SYNTHESES AND APPLICATIONS OF CARBON NANOTUBES AND THEIR COMPOSITES

Edited by **Satoru Suzuki**

Syntheses and Applications of Carbon Nanotubes and Their Composites
http://dx.doi.org/10.5772/3377
Edited by Satoru Suzuki

Contributors

Jijun Zhao, Lizhao Liu, Steve Francis Albert Acquah, Darryl Ventura, Samuel Rustan, Harold Walter Kroto, Mou'ad Al-Tarawneh, Natalia Vladimirovna Kamanina, Elisabeth Lojou, Anne De Poulpiquet, Alecxandre Ciaccafava, Marie-Thérèse Giudici-Orticoni, Saïda Benomar, Alexandre Ciaccafava, Vlad Popa-Nita, Srinivasan Karthikeyan, Ponnusamy Mahalingam, Tawfik A. Abdo Saleh, Emmanuel Beyou, Philippe Cassagnau, Philippe Chaumont, Sohaib Akbar, James Rohan, Mohsen Jahanshahi, Cabrera, Fangchang Tsai, Ning Ma, Chi-Min Shu, Lungchang Tsai, Tai-Chin Chiang, Chi Zhang, Sheng Wen, Hung-Chen Chang, Han-Wen Xiao, Shih-Hsin Chen, Yao-Chi Shu, Tao Jiang, Gang Chang, Yung-Chuan Chu, Igor Levchenko, Zhaojun Han, Shailesh Kumar, Samuel Yick, Jinghua Fang, Ken Ostrikov, Heon Sang Lee, Jing Sun, Ranran Wang, Jun Chen, Tetsu Mieno, Carlos Alberto Avila Orta, Juan Guillermo Martínez Colunga, Aidé Sáenz Galindo, Pablo González Morones, Carlos José Espinoza González, María Neira-Velázquez, Lluvia Itzel López López, Veena Choudhary, Satya Prakash, Wei Shao, Arghya Paul, Vesselin Shanov

Published by InTech

Janeza Trdine 9, 51000 Rijeka, Croatia

Edition 2014

Publishing Process Manager Ana Pantar
Technical Editor InTech DTP team
Cover InTech Design team

Additional hard copies can be obtained from orders@intechopen.com

Syntheses and Applications of Carbon Nanotubes and Their Composites, Edited by Satoru Suzuki
p. cm.
ISBN 978-953-51-1125-2

INTECH

open science | open minds

Contents

Preface IX

Section 1 Syntheses of Carbon Nanotubes and Their Composites 1

Chapter 1 **Production of Carbon Nanotubes and Carbon Nanoclusters by the JxB Arc-Jet Discharge Method 3**
Tetsu Mieno and Naoki Matsumoto

Chapter 2 **Large Arrays and Networks of Carbon Nanotubes: Morphology Control by Process Parameters 19**
I. Levchenko, Z.-J. Han, S. Kumar, S. Yick, J. Fang and K. Ostrikov

Chapter 3 **Classification of Mass-Produced Carbon Nanotubes and Their Physico-Chemical Properties 39**
Heon Sang Lee

Chapter 4 **Fabrication, Purification and Characterization of Carbon Nanotubes: Arc-Discharge in Liquid Media (ADLM) 55**
Mohsen Jahanshahi and Asieh Dehghani Kiadehi

Chapter 5 **Polymer Nanocomposites Containing Functionalised Multiwalled Carbon NanoTubes: a Particular Attention to Polyolefin Based Materials 77**
Emmanuel Beyou, Sohaib Akbar, Philippe Chaumont and Philippe Cassagnau

Chapter 6 **Characterization and Morphology of Modified Multi-Walled Carbon Nanotubes Filled Thermoplastic Natural Rubber (TPNR) Composite 117**
Mou'ad A. Tarawneh and Sahrim Hj. Ahmad

Chapter 7 **Mixtures Composed of Liquid Crystals and Nanoparticles 145**
Vlad Popa-Nita, Valentin Barna, Robert Repnik and Samo Kralj

Chapter 8 **Toward Greener Chemistry Methods for Preparation of Hybrid Polymer Materials Based on Carbon Nanotubes 167**
Carlos Alberto Ávila-Orta, Pablo González-Morones, Carlos José Espinoza-González, Juan Guillermo Martínez-Colunga, María Guadalupe Neira-Velázquez, Aidé Sáenz-Galindo and Lluvia Itzel López-López

Chapter 9 **Carbon Nanotubes and Their Composites 193**
Veena Choudhary, B.P. Singh and R.B. Mathur

Chapter 10 **Kinetics of Growing Centimeter Long Carbon Nanotube Arrays 223**
Wondong Cho, Mark Schulz and Vesselin Shanov

Chapter 11 **Carbon Nanotubes from Unconventional Resources: Part A: Entangled Multi-Walled Carbon Nanotubes and Part B: Vertically-Aligned Carbon Nanotubes 239**
S. Karthikeyan and P. Mahalingam

Chapter 12 **Toroidal and Coiled Carbon Nanotubes 257**
Lizhao Liu and Jijun Zhao

Section 2 Electrical and Biomedical Applications of Carbon Nanotubes 283

Chapter 13 **Carbon Nanotubes for Use in Medicine: Potentials and Limitations 285**
Wei Shao, Paul Arghya, Mai Yiyong, Laetitia Rodes and Satya Prakash

Chapter 14 **Carbon Nanotube Transparent Electrode 313**
Jing Sun and Ranran Wang

Chapter 15 **Latest Advances in Modified/Functionalized Carbon Nanotube-Based Gas Sensors 337**
Enid Contés-de Jesús, Jing Li and Carlos R. Cabrera

Chapter 16 **Carbon Nanotube Composites for Electronic Interconnect Applications 367**
Tamjid Chowdhury and James F. Rohan

Section 3 Carbon Nanotubes for Green Technologies 395

Chapter 17 **Carbon Nanotubes Influence on Spectral, Photoconductive, Photorefractive and Dynamic Properties of the Optical Materials 397**
Natalia V. Kamanina

Chapter 18 **Interconnecting Carbon Nanotubes for a Sustainable Economy 413**
Steve F. A. Acquah, Darryl N. Ventura, Samuel E. Rustan and Harold W. Kroto

Chapter 19 **Carbon Nanotube-Enzyme Biohybrids in a Green Hydrogen Economy 433**
Anne De Poulpiquet, Alexandre Ciaccafava, Saïda Benomar, Marie-Thérèse Giudici-Orticoni and Elisabeth Lojou

Chapter 20 **Adsorption of Methylene Blue on Multi-Walled Carbon Nanotubes in Sodium Alginate Gel Beads 467**
Fang-Chang Tsai, Ning Ma, Lung-Chang Tsai, Chi-Min Shu, Tao Jiang, Hung-Chen Chang, Sheng Wen, Chi Zhang, Tai-Chin Chiang, Yung-Chuan Chu, Wei-Ting Chen, Shih-Hsin Chen, Han-Wen Xiao, Yao-Chi Shu and Gang Chang

Chapter 21 **The Role of Carbon Nanotubes in Enhancement of Photocatalysis 479**
Tawfik A. Saleh

Chapter 22 **Carbon Nanotubes for Energy Applications 495**
Dennis Antiohos, Mark Romano, Jun Chen and Joselito M. Razal

Preface

Preface

Carbon nanotubes are rolled up graphene sheets with a quasi-one-dimensional structure of nanometer-scale diameters. More than twenty years have passed since the pioneering work on carbon nanotubes by Prof. Iijima in 1991. During that time, carbon nanotubes have attracted much attention from physicists, chemists, material scientists, and electronic device engineers, because of their excellent structural, electronic, optical, chemical, and mechanical properties. Most of these unique properties mainly originate in the parent material, graphene, which has also been very intensively studied as a Dirac Fermion system. More recently, demand for innovative industrial applications of carbon nanotubes is increasing.

This book contains recent research topics covering syntheses techniques of carbon nanotubes and nanotube-based composites, and their applications. All of the chapters were written by researchers who are active on the front lines. This book consists of three parts. Part 1 mainly focuses on novel syntheses techniques for single- and multi-wall carbon nanotubes, nanocoils, and their composites. Some chapters in Part 1 describe theoretical aspects of nanotube composite formation. In Part 2, electrical and medical applications of carbon nanotubes are described. This part covers applications for gas sensors, transparent electrodes, and interconnects. Nanotube-based therapeutics and biological detection techniques are also reviewed. Part 3 mainly focuses on applications for green technologies, with much attention paid to energy storage and decontamination technologies. However, note that the above categorization is not rigorous: Some chapters are quite broad in scope and cover topics in more than one category.

I believe that this book will be of interest to physicists, chemists, material scientists, engineers, and students who are working on carbon nanotubes both in the academic and industrial domains

Satoru Suzuki
Senior Research Scientist
Low-Dimensional Nanomaterials Research Group
Materials Science Laboratory
NTT Basic Research Laboratories, Japan

I express my deepest appreciation to all of the authors for their excellent contributions and endeavors in the publication of this book.

Section 1

Syntheses of Carbon Nanotubes and Their Composites

Chapter 1

Production of Carbon Nanotubes and Carbon Nanoclusters by the JxB Arc-Jet Discharge Method

Tetsu Mieno and Naoki Matsumoto

Additional information is available at the end of the chapter

http://dx.doi.org/10.5772/51964

1. Introduction

Since the discovery of methods for the mass production of single-walled carbon nanotubes (SWNTs) [1, 2], applications of SWNTs such as transistor devices, biosensing devices, double-layer-type capacitors, transparent electrode films, radio wave absorbents and material hardeners have been studied [3-5]. Large-scale production and improvement of purity of SWNTs by the electric-arc techniques have been developed [6, 7]. However, the efficient production of high-quality and defect-free SWNTs, and metal/semiconductor selected or diameter-controlled production of SWNTs have not yet been achieved. Therefore, basic study of the various methods of producing SWNTs is still important, by which new high-performance routes to producing desired SWNTs are expected to be found.

Here, the production of SWNTs and carbon nanoclusters by the arc discharge method utilizing a magnetic field, known as the *JxB* arc-jet discharge method, has been studied [8-10]. Although the application of a steady-state magnetic field to arc discharge is not such a popular method, electromagnetic force can change the flow of hot gas in the arc region and thus control the production process of carbon clusters. To realize the large-scale production of carbon clusters by the arc discharge method, a revolver-injection-type *JxB* arc-jet producer was successfully developed by our group, by which the continuous mass production of SWNTs and other carbon clusters can be carried out.

As a result, the more efficient production of SWNTs and other carbon clusters compared with conventional arc discharge methods has been achieved. Here, the development of the *JxB* arc-jet discharge method and results obtained using the method are described.

2. Theoretical model of the *J*x*B* arc jet discharge method

By applying a steady-state weak magnetic field (B_0= 1 – 5 mT) perpendicular to the discharge current in the arc discharge, the Lorentz force (*J*x*B* force) causes the ejection of the arc plasma and surrounding gas in the *J*x*B* direction as shown in Fig. 1 [8, 11]. In the 1960s, this force in a pulsed discharge was actively studied in relation to the electric propulsion engine of rockets [12].

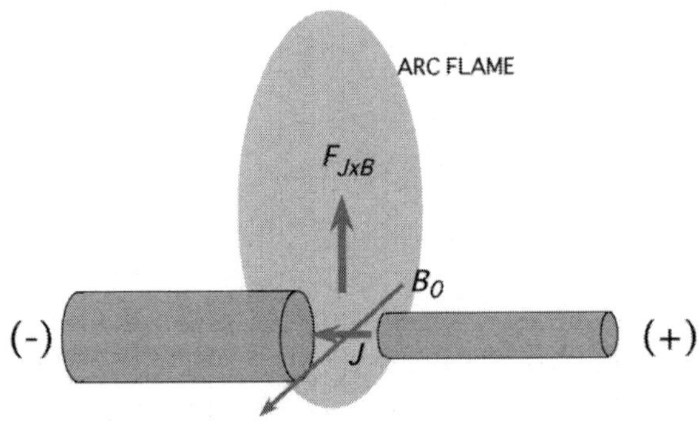

Figure 1. Schematic diagram of the *JxB* arc-jet discharge.

Here, this effect is used to eject sublimated carbon atoms in a selected direction. By controlling the magnetic field, control of the hot gas including the carbon material is possible, and suitable conditions to do hot gas reactions for the production of SWNTs and other carbon clusters can be selected. This method can also reduce the influence of the electrode direction and chamber configuration.

When the discharge current density and applied magnetic field are 50 A/cm^2 and 5 mT, respectively, the Lorentz force causing acceleration of electrons and ions is 0.25 N/cm^3. When the gas pressure and the gas temperature around the arc are 30 kPa and 5000 K, respectively, the mean free path and collision frequency of electrons are about 0.01 mm and 10 GHz, respectively. Because of this high collision frequency, electrons frequently collide with neutral gas atoms and accelerate them in the *J*x*B* direction, resulting in the ejection of hot gas from the arc region. The acceleration time is related to the electron lifetime in the plasma.

3. Production of carbon clusters by the *J*x*B* arc-jet discharge method

3.1. Heat flux

To investigate the *J*x*B* arc-jet discharge reaction, several types of arc reactors are used. Figure 2 shows a schematic of the reactor used to measure the heat flux of the arc plasma [11]. The reactor is made of stainless steel (18 cm diameter, 20 cm height) and has a carbon anode

http://dx.doi.org/10.5772/51964

(8.0 mm), a carbon cathode (15 mm), a viewing port and a movable calorimetric probe. The reactor is evacuated by a rotary pump to a pressure of less than 10 Pa and then closed. After introducing He gas with p(He) = 10 – 80 kPa, discharge starts, where the discharge current is I_d= 20 – 80 A and voltage between the electrodes is V_{rod}= 20 - 35 V. At the front and back of the reactor, solenoid coils (20 cm inner diameter) are installed to produce a steady state magnetic field of B_0= 0 - 5 mT.

When a magnetic field is applied during the discharge, the shape of the arc flame dramatically changes, and a strong plasma flow in the *J*x*B* direction can be observed. Figures 3(a) and (b) respectively show side views of the arc flames for B_0= 0 and B_0= 2.0 mT, where p(He)= 40 kPa and I_d= 80 A. The upper direction is the *J*x*B* direction. By applying a magnetic field, the plasma and the hot gas are ejected in the vertically upward direction. The upward flow of carbon particles can sometimes be clearly observed. By developing a calorimeter [11] in which flowing water absorbs the heat flux, the local heat flux is measured and the results are shown in Figs. 4 (a) and (b) [11]. By increasing the magnetic field, the heat flux is localized in the upper part of the arc plasma (FWHM value of about 50 mm). Above the arc plasma, the heat flux monotonically increases.

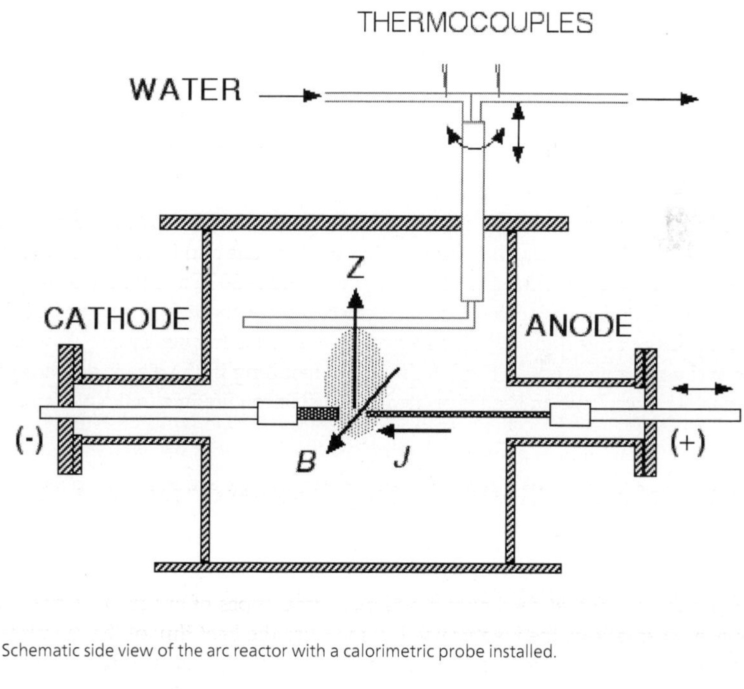

Figure 2. Schematic side view of the arc reactor with a calorimetric probe installed.

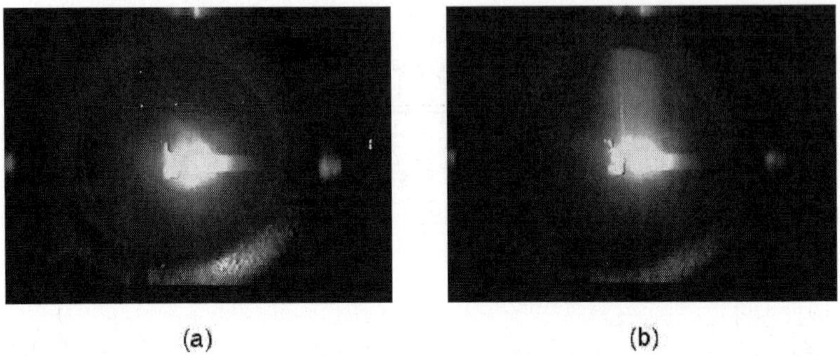

Figure 3. Arc flames for B_0= 0 (a) and B_0= 2.0 mT (b) (side views), where p(He)= 40 kPa and I_d= 80 A.

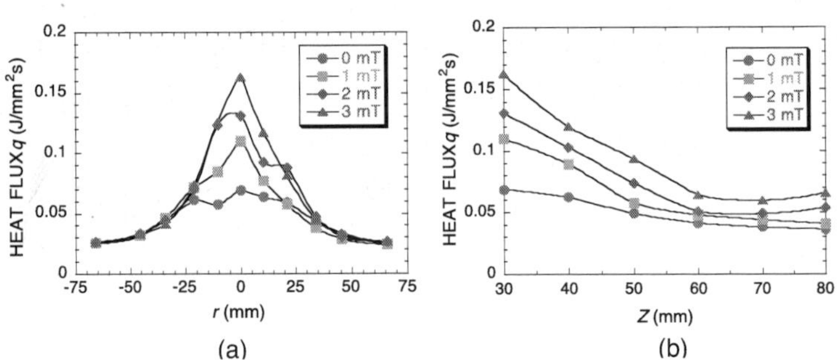

Figure 4. Radial profiles (a) and vertical profiles (b) of heat flux above the arc plasma for B_0= 0, 1.0, 2.0 and 3.0 mT. p(He)= 40 kPa and I_d= 60 A.

The heat flux above the arc plasma as a function of He gas pressure is measured and shown in Fig. 5(a), where I_d= 60 A, d_G= 5 mm, and the distance from the arc center is 40 mm. The heat flux increases monotonically with the pressure, which is particularly in the case of B_0= 2.0 mT. Figure 5(b) shows the heat flux above the arc plasma as a function of the gap distance between the two electrodes d_G, where p= 40 kPa, I_d= 60 A and z= 40 mm. By changing d_G, the effect of the arc jet changes, which can be observed from the viewing port. The heat flux gradually increases with the gap distance, and this effect is greatly enhanced when B_0= 2.0 mT.

To summarize these results, that the *J*x*B* arc jet is enhanced by increasing the applied magnetic field (B_0= 0 - 3.0 mT), the He pressure and the gap distance. However, under a stronger magnetic field of B_0> 4 mT, the discharge tends to be extinguished easily by fluctuation in the discharge.

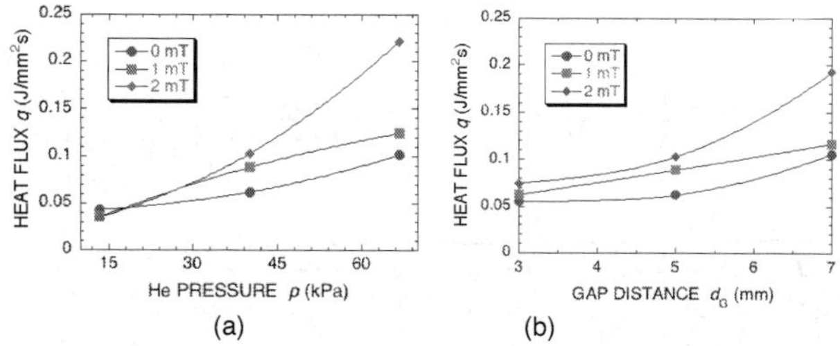

Figure 5. (a) He pressure dependence of the heat flux, where I_d= 60 A, d_G= 5 mm and z= 40 mm from the arc center. (b) Gap distance dependence of the heat flux, where p= 40 kPa, I_d= 60 A and z= 40 mm.

3.2. Relations among directions of the discharge current, magnetic field and gravity

In the case of gas arc discharge, gravity induces strong heat convection. Therefore, by changing the current direction relative to that of gravity, different production characteristics of carbon can be expected [13]. Direction of the ***JxB*** force compared with that of gravity should also be an important parameter. To examine the relations among the directions of the discharge current, magnetic field and gravity for the production of fullerenes, five experimental configurations are prepared and a discharge experiment is carried out. The 5 configurations are shown in Fig. 6. Here, the carbon anode is 6.5 mm in diameter and the carbon cathode is 15 mm in diameter.

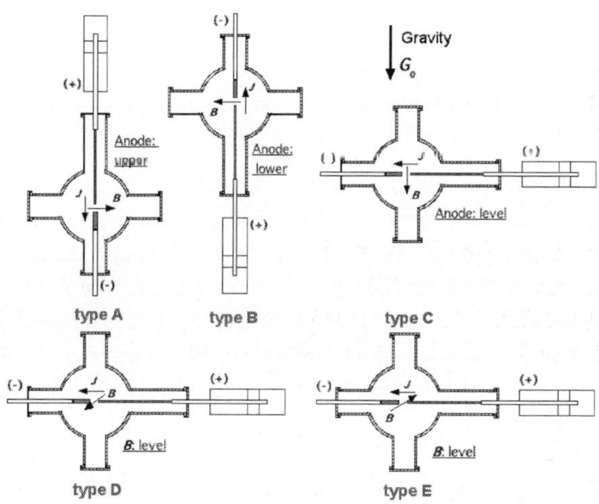

Figure 6. Schematic of five experimental configurations. The directions of the discharge current J and magnetic field B relative to that of gravity G_0 are changed.

The production rates of carbon soot W_{soot} (g/h) as a function of B_0 for configurations (types A – E) are obtained and the results are shown in Fig. 7, where p(He) = 40 kPa, I_d= 70 A and d_G~ 5 mm [13]. Generally, the soot production rate increases steadily with the magnetic field. However, for type A, W_{soot} is very low for B_0 =0 and it rapidly increases with increasing magnetic field. When B_0= 4.0 mT, the differences in W_{soot} are very small among the five configurations.

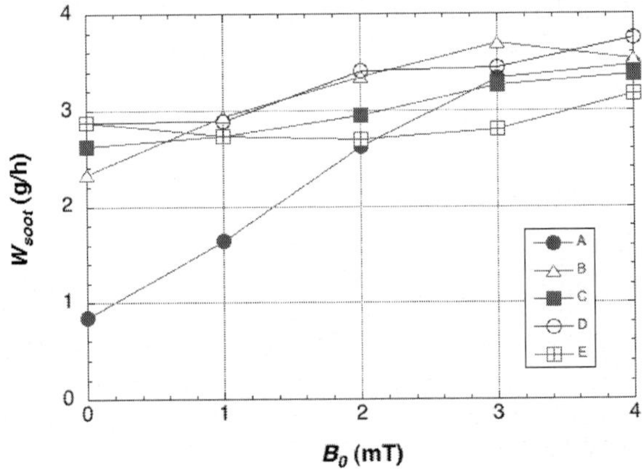

Figure 7. Production rate of carbon soot versus B_0 for the five configurations. p= 40 kPa, I_d= 70 A and discharge time T_d= 60 min

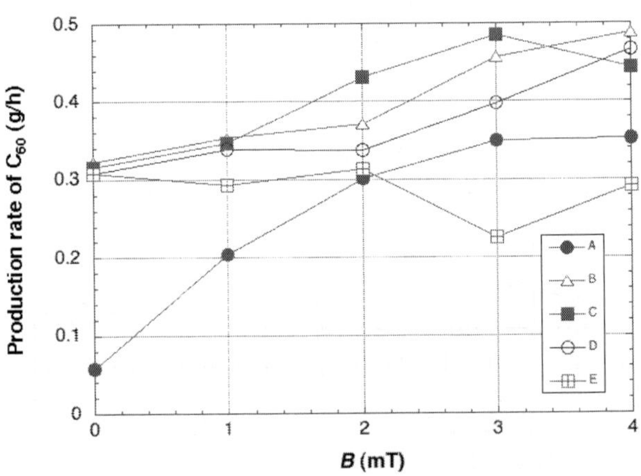

Figure 8. Production rate of C_{60} versus B_0 for the five configurations. p= 40 kPa, I_d= 70 A and discharge time T_d= 60 min.

The production rate of C_{60} as a function of B_0 for configurations (types A – E) is obtained and the results are shown in Fig. 8, where the conditions are the same us those of Fig. 7. Similarly to in Fig. 7, the C_{60} production rate generally increases with B_0, except for type A. C_{60} production rate of type A is very low at B_0= 0. Moreover, for type E, the C_{60} production rate does not increase monotonically with B_0 and the magnetic field does not have a positive effect on the production rate.

From these results, it can be concluded that the directions of the discharge current and magnetic field compared with that of gravity affect the production of carbon soot and fullerenes. The *J*x*B* force tends to reduce the effect of gravity when B_0 is sufficiently large. The type A and the Type E are less suitable for the production of fullerenes.

3.3. Production of SWNTs

The production of high-quality SWNTs is one of the most important targets in advancing nanomaterial development. The growth model of SWNTs in the arc-discharge reactors has not been confirmed. Several models show importance of catalyst particles in the hot gas, carbon density and catalyst temperature. [14, 15] Here, the production of SWNTs is examined using the *J*x*B* arc-jet method, which could modify the growth reactions in the hot gas. In this case, a Ni/Y catalyst is included in the carbon material rods (6.0 x 6.0 mm, rectangular type, 4.2 W% of Ni and 0.9 W% of Y included), and *p*(He)= 60 kPa and I_d= 50 A. The soot production rate as a function of applied magnetic field is shown in Fig. 9(a) [16, 17]. By increasing the magnetic field, the production rate monotonically increases. However, further increasing B_0 to above 3.5 mT makes the discharge unstable. Figure 9(b) shows the pressure dependence of the soot production rate for B_0= 0 and B_0= 2.2 mT. As the pressure increases and the collisional effect of He gas increases, the *J*x*B* force clearly affects soot production.

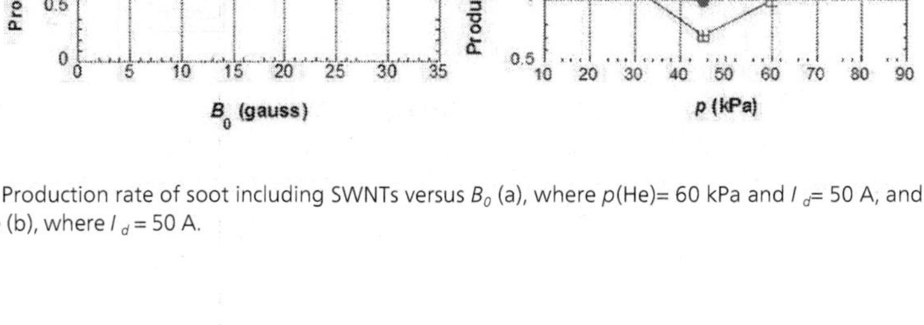

Figure 9. Production rate of soot including SWNTs versus B_0 (a), where p(He)= 60 kPa and I_d= 50 A, and pressure dependence (b), where I_d = 50 A.

Using this *JxB* arc-jet method, a large amount of SWNTs is produced from carbon rods including a Ni/Y catalyst. Figure 10 shows a typical TEM image of the produced soot, in which many bundles of SWNTs are included. There are also carbon nanoparticles and catalyst nanoparticles in the soot, which should be removed during the purification of the SWNT products. The quality of the products is measured by a Raman spectrometer (Jasco Co., NR-1800. An Ar ion laser of λ= 488.0 nm is used.), and the results are shown in Figs. 11 (a) and 11(b) for B_0= 0 and B_0= 3.2 mT. In both cases, there are very small D(disorder) band peaks and large G (graphite) band peaks, from which we can estimate the content and quality of SWNTs in the produced carbon soot. These figures show that the *JxB* arc discharge does not degrade the quality of the SWNTs. From the signals of the radial breathing mode in Fig. 11(a), we can evaluate the SWNT diameters [18]. The major diameter is 1.40 nm, and SWNTs with a diameter of 1.26 nm also exist in the case of B_0= 3.2 mT. SWNTs with diameters of 1.70 nm, 1.16 nm and 1.0 nm also exist in the case of B_0= 0.

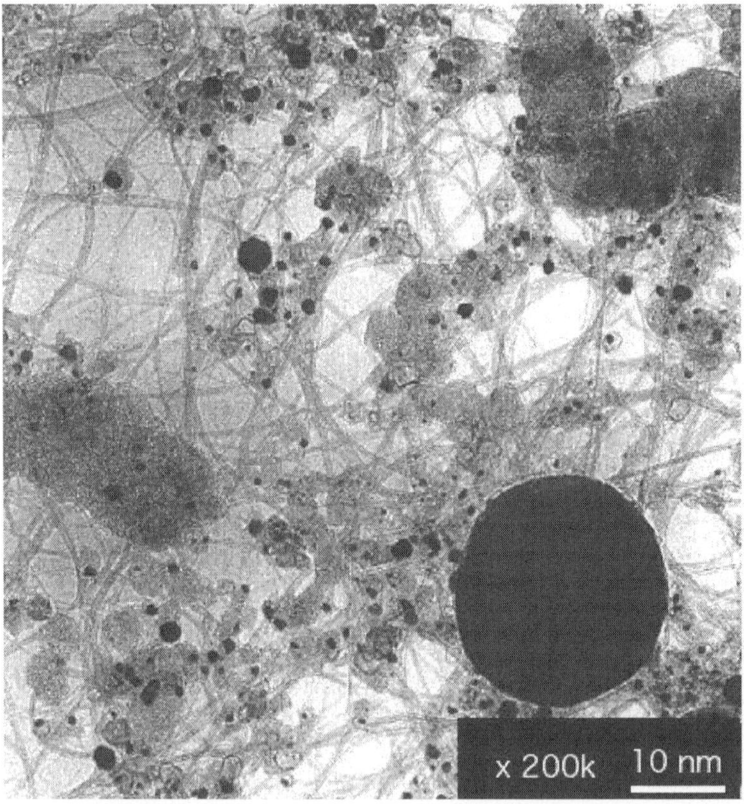

Figure 10. Typical TEM image of SWNTs produced by the *JxB* arc-jet discharge method.

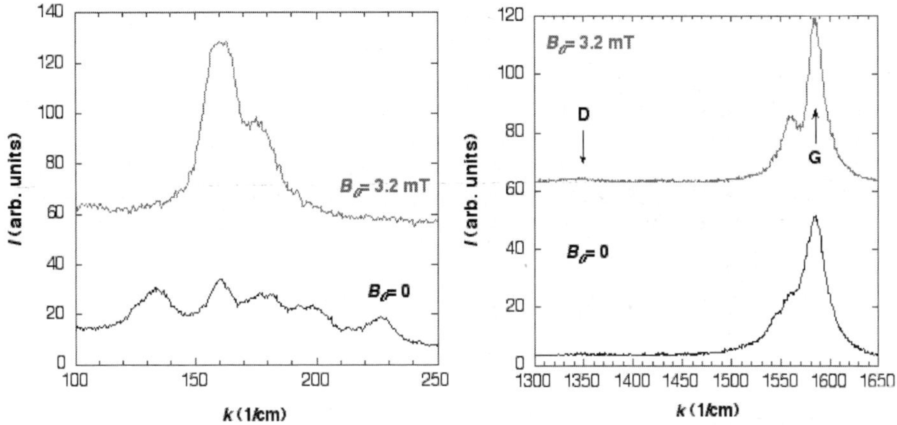

Figure 11. Raman spectra of the produced samples for magnetic fields of B_0= 3.2 mT and 0 T. p(He)= 60 kPa and I_d= 50 A. λ = 488.0 nm.

Figure 12. Photograph of SWNTs dispersed in pure water.

In the arc production of SWNTs, effect of gas species is examined. Ar, Ne or N_2 gas is used instead of He gas, all of which degrade production of SWNTs. When amount of H_2 gas is included in He gas, the SWNT production rate considerably decreases, which is not consistent with the previous report [19]. It is conjectured that He atom has high ionization potential, and it causes almost no chemical reactions and less emission loss. When Co or Fe particles are used as catalyst material instead of Ni/Y particles, the SWNT production rate considerably decreases. However, in case of of Co catalyst, a little amount of very long bundles of SWNTs is obtained, which is about 5 cm long. Improvement of the long-SWNT yield by this method is one of our study targets.

Usually SWNTs have poor dispersibility in water, which limits their potential applications. Therefore, the development of water-dispersible SWNTs is very important. Here we attempt to dissolve a SWNT sample in pure water. First, a small amount of SWNTs is placed in 20 mL of pure water, which is then mixed using a supersonic homogenizer (Sonics Co., VC-130, 25 W) for about 40 min. Then, a small amount of surfactant is added, which is one of bio-polymers [20, 21]. And it is mixed by sonication again for about 40 min. As a result, the SWNTs are well dispersed in water, and the dispersion remains very stable for more than 1 month. Figure 12 shows a photograph of SWNTs dispersed in water after 50 times dilution. The study of SWNTs dispersed in water is continuing with the aim of realizing biological applications.

3.4. Production of endohedral metallofullerenes

Using the above arc-jet discharge methods, endohedral metallofullerenes (such as Gd@C_{82} and La@C_{82}) [22] and carbon nanocapsules are efficiently produced. Applications of these materials are expected.

By performing arc discharge using a Gd_2O_3-containing carbon rod (6.0 x 6.0 mm, rectangular), metallofullerenes are produced, where p(He)= 50 kPa and I_d= 58 A. The production rate of soot is about 2.5 g/h. Figure 13(a) shows a typical TEM image of the sample obtained, in which gadolinium nanoparticles with a diameter on the order of 10 nm are covered with carbon atoms, resulting in the formation of carbon capsules with a size of 10 – 50 nm, which are very stable in air. By refluxing the sample with toluene, fullerenes can be extracted from the soot. After 4 h of reflux, the liquid is filtered and a reddish liquid is obtained. A mass spectrum of this sample obtained using a laser-desorption time-of-flight mass spectrometer (LD-TOF-MS) (Bruker Co., Autoflex, + ion mode, 50 shots averaged) is shown in Fig. 13(b). We can confirm the existence of not only C_{60} and higher fullerenes but also endohedral metallofullerenes Gd@C_{82}. Although the peak intensities are not quantitative, the relative yield of Gd@C_{82} compared with that of C_{60} is several mol%. The Gd@C_{82} is expected to be used as a contrast agent in MRI [23].

(a) (b)

Figure 13. (a) TEM image of gadolinium-continuing carbon nanoparticles produced by the arc-jet discharge method. (b) LD-TOF-MS spectrum of a carbon sample extracted from soot using toluene.

3.5. Production of magnetic nanoparticles

Using the *JxB* arc-jet discharge method, many types of metal-particle-encapsulated carbon nanoparticles [24, 25] can be easily produced. As examples, iron-encapsulated carbon nanoparticles and cobalt-encapsulated carbon nanoparticles have been produced. Both are ferromagnetic nanoparticles with a size of 10 - 100 nm, and are very stable.

Using iron-containing carbon rods (6.0 x 6.0 mm, rectangular), iron-encapsulated carbon nanoparticles are produced, where p(He)= 50 kPa and I_d= 50 A. The soot production rate is about 0.43 g/h. Figure 14 (a) shows a photograph of the produced soot suspended by a magnet, demonstrating the good ferromagnetic property. A typical TEM image of the sample is shown in Fig. 14 (b). Iron particles with a size of 1 nm to 20 nm are covered with carbon atoms, resulting in the formation of carbon particles with a size of 10 – 100 nm. These particles are very stable in air and inactive in hydrochloric acid.

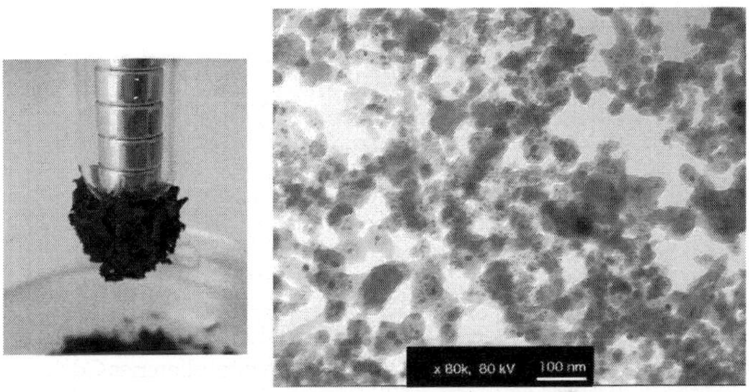

Figure 14. (a) Photograph of iron-containing carbon nanoparticles suspended by a magnet. (b) TEM image of iron-containing carbon nanoparticles.

Figure 15. Photograph of iron-containing carbon nanoparticles dispersed in water.

Cobalt-encapsulated carbon nanoparticles, which also have ferromagnetic properties, are produced by the arc-jet discharge method. They are dispersed in pure water with a small amount of surfactant (gelatin *etc.*) and mixed using a supersonic homogenizer (Sonic Co., VX-130) for 1 h. Finally, a black inklike liquid is obtained. The dispersion is homogeneous and stable, and most of the particles do not precipitate even after one month. These water-soluble magnetic nanoparticles potentially have many applications in the fields of liquid sealing, medical diagnostics and medical treatment [26]. Figure 15 shows a photograph of the stable iron-containing carbon nanoparticles dispersed in water.

4. Development of automatic *JxB* arc-jet producer

To produce SWNTs and carbon nanoclusters at a commercial scale by the *JxB* arc-jet discharge method, a revolver-injection-type arc jet producer (RIT-AJP) has been developed by collaboration with Daiavac Ltd. (Japan) [9].

A schematic and photograph of RIT-AJP are shown in Fig. 16. The left side of the machine is an arc discharge chamber, which consists of a vacuum vessel made of stainless steel 25 cm in diameter and 70 cm high that is uniformly cooled by water jackets. About 5 L of water is stored in the jackets and cooling water is slowly supplied to the jackets. In the central part of the chamber, a cathode electrode (30 mm in diameter), an anode electrode, an exhaust port, a viewing port and an electrode-cleaning hand are mounted. The top and bottom parts of the chamber are soot collectors, with an inner diameter of about 25 cm and a height of 24 cm, in which produced soot is deposited. Using these collectors, as much as 25 L of soot can be easily collected after a single operation.

The right side of the apparatus is a revolver-type carbon rod magazine. In the cylindrical metal vacuum vessel, which is 34 cm in diameter and 49 cm long, there is a rotatable cylindrical magazine, in which as many as 50 carbon rods of 6 - 10 mm diameter and 300 mm length can be loaded. A schematic figure of the material-feeding mechanism of the magazine is shown in Fig. 17(a), and a photograph of a rotatable carbon rod magazine (for 50 carbon rods) is shown in Fig. 17(b). Under the vacuum chamber, there is a vacuum pump, an electrical controller and a microcomputer. Discharge power is supplied by an inverter-type DC power supply (Daihen Co., ARGO-300P).

The production sequence is as follows. First, up to 50 carbon rods are loaded in the magazine, and the chamber is evacuated by the vacuum pump. After evacuation to a pressure of less than 10 Pa, pure helium gas is introduced and the chamber is closed. Upon turning on the electrical controller, a metal striker pushes one of the carbon rods towards the cathode, and the discharge starts upon electrical contact. As the discharge conditions are determined by the discharge voltage and discharge current, the carbon rod is automatically moved until both parameters match the set values. Once the carbon rod is consumed, the cylindrical carbon magazine rotates 1/50 of a turn and the next carbon rod is inserted by the striker. A magnetic field can be applied by a block-type ferrite magnet located outside the chamber, by which a magnetic field of about 2 mT is applied horizontally to the discharge space. Carbon

http://dx.doi.org/10.5772/51964

deposited on the cathode is removed by a cathode-cleaning hand. After the discharge, produced soot that has been deposited is carefully collected.

As an example of the continuous production of carbon clusters, fullerenes are produced. Using 134 carbon rods of 8 mm diameter, continuous *J*x*B* arc-jet discharge is carried out, where *p*(He)= 40 kPa, the discharge current is I_d= 120 A, the voltage between electrodes is V_{rod}~ 33 V and the gap distance is d_G~ 5 mm. The insertion speed of the carbon rods is about 30 cm/h. After the discharge, carbon soot from three parts (the top collector, central chamber and bottom collector) is collected separately and their masses are measured. The amount of soot deposited on the top wall is considerably increased by applying the magnetic field, because the carbon molecules are blown upward onto the top wall. After sufficient mixing, the C_{60} content in the soot is measured by a UV/visible spectrometer (Shimadzu Co. UV-1200). At the top collector, the C_{60} content is the highest and about 14 W% of C_{60} is present, whereas, 4.2 W% is present on the center wall and 2.9 W% is present on the bottom wall. In total, about 105 g of soot containing about 7 g of C_{60} is produced in 12 h.

The contents of higher fullerenes in the soot are measured using a high-pressure liquid chromatograph (HPLC) (Jasco Co., Gulliver Series, PU980) [27]. The collection rates of C_{60}, C_{70}, C_{76}, C_{78} and C_{84} for two different magnetic fields are shown in Fig. 18. White rectangules in the graphs show the measurement errors. By applying a magnetic field, the collection rates of these fullerenes considerably increase.

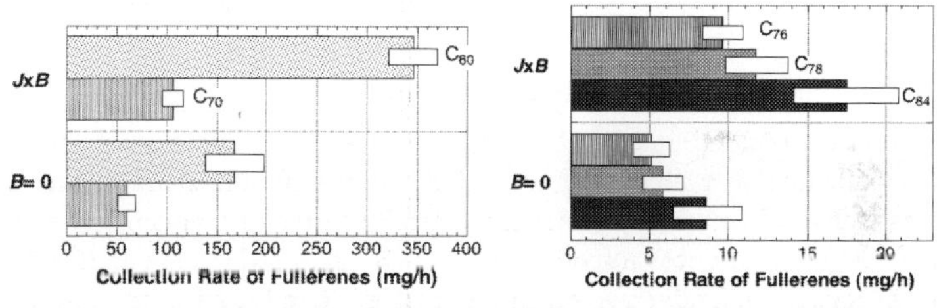

Figure 18. Collection rates of C_{60} and C_{70} (a), and C_{76}, C_{78} and C_{84} (b) for two different magnetic fields.

5. Summary

1. By applying a steady state magnetic field perpendicular to the discharge current, *J*x*B* arc-jet discharge is successfully produced. The flow of hot gas and the heat flux are concentrated in the *J*x*B* direction.

2. Carbon atoms sublimated from the anode are continuously ejected from the arc plasma in the *J*x*B* direction. As a result of the *J*x*B* force, the effect of gravity (heat convection) can be reduced.

3. By increasing the magnetic field, the production rate of carbon soot including SWNTs considerably increases, in which the quality of the SWNTs remains high. Water-soluble SWNTs can be obtained by additional processing.

4. Using the *JxB* arc-jet discharge, endohedral metallofullerenes, Gd@C_{82} and magnetic nano-particles (iron-encapsulateted carbon nanoparticles and cobalt-encapsulated carbon nano-particles) are sucessfully produced.

5. For the continuous and large-scale production of carbon clusters, a revolver-injection-type arc-jet producer (RIT-AJP) has been developed. Using this machine, the automatic mass production of SWNTs and carbon clusters is realized. We are currently attempting to fabricate many new types of carbon clusters using this machine.

Acknowledgments

We thank H. Inoue of Daiavac Co. (Chiba, Japan) for his technical support during the development of the RIT-AJP machine. We also thank W. Tomoda, Md. K. H. Bhuiyan and S. Aoyama of Shizuoka University for their technical assistance.

Author details

Tetsu Mieno and Naoki Matsumoto

Department of Physics, Shizuoka University, Japan

References

[1] Iijima, S. Single-Shell Carbon Nanotubes of 1-nm Diameter, Nature 1993; 363, 603-605.

[2] Bethune, D. S., Klang, M. S., de Vries, M. S., Gorman, G., Savoy, R., Vazquez, J., Beyers, R., Cobalt-catalysed growth of carbon nanotubes with single-atomic-layer walls, Nature 1993; 363, 605-607.

[3] Dresselhaus, M. S., Dresselhaus, G., Avouris, Ph., (eds.) Carbon Nanotubes, Springer, Berlin, 2000, ISBN: 3-540-41086-4.

[4] Jorio. A., Dresselhaus, M. S., Dresselhaus, G., (eds.) Carbon Nanotubes, Springer, Berlin, 2008, ISBN : 978-3-540-72864-1.

[5] Harris, P. J. Carbon Nanotubes Science, Cambridge University Press, Cambridge, 2009, ISBN: 978-0-521-53585-4.

[6] Journet, C., Maser, W. K., Bernier, P., Loiseau, A. Lamy de la Chapele, M., Lefrant, S., Deniard, P., Lee, R., Fischer, J. E., Large-scale production of single-walled carbon nanotubes by the electric-arc technique, Nature, 388, pp. 756-758.

[7] Mansour, A., Razafinimanana, M., Monthoux, M., Pachco, M., Gleizes, A., A siginificant improvement of both yield and purity during SWNT synthesis via the electric arc process, Carbon, 45, 2007, pp. 1651-1661.

[8] Mieno, T. Automatic Production of Fullerenes by a *J*x*B* Arc Jet Discharge Preventing Carbon Vapor from Depositing on the Cathode, Fullerene Science & Technology 1995; 3 (4), 429-435.

[9] Mieno, T. *J*x*B* Arc Jet Fullerene Producer with a Revolver Type Automatic Material Injector, Fullerene Science & Technology 1996; 4 (5), 913-923.

[10] Mieno, T. Production of Fullerenes from Plant Materials and Used Carbon Materials by Means of a Chip-Injection-Type *J*x*B* Arc Reactor, Fullerene Science & Technology 2000; 8 (3), 179-186.

[11] Matsumoto, N., Mieno, T. Characteristics of Heat Flux of *J*x*B* Gas-Arc Discharge for the Production of Fullerenes, Vacuum 2003; 69, 557-562.

[12] Jahn, R. G. Physics of Electric Propulsion, McGraw-hill, New York, 1968, 257-316, (NC)ID: BA18312459.

[13] Aoyama, S., Mieno, T. Effects of Gravity and Magnetic Field in Production of C_{60} by a DC Arc Discharge, Japanese Journal of Applied Physics 1999; 38 (3A), L267- L269.

[14] Kanzow, H., Ding, A, Formation mechanism of single-wall carbon nanotubes on liquid-metal particles, Physical Review B 1999 ; 60 (15) 11180-11186.

[15] Gavillet, J. *et al.*, Microscopic mechanisms for the catalyst assisted growth of single-wall carbon nanotubes, Carbon 2002 ; 40, 1649-1663.

[16] Mieno, T., Tan, G.-D. Effect of Gravity and Magnetic Field on Production of Single-Walled Carbon Nanotubes by Arc-Discharge Method, In: Yellampalli, S. (ed.) Carbon Nanotubes- Synthesis, Characterization, Application; InTech; 2011, 61-76, ISBN: 978-953-307-497-9.

[17] Mieno, T., Matsumoto N., Takeguchi, M. Efficient Production of Single-Walled Carbon Nanotubes by *J*x*B* Gas-Arc method, Japanese Journal of Applied Physics 2006; 43 (12A), L1527- L1529.

[18] Kataura, H. *et al.* Optical Properties of Single-Walled Carbon Nanotubes, Synthetic Metals 1999; 103, 2555-2558.

[19] Zhao, X., Inoue, S., Jinno, M., Suzuki T., Ando Y., Macroscopic oriented web of single-wall carbon nanotubes, Chemical Physics Letters 2003; 373, 266-271.

[20] Takahashi, T., Tsunoda, K., Yajima, H., Ishii, T., Purification of Single Wall Carbon Nanotubes Using Gelatin, Japanese Journal of Applied Physics 2004; 43 (3), 1227-1230.

[21] Takahashi, T., Luculescu, C. R., Uchida, K., Tsunoda, K., Yajima, H., Ishii, T., Dispersion Behavior and Spectroscopic Properties of Single-Walled Carbon Nanotubes in Chitosan Acidic Aqueous Solutions, Chemistry Letters 2005; 34 (11), 1516-1517.

[22] Shinohara, H., Endohedral metallofullerenes, Reports on Progress in Physics 2000; 63, 843-892.

[23] Mikawa, M. *et al.* Paramagnetic Water-Soluble Metallofullerenes Having the Highest Relaxivity for MRI Contrast Agents, Bioconjugate Chemistry 2001; 377, 510-514.

[24] Saito, Y., Nanoparticles and Filled Nanocapsules, Carbon 1995; 33 (7) 979-988.

[25] Saito, Y. *et al.* Iron particles nesting in carbon cages grown by arc discharge, Chemical Physics Letters 1993; 212 (3-4) 379-383.

[26] Pankhurst, Q. A., Connolly, J., Jones, S. K., Dobson, J. Applications of Magnetic Nanoparticles in Biomedicine 2003; 36, R167-R181.

[27] Bhuiyan, Md. K. H., Mieno, T. Production Characteristics of Fullerenes by Means of the *J*x*B* Arc Discharge Method, Japanese Journal of Applied Physics 2002; 41 (1), 314-318.

Chapter 2

Large Arrays and Networks of Carbon Nanotubes: Morphology Control by Process Parameters

I. Levchenko, Z.-J. Han, S. Kumar, S. Yick, J. Fang and K. Ostrikov

Additional information is available at the end of the chapter

http://dx.doi.org/10.5772/52674

1. Introduction

Large arrays and networks of carbon nanotubes, both single- and multi-walled, feature many superior properties which offer excellent opportunities for various modern applications ranging from nanoelectronics, supercapacitors, photovoltaic cells, energy storage and conversation devices, to gas- and biosensors, nanomechanical and biomedical devices etc. At present, arrays and networks of carbon nanotubes are mainly fabricated from the pre-fabricated separated nanotubes by solution-based techniques. However, the intrinsic structure of the nanotubes (mainly, the level of the structural defects) which are required for the best performance in the nanotube-based applications, are often damaged during the array/network fabrication by surfactants, chemicals, and sonication involved in the process. As a result, the performance of the functional devices may be significantly degraded. In contrast, directly synthesized nanotube arrays/networks can preclude the adverse effects of the solution-based process and largely preserve the excellent properties of the pristine nanotubes. Owing to its advantages of scale-up production and precise positioning of the grown nanotubes, catalytic and catalyst-free chemical vapor depositions (CVD), as well as plasma-enhanced chemical vapor deposition (PECVD) are the methods most promising for the direct synthesis of the nanotubes.

On the other hand, these methods demonstrate poor controllability, which results in the unpredictable properties, structure and morphology of the resultant arrays. In our paper we will discuss our recent results obtained by the application of CVD and PECVD methods. Specifically, we will discuss carbon nanotube arrays and networks of very different morphology. The fabrication of the arrays of vertically aligned and entangled nanotubes, as well as arrays of arbitrary shapes grown directly on the pre-patterned substrates will be considered with a special attention paid to the fabrication methods and the influence of the process parameters on the array growth morphology (see Figure 1). Besides, the possibility of creating the 3D structures of carbon nanotubes through post-processing of the arrays by liquids will be discussed.

The fabrication methods involved are the conventional CVD utilizing various gases such as methane, ethane, acetylene, argon, and hydrogen; plasma-enhanced CVD based on inductively-coupled low-temperature plasma reactor; microwave PECVD. The advantages of the plasma-based CVD process will be shown and discussed with a special attention. We will also discuss the influence of the process parameters such as process temperature, pressure, gas composition, discharge power etc. on the morphology of the nanotube arrays and networks, and demonstrate that the proper selection of the parameters ensures very high level of the process controllability and as a result, sophisticated control and tailoring of the growth structure and morphology of the carbon nanotube arrays.

Characterization technologies used are scanning and transmission electron microscopy (SEM and TEM), as well as atomic force microscopy (AFM), Raman and X-ray photoelectron spectroscopy techniques. The results of the numerical simulations will also be used to support the growth models and proposed growth mechanisms.

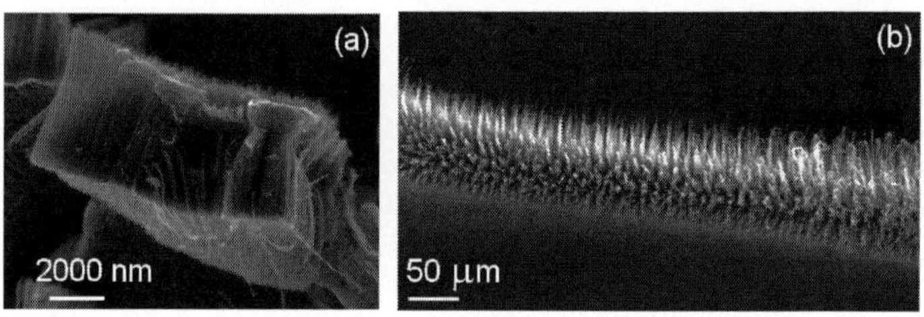

Figure 1. Morphologies of the representative CNT arrays grown by CVD (a) and PECVD (b).

2. CVD versus PECVD: Morphology control issues

2.1. CVD and PECVD: General

The term 'chemical vapor deposition', or 'CVD', is commonly used for describing the processes and chemical reactions which occur in a solid material deposited onto a heated substrate using a gaseous precursor. However, more complicated process than the 'common' CVD takes place [1,2] during the growth of CNTs. In this case, the carbon-containing gaseous precursors (e.g., CH_4, C_2H_4, C_2H_2, CO) firstly dissociate into atomic or molecular carbon species on the surface of catalyst nanoparticles, and then the nucleation occurs as these carbon species diffuse into the catalyst nanoparticles, reach a supersaturated state, and then segregate from the surface of nanoparticles to form a nanotube cap. Subsequently, the growth of nanotubes is sustained by the continuous incorporation of carbon atoms *via* bulk and/or surface diffusion. Figure 2 shows the SEM image of randomly-oriented SWCNTs with a unique 'bridging' morphology in catalytic

http://dx.doi.org/10.5772/52674

CVD. This array was grown by using $Ar/H_2/CH_4$ gas mixture on Fe/Al_2O_3 catalyst. By tuning the growth condition (e.g., temperature and pressure), it was demonstrated that the SWCNTs could be of a high quality (a high I_G/I_D in the Raman spectra) and could contain a significantly higher content of metallic nanotubes as compared to the 'standard' metallic nanotube content of 33% (1/3 metallic and 2/3 semiconducting) produced in many CVD processes [3].

On the other hand, PECVD refers to the CVD process that uses plasma environment as an extra dimension to control the growth of CNTs. Plasma by definition contains ionized species and is generally considered as the fourth state of matter along with solid, liquid and gas. Recent advances in the plasma-based nanofabrication offer unprecedented control over the structure and surface functionalities of a range of nanomaterials [4]. One of the major advantages, as compared to the conventional CVD processes, is that nanostructures can grow vertically-aligned due to the electrical field in the vicinity of surface [5]. Another benefit of using plasma is that the temperature required to dissociate carbon feedstock could be greatly reduced [6]. Figure 3 illustrates the isolated CNTs grown in a PECVD system using Ni/SiO_2 as the catalyst, C_2H_2/NH_3 as the gas precursors, and a DC glow discharge. It can be seen clearly that these nanotubes are aligned vertically to the substrate surface, due to the plasma sheath-directed growth. These freestanding nanotubes could give many opportunities to custom-design novel functional devices.

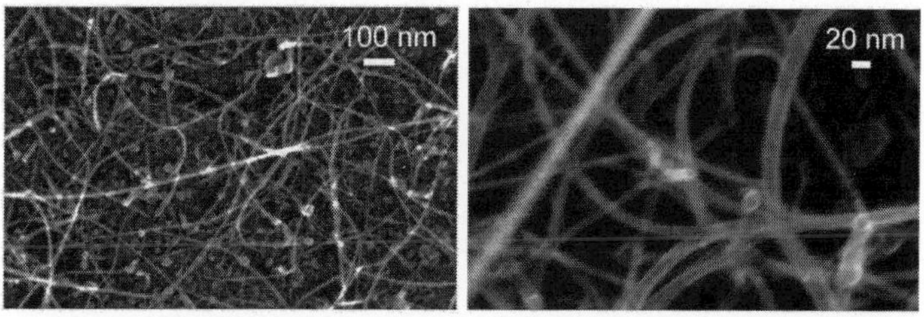

Figure 2. Typical randomly-oriented SWCNT networks with a unique "bridging" morphology grown in catalytic CVD [3].

Figure 3. Low- and high-resolution SEM images of the typical arrays of vertically-aligned CNTs grown in PECVD process with a glow discharge. The growth followed the 'tip-growth' mode as the catalyst nanoparticles are noticeable on the top of each nanotube [4].

2.2. Morphologies of nanotube arrays

In general, there are three types of morphologies observed in the directly-grown nanotube arrays: entangled, horizontally aligned, and vertically aligned. Each of these morphologies has their specific functionalities and can be desirable for different applications. In this work, we will briefly describe the first two morphologies and then pay the most attention to the arrays of vertically aligned nanotubes.

The horizontally-aligned CNT arrays were usually grown on the quartz wafers using CVD. The alignment could in such arrays be controlled by two factors: gas flow direction and substrate lattice. These arrays could have a very high density (up to 50 SWNT/μm) over large area. These nanotubes have also a large diameter and good electrical properties desirable for the nanoelectronic applications [7].

On the other hand, the vertically aligned CNTs could be grown using both CVD and PECVD. *Hata et al.* demonstrated that by using Fe/Al_2O_3 as the catalysts, C_2H_4 as the feedstock and a trace amount of water vapor (100 – 300 ppm) as the growth enhancer, high-yield, milli-meter long vertically aligned SWCNTs could be produced [8]. Water in this process was used to etch the possible amorphous carbon phase deposited onto the catalyst during the growth, therefore enhancing the lifetime and activity of the catalyst. The vertical alignment was supported by the collective van der Waals' interactions among the nanotubes [9]. In contrast, the CNTs grown in PECVD process do not require such growth enhancer to align them vertically, since the electrical field in the plasma-surface sheath at the vicinity of the substrate could easily direct the growth.

The third type of CNT arrays is the entangled network consisting of interconnected randomly-oriented nanotubes. In some cases, these networks are not entirely 'random'; instead, they may form certain unique features such as the 'Y-junctions', as well as 'knotted' and 'bridging' structures. *Sun et al.* demonstrated that by using a porous membrane filter to collect the nanotubes at room temperature, a unique 'Y-junction' with high electronic performance could be induced in an aerosol CVD process [10]. Similar to VACNTs, they can be produced in both CVD and PECVD processes. Figure 4 illustrates both the horizontally and vertically aligned morphologies obtained by our group.

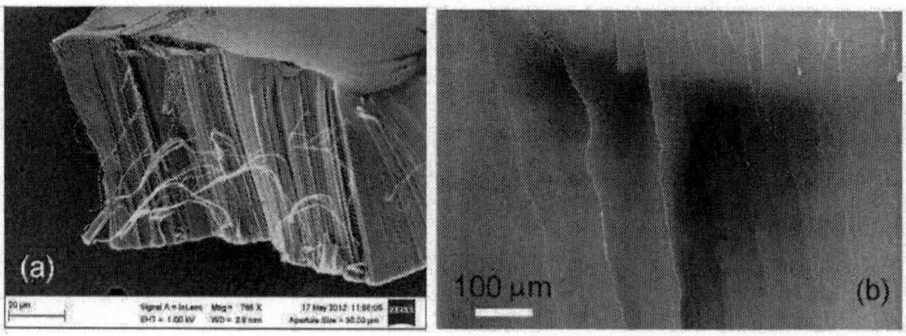

Figure 4. Highly uniform, dense array of vertically aligned single-walled carbon nanotubes (SWCNTs) grown on trilayered $Fe/Al_2O_3/SiO_2$ catalyst (a). Horizontally-aligned nanotubes (b).

2.3. Vertically-aligned arrays of carbon nanotubes

Vertically-aligned CNTs not only preserve the excellent intrinsic properties of individual nanotubes, but also show a high surface-to-mass ratio owing to their three-dimensional microstructure. Moreover, the surface of the vertically-aligned CNTs could be easily functionalized. These advantageous features have placed the vertically-aligned CNT arrays among the most promising materials for a variety of applications ranging from field emitters, heat sinks, nanoelectrochemical systems, gas- and bio-sensors, drug delivery systems, to molecular/particular membranes. For example, Wu et al. used the functionalized vertically-aligned CNTs to deliver nicotine for therapeutic purposes [11]; Han et al. studied the release behaviors of bone morphogenetic protein-2 (BMP-2; a growth factor for human mesenchymal stem cells) on the vertically-aligned CNTs with different surface wettability, in attempting to control the differentiation and proliferation of these stem cells [12].

Growth of the vertically-aligned CNTs can be easily obtained in PECVD. Figure 5 shows SEM images (high and low magnification) of the vertically-aligned nanotubes grown in the low-temperature plasma [13]. These CNTs have a diameter of 50-200 nm, a height of several micrometers, and followed a "tip-growth" mechanism. Interesting, they collapse upon liquid wetting (this will be discussed in more detail in the next section).

Figure 5. Dense array of vertically aligned single-walled carbon nanotubes.

The the vertically-aligned CNTs grown using CVD process are much denser and longer, and have more uniform distribution of diameters. In such arrays, very strong Van der Waals forces are present. The CVD process is therefore suitable for mass production of CNTs, and may contribute to lowering the price of CNTs. We have recently demonstrated that highly uniform and dense arrays of SWCNTs with more than 90% population of thick nanotubes (>3 nm in diameter) could be obtained by tailoring the thickness and microstructure of the catalyst supporting SiO2 layer [14].

2.4. Entangled arrays of carbon nanotubes

Networks of entangled nanotubes consist of randomly-oriented nanostructures. The thickness of the entangled array may vary from sum-monolayer to a few monolayers. Advantages of such morphology, as compared to individual nanotubes, are scalability, stability, reproducibility, and low cost of the CNT-based devices. They are therefore widely used as thin film transistors, transparent conductive coatings, solar cells, gas and biosensors. The electrical resistivity in entangled SWCNTs is determined by the nanotube-nanotube junctions in the network, and the nanotube-metal junctions at the electrodes (so-called Schottky barrier). The intrinsic resistance of the nanotubes usually plays a minor role if the array density is not far away from the percolation threshold [15]. In addition, it is generally perceived that for the CNT-based device to deliver outstanding performance, chirality-selected growth of CNT is a pre-requisite. However, for entangled SWCNTs, this stringent requirement may be avoided if the density is within a certain range (usually 1–3 nanotubes/μm^2) [16,17,18].

There are many parameters of the CVD process that should be controlled to grow entangled CNTs with some special patterns. For example, the length of the nanotubes could be determined by the exposure time of the carbon feedstocks. Recently, we have demonstrated that the density of entangled SWCNTs, which is a critical factor in device performance, could be controlled over 3-order-of-magnitude in acetylene-modulated CVD processes (Figure 6a) [2]. In addition, we also obtained a special 'knotted' morphology of the CNT network by using porous silica as the catalyst-supporting layer (Figure 6b) [19]. In contrast to this morphology, a much lower density of nanotubes was observed on flat silica surface.

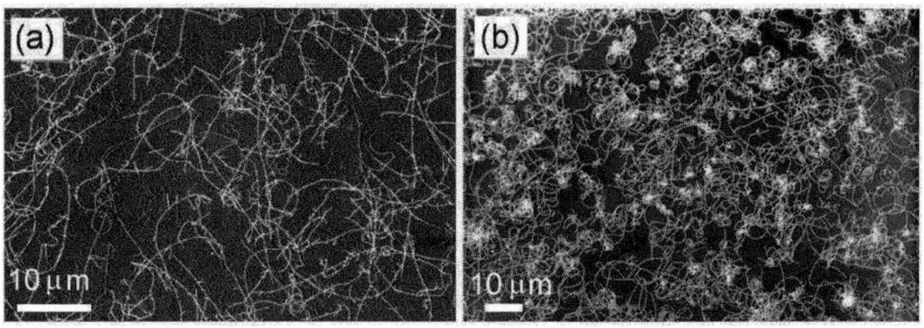

Figure 6. Representative arrays of entangled carbon nanotubes [2,19].

3. Complex catalyst-free arrays by mechanical writing

Plasma-based techniques yet being capable of producing high-quality nanotube arrays, still require metal catalyst to initialize the control the nanotube growth process. However, there is a strong demand in metal-free CNTs, i.e. the CNTs not containing a catalyst metal which is usually incorporated in the nanotube structure (from the nanotube top or bottom, depending on the process used). Removal of metal catalyst from CNTs implies a complex post-processing [20] which results in significant disadvantages, such as essential change in electronic properties or degradation of the nanotube ordering or orientation (in particular, post-processing deteriorates the vertical orientation of the nanotubes), damages the substrate structure in high temperature annealing process, etc. Thus, removal of the metallic catalyst by after-growth post-processing is feasible only for limited small-scale experimental production [21]. Hence, the development of the catalyst-free methods for growing arrays of high quality, dense vertically aligned nanotubes is a pressing demand now. The metal-free nucleation and growth of carbon nanotubes is possible, yet with the use of other catalytic material, and with a low quality outcome. For example, the nucleation and growth on semiconductor nanoparticles in CVD process was recently reported [22,23,24]. In these works, the nanotubes were catalyzed and grown without metal catalyst, but those nanotubes are not vertically aligned but highly tangled, tousled, and the surface density is quite low. Therefore, obtaining high quality arrays of CNTs on a catalyst-free silicon substrate still remains elusive.

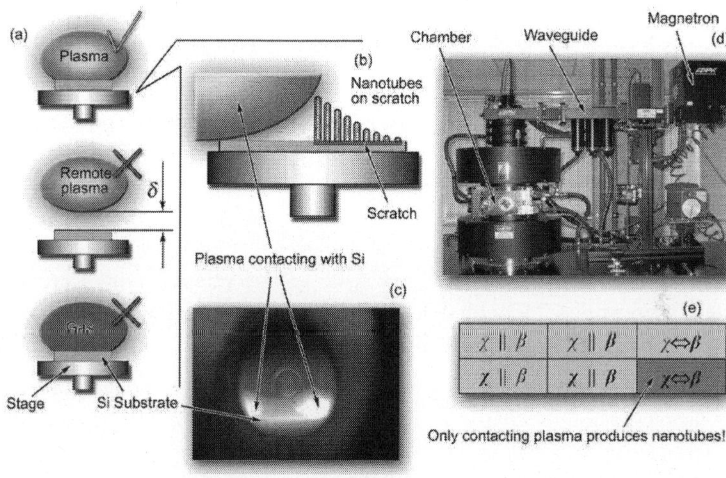

Figure 7. a) Three typical process configurations: localized plasma, remote plasma, gas environment; (b) nanotubes growth on Si substrate contacting with plasma: dense nanotubes as-grown on a doted spot; (c) photo of the plasma above substrate and (d) photo of the microwave reactor; (e) complete experiment matrix, which indicates the substrate condition (for scratched or non-scratched surface), and the process environment condition; remote gas/plasma and contacting gas/plasma. Among all possible 6 variants tested, only localized plasma process have produced nanotubes on substrate [25].

Figure 8. a,b) Top-view of CNTs on doted spots (SEM images); (c) microwave reactor; (d, e) tilted SEM image of CNT arrays showing a high number density of CNTs. Insets illustrate the process of making pattern and TEM image of the carbon nanotube [25].

Here we describe a novel plasma-based catalyst-free growth technique that is capable of producing very dense, strongly aligned arrays of extremely long (up to several hundred μm) CNTs on Si wafer surface in very fast process (with growth rate achieving 50 μm/sec), with experimentally proven possibility to arrange the nanotubes into complex arrays of various shapes such as separate nests and linear strands.

The six different experimental variations were used, with respect to the plasma/gas environments and plasma location relative to the substrate, as shown in Figure 7. We did not observe the nanotube nucleation in gas environment, on both smooth and patterned surfaces; we also did not observe the nucleation on both smooth and patterned surfaces with the remotely located plasma, and only the process conducted in plasma contacting the patterned surface resulted in the nucleation and growth of CNTs. The process starts by applying a special notch pattern (NP) on the prepared Si(100) wafers.

Then, the substrates with a specific NP (we used a linear NP consisting of parallel notches, and spot pattern of small pits) was treated in a chemical vapor deposition (CVD) reactor (Figure 8) where a microwave discharge was ignited in gas mixture of CH_4 and N_2, at pressure of 13 Torr and power density of 1.28 W/cm^3, typically for 3 min. The substrates were heated up to ~800 °C only by the plasma. The plasma localization relative to the substrate was varied to study in detail the plasma effect on CNT growth process (Figure 1e); namely, the process conducting with plasma contacting the wafer surface was effective for nucleation and growth of CNTs. More details on making the mechanical pattern are shown in Figure 8.

The scanning electron microscopy (SEM) investigations (Figure 9) show that a fast growth of a high-density, highly aligned CNTs are produced exactly replicating the pattern configuration (a complex pattern configuration which consists of a linear notch and spot applied directly on the notch have been achieved). The SEM images clearly show that the complex pattern of CNTs was perfectly replicated by nanotubes and the longest nanotubes reach ~140 μm in length. The growth sites are very densely occupied, and the rest wafer surface is absolutely free of nanotubes. Notably, these very dense arrays were formed in a very fast process, such unusual growth rates (up to ~48 μm/minute) were not reported previously in the absence of metal catalyst. Figure 9a shows the high-magnification SEM image of the nanotube array.

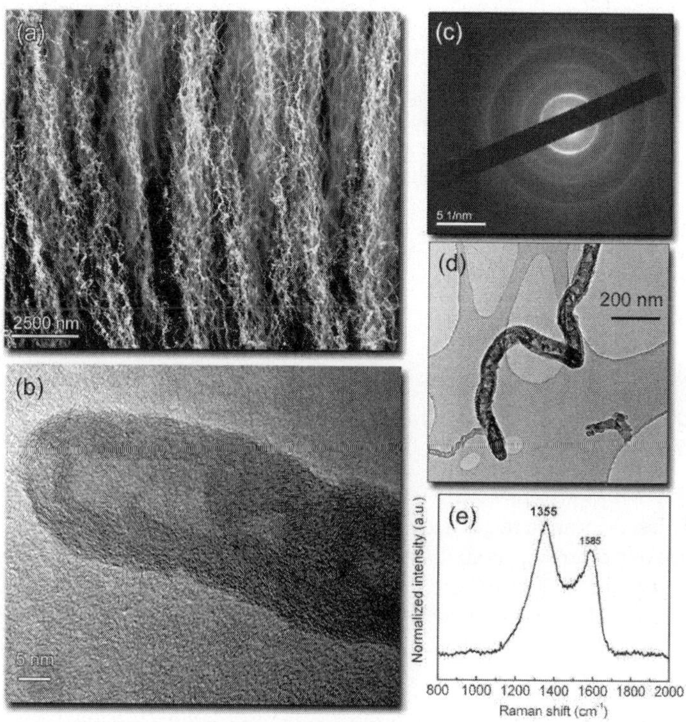

Figure 9. a) High-magnification SEM image of the vertically-oriented CNTs; (b) a high-resolution TEM image showing the planes in CNTs, with the inter-planer distance of ~0.34 nm; (c) the electron diffraction pattern of CNTs; (d) low-resolution TEM image showing the nanotube diameter of about 10 nm; (e) micro-Raman spectrum of the carbon nanotubes [25].

Further characterization of the nanotube structure was done with a high-resolution transmission electron microscopy (TEM) and Raman techniques (Figure 9). The TEM images (inset in Figure 8, Figures 9b and 9d) clearly show the absence of catalyst particle at the closed end tip of the CNTs, this reveals that the nanotubes were following in a "base-growth" mode [25]. As follows from TEM images, the diameters of the nanotubes are in the range of 10-80 nm, with up to 25 walls. Figure S13c shows the electron diffraction pattern of multi-wall nanotubes. Raman spectrum of as-grown nanotubes obtained at a room temperature (Figure 9e) shows a Raman broad-band peak at 1585 cm^{-1}, which is the characteristic of in-plane C-C stretching E2g mode of the hexagonal sheet. The appearance of a broad-band peak at 1355 cm^{-1} indicates the disordered graphitic nature of the nanotubes.

Thus, the nanotubes in our experiments were grown on the features mechanically written on the surface of Si wafer, and no nanotubes were formed on the intact silicon. To explain this, we propose a mechanism based on the key role of nano-elements on Si surface, so-called 'nano-hillocks'. These hillocks are formed on the surface when writing pattern, they establish a strong covalent bond to the Si surface at a temperature of ~800 °C during the process of CNT nucleation, and thus remain on Si surface, and hence at the bottom of nanotube during the whole growth process. Indeed, the solubility of carbon in Si is very low (10^{-3} %) [26] as compared to the conventional metallic catalyst such as Fe, Co, Ni etc., and thus the extremely high (up to 1 μm/s) growth rate observed in these experiments indicates that the nanotubes were grown via a surface diffusion, without involving very slow bulk-Si diffusion. Thus, a vapor-liquid-solid (VLS) mechanism was not involved, and the plasma played a key role in this process. We propose the following mechanism, so-called *reshaping-enhanced surface catalyzed* (RESC)growth. During the first stage, the tip region of a Si nano-hillock was heated up by plasma due to increased current density to the nano-sized tip (Figure 10).

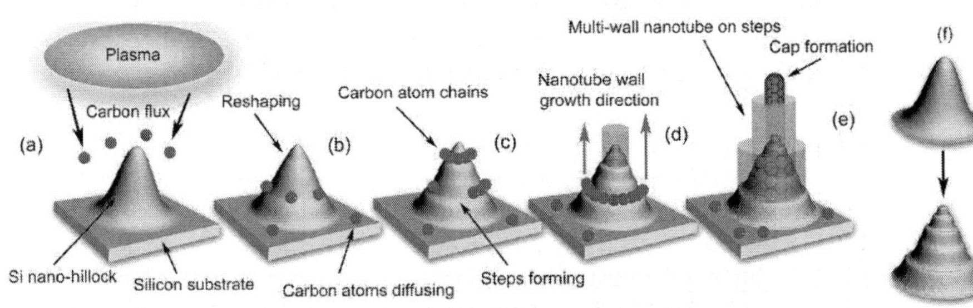

Figure 10. Scheme of the proposed mechanism of carbon nanotube nucleation and growth on silicon nano-hillocks in the plasma environment. (a) Si nano-hillock (with the shape 'as-produced' by mechanical patterning) is locally (mainly at the top) heated by the plasma; (b) heated Si nano-hillock starts reshaping – multiple step-like features are formed due to thermal re-arrangement and carbon saturation of the upper (overheated) Si layer; single carbon atoms incorporate into the steps; (c) reshaping continues, the steps become well-shaped, carbon atoms form chains (nanotube nuclei) along the multiple steps; (d) carbon chains close, nanotube start growing; (e) nanotube grow and close; (f) supposed reshaping of the silicon nano-hillock during plasma heating and nanotube nucleation [25].

Further, the heated silicon nano-hillock starts reshaping [27] by forming multiple step-like features due to thermal re-arrangement of silicon (to minimize the surface energy), and partially due to the possible carbon solution and saturation in the upper overheated Si layer. Then, carbon atoms incorporate into the steps and form closed chains. Simultaneously, the steps become well-shaped and thus carbon atoms assist the nanotube nucleation along the multiple steps. Later, multi-walled nanotubes start growing. Eventually, when the nanotube reaches 100-150 μm in length, the tip of carbon nanotube closes.

Thus, in this process the carbon catalization proceeds by the minimization of surface energy at the nano-hillock steps [28], since the adatom adsorbed in the step can be considered as 'partially dissolved'. As a result, this process leads to the formation of very dense array of very long multiwall nanotubes on the mechanically patterned areas. Thus, the proposed growth mechanism explains all the observed features; it is noteworthy that just the effect of plasma on the patterned surface explains several important characteristics due to plasma-related heating and high rate of material delivery. As a result, the catalyst-free, very dense arrays of long (up to 150 μm) vertically oriented multiwall carbon nanotubes were grown on the mechanically patterned silicon wafers in a low-temperature microwave discharge. These experiments have demonstrated an extremely high (up to 48 μm/min) growth rate.

4. Three-dimensional CNT arrays by post-processing with liquids

The above-described method can be used to produce 'planar', drawing-like arrays of the vertically-aligned carbon nanotubes on silicon surfaces. When a need in a complex three-dimensional array arises, post-processing of the uniform array (array-precursor) can be used. Among others, the post-treatment with a liquid is the most cheap and convenient [29-38]. Nevertheless, this technique still lacks controllability. In this section we show several possible ways of enhancing controllability of the fabrication of three-dimensional structures of the vertically-aligned carbon nanotube arrays. Specifically, we show that the array structure can be a key factor of the resultant structure fabricated by immersing the CNT array into liquid.

Figure 11a is an SEM image of the cross-section of the array of vertically-aligned CNTs grown using the CVD technique. This array exhibits super hydrophobic properties and thus, it cannot be wetted by water. After immersing into water, only weakly-collapsed irregular structure was produced (Figure 11b). In contrast, this array does not show super-hydrofobicity to acetone, and thus, highly-regular completely collapsed pattern was produced by immersing this array into acetone (Figures 11c, 11d). As one can see in this figure, this pattern exhibits very high surface area of the 'sponge', produced by carbon nanotubes (and hence, the walls of this sponge can be highly-conductive or semi-conductive). Such structures could be very useful for the fabrication of gas and bio-sensors, gas storage devices, as well as energy-transforming applications requiring very high levels of the light absorbance.

It is apparent that the control over the resultant structure of such patterns is a key issue for the above applications. Using different growth conditions, we have grown a similar CNT array with denser structure (see Figure 12a), which does not exhibit super-hydrophobic properties.

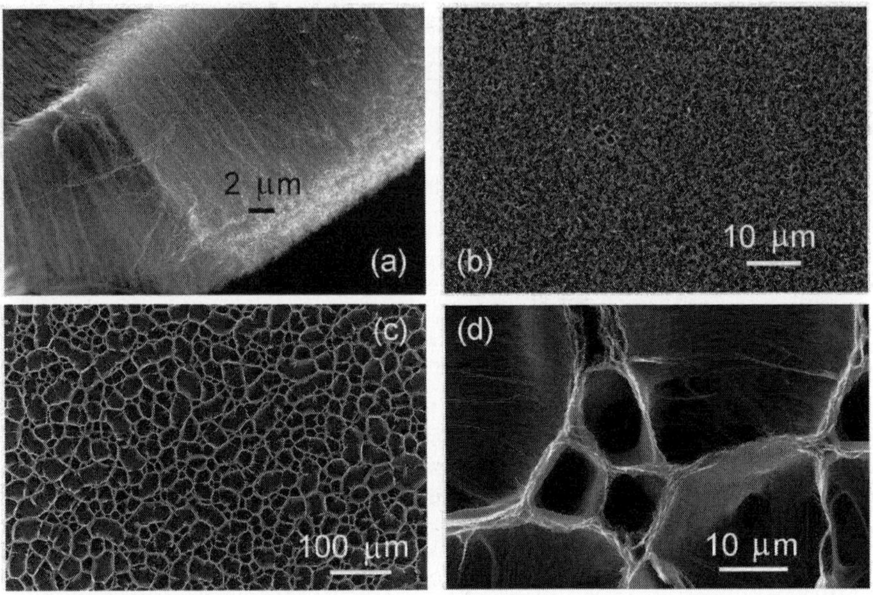

Figure 11. Treatment of super-hydrophobic sample with water and acetone. (a) Cross-section of the arrays of vertically-aligned nanotubes. (b) Irregular structure was produced after treating with water. (c, d) Regular structure (completely collapsed pattern) was produced by acetone, low and high magnifications.

Post-treatment of this array with ethanol and acetone has produced the sponge-like structure consisting of separated fine-porous island (Figures 12b, c, d). This structure is significantly different of that fabricated using super-hydrophobic sample (Figure 11). Moreover, variation of the dosage of liquid (we used 5 and 10 droplets of acetone, applied in sequence after complete drying of the preciously-applied drop) can be used to slightly change the structure. A comparison of the structures produced by 5 and 10 drops (Figures 12c and d) reveals a slight change in the pore sizes.

The use of water to treat this weakly hydrophobic CNT array produces slightly different pattern (Figure 13) consisting of smaller islands, which still demonstrate the sponge-like structure, i.e., each island is not a solid, intact array of the vertically-aligned carbon nanotubes but also consists of collapsed CNTs forming fine pores. One can expect that the fine-sponged structures produced using weakly hydrophobic CNT arrays can be very promised for the gas storage applications, whereas the highly-collapsed patterns may be more promising for sensing and other applications requiring control of the electrical resistivity of the surface. Thus, different internal structure of the vertically-aligned CNT array, together with the type of liquid and dosage, can be control parameters for the production of CNT patterns with a high level of the controllability.

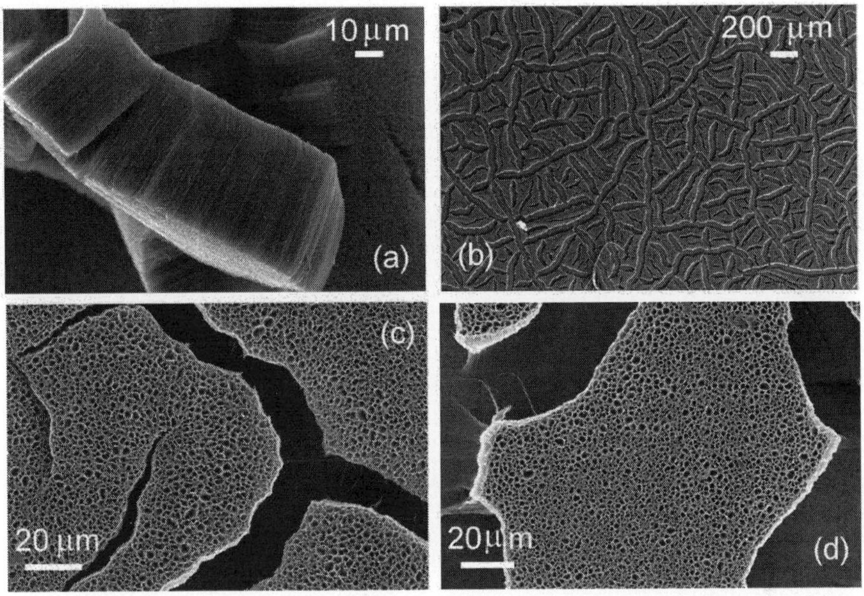

Figure 12. Treatment of weakly hydrophobic CNT array (a) with ethanol (b) and acetone, application of 5 drops (c) and 10 drops (d). Acetone produces a patterned sponge-like structure consisting of fine-porous island.

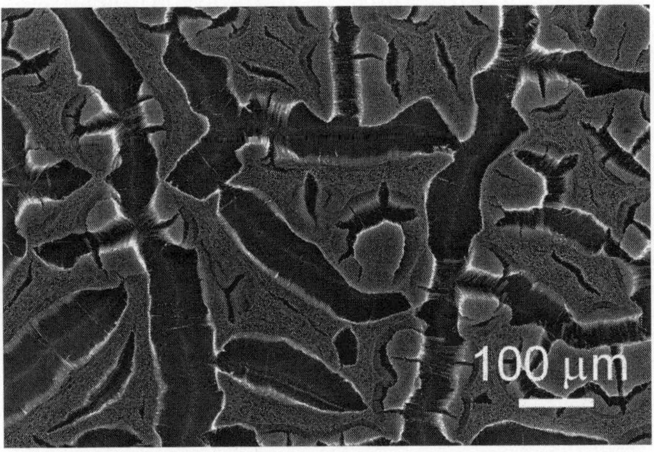

Figure 13. Treatment of weakly hydrophobic sample with water. Smaller islands demonstrate the sponge-like structure.

5. Perspective approaches for the control of array morphology

Dense arrays of highly ordered surface-bound vertically aligned nanotubes on silicon and metal oxides have a great potential for the fabrication of various advanced nano- and micro-devices such as fuel cells, sensors, and field emission element [39,40,41,42] One possible way to integrate the carbon nanotube array in the silicon platform is the use of anodized aluminum oxide (AAO) membranes to grow the pre-structured CNT patterns, bonded to the template surface. Indeed, the use of AAO membranes as growth templates was successful for the fabrication of, i.e., electron emitters [43]. Synthesis of carbon nanotubes on AAO templates allows precise and reproducible control of the dimensions of nanotubes [44, 45]. In this section we will review in short the AAO template characteristics important for growing the carbon nanotube arrays, and discuss the most important control parameters.

An AAO template can be prepared by the anodic oxidation of aluminum in various acid solutions. The thickness, pore size and interpore distance can be easily controlled by varying conditions of anodization such as composition of electrolyte, process temperature, applied voltage, process time and pore widening time [46,47]. Figure 14 shows SEM images of the free standing AAO templates fabricated by the two-step anodization.

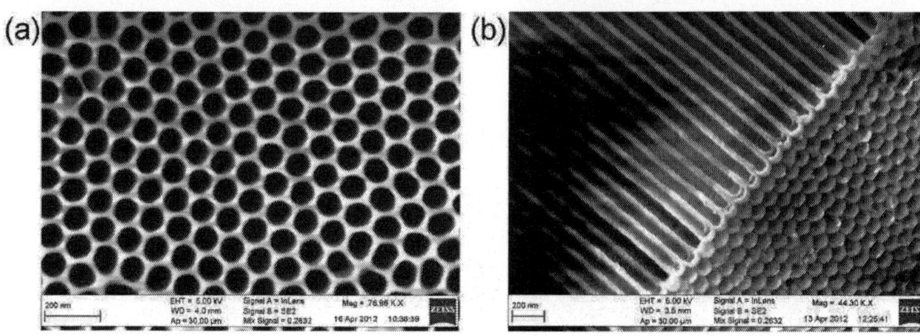

Figure 14. A Free-standing AAO fabricated using a two-step anodization. (a) Top view and (b) side view.

However, the free-standing AAO templates and membranes fabricated on aluminum foil are not be suitable for growing carbon nanotube arrays due to the thermal instability. Under thermal treatment, which is inevitable in the nanotube fabrication process, the AAO templates fabricated on aluminum foil easily crack due to the difference in thermal expansion coefficient of the alumina oxide and underlying aluminum. Moreover, the growth temperature cannot exceed the aluminum melting point e. In addition, a free-standing AAO template easily cracks due to its ceramic nature. Therefore, the conventional approach based on the use of the aluminum foil is not suitable for the CNT growth and fabrication of the carbon nanotube-based electronic devices.

To avoid this problem, it is necessary to fabricate AAO templates on other functional substrates. The alternative materials include silicon, quartz and ITO glass, on which the highly ordered structure of very thin AAO templates can be fabricated. For example, AAO templates fabricated on silicon wafers have already been used to fabricate highly ordered carbon nanotubes [48]. The AAO templates fabricated on non-aluminum substrates can be compatible with much higher processing temperatures, well above the 700 °C. Silicon substrates may be also useful for protecting AAO from distortion during the CNT growth.

However, quartz is more advantageous substrate for AAO membranes to be used as templates for the CNT synthesis. Quartz has a very high melting point, allowing for much higher temperatures, and can protect the AAO templates from cracking during the thermal treatment. Another advantage of quartz is transparency enabling the use of AAO templates as photonic crystals, and thus significantly broadening the application of fabricated AAO templates in other optics related applications.

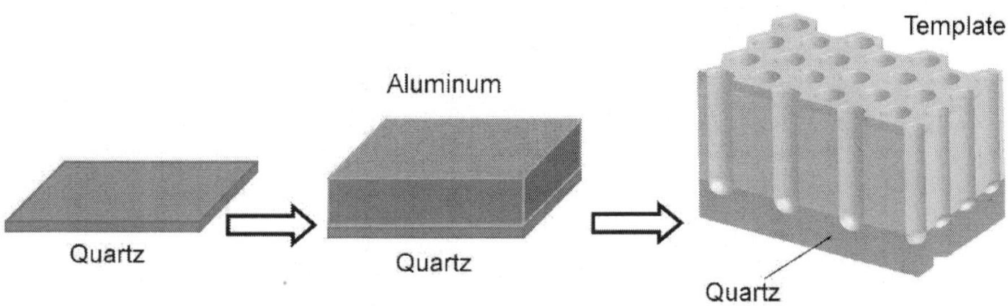

Figure 15. Schematic of the fabrication of AAO on quartz. High purity aluminum is deposited onto cleaned quartz, and the template is fabricated by anodization.

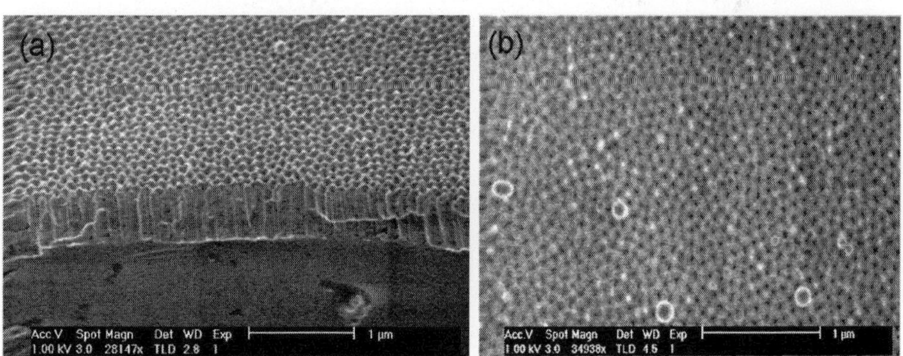

Figure 16. SEM images of the quartz-based AAO template suitable for the fabrication of carbon nanotube arrays. (a) Side view and (b) top view.

The high quality AAO templates were fabricated directly on quartz, using a two-step anodization without using any inter-layers between the deposited aluminum and quartz substrate. Figure 15 illustrates the schematic of this process. Prior to the fabrication of AAO template, quartz samples are cleaned in boiling solution of 30% w.t. of H_2SO_4 and 70% W.t. of H_2O_2. After that, the samples were etched in HF solution (0.1 w.t.% for 30 seconds), placed into an e-beam evaporator and coated with high purity (99.999%) aluminum to a thickness of 1.0 μm at a deposition rate of ≈1.5 nm×s^{-1}. The deposited Al film was then anodized to produce porous alumina templates in an electrolytic cell using a two-step anodization process. As a result, high quality AAO templates on quartz were fabricated. Figure 16 shows the SEM image of the AAO templates.

Above results demonstrate that AAO template technology not only can be used in a piece of aluminum foil, but also can be combined with silicon and other functional substrate technology. AAO template on functional substrate were used in the fabrication of CNT arrays can be realize the field emitters or possible become optical devices when CNT in quartz-AAO. Moreover, since the crystallinity of CNTs increase with the synthesis temperature, the emission current density increases with the synthesis temperature of CNTs. This again demonstrated that only AAO on functional substrate can realize high quality of CNTs array fabrication.

Author details

I. Levchenko[1,2], Z.-J. Han[1,2], S. Kumar[1,2], S. Yick[1,2], J. Fang[1,3] and K. Ostrikov[1,2]

1 Plasma Nanoscience Centre Australia (PNCA), CSIRO Materials Science and Engineering, Lindfield, Australia

2 Plasma Nanoscience, School of Physics, The University of Sydney, Sydney, Australia

3 School of Physics, The University of Melbourne, Parkville, VIC, Australia

References

[1] C. A. Crouse, B. Maruyama, R. Colorado Jr., T. Back, A. R. Barron. Growth, new growth, and amplification of carbon nanotubes as a function of catalyst composition. J. Am. Chem. Soc. 2008;130(25), 7946–7954.

[2] Z.-J. Han, I. Levchenko, S. Yick, K. Ostrikov. 3-Orders-of-magnitude density control of single-walled carbon nanotube networks by maximizing catalyst activation and dosing carbon supply. Nanoscale 2011;3, 4848-4853.

[3] Z.-J. Han, S. Yick, I. Levchenko, E. Tam, M. M. A. Yajadda, S. Kumar, P. J. Martin, S. Furman, K. Ostrikov. Controlled synthesis of a large fraction of metallic single-wal-

led carbon nanotube and semiconducting carbon nanowire networks. Nanoscale 2011;3, 3214-3220.

[4] Z. J. Han, B. K. Tay, M. Shakerzadeh, K. Ostrikov. Superhydrophobic amorphous carbon/carbon nanotube nanocomposites. Appl. Phys. Lett. 2009;94, 223106.

[5] I. Levchenko and K. Ostrikov, M. Keidar, S. Xu. Deterministic nanoassembly: Neutral or plasma route? Appl. Phys. Lett. 2006;89, 033109.

[6] M Meyyappan. A review of plasma enhanced chemical vapour deposition of carbon nanotubes. J. Phys. D: Appl. Phys. 2009;42, 213001.

[7] W. Zhou, L. Ding, S. Yang, J. Liu. Synthesis of high-density, large-diameter, and aligned single-walled carbon nanotubes by multiple-cycle growth methods. ACS Nano 2011;5, 3849–3857.

[8] K. Hata, D. N. Futaba, K. Mizuno, T. Namai, M. Yumura, S. Iijima. Water-assisted highly efficient synthesis of impurity-free single-walled carbon nanotubes. Science 12004;306, 1362-1364.

[9] G. Zhang, D. Mann, L. Zhang, A. Javey, Y. Li, E. Yenilmez, Q. Wang, J. P. McVittie, Y. Nishi, J. Gibbons, H. Dai. Ultra-high-yield growth of vertical single-walled carbon nanotubes: Hidden roles of hydrogen and oxygen. PNAS 2005;102, 16141-16145.

[10] D. Sun, M. Y. Timmermans, Y. Tian, A. G. Nasibulin, E. I. Kauppinen, S. Kishimoto, T. Mizutani, Y. Ohno. Flexible high-performance carbon nanotube integrated circuits. Nature Nanotech. 2011;6, 156–161

[11] J. Wu, K. S. Paudel, C. Strasinger, D. Hammell, A. L. Stinchcomb, and B. J. Hinds, Programmable transdermal drug delivery of nicotine using carbon nanotube membranes. Proc. Natl. Acad. Sci. USA 2010;107, 11698.

[12] Z. J. Han, K. Ostrikov, C. M. Tan, B. K. Tay, S. A. F. Peel. Effect of hydrophilicity of carbon nanotube arrays on the release rate and activity of recombinant human bone morphogenetic protein-2. Nanotechnology 2011;22, 295712.

[13] S. Kumar, I. Levchenko, Q. J. Cheng, J. Shieh, K. Ostrikov. Plasma enables edge-to-center-oriented graphene nanoarrays on Si nanograss. Appl. Phys. Lett. 2012;100, 053115.

[14] Z. J. Han, K. Ostrikov. Uniform, dense arrays of vertically aligned, large-diameter single-walled carbon nanotubes. J. Am. Chem. Soc. 2012;134, 6018–6024.

[15] M. P. Garrett, I. N. Ivanov, R. A. Gerhardt2,3, Alex A. Puretzky2, and David B. Geohegan. Separation of junction and bundle resistance in single wall carbon nanotube percolation networks by impedance spectroscopy Appl. Phys. Lett. 97, 163105 (2010); http://dx.doi.org/10.1063/1.3490650 (3 pages)

[16] Q. Cao, H. Kim, N. Pimparkar, J. P. Kulkarni, C. Wang, M. Shim, K. Roy, M. A. Alam, J. A. Rogers. Medium-scale carbon nanotube thin-film integrated circuits on flexible plastic substrates. Nature 2008;454, 495-500.

[17] M. A. Topinka, M. W. Rowell, D. Goldhaber-Gordon, M. D. McGehee, D. S. Hecht, G. Gruner. Charge transport in interpenetrating networks of semiconducting and metallic carbon nanotubes. Nano Letters 2009;9, 1866–1871.

[18] K. Ostrikov. Reactive plasmas as a versatile nanofabrication tool. Rev. Mod. Phys. 2005;77, 489–511.

[19] Z. J. Han, H. Mehdipour, X. Li, J. Shen, L. Randeniya, H. Y. Yang, K. Ostrikov. SWCNT networks on nanoporous silica catalyst support: morphological and connectivity control for nanoelectronic, gas-sensing, and biosensing devices. ACS Nano 2012;6, 5809–5819.

[20] T. W. Ebbesen, P. M. Ajayan, H. Hiura, K. Tanigaki. Purification of nanotubes. Nature 1994;367 519-page.

[21] X. M. H. Huang, R. Caldwell, L. Huang, S. C. Jun, M. Huang, M. Y. Sfeir, S. P. O'Brien, J. Hone. Controlled Placement of Individual Carbon Nanotubes Nano Letters 2005;5 1515.

[22] D. Takagi, H. Hibino, S. Suzuki, Y. Kobayshi, and Y. Homma. Carbon Nanotube Growth from Semiconductor Nanoparticles. Nano Letters 2007;7(8) 2272-page.

[23] B. Liu, W. Ren, L. Gao, S. Li, S. Pei, C. Liu, C. Jiang and H. M. Cheng. Metal-Catalyst-Free Growth of Single-Walled Carbon Nanotubes. J. Am. Chem. Soc. 2009;131 2082.

[24] S. Huang, Q. Cai, J. Chen, Y. Qian, and L. Zhang. Metal-Catalyst-Free Growth of Single-Walled Carbon Nanotubes on Substrates. J. Am. Chem. Soc. 2009;131, 2094-page.

[25] S. Kumar, I. Levchenko, K. Ostrikov and J. McLaughlin. Plasma-enabled, catalyst-free growth of carbon nanotubes on mechanically-written Si features with arbitrary shape. Carbon 2012, 50, 321.

[26] P. Laveant, P. Werner, G. Gerth, and U.Gosele. Incorporation, diffusion and agglomeration of carbon in silicon. Solid State Phenomeona 2002;82-84, 189.

[27] S. Helveg, C. Lopez-Cartes, J. Sehested, P. L. Hansen, B. S. Clausen, J. R. Rostrup-Nielsen, F. Abild-Pedersen, J. K. Norskov. Atomic-scale imaging of carbon nanofibre growth. Nature 2004;427, 426.

[28] J. Pezoldt, Y. V. Trushin, V. S. Kharlamov, A. A. Schmidt, V. Cimalla, O. Ambacher. Nuclear Instr. Meth. in Phys. Res. B 2006;253, 241.

[29] E.M. Kotsalisa, E. Demosthenousb, J.H. Walthera, c, S.C. Kassinosb, P. Koumoutsakos. Wetting of doped carbon nanotubes by water droplets. Chem. Phys. Lett. 2005;412, 250–254.

[30] M. A. Correa-Duarte, N. Wagner, J. Rojas-Chapana, C. Morsczeck, M. Thie, M. Giersig. Fabrication and Biocompatibility of Carbon Nanotube-Based 3D Networks as Scaffolds for Cell Seeding and Growth. Nano Lett. 2004;4 2233-2236.

[31] B. Q. Wei, R. Vajtai, Y. Jung, J. Ward, R. Zhang, G. Ramanath, P. M. Ajayan. Organized assembly of carbon nanotubes. Nature 2002; 416, 495-496.

[32] C. L. Pint, Y.-Q. Xu, M. Pasquali, R. H. Hauge. Formation of Highly Dense Aligned Ribbons and Transparent Films of Single-Walled Carbon Nanotubes Directly from Carpets. ACS Nano 2008;2, 1871–1878.

[33] B. Pokroy, S. H. Kang, L. Mahadevan, J. Aizenberg. Self-Organization of a Mesoscale Bristle into Ordered, Hierarchical Helical Assemblies. Science 2009; 323, 237-240.

[34] D. N. Futaba, K. Hata, T. Yamada, T. Hiraoka, Y. Hayamizu, Y. Kakudate, O. Tanaike, H. Hatori, M. Yumura, S. Iijima. Shape-engineerable and highly densely packed single-walled carbon nanotubes and their application as super-capacitor electrodes. Nature Materials 2006;5, 987-994.

[35] S. Li, H. Li, X. Wang, Y. Song, Y. Liu, L. Jiang, D. Zhu. Super-hydrophobicity of large-area honeycomb-like aligned carbon nanotubes. J. Phys. Chem. B 2002;106, 9274-9276.

[36] W. Wang, M. Tian, A. Abdulagatov, S. M. George, Y.-C. Lee, R. Yang. Three-dimensional Ni/TiO2 nanowire network for high areal capacity lithium ion microbattery applications. Nano Lett. 2012;12, 655–660.

[37] H. Liu, J. Zhaia, L. Jiang. Wetting and anti-wetting on aligned carbon nanotube films. Soft Matter. 2006;2, 811–821.

[38] C. T. Wirth, S. Hofmann, J. Robertson. Surface properties of vertically aligned carbon nanotube arrays. Diam. Relat. Mater. 2008;17, 1518–1524.

[39] Z. Wang, L. Ci, L. Chen, S. Nayak, P. M. Ajayan and N. Koratkar. Polarity-Dependent Electrochemically Controlled Transport of Water through Carbon Nanotube Membranes. Nano Letters 2007;7, 697.

[40] K. Gong, F. Du, Z. Xia, M. Durstock and L. Dai. Nitrogen-doped carbon nanotube arrays with high electrocatalytic activity for oxygen reduction. Science 2009;323, 760.

[41] I. Levchenko, S.Kumar, M. M. A. Yajadda, Z. J. Han, S. Furman, K. Ostrikov. Self-organization in arrays of surface-grown nanoparticles: characterization, control, driving forces. J. Phys. D: Appl. Phys. 2011;44, 174020.

[42] M. E. Itkis, A. Yu and R. C. Haddon. Single-Walled Carbon Nanotube Thin Film Emitter–Detector Integrated Optoelectronic Device. Nano Letters 2008;8, 2224.

[43] S.-K. Hwang, J. Lee, S.-H. Jeong, P.-S. Leen, K.-H. Lee. Fabrication of carbon nanotube emitters in an anodic aluminum oxide nanotemplate on a Si wafer by multi-step anodization. Nanotechnology 2005;16, 850–858.

[44] K. Liu, M. Burghard and S. Roth. Conductance spikes in single-walled carbon nanotube field-effect transistor. Appl. Phys. Lett. 1999;75, 2494.

[45] J. S. Suh, J. S. Lee. Highly ordered two-dimensional carbon nanotube arrays. Appl. Phys. Lett. 1999;75 2047.

[46] J. Fang, P. Spizzirri, A. Cimmino, S. Rubanov, and S. Prawer, Nanotechnology, 20, 065706 (2009). Extremely high aspect ratio alumina transmission nanomasks: their fabrication and characterization using electron microscopy.

[47] A. P. Li, F. Müller, A. Birner, K. Nielsch, U. Gösele. Hexagonal pore arrays with a 50–420 nm interpore distance formed by self-organization in anodic alumina. J. Appl. Phys. 1998;84, 6023.

[48] S. Chu, K.Wada, S. Inoue, S. Todoroki. Synthesis and characterization of titania nanostructures on glass by Al anodization and sol-gel process. Chem. Mat. 2002;14(1), 266.

Chapter 3

Classification of Mass-Produced Carbon Nanotubes and Their Physico-Chemical Properties

Heon Sang Lee

Additional information is available at the end of the chapter

http://dx.doi.org/10.5772/52613

1. Introduction

Carbon nanotube (CNT) may be classified into single walled (SWCNT), double walled (DWCNT), and multiwalled carbon nanotube (MWCNT) according to the number of graphene layers. In some cases, bamboo-shaped multiwalled carbon nanotubes were also synthesized. Among these carbon nanotubes, multiwalled carbon nanotubes have been mass-produced in hundreds metric tons level. Many researchs on multiwalled carbon nanotubes point to an electrode, polymer composites, coating, and others. The number of graphene layers, purity, and crystallinity are the main features of multiwalled carbon nanotubes, which need to be characterized. We have proposed one important characteristic of multiwalled carbon nanotubes, the mesoscopic shape of MWCNT, of which many industrial applications may be comprised. According to our suggestion, one can determine the degree of tortuousness of MWCNT, quantitatively. (*see ref.1-5 and sections 1-6 in this chapter*)In this chapter, we will describe the mesoscopic shape factor of MWCNT in detail. Various physical properties as well as toxicity may strongly depend on the mesoscopic shape factor of MWCNT. Our suggestion has also been published as an international standard ISO/TS11888 by international organization for standardization (ISO) in 2011.

I hope readers enjoy the concepts and expressions shown in this chapter. Especially, this chapter shall be helpful to whom may want to develop a commercial application by selecting a proper CNT.

2. Static bending persistence length (SBPL, l_{sp})

If MWCNTs have no defect along their axis, their appearance would be straight to several hundred micro meter. Persistence length is the maximum straight length that is not bent by thermal energy. The persistence length of MWCNT is expected to be several hundred micro meter due to its exceptional high modulus. Static bending persistence length (SBPL) has been proposed in our earlier work to quantify the mesoscopic shape of MWCNT. SBPL is the maximum straight length that is not bent by permanent deformation. Fig. 1 shows the concept of SBPL. When a length considered is longer than SBPL, the shape of MWCNT looks tortuous. On the contrary, the shape of MWCNT looks straight as a length considered is shorter than SBPL.

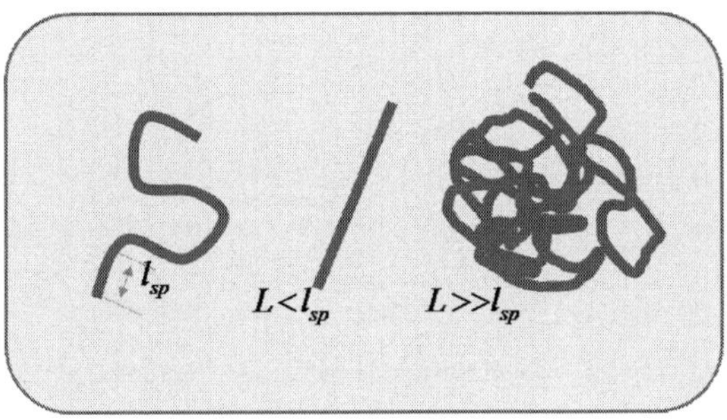

Figure 1. The concept of static bending persistence length of MWCNT (l_{sp}).

If length considered is longer than SBPL, the shape of MWCNT looks tortuous. On the contrary, the shape of MWCNT looks straight as a length considered is shorter than SBPL.

3. Mathematical expression of SBPL (l_{sp})

The end-to-end vector can be obtained such as eq 1 when the distribution of bending points ($\{\varphi\} \equiv (\varphi_1, \varphi_2, \ldots, \varphi_k)$) is given.

$$\mathbf{R} = N \sum_{i=1}^{k} \varphi_i \mathbf{r}_i \tag{1}$$

The spatial average of end-to-end distance $\langle R \rangle$ should be zero, since probability to bend to one direction is the same as that to the opposite direction. Then spatial average of square end-to-end vector is obtained as eq 2

$$\left\langle \mathbf{R}^2 \right\rangle = N^2 \sum_{i=1}^{k}\sum_{j=1}^{k} (\varphi_i \mathbf{r_i}) \cdot (\varphi_j \mathbf{r_j}) = N^2 b^2 D_b \tag{2}$$

$$D_b \equiv \left\langle \mathbf{R}^2 \right\rangle / L^2 \cong \sum_{i=1}^{k} \varphi_i^2 \tag{3}$$

where D_b is a bending ratio, $\varphi_i = N_i / N$, N_i is the number of unit segment in i-direction segment, N is the total number of unit segment, $k = m + 1$, m is the number of static bending points on a coil, and r_i is i-direction segment vector with the length of b. The expression shown in eq 3 is significant. This indicates that we can obtain the distribution function when we have enough data. This is often called as ill-posed problem. Regularization method in applied mathematics gives us the solution for solving the problems. Equation 3 holds only if a probability of the fold-back conformation is the same as that of the straight conformation.By using the scaling law, the coil expressed in eq 2 and 3 can be renormalized into the coil that has constant segment length, $2l_{p0}$. Then we can obtain eq 3 with $\varphi_i = 2l_{p0} / L$ and $k = L \ / 2l_{p0}$. We can also consider a case where the bent angle (θ) between the *ith* and *(i+1)*th segments is a fixed small angle. The spatial average of the square end-to-end vector is obtained as following

$$\left\langle \mathbf{r}_n \cdot \mathbf{r}_m \right\rangle = N \left(\sum_{i=1}^{k} \varphi_i^2 \right) b^2 (\cos\theta)^{|n-m|} \tag{4}$$

$$\left\langle \mathbf{R}^2 \right\rangle = \sum_{n=1}^{k}\sum_{m=1}^{k} \left\langle \mathbf{r}_n \cdot \mathbf{r}_m \right\rangle = \sum_{n=1}^{k}\sum_{p=-n+1}^{k-n} \left\langle \mathbf{r}_n \cdot \mathbf{r}_{n+p} \right\rangle \cong \sum_{n=1}^{k}\sum_{p=-\infty}^{\infty} \left\langle \mathbf{r}_n \cdot \mathbf{r}_{n+p} \right\rangle \tag{5}$$

$$\sum_{p=-\infty}^{\infty} \left\langle \mathbf{r}_n \cdot \mathbf{r}_{n+p} \right\rangle = N \left(\sum_{i=1}^{k} \varphi_i^2 \right) b^2 \left(1 + 2\sum_{p=1}^{\infty} \cos^p \theta \right) = N \left(\sum_{i=1}^{k} \varphi_i^2 \right) b^2 \left(\frac{1+\cos\theta}{1-\cos\theta} \right) \tag{6}$$

$$\left\langle \mathbf{R}^2 \right\rangle = (N^2 b^2)(\sum_{i=1}^{k} \phi_i^2)(\frac{1+\cos(\theta)}{1-\cos(\theta)}) = L^2 D_b \tag{7}$$

$$D_b \equiv \frac{\left\langle \mathbf{R}^2 \right\rangle}{L^2} \cong \left(\sum_{i=1}^{k} \varphi_i^2 \right) \left(\frac{1+\cos(\theta)}{1-\cos(\theta)} \right) \tag{8}$$

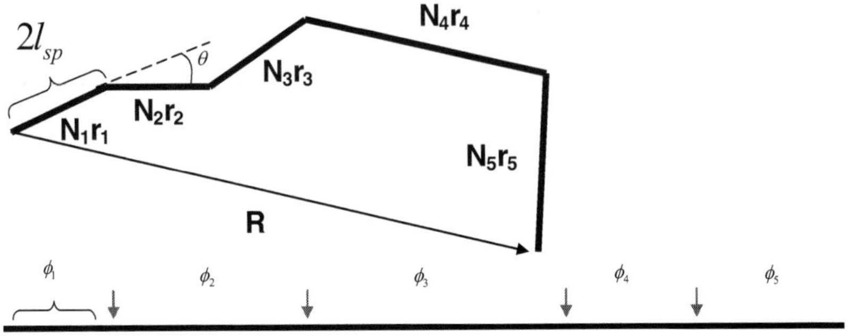

Figure 2. Tortuous MWCNT; bent points are distributed randomly along MWCNT axis.

Equation 7 can also be renormalized into the coil that has a constant segment length, $2l_{sp}$. The bending ratio (D_b) is expressed as eq 9

$$D_b \equiv \frac{\langle \mathbf{R}^2 \rangle}{L^2} \cong \left(\frac{2l_{p0}}{L}\right)\left(\frac{1+\cos(\theta)}{1-\cos(\theta)}\right) = C\left(\frac{2l_{p0}}{L}\right) = \frac{2l_{sp}}{L} \tag{9}$$

where $l_{sp}=Cl_{p0}$ is the static bending persistence length and Cshould be a constant for a fixed bent angle. The static bending persistence length is a statistical quantity, representing the maximum straight length that is not bent by static bending. In the case of continuous curvature, a more accurate statement is that the static bending persistence length is the mean radius of curvature of the rigid random-coil due to static bending. The same quantity arising from dynamic bending instead of static bending is dynamic bending persistence length (l_p). The dynamic bending persistence length represents the stiffness of the molecules as determined by the effective bending modulus against thermal energy in Brownian motion. Equation 5 is valid when $L >> l_{sp}$, the coil limit. $D_b = \langle \mathbf{R}^2 \rangle / L^2 = 1$ when $L < l_{sp}$, the rod limit.If we know the values of end-to-end distance and contour length, the bending ratio can be obtained from the mean-squared end-to-end distance divided by the mean-squared contour length. The end-to-end distance of RRC varies with the change of bending angle. The difference can be compromised by using an arbitrary unit segment length which is similar to the scaling of polymer chain. The mean-squared end-to-end distance by the Kratky-Porod (KP) expression is given by eq 10 when the dynamic bending persistence length (l_p) is replaced by the static bending persistence length (l_{sp}) and the twice l_{sp} equals to Kuhn length.

$$\langle \mathbf{R}^2 \rangle = 2l_{sp}L + 2l_{sp}^2\left(e^{-L/l_{sp}} - 1\right) \tag{10}$$

http://dx.doi.org/10.5772/52613

4. Measurement methods for SBPL

The plot of eq 10 is presented in Fig. 3. Given data, the SBPL can be obtained by eq 11.

$$l_{sp} = \lim_{L\to\infty} \frac{1}{2} \frac{dD_b}{d\ln L} \tag{11}$$

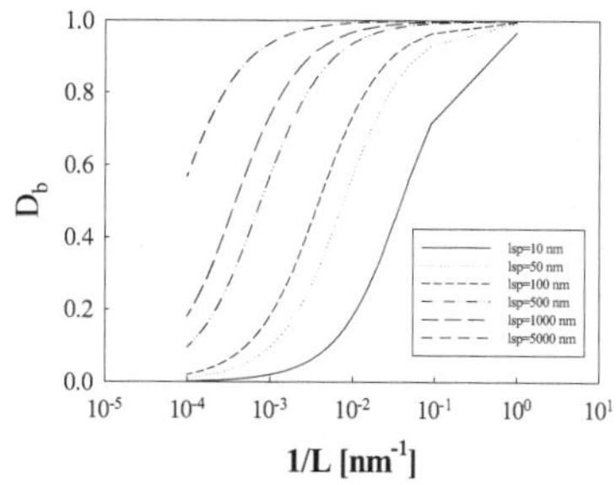

Figure 3. Bending ratio (D_b) with respect to reciprocal contour length.

In this method, one need to have experimental data for $\langle R^2 \rangle$ and L. In order to obtain these data, one have to cut MWCNTs into pieces with various L. Acid cutting or mechanical cutting method may be applied to obtain pieces of MWCNT. It is worth to note that $\langle R^2 \rangle$ are Gaussian, given contour length (L). That is, various end-to-end distances may be measured for a constant L. This method is exact, but hard to obtain the experimental data.

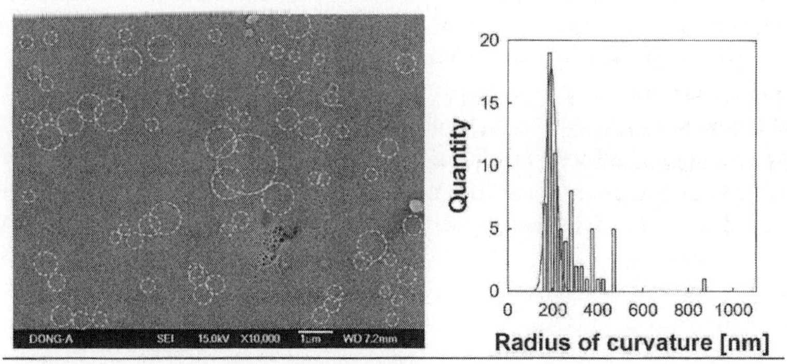

Figure 4. Approximation method to determine SBPL; the mean radius of curvature approximate SBPL.

The mean radius of curvature approximates the SBPL. One can easily obtain the mean value of the radius of curvatures of MWCNTs from any SEM images as seen in Fig. 4. The approximation method is convenient because SEM images of as-synthesized or as-received MWCNT can be directly used. The SBPL obtained by the approximation could have an error up to 200% compared to those obtained by exact method. However, the approximated value of SBPL still has physical significant in many applications, since many applied properties depend on the order of magnitude of SBPL.

5. Intrinsic viscosity of MWCNTs

From the molecular weight, the contour length, and the persistence length, the intrinsic viscosity of MWCNTs can be calculated. If we apply the intrinsic viscosity model of a worm-like coil to the rigid random-coil, the following expressions are obtained,

$$[\eta] = 2.20x10^{21}\frac{\langle R^2\rangle^{3/2}}{M}f \tag{12}$$

$$f \cong \left[1+0.926\theta\left(D_b\right)^{1/2}\right]^{-1} \tag{13}$$

$$\theta = \ln\left(\frac{2l_{sp}}{e}\right) - 2.431 + \left(e/a\right) \tag{14}$$

where M is molecular weight, $e$is spacing between frictional elements along the contour, $a=\zeta/3\pi\eta_s$, $\zeta$is the friction factor for a single frictional element, and η_s is the solvent viscosity. For the non-draining limit for the random coil, $f=1$, giving the maximum value of intrinsic viscosity in the model. When we take the static bending persistence length (l_{sp}) as the length of a single frictional element, the friction factor of the element in eq 14 may follow the rigid-rod model such that $\zeta_T = 3\pi\eta_s l_{sp}/(\ln(l_{sp}/d)+0.3)$ for the translational motion and may be $\zeta_r = \pi\eta_s l_{sp}{}^3/(3(\ln(l_{sp}/d)-0.8))$ for the end-over-end rotational motion. Translational-rotational coupling and hydrodynamic shielding may also be considered for the evaluation of friction factor in eq 14.In this case, we can surmise that friction factor in eq 14 is scaled with l_{sp}^s, where s is larger than unit value. We can reasonably neglect e/a in eq 14. The measurement of intrinsic viscosity assumes the deformation rate is slow enough. The intrinsic viscosity is determined by the competition of tendency of orientation toward flow direction and tendency to random orientation due to thermal motion (Brownian motion). The measurement often performed at shear rate of several hundreds reciprocal second. At this regime, CNTs may be extended to the static shape by shear force where peclet number

($Pe = \dot{\gamma}(2R_h)^2 / D_T$) is over 10.It is worth noting that the static bending persistence length determined from intrinsic viscosity is consistent with that determined from 3-D SEM analysis in dried state.

6. Diffusions of MWCNTs

Not only the toxicological issues but also researches on novel hybrid materials or nano-scale devices points to the need for the understanding of overall shape and mobility of carbon nanotube particles in a solution or in atmosphere.The degree of flexibility of carbon nanotubes is the major ingredient for the shape and mobility, however it is also puzzling.The persistence lengths of single-walled carbon nanotubes are expected to be in the order of tens to hundreds of micrometers due to their exceptionally large modulusand to have longer persistence lengths for muliwalled nanotubes, indicating currently prepared several-micrometer long nanotubes behave like rigid rods. Elastic fluctuations of semi-rigid particles by thermal energy have been described exactly by the worm-like coil model proposed more than 50 years ago by Kratky and Porod. The model describes the stiffness of molecules by dynamic bending persistence lengths (mean radius of curvatures) which are determined by effective bending modulus (E_{eff}) against thermal energy (kT) in a solution. Theoretical calculationshave shown that the dynamic bending persistence lengths (l_ps) of carbon nanotubes (CNTs) are up to several millimetersdue to their exceptionally large Young's modulus of about 1.5 TPa. Real-time visualization technique revealed that l_p s of singlewalled carbon nanotubes (SWCNTs) are between 32 and 174μm, indicating SWCNTs shorter than l_p(=32μm) may be rigid around room temperature in a solution. However, rippling developed on the compressive side of the tube leading to a remarkable reduction of the effective bending modulus, which is more pronounced for multiwalled carbon nanotubes (MWCNTs). Theoretical calculationshave shown that the effective bending moduli of MWCNTs are around 0.5 $nN\ nm^2$ when the radii of curvatures are around 150 ~ 500nm. This indicates MWCNTs longer than 0.5 μmmight be flexible in a solution around room temperature, since thermal energy is about 4.1 x $10^{-3}nN\ nm$. It seems not likely that van der Waals interaction between graphene layers is the only reason that makes the effective bending moduli of MWCNTs more than 100 times smaller than SWCNT.

Both MWCNTs and SWCNTs discussed above are no more than worm-like coils (WLCs) where ensemble average of overall size (end-to-end distance) scales with the square root of molecular weight (contour length) in asymptotic limit. Our recent work has revealed that the spatial average of overall size of MWCNTs also follows the same scaling as WLCs in spite of their static bent points. We designated these MWCNTs as rigid random-coils (RRCs).The only difference between RRCs and WLCs is whether the bending points are static or dynamic by thermal energy.The relationship between the shape and size of RRCs has been characterized by static bending persistence lengths (l_{sp}s). Because both RRCs and WLCs are Gaussian, the models for the mobility of WLCs may also work for RRCs.

Translational diffusion coefficient is defined by the mobility of particle against thermal energy as Einstein relation, eq 15.

$$D = \frac{kT}{\zeta} \tag{15}$$

where k is Boltzman constant, T is temperature, and $1/\zeta$ is the mobility. By analogy to macromolecule, a MWCNT with static bend points can also be considered to be made up N identical structural elements with a frictional factor ζ_e per unit element and a spacing e between elements along the contour of the coil.In this case, the mobility in eq 15 may be expressed as the sum of free-draining contribution $(1/N\zeta_e)$ and hydrodynamic interaction contribution which is called non-draining term.

$$\frac{1}{\zeta} = \frac{1}{N\zeta_e} + \frac{1}{2!N^2}\sum_i\sum_j\frac{1}{\zeta_{ij}} + \cdots \tag{16}$$

where ζ_{ij} is the frictional factor by interaction between ith and jth element and $i \neq j$.

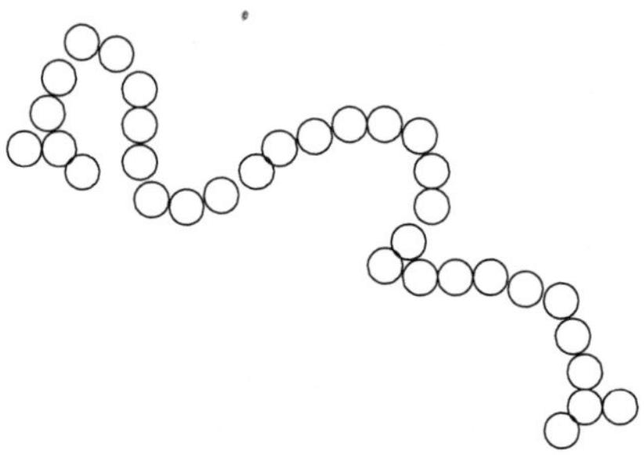

Figure 5. CNT made up N identical frictional element.

When we choose a spherical bead having diameter of a as a frictional element, the frictional factor of each element follows Stokes-Einstein relation, $\zeta_e = 3a\pi\eta_s$ where η_s is the viscosity of solvent. The frictional factor by interaction between ith and jth element may also follow Stokes-Einstein relation since a mean value of distance between elements i and $j$is small. Then, Kirkwood expression is obtained such as eq 17.

$$\frac{1}{\zeta} = \frac{1}{N3a\pi\eta_s}\left[1 + \frac{a}{2Ne}\sum_i\sum_j\left(1-\delta_{ij}\right)\left\langle r_{ij}^{-1}\right\rangle\right] \tag{17}$$

$$\frac{1}{\overline{R_{ij}}} = \frac{1}{e\overline{r_{ij}}} = \left(\frac{1}{e}\right)\sum_i\sum_j\left\langle\left(1-\delta_{ij}\right)\left\langle r_{ij}^{-1}\right\rangle\right\rangle \tag{18}$$

where $r_{ij} \equiv R_{ij}/e$ and R_{ij} is the distance between element i and j,e is solvated diameter of molecule, and N is number of frictional element, $N = L\ /e$. Here, we can see that mathematical expression for the mobility of RRCs is similar to that of WLCs. Equation 17 is widely used for the estimation of translational diffusion coefficient of macromolecules. Equation 18 has been solved by Hearst and Stockmayer using Riseman-Kirkwood theory and Daniel distribution. We notice that $\bar{r}_{ij}$ is no more than a mean value of distance between the frictional element i and j. Then, $\bar{r}_{ij}$ must depend on the conformation of the carbon nanotubes. Hearst and Stockmayer obtained $\bar{r}_{ij}$ by the Kirkwood-Riseman theory as eq 19.

$$\frac{1}{2}\sum_x\sum_n\left\langle\left(1-\delta_{xn}\right)\left\langle r_n^{-1}\right\rangle_x\right\rangle = \int_1^N dl\int_1^l dn\int_0^\infty F(r,n)rdr = \int_1^N dl\int_1^l dn\int_0^\infty\left[F(r,n)-f(r,n)\right]rdr + \int_1^N dl\int_1^l dn\int_0^\infty f(r,n)rdr \tag{19}$$

where $F(r,\ n)$ is the unknown distribution for all n, $f(r,\ n)$ is the known distribution, x is the contour distance of the point of interest from one end of the carbon nanotube, $n$is the contour distance from the point of interest to the frictional element n, and r is the displacement of frictional element n from the point of interest. Hearst and Stockmayer chose the Daniels distribution which includes a first-order correction to a Gaussian distribution as $f(r,\ n)$, and obtained $\bar{r}_{ij}$ as eq 20.

$$\bar{r}_{ij} = \left[2N\left(\ln\left(2L_p\right)-2.431+1.843\left(N/2L_p\right)^{1/2}+0.138\left(N/2L_p\right)^{-1/2}-0.305\left(N/2L_p\right)^{-1}\right)\right]^{-1} \tag{20}$$

where $N = L\ /e$, L is contour length, $e$is spacing between frictional elements along the contour, $L_p = l_p/e$, and l_p is persistence length. The translational diffusion coefficient of worm-like coil can be estimated by eqs 15, 17, and 20. Rotational diffusion coefficientis expressed as eq 21.

$$D_r = \left(\frac{kT}{\eta_s}\right)\left(\frac{2}{l_pL^2}\right)\left[0.253\left(\frac{L}{4l_p}\right)^{1/2}+0.159\ln\left(2L_p\right)-0.387+0.160\right] \tag{21}$$

Equations 20 and 21 are valid for a semi-flexible rod when the contour length of rod is much longer than its persistence length such that the mean squared end-to-end distance follows random-coil scaling, $\langle R^2\rangle = Nb^2$ where $b = 2l_p$.The semi-flexible rods in this coil limit, $L >> l_p$, are so-called worm-like coils (WLCs). We see that the mobility is determined solely by the average conformation of particle with a given solvent viscosity and a contour length in eqs 13 and 14. We can reasonably surmise that the diffusion coefficients of RRCs are similar to those of WLCs with a given contour length, if the values of static bending persistence lengths (l_{sp}s) of RRCs are the same as those of the dynamic bending persistence lengths (l_ps) of WLCs.When hydrodynamic shielding effect is taken into account, the diffusion coefficients of RRCs might be slightly larger than those of WLCs due to the static bent points. The root mean-squared end-to-end distance of RRCs are given by eq 22.

$$\langle \mathbf{R}^2 \rangle = \left(N^2 b^2\right)\left(\sum_{i=1}^{k}\varphi_i^2\right)\left(\frac{1+\cos(\theta)}{1-\cos(\theta)}\right) = L^2 D_b \tag{22}$$

$$D_b \equiv \frac{\langle \mathbf{R}^2 \rangle}{L^2} \cong \left(\sum_{i=1}^{k}\varphi_i^2\right)\left(\frac{1+\cos(\theta)}{1-\cos(\theta)}\right) \cong \left(\frac{2l_{p0}}{L}\right)\left(\frac{1+\cos(\theta)}{1-\cos(\theta)}\right) = C\left(\frac{2l_{p0}}{L}\right) = \frac{2l_{sp}}{L} \tag{23}$$

where D_b is bending ratio, l_{sp} is static bending persistence length, l_{p0} is an arbitrary constant segment length, $\theta$is static bent angle from the MWCNT axis, $\varphi_i = N_i / N$, $N_i$is the number of unit segment in i-direction segment, N is the total number of unit segment, $k = m + 1$, m is the number of static bending points on a coil. When RRCs have semi-flexibility by thermal energy, the ensemble average of bent angle (θ_2) always becomes larger in amount of $\Delta\theta$ than the static bent angle $\theta;\theta_2 = \theta + \Delta\theta$. This is due to the fact that the effective bending modulus toward the bent direction is smaller than that toward the opposite direction.This indicates that the overall size of RRCs may be decreased when they are fluctuated by thermal energy. Because frictional elements of RRCs have Gaussian distribution by the definition of RRCs, those of semi-flexible RRCs also have Gaussian distrbution. Therefore, eqs 20 and 21 are also valid for semi-flexible RRCs when the persistence length is replaced by an apparent persistence length. The apparent persistence length (l_{ap}) is determined by the static bent angle (θ) and dynamic bent angle due to thermal energy ($\Delta\theta$) as following.

$$l_{ap} = l_{sp}\left(\frac{1+\cos(\theta+\Delta\theta)}{1-\cos(\theta+\Delta\theta)}\right)\left(\frac{1-\cos(\theta)}{1+\cos(\theta)}\right) \tag{24}$$

Expression for the translational diffusion coefficient of RRCs can be obtained from eqs 15, 17,18, and 24.

$$D_T = \frac{kT}{3\pi\eta_s L}\left[1+\ln\left(2L_{ap}\right) - 2.431 + 1.843\left(N/2L_{ap}\right)^{1/2} + 0.138\left(N/2L_{ap}\right)^{-1/2} - 0.305\left(N/2L_{ap}\right)^{-1}\right] \tag{25}$$

Similary, expression for the rotational diffusion coefficient of RRCs can be obtained as following.

$$D_r = \left(\frac{kT}{\eta_s}\right)\left(\frac{2}{l_{ap}L^2}\right)\left[0.253\left(\frac{L}{4l_{ap}}\right)^{1/2} + 0.159\ln\left(2L_{ap}\right) - 0.387 + 0.160\right] \quad (26)$$

where $L_{ap} = l_{ap} / e$.Equations 25 and 26 are promising for the estimation of diffusion coefficients of MWCNTs synthesized by a CVD method. In other words, eqs 25 and 26 give us the information of the shape and size of MWCNTs if we have the measured values of diffusion coefficients.

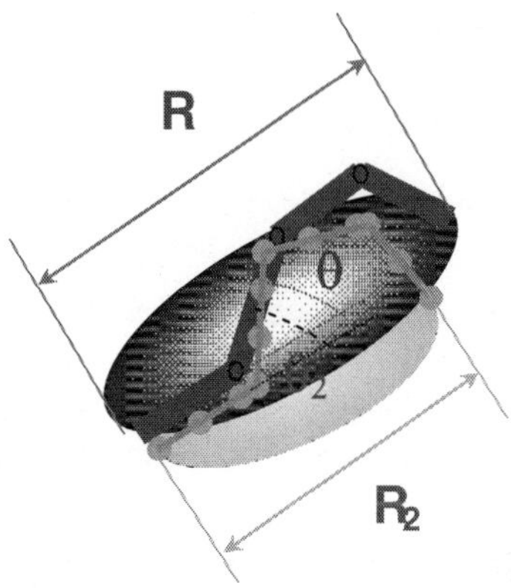

Figure 6. Size of carbon nanotube (R) is decreased to R2 at the elevated temperature.

7. Dynamic light scattering

The translational and rotational Brownian motions lead to fluctuation in the intensity of scattered light.The velocity of particles in Brownian motion can be directly measured by using dynamic light scattering (DLS) method, since time is correlated to obtain the intensity autocorrelation function ($g^{(2)}(t)$). The intensity autocorrelation function is connected to the electric field autocorrelation function ($g^{(1)}(t)$) which is given by eq 27 for a monodisperse solution.

$$\left|g^{(1)}(t)\right| = \exp(-\Gamma t) \quad (27)$$

$$\int_0^\infty G(\Gamma)d\Gamma = 1$$

where $\Gamma = Dq^2$ with D, the translational diffusion coefficient of the molecules, and q, the scattering vector magnitude ($q = 4\pi n \sin(\theta/2)/\lambda_0$ where n is the solution refractive index, θ is the scattering angle, and λ_0 is the incident light wavelength *in vacuo*). For polydisperse solutions, the electric field correlation function is given by a sum or distribution of exponentials,

$$\left|g^{(1)}(t)\right| = \int_0^\infty G(\Gamma)\exp(-\Gamma t)d\Gamma \tag{28}$$

where

In many cases, a single exponent obtained from an average translational diffusion coefficient value fits the decay rate of the electric field correlation function such as eq 29

$$K_1 = \langle\Gamma\rangle_{av} = \int_0^\infty \Gamma G(\Gamma)d\Gamma \tag{29}$$

where K_1 is the first cumulant. When a solution is dilute enough neglecting the interaction between particles, the effective diffusion coefficient can be obtained by the intensity average. The intensity of light scattered by macromolecule species i is often proportional to the molecular weight (M_i) times the weight concentration (c_i). In this case, the intensity average diffusion coefficient equals to z average diffusion coefficient.

The average decay rate (Γ) of the electric field autocorrelation function can be obtained by using conventional DLS.The first cumulant generally fits the data well for carbon nanotube solutions. When the incident light and detector are both vertical, V_v, translational diffusion are characterized by eq 30.

$$\Gamma_{Vv} = q^2 D_T \tag{30}$$

When the incident light vertical and detector horizontal, H_v , the diffusion of anisotropic particle are characterized by eq 31

$$\Gamma_{Hv} = q^2 D_T + 6D_R \tag{31}$$

where D_T is translational diffusion coefficient and D_R is rotational diffusion coefficient. This equation is valid if the particle rotates many times while diffusing a distance comparable to q^{-1} or if there is little anisotropy in particle dimension. MWCNTs solution meets with the former case in this work. Now we have three independent mathematical model eqs 25, 26, and 12. And two equations for DLS measurement.Three unknown shape factors of static bending persistence length, contour length, and thermal fluctuation angle can be determined from the measured diffusion coefficients and intrinsic viscosity using eqs 25, 26, and 12.This is uniqueness of carbon nanotubes compared to macromolecules, since macromolecules have only two unknown shape factors of persistence length and contour length.

8. Micro rheology

The terminology of "microrheology" is used, to distinguish the technique from conventional (macro) rheology. In the microrheology, colloidal particles are used for probing the rheology of material of interest. The starting point is the Stokes-Einstein relation.

$$D_T = \frac{kT}{6a\pi\eta_s} \tag{32}$$

If we have measured values of translational diffusion coefficient, the viscosity of material of interest can be easily obtained by

$$\eta_s = \frac{kT}{6a\pi D_T} \tag{33}$$

where a is the radius of spherical colloidal particle; colloidal particles have usually average diameter between 1 nm and 1000 nm. It is comparable the ISO definition of nanoparticles those have average diameter between 1 nm and 100 nm. In this sense, nanoparticles are just some kinds of colloidal particles. When we have measured value of mean-squared displacement of probe particle, the following equation can be utilized.

$$\left\langle r^2(t) \right\rangle = 6D_T t = \frac{kT}{\pi a \eta_s} \tag{34}$$

This seemingly simple idea has done a great impact on various research fields, indeed. One example is the nanoparticles dispersed in a polymer melt. It is often reported that nanoparticles seems diffuse faster than expected. The origin of this phenomenon lies in the "Nano" size. The viscosity of polymer melt is well described by integral constitutive equations such as reptation model. In this model, the viscosity is determined by the stress relaxation time of polymer chain from the constraint of entanglement. When the observation time is much shorter than any relaxation time of polymer in rheometry of frequency sweep, the polymers behave like a crosslinked rubber, exhibiting a plateau modulus. The plateau modulus of polymers is determined from the entanglement lengths of polymer such as

$$G_o^N = \frac{\rho RT}{M_e} \tag{35}$$

The plateau modulus of polymer is usually reported in the order of $10^{6\sim7}$Pa. Entanglement molecular weight of polymer is about 1000~2000 g/mole. The entanglement length is about 10~100 nm.The particles having comparable size to the entanglement length of a polymer would feel less frictional force than expected from the melt viscosity in macrorheology. Therefore, viscosity of polymer melt is much lower for the nanoparticles. This may lead to the faster thermal motion of nanoparticle compared to a larger particles.

9. Applications

In our previous works, we demonstrated that MWCNT having shorter SBPL have a certain merit in a polymer composite for electrical conductive application. When MWCNTs are needle-like, polymer composites comprised of them exhibit higher electrical conductivity compared to those comprised of tortuous MWCNTs. The situation changed drastically when the composites were molded into a specimen by injection molding machine. The needle like CNTs aligned to the flow direction, which broke the electrical conductive networks, then the composites lose the electrical conductivity. However, this problem was not observed when the composites contained tortuous MWCNTs which have a short SBPL. We also showed that the electrical percolation threshold depends on the length of MWCNT when MWCNT are needle-like. But, the electrical percolation threshold depends on the SBPL for a tortuous MWCNT. Thermal conductivity and linear thermal expansivity are also strongly dependant properties on the SBPL of MWCNT. Especially, thermal shrinkable material can be fabricated as well as thermal expansive material by controlling SBPL of MWCNT.

Author details

Heon Sang Lee

Address all correspondence to: heonlee@dau.ac.kr

Dong-A University, Department of Chemical Engineering, Sahagu, Busan, South Korea

References

[1] Lee, H. S.; Yun, C. H.; Kim, S. K.; Choi, J. H.; Lee, C. J.; Jin, H. J.; Lee, H.; Park, S. J.; Park, M. Appl. Phys. Lett. 2009, 95, 134104.

[2] Lee, H. S.; Yun, C. H. J. Phys. Chem. C 2008, 112, 10653-10658.

[3] Lee, H. S.; Yun, C. H.; Kim, H. M.; Lee, C. J. J. Phys. Chem. C 2007, 111, 18882- 18887.

[4] Han, M. S.; Lee, Y. K.; Yun, C. H.; Lee, H. S.; Lee, C. J.; Kim, W. N. Synthetic Metals 2011, 161, 1629-1634.

[5] Heo, Y. J.; Yun, C. H.; Kim, W. N.; Lee, H. S. Curr. Appl. Phys. 2011, 11, 1144-1148.

[6] Hearst, J. E.; Stockmayer, W. H. J. Chem. Phys. 1962, 37, 1425-1433.

[7] Squires, T. M.; Mason, T. G. Ann. Rev. Fluid Mech. 2010, 42, 413-438.

[8] Hearst, J. E. J. Chem. Phys. 1963, 38, 1062-1065.

[9] Hearst, J. E. J. Chem. Phys. 1964, 40, 1506-1509.

[10] Kratky, O; Porod, G. Recl. Trav. Chim. Pays-Bas 1949, 68, 1106-1122.

[11] Doi, M.; Edwards, S. F. The theory of polymer dynamics; Oxford University Press: New York, U.S., 1986.

[12] Berne, B. &Pecora, R. Dynamic Light Scattering; Wiley: New York, 1976.

[13] Tanford, C. Physical Chemistry of Macromolecules; John Wiley & Sons: New York, 1961.

[14] Koppel, D. E. J. Chem. Phys. 1972, 57, 4814-4820.

[15] Cush, R.; Russo, P. S.; Kucukyavuz, Z.; Bu, Z.; Neau, D.; Shih, D.; Kucukyavuz, S.; Ricks, H. Macromolecules1997, 30, 4920-4926.

Chapter 4

Fabrication, Purification and Characterization of Carbon Nanotubes: Arc-Discharge in Liquid Media (ADLM)

Mohsen Jahanshahi and Asieh Dehghani Kiadehi

Additional information is available at the end of the chapter

http://dx.doi.org/10.5772/51116

1. Introduction

Carbon nanotubes (CNTs) were first discovered by Iijima in 1991 [1]. CNTs have sparked great interest in many scientific fields such as physics, chemistry, and electrical engineering [2, 3]. CNTs are composed of graphene sheets rolled into closed concentric cylinders with diameter of the order of nanometers and length of micrometers. CNTs are in two kinds, based on number of walls, the single-walled and multi-walled.

The diameter of single walled carbon nanotubes (SWNTs) ranges from 0.4 nm to 3nm and the length can be more than 10 mm that makes SWNTs good experimental templates to study one-dimensional mesoscopic physics system [3]. These unique properties have been the engines of the rapid development in scientific studies in carbon based mesoscopic physics and numerous applications such as high performance field effect transistors [4-9], single-electron transistors [10, 11], atomic force microscope tips [12], field emitters [13, 14], chemical/biochemical sensors [15-18], hydrogen storage [19].

There are three important methods to produce high quality CNT namely laser [20], arc discharge [21, 22], and Chemical Vapor Deposition (CVD) [23, 24]. Recently, arc discharge in liquid media has been developed to synthesize several types of nano-carbon structures such as: carbon onions, carbon nanohorns and carbon nanotubes. This is a low cost technique as it does not require expensive apparatus [25,26].

However, several techniques such as oxidation, nitric acid reflux, HCl reflux, organic functionalization, filtration, mechanical purification and chromatographic purification have been developed that separate the amorphous carbons and catalyst nanoparticles from the CNTs while a significant amount of CNTs are also destroyed during these purification processes [27].

In this review paper, synthesis, purification and structural characterization of CNTs based on arc discharge in liquid media are reviewed and discussed. In addition, several parameters such as: voltage difference between electrodes, current, type and ratio of catalysts, electrical conductivity, concentration, type and temperature of plasma solution, as well as thermal conductivity on carbon nanotubes production are investigated.

2. Synthesis of CNTs

2.1. Laser vaporization

The laser vaporization method was developed for CNT production by Smalley's group [28, 29]. The laser is suitable for materials with high boiling temperature elements such as carbon because of its high energy density. The quantities of CNTs, in this method are large. Smalley's group further developed the laser method also known as the laser-furnace method [28]. Fullerenes with a soccer ball structure are produced only at high furnace temperatures, underlining the importance of annealing for nanostructures [28]. These discoveries were applied to produce CNTs [29] in 1996, especially SWNTs. A beam of high power laser impinges on a graphite target sitting in a furnace at high temperature as Figure1 shows.

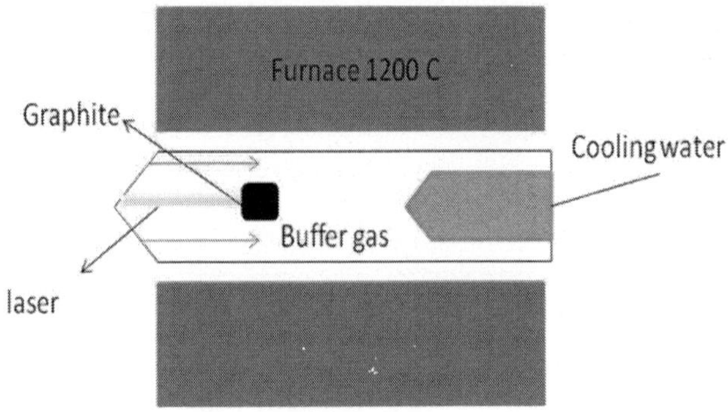

Figure 1. Schematic drawing of a laser obtain.

The target is vaporized in high-temperature buffer gas like Ar and formed CNTs. The produced CNTs are conveyed by the buffer gas to the trap, where they are collected. Then CNTs can be found in the soot at cold end.

This method has several advantages, such as high-quality CNTs production, diameter control, investigation of growth dynamics, and the production of new materials. High-quality CNTs with minimal defects and contaminants, such as amorphous carbon and catalytic metals, have been produced using the laser-furnace method together with purification processes [30-32] but the laser has sufficiently high energy density to vaporize target at the molecular level. The graphite vapor is converted into amorphous carbon as the starting material of

CNTs [33-35]. However, the laser technique is not economically advantageous, since the process involves high purity graphite rods, high power lasers and low yield of CNTs.

2.2. Chemical vapor deposition (CVD)

The chemical vapor deposition (CVD) is another method for producing CNTs in which a hydrocarbon vapor is thermally decomposed in the presence of a metal catalyst. In this method, carbon source is placed in gas phase in reaction chamber as shown in Figure 2. The synthesis is achieved by breaking the gaseous carbon molecules, such as methane, carbon monoxide and acetylene, into reactive atomic carbon in a high temperature furnace and sometime helped by plasma to enhance the generation of atomic carbon [36].

This carbon will get diffused towards substrate, which is coated with catalyst and nanotubes grow over this metal catalyst. Temperature used for synthesis of nanotube is 650 – 900 ^{0}C range and the typical yield is 30% [36].

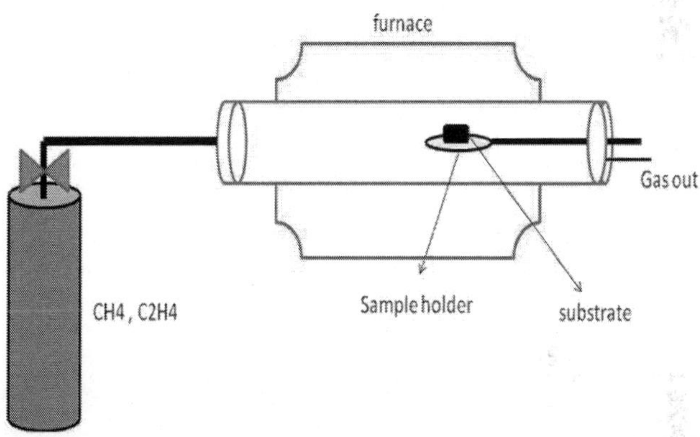

Figure 2. Schematic diagram of a CVD setup.

In fact, CVD has been used for producing [37-39] carbon filaments and fibers since 1959. Figure 2 shows a schematic diagram of the setup used for CNT growth by CVD in its simplest form. CNTs grow over the catalyst and are collected upon cooling the system to room temperature. The catalyst material may be different, solid, liquid, or gas and can be placed inside the furnace or fed in from outside [40, 41].

2.3. Arc Discharge

A schematic diagram of the arc discharge apparatus for producing CNTs is shown in Figure 3. In this method, two graphite electrodes are installed vertically, and the distance between the two rod tips is usually in the range of 1–2 mm. The anode and cathode are made of pure graphite (those are, with a purity of 99.999%).

The anode is drilled, and the hole is filled with catalyst metal powder then the chamber is connected to a vacuum line with a diffusion pump and to a gas supply [43]. Like the anode in a DC electric arc discharge reactor, CNT is synthesized of graphite rod. After the evacuation of the chamber by a diffusion pump, rarefied ambient gas is introduced [43].

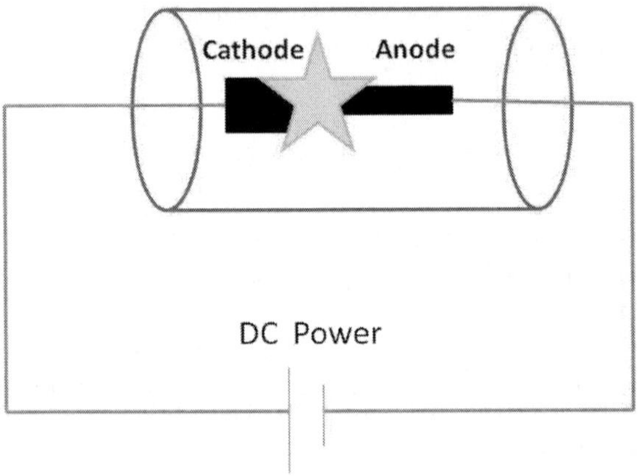

Figure 3. Schematic diagram of apparatus for preparing CNTs.

When a dc arc discharge is applied between the two graphite rods, the anode is consumed, and fullerene is formed in the chamber soot [43]. The mass production of multi wall carbon nanotubes (MWCNTs) by this dc arc discharge evaporation was first achieved by Ebbesen and Ajayan [44].

3. Arc-Discharge in Liquid Media (ADLM)

The traditional arc-plasma growth method for CNTs necessitates complex gas-handling equipment, a sealed reaction chamber, a liquid-cooled system and time consuming purge cycles. The act of extraction of nanotube's product is so complicated [45]. To be compared, the growth of arc plasma in water requires simple operation and equipment which had made the process of CNTs production noticeable [47].

Ishigami et al. [48] developed a simple arc method in liquid nitrogen for the first time that allow for the continuous synthesis of high-quality CNTs. The materials obtained are mainly MWCNTs, amorphous carbon, graphitic particles and carbonaceous material [48, 49]. Subsequently, an aqueous arc-discharge (arc-water) method was developed. Lange et al. [50] generated onions, nanotubes and encapsulates by arc discharge in water. Figure 4 is shown produced CNTs in LiCl 0.25 N media [51].

http://dx.doi.org/10.5772/51116

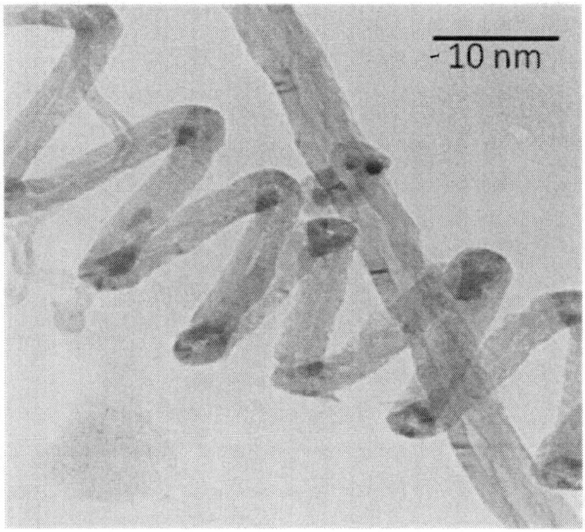

Figure 4. Carbon nanotubes produced in LiCl 0.25 N [51].

The arc discharge in liquid is initiated between two high purity graphite electrodes. Figure 5 is shown schematic device of arc discharge in liquid. Both electrodes are submerged in the liquid in a beaker. At first, the electrodes touch each other and are connected with a direct current (DC) power supply.

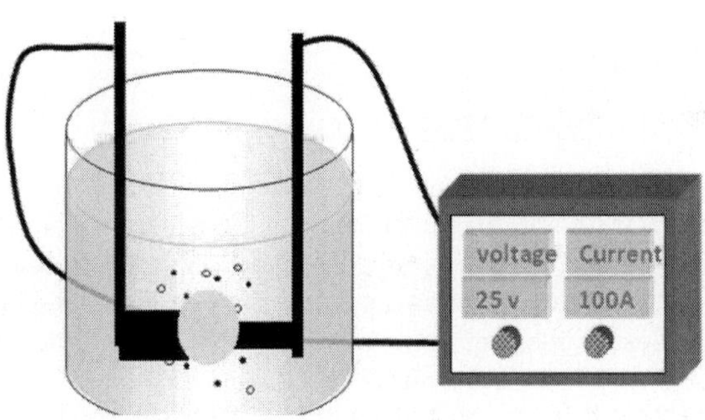

Figure 5. Schematic device of arc discharge in liquid.

The cathode is usually 20 mm in diameter, while the anode is 6 mm in diameter. Then the arc discharge is initiated by slowly detaching the moveable anode from the cathode. The arc gap is kept at the proper value (about 1 mm) that the continuous arc discharge could be obtained [52].

Recently carbon nanotubes (CNTs) were fabricated successfully with arc discharge in solution by a novel full automatic set up [51].

The arc discharge and consequently consumption of anode result to increase the distance between the two electrodes and degrees the voltage between them as well. In order to remain constant voltage and gap between the two electrodes, the program automatically will compare the initial voltage with the voltage of two electrodes with an accuracy of 0.1 V. Based on the calculated difference the program calculates the proportional coefficient for the proportional controller. Figure 6 is schematic of the apparatus used for automatic arc discharge in solution [51].

During the arc discharge, the gap between the two electrodes is maintained at approximately 1 mm, while the synthesis time is 60 s [51].

Figure 6. Schematic of the apparatus used for automatic arc discharge in solution [51].

3.1. Catalyst Materials and their ratio

A metal catalyst is necessary for the growth of the CNTs in all the methods used for synthesis of CNTs. Catalysts use to prepare CNTs usually include transition metals as a single or mixture of two catalysts such as a single, Fe, Co, Ni or Mo [53] or mixture of two catalysts such as FeNi [54], PtRh [55] and NiY [22]. Hsin et al. [57] firstly reported the production of metal-containing CNTs by arc discharge in solution.

The catalyst activation is determined in relation to the melting temperature and the boiling temperature thus the melting and boiling temperature of a catalyst can be one of the vital factors in the synthesis of SWCNTs [51].

CNTs are synthesized while the anode is filled by divergent single or bimetallic catalysts [51]. Scanning electron micros copies (SEM) show that the fabricated CNTs without any catalyst, figure 7, is in a very short long, disordered and is faulty grown.

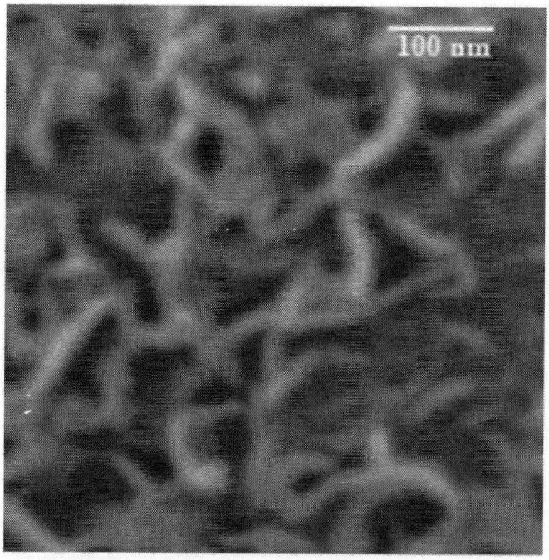

Figure 7. SEM image of the product which was fabricated without catalyst [51]

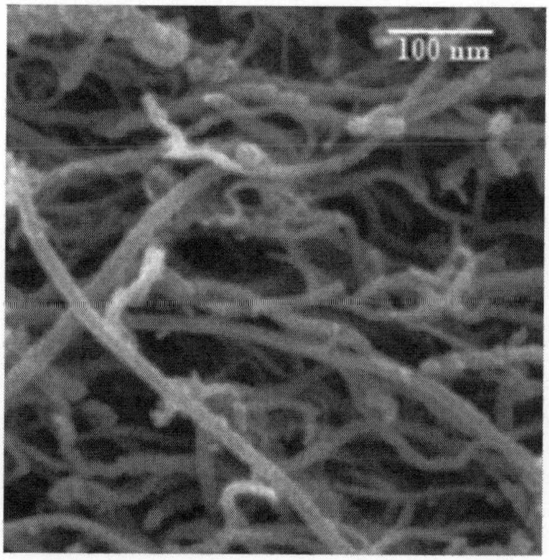

Figure 8. SEM image of the product which is fabricated with 5% Ni as catalyst [51].

Ni catalyst, figure 8, motivate a production of elongated CNTs and springy CNTs with a relatively good yield while Fe catalysts, figure 9, promote a production of CNTs with short length and defect structure and the yield is moderate.

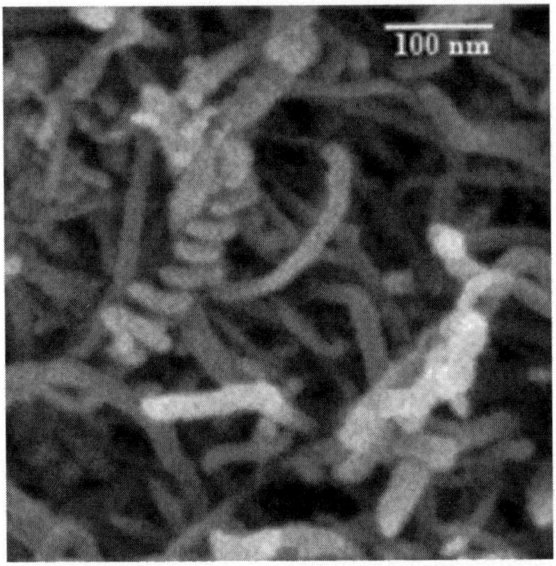

Figure 9. SEM image of the product which was fabricated with 5% Fe as catalyst [51].

Jahanshahi et al. showed that Mo catalysts motivate a production of CNTs with long length and high crystalline structure but with a wide diameter while it has a relatively good yield. In contrast, Mo-Ni bimetallic catalyst cause the production of CNTs with long length, narrow distribution diameters and crystalline structure without any defect and follow with a good yield [51].

3.2. Plasma Solution Temperature

The effect of solution temperature on the synthesis of CNTs and the structure of fabricated CNTs was investigated by Dehghani et al [57]. Scanning electron microscopes and transient electron microscopes (TEMs), figure 10, shows that the fabricated CNTs below zero as thermal condition is not suitable for synthesizing CNTs by the arc discharge method in liquid and CNTs cannot grow under low temperatures, especially below zero. High temperature is also not suitable for synthesizing CNTs by the arc discharge method in liquid media.

In contrast, observations show that in the environment with (25 ^{0}C) temperature, long CNTs are formed with narrow distribution diameter, complete clean, flat surface and arranged structure. Constant temperature around 25^{0}C is the best thermal condition for synthesizing CNTs by the arc discharge method in liquid [57].

http://dx.doi.org/10.5772/51116

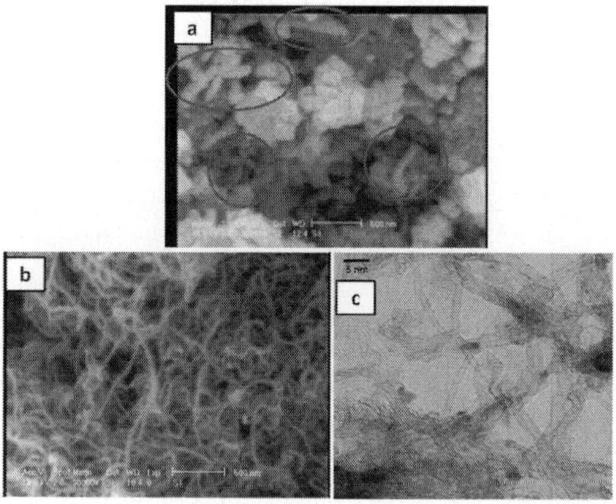

Figure 10. SEM image of the product which was fabricated a) below zero temperature b) at a high temperature (80°C), c) at the environment temperature (25 °C)[57].

3.3. Voltage difference between electrodes

The voltage effects on the production of the nanostructures by applying a variety of voltage values in different experiments were investigated by Jahanshahi et al [58]. The SEMs of the synthesized materials, figure 11 shows the formation of fullerene at a voltage of 10 V, while both CNTs and fullerenes are fabricated at a voltage of 20 V.

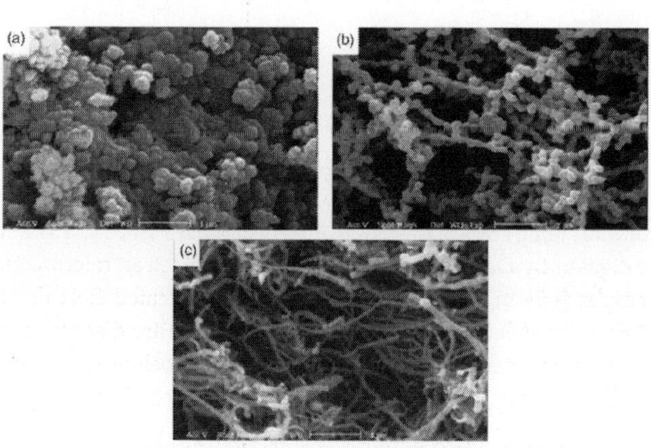

Figure 11. a) SEM images of the produced sample by arc discharge at a voltage of 10 V. (b) SEM images of the produced sample by arc discharge at a voltage of 20 V. (c) SEM images of the produced sample by arc discharge at a voltage of 30 V [58].

On the contrary, the elongated CNTs were synthesized with high quality at a voltage of 30 V. The results show that the rate of production efficiency and anode consumption is increased by increasing the voltage amount [58].

3.4. Plasma Solution Concentration

Liquid nitrogen provides a good environment for the MWCN synthesis, but the strong evaporation cause by the operation of the arc discharge does not allow a good thermal exchange between the synthesized material and its surroundings.

Arc discharge in deionised water and liquid nitrogen are erratic due to their electrical insulation [50]. The electrical conductivity of LiCl solution is also better than deionised water and liquid nitrogen [59].

Figure 12 shows TEM image of the as-grown MWCN synthesized in LiCl. Investigators have demonstrated the possibility of producing carbon nanostructures in the liquid phase (water, hydrocarbons, dichloromethane, CCl_4, in liquid gases) [61].

Figure 12. TEM image of the as-grown MWCN synthesized in LiCl [58].

In contrast, liquid water besides providing a suitable environment also provides the thermal conditions necessary to retain good quality CNTs in the raw material, while the reactivity of the water with hot carbon does not appear to have any major effects on the reaction [59].

Figure 13 shows CNTs are produced in NaCl [62]. Nevertheless, arc discharge in NaCl solution is extremely stable owing to the excellent electrical conductivity induced by Cl^- and Na^+ ions. Too many Na^+ ions would hinder carbon ions flying from anode towards the center of cathode. Researchers found that perhaps this is another reason that the length of SWCNT is short and only a single SWCNT [61].

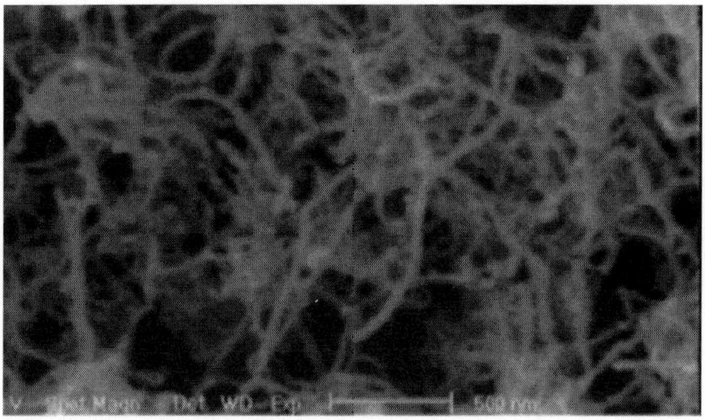

Figure 13. Production of CNTs in NaCl solution [62].

The optimized conditions to synthesize large quantities of SWCNT by applying arc discharge in NaCl solution deserve further investigation. Arc discharge in NaCl solution provides a very simple and cheap method to synthesize CNTs [60].

3.5. Discharge current

Discharge current is another important parameter influencing the products of arc discharge. If the catalyst percentage is 1 mol% Fe, and the discharge current is intentionally reduced to 20 A, the arc became very unstable, and disappear when the voltage is increased to 28 V [63].

3.6. The solution electrical conductivity effect

The effect of electrical conductivity of liquid on CNTs production might be important. A series of experiments carried out and the products were fabricated using arc discharge between two graphite electrodes submerged in different aqueous solutions of NaCl, KCl as well as LiCl. In comparative studies, CNTs were synthesized under different electrical conductivity conditions, and the results were analyzed, compared and discussed. The scanning electron microscopy (SEM), transmission electron microscopy (TEM) and Raman spectroscopy were employed to study the morphology of these carbon nanostructures and reported. LiCl 0.25 N (with 22.7mS as electrical conductivity) media when applied as solution, high-crystalline and a longed multi walled carbon nanotubes, single walled carbon nanotubes and springy carbon nanotubes (SCNTs) were synthesized. This study is one of the first one have demonstrated application of an arc discharge in liquid media with electrical conductivity effects upon CNT preparation and deserves further study (Dehghani and Jahanshahi, (2012); unpublished data).

4. Purification of fabricated CNTs

CNTs usually contain a large amount of impurities such as metal particles, amorphous carbon, and multi shell. There are different steps in purification of CNTs. Purification of CNTs is a process that separates nanotubes from non-nanotube impurities included in the raw products, or from nanotubes with undesired numbers of walls. Purification has been an important synthetic effort since the discovery of carbon nanotubes, and there are many publications discussing different aspects of the purification process. Good review articles on the purification of CNTs are available in the recent literature [64, 65].

The current industrial methods applied oxidation and acid-refluxing techniques that affect the structure of tubes. Purification difficulties are great because of insolubility of CNT and the limitation of liquid chromatography.

CNT purification step (depending on the type of the purification) removes amorphous carbon from CNTs, improves surface area, decomposes functional groups blocking the entrance of the pores or induces additional functional groups.

Most of these techniques are combined with each other to improve the purification and to remove different impurities at the same time. These techniques are as follow:

4.1. Oxidation

Oxidation is a way to remove CNTs impurities. In this way CNTs and impurities are oxidized. The damage to CNTs is less than the damage to the impurities. This technique is more preferable with regard to the impurities that are commonly metal catalysts which act as oxidizing catalysts [66, 67].

Altogether, the efficiency and yield of the procedure are highly depending on a lot of factors, such as metal content, oxidation time, environment, oxidizing agent and temperature [67].

4.2. Acid treatment

Refluxing the sample in acid is effective in reducing the amount of metal particles and amorphous carbon. Different used acids are hydrochloric acid (HCl), nitric acid (HNO_3) and sulphuric acid (H_2SO_4), while HCl is identified to be the ideal refluxing acid. When a treatment in HNO3 had been used the acid had an effect on the metal catalyst only, and no effects was observed on the CNTs and the other carbon particles. [66-69]. Figure 14 shows the SEM images of CNTs after and before purification stage with HCl [51].

If a treatment in HCl is used, the acid has also a little effect on the CNTs and other carbon particles [69, 70]. A review of literature demonstrates the effects that key variables like acid types and concentration & temperature have on the acid treatment [69, 70].

http://dx.doi.org/10.5772/51116

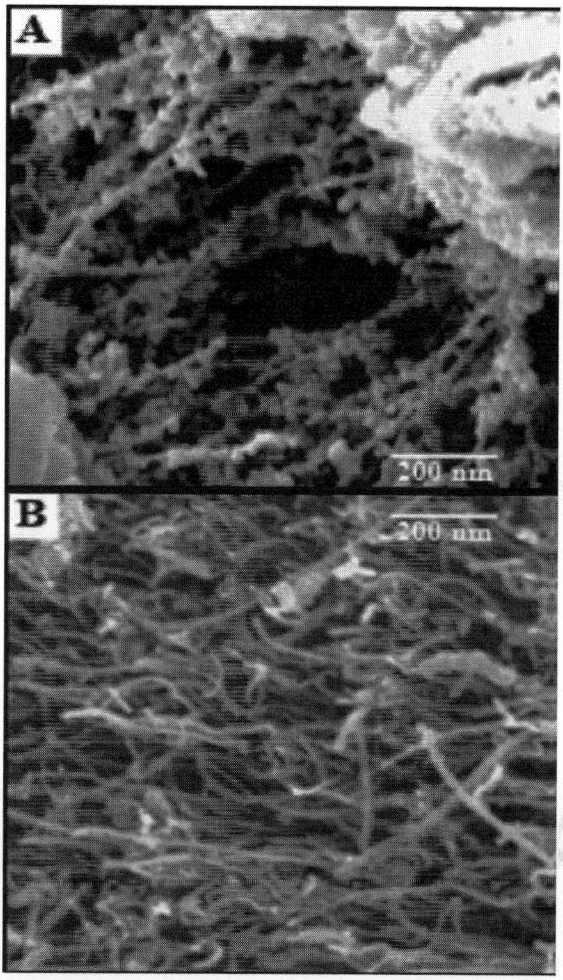

Figure 14. The SEM images of CNTs (A) after (B) before purification stages with HCl [51].

4.3. Annealing and thermal treatment

High temperature has effect on the productions and paralizes the graphitic carbon and the short fullerenes. When high temperature is used, the metal will be melted and can also be removed [69].

4.4. Ultrasonication

This technique is based on the separation of particles due to ultrasonic vibrations and also agglomerates of different nanoparticles will be more dispersed by this method. The separation of the particles is highly dependable on the surfactant, solvent and reagents which are used [67-70].

When an acid is used, the purity of the CNTs depends on the sonication time. During the tubes vibration to the acid for a short time, only the metal is solvated, but in a more extended period, the CNTs are also chemically cut. [69].

4.5. Micro-filtration

Micro-filtration is based on particle size. Usually CNTs and a small amount of carbon nanoparticles are trapped in a filter. The other nanoparticles (catalyst metal, fullerenes and carbon nanoparticles) are passing through the filter [65, 69, 70, 72].

A special form of filtration is cross flow filtration. Through a bore of fiber, the filtrate is pumped down at head pressure from a reservoir and the major fraction of the fast flowing solution is reverted to the same reservoir in order to be cycled through the fiber again. A fast hydrodynamic flow down the fiber bore sweeps the membrane surface and prevents building up of a filter cake [67].

5. Morphological and structural characterizations

To investigate the morphological and structural characterizations of the CNTs, a reduced number of techniques can be used. It is very important to characterize and determine the quality and properties of the CNTs, since its applications will require certification of properties and functions [74].

However, only few techniques are able to characterize CNTs at the individual level such as scanning tunneling microscopy (STM) and transmission electronic microscopy (TEM). X-ray photoelectron spectroscopy is required to determine the chemical structure of CNTs in spite of the fact that Raman spectroscopy is mostly introduced as global characterization technique.

5.1. Electron microscopy (SEM & TEM)

The morphology, dimensions and orientation of CNTs can be easily revealed by using scanning (SEM) and Transmission Electron Microscopes (TEM) which have high resolution. [70-75] (Figs. 15).

Therefore, the TEM technique is applied as a method for measurement of the outer and inner radius and linear electron absorption coefficient of CNTs [76].This method is used to study CNTs before and after annealing and notice a significant increase of the electron absorption coefficient. The inter shell spacing of MWNTs was studied by Kiang et al. [77] using high resolution TEM images.

http://dx.doi.org/10.5772/51116

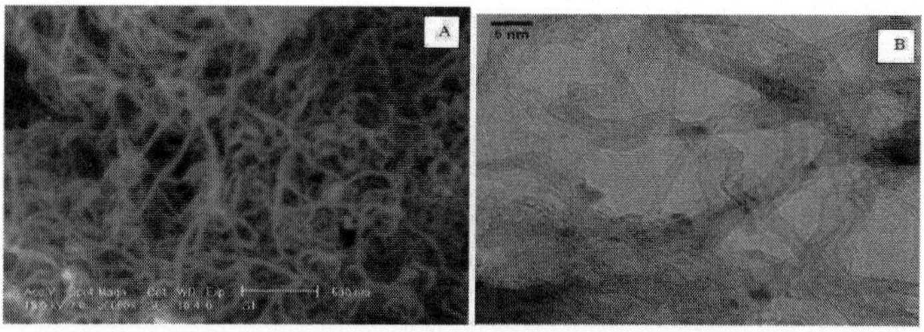

Figure 15. Electron micrographs of CNT (A) SEM of the CNT. (B) TEM of the CNT [57].

5.2. X-ray diffraction (XRD)

This technique is used to obtain some information on the interlayer spacing, the structural strain and the impurities. However, in comparing CNTs with x-ray incident beam, CNTs have multiple orientations. This leads to a statistical characterization of CNTs [78].

5.3. Raman spectroscopy

Raman spectroscopy is one of the most powerful tools for characterization of CNTs. Without sample preparation, a fast and nondestructive analysis is possible. All allotropic forms of carbon are active in Raman spectroscopy [79]. The position, width, and relative intensity of bands are modified according to the carbon forms [80].

A Raman spectrum of a purified sample (after applying the purification procedure) is shown in figure 16. The peaks at 1380 cm^{-1} and 1572 cm^{-1} correspond to disorder (D-band) and graphite (G-band) bands, respectively. The former is an indication of the presence of defective material and the latter one refers to the well-ordered graphite [62].

The most characteristic features are summarized as following:

1. Low-frequency peak <200 cm-1 characteristic of the SWNT, whose frequency is depended on the diameter of the tube mainly (RBM: radical breathing mode).

2. D line mode (disorder line), which is a large structure assign of residual ill-organized graphite.

3. High-frequency bunch that is called G band and is a characteristic of CNTs. This bunch has the ability to be superimposed with the G-line of residual graphite [81].

Raman spectroscopy is considered an extremely powerful tool for characterizing CNT, which gives qualitative and quantitative information on its diameter, electronic structure, purity and crystalline, and distinguishes metallic and semiconducting material as well as chirality.

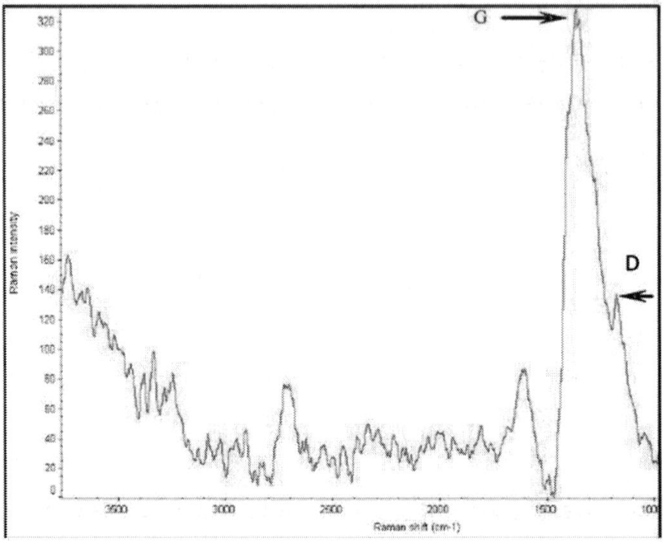

Figure 16. Raman spectrum showing the most characteristic features of CNTs produced by arc discharge method in liquid followed by acid treatment [62].

6. Conclusion

Carbon nanotubes (CNTs), a new structure of carbon element, are composed of graphen sheets rolled into closed concentric cylinders with diameter of the order of nanometers and length of micrometers. CNTs are attracted significant attention because of their unique physical and mechanical properties. These properties have been the engines of the rapid development in scientific studies in numerous applications such as in fuel cell and electrocatalyst, nanobiosensors, gas adsorptions and membrane separation [82-88].

Three methods, laser, arc discharge and chemical vapor deposition are used to synthesize CNTs. The laser method is also known as the laser-furnace method. The quantities CNTs in this method are large but this technique is not economically advantageous, since the process demanded considerable power. The chemical vapor deposition is another method for producing CNTs. It could produce CNTs at temperatures above 700 ^{0}C in large quantities, but the walls of the CNTs frequently contain many defects. Traditional arc discharge requires a complicated vacuum and heat exchange system. The yields of the laser and traditional arc discharge methods are very low (mg/h). From the application perspective, researchers are continuously trying to devise improved methods for CNTs fabrication.

Arc discharge in liquid media is a new method of synthesizing CNTs developed recently. All that is required is a dc power supply and an open vessel full of liquid nitrogen, deionized water or aqueous solution. This method is not requiring vacuum equipment, reacted gases, a high temperature furnace and a heat exchange system. Consequently, this method is extremely simple and cheap.

As it has been deeply investigated above, synthesis, purification and characterization of CNTs based on arc discharge in liquid media were described and discussed in this review paper. The observations of CNT growth under electron microscopy and other analytical techniques by different groups suggested that the mechanism are extremely sensitive to each fabrication parameter such as voltage difference between electrodes, current, type and ratio of catalysts, electrical conductivity, concentration, type and temperature of plasma solution and thermal conductivity. All these parameters were reviewed and studied herein. To the best of our knowledge the current review is the first one has discussed all aspects of arc discharge method in liquid media for CNT preparation and this technique deserves further attention.

Acknowledgements

The authors are grateful to Prof. M. Shariaty Niasar (Tehran Uiversity, Iran), Prof. J. Raoof (The University of Mazandaran, Iran), Dr. H. Molavi and PhD student Mrs R. Jabari Sheresht for their previous collaborations and productive discussion during preparation of this paper.

Author details

Mohsen Jahanshahi* and Asieh Dehghani Kiadehi

*Address all correspondence to: mjahan@nit.ac.ir

Nanotechnology Research Institute and Faculty of Chemical Engineering Babol University of Technology, Babol,, Iran

References

[1] Iijima, S. (1991). Helical Microtubules of Graphitic Carbon. *Nature*, 354, 56-58.

[2] Dresselhaus, M. S., Dresselhaus, G., & Eklund, P. C. (1996). Science of fullerenes and carbon nanotubes., Academic Press, New York.

[3] Dekker, C. (1999). Carbon nanotubes as molecular quantum wires. *Physics Today*, 52, 22-30.

[4] Tans, S. J., Verschueren, A. R. M., & Dekker, C. (1998). Room-temperature transistor based on a single carbon nanotube. *Nature*, 393, 49-52.

[5] Javey, A., Kim, H., Brink, M., Wang, Q., Ural, A., Guo, J., Mc Intyre, P., Mc Euen, P., Lundstrom, M., & Dai, H. (2002). *Nature Materials*, 1, 241-246.

[6] Javey, A., Guo, J., Wang, Q., Lundstrom, M., & Dai, H. J. (2003). Ballistic Carbon Nanotube Transistors. *Nature*, 424, 654-657.

[7] Rosenblatt, S., Yaish, Y., Park, J., Gore, J., Sazonova, V., & Mc Euen, P. L. (2002). High Performance Electrolyte Gated Carbon Nanotube Transistors. *Nano Letters*, 2, 869-872.

[8] Lu, C., Fu, Q., Huang, S., & Liu, J. (2004). *Nano Letters*, 4-623.

[9] Huang, S. M., Woodson, M., Smalley, R., & Liu, J. (2004). Growth Mechanism of Oriented Long Single Walled Carbon Nanotubes Using "Fast-Heating" Chemical Vapor Deposition Process. *Nano Letters*, 4, 1025-1028.

[10] Tan, S. J., Devoret, M. H., Dai, H. J., Thess, A., Smalley, R. E., Geerligs, L. J., & Dekker, C. (1997). *Nature*, 386-474.

[11] Bockrath, M., Cobden, D. H., & Mc Euen, P. L. (2000). Individual Single-Wall Carbon Nanotubes As Quantum Wires. *Science*, 290, 1552-1555.

[12] Dai, H. J., Hafner, J. H., Rinzler, A. G., Colbert, D. T., & Smalley, R. E. (1996). Nanotubes As Nanoprobes in Scanning Probe Microscopy. *Nature*, 384, 147-150.

[13] Wong, S. S., Harper, J. D., Lansbury, P. T., & Lieber, C. M. (1998). *Am J. Chem. Soc.*, 120, 603.

[14] Rinzler, A., & Hafner, J. (1995). Unraveling Nanotubes- Field-Emission from an Atomic Wire. *Science*, 269, 1550-1553.

[15] Collins, P. G., Bradley, K., Ishigami, M., & Zettl, A. (2000). Extreme oxygen sensitivity of electronic properties of carbon nanotubes. *Science*, 287, 1801-1804.

[16] Kong, J., Franklin, N. R., Zhou, C., Peng, S., Cho, J. J., & Dai, H. (2000). Nanotube molecular wires as chemical sensors. *Science*, 287, 622-625.

[17] Chen, R. J., Bangsaruntip, S., Drouvalakis, K. A., Kam, N. W. S., Shim, M., Li, Y. M., Kim, W., Utz, P. J. H., Dai, J., & Natl, P. (2003). Nanotube molecular wires as chemical sensors. *Acad. Sci. USA*, 100, 4984-4989.

[18] Star, A., Gabriel, J. C. P., Bradley, K., & Grüner, G. (2003). Electronic detection of specific protein binding using nanotube FET devices. *Nano Letters*, 3, 459-463.

[19] Kuchta, B., Firlej, L., Pfeifer, P., & Wexler, C. (2010). Numerical Estimation of Hydrogen Storage Limits in Carbon Based Nanospaces. *Carbon*, 48, 223-231.

[20] Scott, C., Arepalli, S., Nikolaev, P., & Smalley, R. E. (2001). *Applied Physics A: Materials Science & Processing*, 72, 573-580.

[21] Bethune, D., Kiang, C., De Vries, M., Gorman, G., Savoy, R., & Beyer, R. (1993). *Nature*, 363, 605-607.

[22] Journet, C., Maser, W. K., Bernier, P., Loiseau, A., Chapelle, M., Lefrant, S., Deniard, P., Lee, R., & Fischer, J. E. (1997). Large scale production of single wall carbon nanotubes by the electric arc technique. *Nature*, 388, 756-758.

[23] Cassell, A. M., Raymakers, J. A., Kong, J., & Dai, H. (1999). Solvation of fluorinated single-wall carbon nanotubes in alcohol solvents. *Journal of Physical Chemistry B*, 103, 6484-6492.

[24] Liu, J., Fan, S., & Dai, H. (2004). *MRS Bulletin*, 4, 224-250.

[25] Jabari, Seresht. R., & Jahanshahi, M. (2010). *Fullerenes Nanotubes Carbon Nanostruct*, 2, 1-12.

[26] Biró, L., Horváth, Z., Szalmás, L., Kertész, K., Wéber, F., Juhász, G., Radnóczi, G., & Gyulai, J. (2003). Continuous carbon nanotube production in underwater ac electric arc. *Chemical physics letters*, 372, 399-402.

[27] Feng, Y., & Zhou, G. (2003). Removal of some impurities from carbon nanotubes. *Chemical physics letters*, 375, 645-648.

[28] Guo, T., Diener, M., & Chai, Y. (1992). Uranium Stabilization of C28- a Tetravalent Fullerene. *Science*, 257, 1661-1664.

[29] Thess, A., & Lee, R. (1996). Crystalline Ropes of Metallic Carbon Nanotubes. *Science*, 273, 483-487.

[30] Bandow, S., & Rao, A. (1997). Purification of single-wall carbon nanotubes by microfiltration. *Journal of Physical Chemistry B*, 101, 8839-8842.

[31] Chiang, I., & Brinson, B. (2001). *Journal of Physical Chemistry B*, 105, 8297-8301.

[32] Ishii, H., & Kataura, H. (2003). Direct observation of Tomonaga-Luttinger-liquid state in carbon nanotubes at low temperatures. *Nature*, 426, 540-544.

[33] Puretzky, A., & Geohegan, D. (2000). *Applied Physics A: Materials Science & Processing*, 70, 153-160.

[34] Sen, R., & Ohtsuka, Y. (2000). Time period for the growth of single-wall carbon nanotubes in the. *Chemical physics letters*, 332, 467-473.

[35] Kokai, F., & Takahashi, K. (2000). *Journal of Physical Chemistry B*, 104, 6777-6784.

[36] Li, Y., & Mann, D. (2004). *Nano Letters*, 4, 317-321.

[37] Walker, P., & Rakszawski, J. (1959). *Journal of Phys. Chem*, 63, 133-140.

[38] Dresselhaus, M., & Dresselhaus, G. (1988). Physical Properties of Carbon Nanotubes : Synthesis, Structure, Properties and Applications. Springer-Verlag, Berlin.

[39] Endo, M. (1988). Grow Carbon Fibers in the Vapor Phase. *Chemtech*, 18, 568-576.

[40] Baker, R. T. K., & Harris, P. S. (1978). Marcel Dekker, New York.

[41] Tibbetts, G. G. (1984). Why are carbon filaments tubular? *Journal of crystal growth*, 66, 632-638.

[42] Seresht, R. J., Jahanshahi, M., & Yazdani, M. (2009). Parametric study on the synthesis of single wall carbon nanotube by gas arc-discharge method with multiple linear regressions and artificial neural network. *International Journal of Nanoscience*, 8, 243-249.

[43] Saito, Y., & Inagaki, M. (1992). Yield of fullerenes generated by contact arc method under He and Ar-dependence on carbon nanotube. *Chemical physics letters*, 200, 643-648.

[44] Ebbesen, T., & Ajayan, P. (1992). *Nature*, 358, 220-222.

[45] Belin, T., & Epron, F. (2005). Characterization methods of carbon nanotubes: a review. *Materials Science and Engineering B*, 119, 105-118.

[46] Jahanshahi, M., Raoof, J., Hajizadeh, S., & Jabari, Seresht. R. (2007). *Nanotechnol. Appl.*, 929, 71-77.

[47] Bera, D., & Johnston, G. (2006). A parametric study on the nanotube structure through arc-discharge in water. *Nanotechnology*, 17, 1722-1730.

[48] Ishigami, M., & Cumings, J. (2000). A simple method for the continuous production of carbon nanotubes. *Chemical physics letters*, 319, 457-459.

[49] Jung, S. H., & Kim, M. R. (2003). Applied Physics A: Materials Science & Processing . 76, 285-286.

[50] Lange, H., Sioda, M., Huczko, A., Zhu, Y. Q., Kroto, H. W., & Walton, D. R. M. (2003). Nanocarbon production by are discharge in water. *Carbon*, 41, 1617-1623.

[51] Jahanshahi, M., & Seresht, R. J. (2009). Catalysts effects on the production of carbon nanotubes by an automatic arc discharge set up in solution. *physica status solidi (c)*, 6, 2174-2178.

[52] Xing, G., & Jia, S. (2007). Influence of transverse magnetic field on the formation of carbon nanotube. *Carbon*, 45, 2584-2588.

[53] Xu, B., Guo, Ju., Wang, X., Liu, X., & Ichinose, H. (2006). Synthesis of carbon nanocapsules containing Fe, Ni or Co by arc discharge in aqueous solution. *Carbon*, 44-2631.

[54] Seraphin, S., & Zhou, D. (1994). *Applied Physics Letters*, 64, 2087-2089.

[55] Saito, Y., & Tani, Y. (1998). High yield of single-wall carbon nanotubes by arc discharge using Rh-Pt mixed catalysts. *Chemical physics letters*, 294, 593-598.

[56] Hsin, Y. L., Hwang, K. C., Chen, R. R., & Kai, J. J. (2001). Production and in-situ Metal Filling of Carbon Nanotubes in Water. *J. Advanced Materials*, 13, 830-833.

[57] Dehghani, Kiadehi. A., Jahanshahi, M., Mozdianfard, M. R., Vakili-Nezhaad, G. H. R., & Jabari, Seresht. R. (2011). Influence of the solution temperature on carbon nano-

tube formation by arc discharge method. *Journal of Experimental Nanoscience*, 4, 432-440.

[58] Jahanshahi, M., Raoof, J., & Seresht, R. J. (2009). Voltage effects on production of nanocarbons by a unique arc-discharge set-up in solution. *Journal of Experimental Nanoscience*, 4, 331-339.

[59] Lide, D. R. (2003). *CRC handbook of chemistry and physics; CRC Pr I Llc.*

[60] Wang, Sh., Chang, M., Lan, K. M., Wu, Ch., Cheng, J., & Chang, H. (2005). *Carbon*, 43, 1778-1814.

[61] Schur, D. V., Dubovoy, A. G., Zaginaichenko, S., Yu, Adejev. V. M., Kotko, A. V., Bogolepov, V. A., Savenko, A. F., & Zolotarenko, A. D. (2007). *Carbon*, 45, 1322-1329.

[62] Jahanshahi, M., Raoof, J., Hajizadeh, S., & Seresht, R. J. (2009). Synthesis and subsequent purification of carbon nanotubes by arc discharge in NaCl solution. *Phys. Status Solidi A*, 1, 101-105.

[63] Wang, S. D., Chang, M. H., Cheng, J. J., Chang, H. K., & Lan, K. M. D. (2005). *Carbon*, 43, 1317-1339.

[64] Park, T. J., Banerjee, S., Hemraj-Benny, T., & Wong, S. S. (2006). Mater J. Chem. , 16, 141-154.

[65] Haddon, R. C., Sippel, J., Rinzler, A. G., & Papadimitrakopoulos, F. (2004). Purification and separation of carbon nanotubes. *MRS Bull.*, 29, 252-259.

[66] Hajime, G., Terumi, F., Yoshiya, F., & Toshiyuki, O. (2002). Method of purifying single wall carbon nanotubes from metal catalyst impurities. Honda Giken Kogyo Kabushiki Kaisha, Japan.

[67] Borowiak-Palen, E., & Pichler, T. (2002). *Chemical physics letters*, 363, 567-572.

[68] Farkas, E., Anderson, M. E., Chen, Z. H., & Rinzler, A. G. (2002). Length sorting cut single wall carbon nanotubes by high performance liquid chromatography. *Chem. Phys. Lett.*, 363, 111-116.

[69] Kajiura, H., Tsutsui, S., Huang, H. J., & Murakami, Y. (2002). High-quality single-walled purification from arc-produced soot. *Chem Phys Lett.*, 364, 586-92.

[70] Chiang, I., & Brinson, B. (2001). *Journal of Physical Chemistry B*, 105, 8297-8301.

[71] Bandow, S., & Rao, A. (1997). *Journal of Physical Chemistry B*, 101, 8839-8842.

[72] Moon, J. M., & An, K. H. (2001). High-yield purification process of singlewalled carbon nanotubes. *Journal of Physical Chemistry B*, 105, 5677-5681.

[73] Jahanshahi, M., Tobi, F., & Kiani, F. (2005). Carbon-Nanotube Based Nanobiosensors:. Paper presented at Electrochemical Pretreatment. The 10th Iranian Chemical Engineering Conference,, Zahedan, Iran.

[74] Mawhinney, D. B., Naumenko, V., Kuznetsova, A., Yates, J. T., Liu, J., & Smalley, R. E. (2000). Surface defect site density on single walled carbon nanotubes by titration. *Chem. Phys. Lett.*, 324, 213-216.

[75] Li, W., Wen, J., & Tu, Y. (2001). *Ren Z. Appl. Phys. A*, 73, 259-264.

[76] Gommes, C., & Blacher, S. (2003). *Carbon*, 41, 2561-2572.

[77] Kiang, C. H., & Endo, M. (1998). Size effect in carbon nanotubes. *Physical review letters*, 81, 1869-1872.

[78] Zhu, W., Miser, D., Chan, W., & Hajaligol, M. (2003). *Mater. Chem. Phys*, 82, 638-647.

[79] Arepalli, S., & Nikolaev, P. (2004). Protocol for the characterization of SWCNT material quality. *Carbon*, 42, 1783-1791.

[80] Ferrari, A., & Robertson, J. (2000). *Physical Review B*, 61, 14095.

[81] Hiura, H., Ebbesen, T. W., Tanigaki, K., & Takahashi, H. (1993). Single-walled carbon nanotubes produced by electric arc. *Chem. Phys. Lett.*, 202, 509.

[82] Raoof, J. B., Jahanshahi, M., & Momeni, Ahangar. S. (2010). Nickel Particles Dispersed into Poly (o-anisidine) and Poly (oanisidine)/Multi-walled Carbon Nanotube Modified Glassy Carbon Electrodes for Electrocatalytic Oxidation of Methanol. *Int. J. Electrochem. Sci.*, 5, 517-530.

[83] Toubi, F., Jahanshahi, M., Rostami, A. A., & Hajizadeh, S. (2009). Voltametric tests on different carbon nanotubes as nanobiosensor devices. *D. Biochem Process Biotech. Mol. Biol.*, 2, 71-74.

[84] Khalili, S., Ghoreyshi, A. A., & Jahanshahi, M. (2012). Equilibrium, kinetic and thermodynamic studies of hydrogen adsorption on multi-walled carbon nanotubes. *Iranica Journal of Energy & Environment*, 1, 69-75.

[85] Delavar, M., Ghoreyshi, A. A., Jahanshahi, M., Khalili, S., & Nabian, N. (2012). The effect of chemical treatment on adsorption of natural gas by multi-walled carbon nanotubes: sorption equilibria and thermodynamic studies. Chemical Industry & Chemical Engineering Quarterly Inpress.

[86] Delavar, M., Ghoreyshi, A. A., Jahanshahi, M., & Nabian, N. (2012). Journal of Experimental Nanoscience Inpress.

[87] Delavar, M., Ghoreyshi, A. A., Jahanshahi, M., Khalili, S., & Nabian, N. (2012). Equilibria and kinetics of natural gas adsorption on multi-walled carbon nanotube material. *Green Chemistry*, 2, 4490-4497.

[88] Rahimpour, A, Jahanshahi, M, Khalili, S, Mollahosseini, A, Zirepour, A, & Rajaeian, B. (2012). Novel functionalized carbon nanotubes for improving the surface properties and performance of polyethersulfone (PES) membrane. *J. Desalination.*, 286, 99-107.

Chapter 5

Polymer Nanocomposites Containing Functionalised Multiwalled Carbon NanoTubes: a Particular Attention to Polyolefin Based Materials

Emmanuel Beyou, Sohaib Akbar,
Philippe Chaumont and Philippe Cassagnau

Additional information is available at the end of the chapter

http://dx.doi.org/10.5772/50710

Introduction

Incorporation of carbon nanotubes (CNTs) into a polymer matrix is a very attractive way to combine the mechanical and electrical properties of individual nanotubes with the advantages of plastics. Carbon nanotubes are the third allotropic form of carbon and were synthesized for the first time by Iijima in 1991 [1]. Their exceptional properties depend on the structural perfection and high aspect ratio (typically ca 100-300). Two types of CNTs are distinguished : single-walled CNTs (SWCNTs) consist of a single graphene sheet wrapped into cylindrical tubes with diameters ranging from 0.7 to 2nm and have lengths of micrometers while multi walled CNTs (MWCNTs) consist of sets of concentric SWCNTs having larger diameters [2-5]. The unique properties of individual CNTs make them the ideal reinforcing agents in a number of applications [6-9] but the low compatibility of CNTs set a strong limitation to disperse them in a polymer matrix. Indeed, carbon nanotubes form clusters as very long bundles due to the high surface energy and the stabilization by numerous of π−π electron interactions among the tubes. Non covalent methods for preparing polymer/CNTs nanocomposites have been explored to achieve good dispersion and load transfer [10-12]. The non-covalent approaches to prepare polymer/CNTs composites via processes such as solution mixing [13,14], melt mixing [14,15], surfactant modification [16], polymer wrapping [17], polymer absorption [18] and in situ polymerization [19, 20] are simple and convenient but interaction between the two components remains weak. Relatively uniform dispersion of CNTs can be achieved in polar polymers such as nylon, polycarbonate and polyimide because of the strong interaction between the polar moiety of the polymer chains and the surface of the CNTs [21-24]. Moreover, it was found that MWNTs disperse well in PS and form a network-like structure due to π-stacking interactions with aromatic groups of the PS chains [25]. However, it is difficult to disperse CNTs within a non polar polymer matrix such as polyolefins. To gain the advantages of CNTs at its best, one needs: (i) high interfacial area between nanotubes and polymer; and, (ii) strong interfacial interaction. Unfortunately the solvent technique does not help much in achieving these targets and, as a result, a nanocomposite having properties much inferior to theoretical expectations are obtained. For example, the mechanical properties of polyethylene (PE) reinforced by carbon nanotubes do not improve significantly because the weak polymer-CNT interfacial adhesion prevents efficient stress transfer from the polymer matrix to CNT [26-28]. A strategy for enhancing the compatibility between nanotubes and polyolefins consists in functionalising the sidewalls of CNT to introduce reactive moieties and to disrupt the rope structure. Functional moieties are attached to open ends and sidewalls to improve the solubility of nanotubes [29-32] while the covalent polymer grafting approaches, including 'grafting to' [33-36] and 'grafting from' [37-39] that create chemical linkages between polymer and CNTs, can significantly improve dispersion and change their rheological behaviour. First, methods used for processing CNTs-based nanocomposites and for the functionalisation of carbon nanotubes (CNTs) with polymers will be described. This is followed by a review of the surface chemistry of carbon nanotubes in order to perform their dispersion in polyolefin matrix. Finally, general trends of the viscoelastic properties of CNTs/ polyolefin composites are discussed.

1. Methods to process polymer/carbon nanotubes composites

Similar to the case of carbon nanotube/solvent suspensions, pristine carbon nanotubes have not yet been shown to be soluble in polymers illustrating the extreme difficulty of overcoming the inherent thermodynamic drive of nanotubes to bundle [40]. Several processing methods available for fabricating CNT/polymer composites based on either thermoplastic or thermosetting matrices mainly include solution mixing, melt blending, and in situ polymerisation (figure 1) [41, 42].

1.1. Solution blending

The most common method for preparing polymer nanotube composites has been to mix the nanotubes and polymer in a suitable solvent before evaporating the solvent to form a composite film (Figure 1a). One of the benefits of this method is that agitation of the nanotubes powder in a solvent facilitates nanotubes' de-aggregation and dispersion. Almost all solution processing methods are based on a general theme which can be summarised as:

1. Dispersion of nanotubes in either a solvent or polymer solution by energetic agitation.
2. Mixing of nanotubes and polymer in solution by energetic agitation.
3. Controlled evaporation of solvent leaving a composite film.

In general, agitation is provided by magnetic stirring, shear mixing, reflux or ultrasonication. Sonication can be provided in two forms, mild sonication in a bath or high-power sonication using a tip or horn. An early example of solution based composite formation is described by Jin *et al* [43]. By this method, high loading levels of up to 50wt% and reasonably good dispersions were achieved. A number of papers have discussed dispersion of nanotubes in polymer solutions [44-46]. This can result in good dispersion even when the nanotubes cannot be dispersed in the neat solvent. Coleman *et al* [44] used sonication to disperse catalytic MWCNT in polyvinylalcohol/H_2O solutions, resulting in a MWCNT dispersion that was stable indefinitely. Films could be easily formed by drop-casting with microscopy studies showing very good dispersion. Cadek *et al* [46] showed that this procedure could also be applied to arc discharge MWCNTs, double walled nanotubes (DWNTs) and High-Pressure CO Conversion (HiPCO) SWCNTs. They also showed that this procedure could be used to purify arc-MWCNTs by selective sedimentation during composite production.

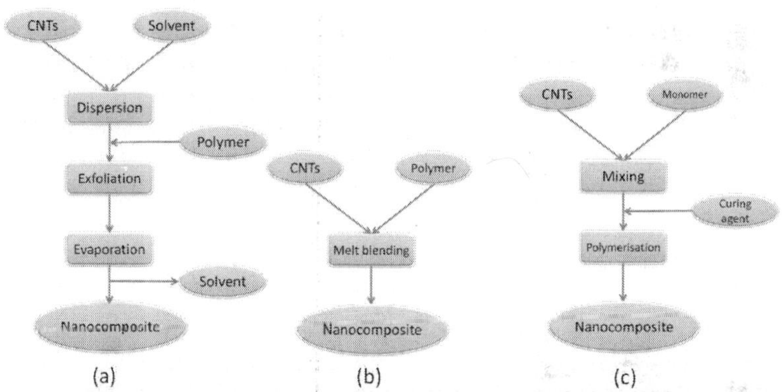

Figure 1. Schematic representation of different steps of polymer/CNTs composite processing: solution mixing (a); melt mixing (b); *in situ* polymerisation (c).

1.2. Melt mixing

While solution processing is a valuable technique for both nanotube dispersion and composite formation, it is completely unsuitable for the many polymer types that are insoluble. Melt processing is a common alternative method, which is particularly useful for dealing with thermoplastic polymers (Figure 1b). This range of techniques makes use of the fact that thermoplastic polymers soften when heated. Amorphous polymers can be processed above their glass transition temperature while semi-crystalline polymers need to be heated above their melt temperature to induce sufficient softening. Advantages of this technique are its speed and simplicity, not to mention its compatibility with standard industrial techniques [47, 48]. Any additives, such as carbon nanotubes can be mixed into the melt by shear mixing. However, Bulk samples can then be fabricated by techniques such as compression moulding, injection moulding or extrusion. However it is important that processing conditions are optimised for the whole range of polymer–nanotube combinations. High temperature and shear forces in the polymer fluid are able to break the carbon nanotubes bundles and CNTs can additionally affect melt properties such as viscosity, resulting in unexpected polymer degradation [49]. Andrews and co-workers [50] showed that commercial polymers such as high impact polystyrene, polypropylene and acrylonitrile–butadiene–styrene (ABS) could be melt processed with CVD-MWCNT to form composites. The polymers were blended with nanotubes at high loading level in a high shear mixer to form master batches. An example of using combined techniques was demonstrated by Tang *et al* [51]. High density polyethylene pellets and nanotubes were melted in a beaker, then mixed and compressed. The resulting solid was broken up and added to a twin screw extruder at 170°C and extruded through a slit die. The resulting film was then compression moulded to form a thin film.

1.3. In Situ Polymerisation

This fabrication strategy starts by dispersing carbon nanotubes in vinyl monomers followed by polymerising the monomers (Figure 1c). This method produces polymer-grafted CNTs mixed with free polymer chains resulting in a homogeneous dispersion of CNTs. In situ radical polymerisation was applied for the synthesis of PMMA-based composites by Jia *et al* [52] using a radical initiator and the authors suggested that π-bonds of the CNT graphitic network were opened by the radical fragments of initiator and therefore the carbon nanostructures could participate in PMMA polymerisation by acting as efficient radical scavengers. Dubois et al [53] applied the in situ polymerization to olefin monomers by anchoring methylaluminoxane, a commonly used co-catalyst in metallocene-based olefin polymerization onto carbon nanotubes surface. Then, the metallocene catalyst was added to the surface-activated CNTs and the course of ethylene polymerization was found to be similar to the one without the presence of pristine MWCNTs. Epoxy nanocomposites comprise the majority of reports using in situ polymerisation methods [54, 55], where the nanotubes are first dispersed in the resin followed by curing the resin with the hardener. Zhu *et al* [56] prepared epoxy nanocomposites by this technique using end-cap carboxylated SWCNTs and an esterification reaction to produce a composite with improved tensile modulus (E is 30% higher with 1 wt % SWCNT).

1.4. Novel methods

Rather than avoid the high viscosities of nanotube/polymer composites, some researchers have decreased the temperature to increase viscosity to the point of processing in the solid state. Solid-state mechanochemical pulverisation processes (using pan milling [57] or twin-screw pulverisation [58]) have mixed MWCNTs with polymer matrices. Pulverisation methods can be used alone or followed by melt mixing. Nanocomposites prepared in this manner have the advantage of possibly grafting the polymer on the nanotubes, which account in part for the observed good dispersion, improved interfacial adhesion, and improved tensile modulus.

An innovative latex fabrication method for making nanotube/polymer composites has been used by first dispersing nanotubes in water (SWCNT require a surfactant, MWCNT do not) and then adding a suspension of latex nanoparticles [59,60]. For example, PEG-based amphiphilic molecule containing aromatic thiophene rings, namely, oligothiophene-terminated poly(ethylene glycol) (TN-PEG) was synthesized, and its ability to disperse and stabilize pristine carbon nanotubes in water was shown.This promising method can be applied to polymers that can be synthesised by emulsion polymerisation or formed into artificial latexes, e.g., by applying high-shear conditions [61].

Finally, to obtain nanotube/polymer composites with very high nanotube loadings, Vigolo et al [62] developed a "coagulation spinning" method to produce composite fibers comprising predominately nanotubes. This method disperses SWCNT using a surfactant solution, coagulates the nanotubes into a mesh by wet spinning it into an aqueous poly(vinyl alcohol) solution, and converts the mesh into a solid fiber by a slow draw process. In addition, Mamedov et al [63] developed a fabrication method based on sequential layering of chemically modified nanotubes and polyelectrolytes to reduce phase separation and prepared composites with SWCNT loading as high as 50 wt %.

2. Surface modifications of carbon nanotubes with polymers

CNTs are considered ideal materials for reinforcing fibres due to their exceptional mechanical properties. Therefore, nanotube–polymer composites have potential applications in aerospace science, where lightweight robust materials are needed [64]. It is widely recognised that the fabrication of high performance nanotube–polymer composites depends on the efficient load transfer from the host matrix to the tubes. The load transfer requires homogeneous dispersion of the filler and strong interfacial bonding between the two components [65]. A dispersion of CNT bundles is called "macrodispersion" whereas a dispersion of individual nonbundled CNT is called a nanodispersion [66, 67]. To address these issues, several strategies for the synthesis of such composites have been developed. Currently, these strategies involve physical mixing in solution, *in situ* polymerisation of monomers in the presence of nanotubes, surfactant-assisted processing of composites, and chemical functionalisation of the incorporated tubes. As mentioned earlier, in many applications it is necessary to tailor the chemical nature of the nanotube's walls in order to take advantage of their unique properties. For this purpose, two main approaches for the surface modification of CNTs are adopted i.e. covalent and noncovalent, depending on whether or not covalent bonding between the CNTs and the functional groups and/or modifier molecules is involved in the modification surface process. Figure 2 depicts a typical representation of such surface modifications.

2.1. Noncovalent attachment of polymers

The noncovalent attachment, controlled by thermodynamic criteria [68], which for some polymer chains is called wrapping, can alter the nature of the nanotube's surface and make it more compatible with the polymer matrix. Non-covalent surface modifications are based mainly on weak interactions, such as van der Waals, π–π and hydrophobic interactions, between CNTs and modifier molecules. Non-covalent surface modifications are advantageous in that they conserve sp^2-conjugated structures and preserve the electronic performance of CNTs. The main potential disadvantage of noncovalent attachment is that the forces between the wrapping molecule and the nanotube might be weak, thus as a filler in a composite the efficiency of the load transfer might be low.

Non-covalent modification approaches typically use organic mediating molecules that range from low molecular weight molecules to supra- molecules to polymers.

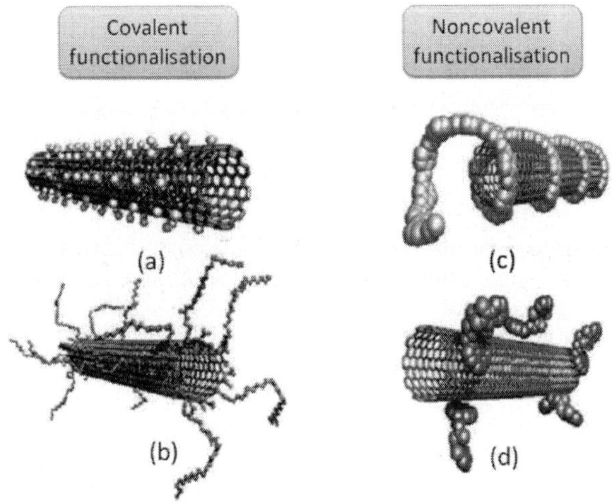

Figure 2. Different routes for nanotubes' functionalisation: sidewall covalent functionalisation (a); defect-group covalent functionalisation (b); noncovalent polymer wrapping (c); noncovalent pi-stacking (d).

2.1.1. Polymer wrapping

O'Connell et al. [68] reported that nanotubes could be reversibly solubilised in water by noncovalently associating them with a variety of linear polymers such as polyvinyl pyrrolidone (PVP) and polystyrene sulfonate (PSS). They demonstrated that the association between the polymer and the nanotubes is robust, not dependent upon the presence of excess polymer in solution, and is uniform along the sides of the tubes (Figure 1c). A general thermodynamic driving force for such wrapping in an aqueous environment has been identified [68].

Conjugated luminescent polymer poly-{(m-phenylenevinylene)-co-[(1,5-dioctyloxy-p-phenylene)-vinylene]} (PmPV) and its derivatives [69-71] have been successfully used for the wrapping around nanotubes on account of stabilising noncovalent bonding interactions, presumably as a result of π–π stacking (Figure 1d) and van der Waals interactions between PmPV and the surfaces of the nanotubes. Star et al [72] also synthesised the Stilbenoid dendrimers, a hyperbranched variant of the PmPV polymer, which exhibits an appropriate degree of branching, and it was found to be more efficient at breaking up nanotube bundles, provided it is employed at higher polymer-to-nanotube ratios than was the "parent" PmPV polymer.

In addition, the behavior of single walled and multi walled carbon nanotubes in aqueous solutions of Gum Arabic, a natural polysaccharide, has been described by Nativ-Roth et al [73]. They observed that while the as-prepared nanotube powders contain highly entangled ropes and bundles, the dispersions are mainly composed of individual tubes suggesting that the ability of Gum Arabic to exfoliate the bundles, and stabilize individual tubes in aqueous dispersions, can be utilized in the preparation of carbon nanotube-polymer composites. In the latter case, the dispersing polymer acts as a compatabilizer and as an adhesion promoter leading to strengthening of the matrix-nanotube interface.

It is clear from these accounts that noncovalent functionalisation of carbon nanotubes can be achieved without disrupting the primary structure of the nanotubes themselves.

2.1.2. Polymer absorption

Xia et al [74] has described a method to prepare polymer-encapsulated MWCNTs : it has been successfully prepared through ultrasonically initiated in situ emulsion polymerisations of n-butyl acrylate (BA) and methyl methacrylate (MMA) in presence of MWCNT. By employing the multiple effects of ultrasound, i.e., dispersion, pulverizing, activation, and initiation, the aggregation and entanglement of carbon nanotubes in aqueous solution can be broken down, while in situ polymerization of monomer BA or MMA on the surface of MWCNTs proceeds and the MWCNTs are coated by the formed polymer.

The hydrophilic regions of surfactants interact with polar solvent molecules, and the hydrophobic regions can adsorb onto nanotube surfaces [75]. Thus, the process of dispersing CNTs from aggregates, bundles, or ropes into separated individual CNTs depends strongly on the length of the hydrophobic regions and the types of hydrophilic groups of the surfactant. A topological, noncovalent solution to improving the dispersion of SWNTs by encasing them in cross-linkable surfactant micelles was demonstrated by Kang and Taton [16]. SWCNTs were dispersed in the dimethylformamide (DMF) solutions of amphiphilic poly(styrene)-block-poly(acrylic acid) copolymer. Water was added to the solutions and the poly(styrene)-block-poly(acrylic acid) copolymer wrapped the SWCNTs and formed micelle. Then the PAA blocks of the micellar shells were permanently crosslinked by addition of a water-soluble diamine linker and a carbodiimide activator. This encapsulation significantly enhances the dispersion of SWCNTs in a wide variety of polar and nonpolar solvents and polymer matrices [76]. Encapsulated SWNTs can be used as an alternative starting material to pure SWNTs for the production and investigation of nanotube composite materials.

2.2. Covalent attachment of polymers

Functionalisation of carbon nanotubes with polymers is a key issue to improve the interfacial interaction between CNTs and the polymer matrix when processing polymer/CNT nanocomposites. The covalent reaction of CNT with polymers is important because the long polymer chains help to dissolve the tubes into a wide range of solvents even at a low degree of functionalisation. There are two main methodologies for the covalent attachment of polymeric substances to the surface of nanotubes, which are defined as "grafting to" and 'grafting from' methods [76, 77]. The former relies on the synthesis of a polymer with a specific molecular weight followed by end group transformation. Subsequently, this polymer chain is attached to the graphitic surface of CNT. A disadvantage of this method is that the grafted polymer contents are limited because of high steric hindrance of macromolecules. The 'grafting from' method involved the immobilisation of initiators onto the substrate followed in situ surface polymerization to generate grafted polymer chains. Because the covalent attachment of the surface modifiers involves the partial disruption of the sidewall sp2 hybridization system, covalently modified CNTs inevitably lose some degree of their electrical and/or electronic performance properties [78].

2.2.1. 'Grafting to' method

Since the curvature of the carbon nanostructures imparts a significant strain upon the sp^2 hybridised carbon atoms that make up their framework, the energy barrier required to convert these atoms to sp^3 hybridisation is lower than that of the flat graphene sheets, making them susceptible to various addition reactions. Therefore, to exploit this chemistry, it is only necessary to produce a polymer-centred transient in the presence of CNT material. Alternatively, defect sites on the surface of oxidized CNTs, as open-ended nanostructures with terminal carboxylic acid groups, allow covalent linkages of oligomer or polymer chains. So, the 'grafting to' method involves the chemical reaction between as-prepared or commercially available polymers with reactive end groups and nanotubes' surface functional groups or the termination of growing polymer radical, cation and anion formed during the polymerization of various monomers in the presence of CNTs or the deactivation of living polymer chain ends with the CNT surface.

For example, oxidized SWCNTs were grafted with amino-terminated poly (N-isopropylacrylamide) (PNIPAAm) by carbodiimide-activated reaction, which yielded a 8wt% polymer content[77]. In a different approach, oxidized MWCNTs were attached ontopolyacrylonitrile(PAN) nanoparticles through the reaction of the reduced cyano-groups of the polymer and the carboxylic moieties of CNT surface [79]. In addition, the amidation reaction was used for grafting of oligo-hydroxyamides to MWCNTs as described in figure 3 [80].

Ester-based linkages have been used by Baskaran et al. [81] by performing the reaction of hydroxy-terminated PS with thionyl chloridetreated MWCNTs, resulting in a hybrid containing 86wt% of CNTs. The esterification reaction was also used for grafting polyethylene glycol(PEG) chains to acylchloride-activated SWCNTs [82]. Silicone-functionalised CNT derivatives were prepared by opening terminal epoxy groups of functionalised polydimethylsiloxanes (PDMS) by the carboxylic groups of acid-treated MWCNTs [83]. Another example of the "grafting to" approach has been reported by Qin et al. [84] through the grafting of polystyrene with azide end group onto SWCNTs (Figure 4).

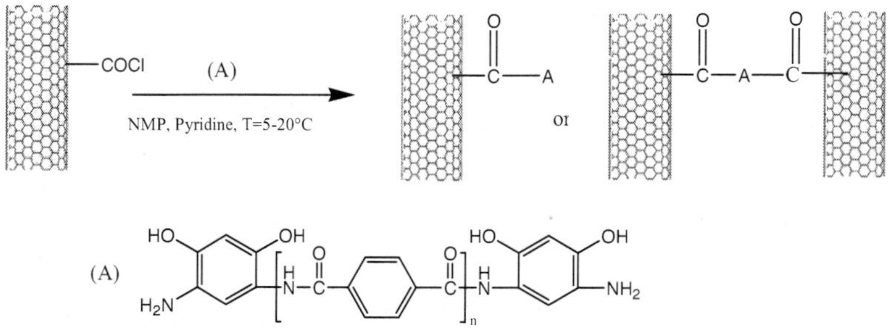

Figure 3. Synthesis of oligo-hydroxyamide-grafted MWNT. Reproduced from [80] with permission of Elsevier.

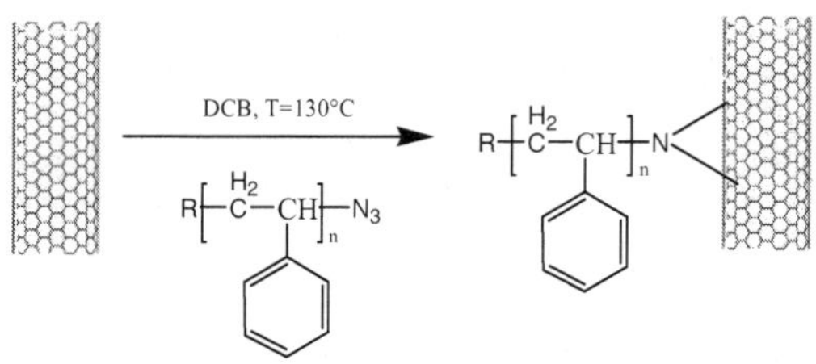

Figure 4. reaction of azide-terminatedpolystyreneonto CNTs surface. Reproduced from [80] with permission of ACS publications.

In an analogous approach, alkyne-decorated SWCNTs and PS-N_3were coupled via [3+1] Huisgen cycloaddition between the alkyne and azide end groups [85]. A new method was developed by Hung et al. [86] for preparing polystyrene-functionalized multiple-walled carbon nanotubes (MWNTs) through the termination of anionically synthesized living poly styryllithium with the acyl chloride functionalities on the MWNTs. The acyl chloride functionalities on the MWNTs were in turn obtained by the formation of carboxyls via chemical oxidation and their conversion into acyl chlorides (Figure 5).

Lou et al. [87] reported the radical grafting of polyvinylpyridine chains onto the surface of nanotubes through the thermolysis of poly (2-vinylpyridine) terminated with a radical-stabilizing nitroxide (Figure 6), resulting in grafting densities up to 12 wt.-%.

2.2.2. *'Grafting from' technique*

Mostly, it involves the polymerisation of monomers from surface-derived initiators on either MWCNTs or SWCNTs. These initiators are covalently attached using the various functionalisation reactions developed for small molecules [77]. Then, the polymer is bound via in situ radical, cationic, anionic, ring opening and condensation polymerizations. The advantage of 'grafting from' approach is that the polymer growth is not limited by steric hindrance, allowing high molecular weight polymers to be efficiently grafted as well as quite high grafting density [9].

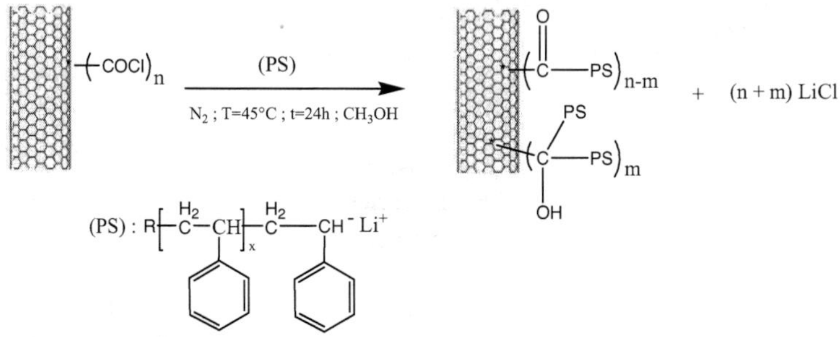

Figure 5. substitution reaction of living polystyryllithium anions with acyl choride-modified CNTs. Reproduced from [86] with permission of John Wiley and Sons.

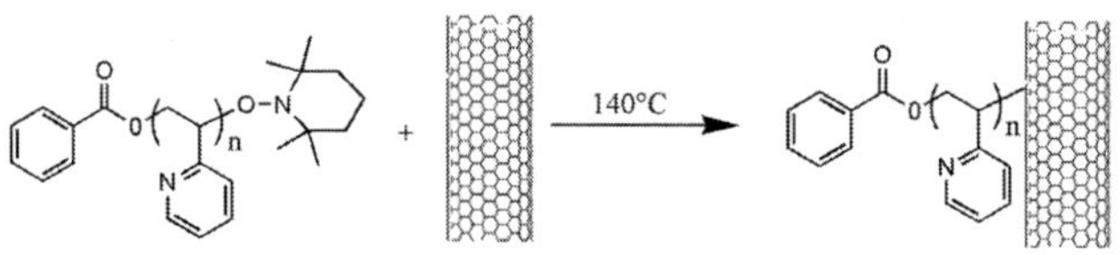

Figure 6. Radical grafting of TEMPO-end capped PVP to MWCNTs. Reproduced from [87] with permission of Elsevier.

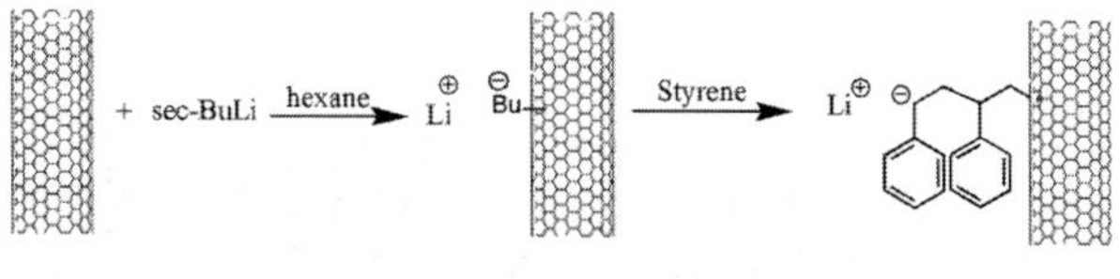

Figure 7. Anionic polymerisation of styrene onto carbon nanotubes

For example, Viswanathan et al [88]. have developed a procedure based on the SWCNT surface treatment with butyllithium providing initiating sites for the anionic polymerization of styrene (Figure 6).

The latter procedure eliminates the need for nanotube pretreatment prior to functionalization and allows attachment of polymer molecules to pristine tubes without altering their original structure.

In addition, polyethylenimine has been grafted onto the surface of MWNTs by performing a cationic polymerization of aziridine in the presence of amine-functionalized MWNTs (NH_2–MWNTs [89]. The grafting of PEI was realized through two mechanisms, the activated monomer mechanism (AMM) or the activated chain mechanism (ACM), by which protonated aziridine monomers or the terminal iminium ion groups of propagation chains, respectively, are transferred to amines on the surface of MWNTs [89].

Bonduel et al. [90] reported the homogeneous surface coating of long carbon nanotubes by in situ polymerization of ethylene as catalyzed directly from the nanotube surface-treated by a highly active metallocene-based complex. It allowed for the break-up of the native nanotube bundles leading, upon further melt blending with HDPE, to high-performance polyolefinic nanocomposites [90]. In another attempt, an easy method for preparing polystyrene-grafted multi-walled carbon nanotubes (MWCNTs) with high graft yields was developed by using free radical graft polymerisation from photoinduced surface initiating groups on MWCNTs [91]. Conventional microscopy, including optical, atomic force, sanning electronic microscopy (SEM), and transmission electronic microscopy (TEM), reveal the dispersion state or quality of CNTs within a very limited area of a given nanofiller composite [67]. High resolution-TEM image of the MWCNTs-PS (Figure 8) shows that the surface of the MWCNTs-PS is covered with 4–5nm thick amorphous PS layers while the wall surface of purified MWCNTs was smooth, without any detectable polymer Layer [91].

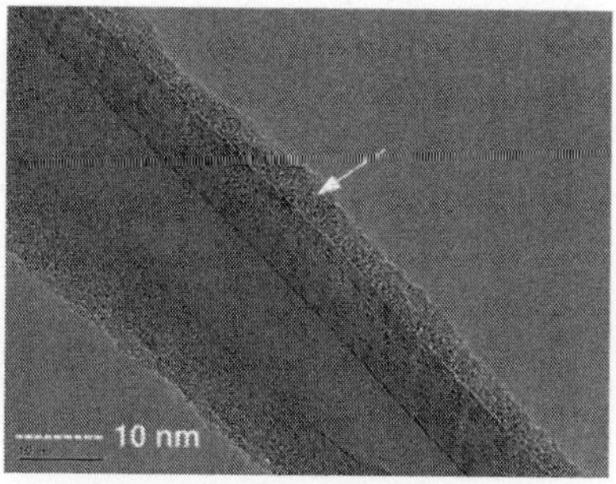

Figure 8. HR-TEM image of PS-g-MWCNT. Reprinted from [91] with permission of Elsevier.

Controlled radical polymerisation techniques such as nitroxide mediated polymerisation (NMP), atom transfer radical polymerisation (ATRP) and radical adition fragmentation transfer (RAFT) have been also used to graft polymer chains from the CNT surface [92-103] (see figure 9 as example).

3. Carbon nanotubes nanocomposites based on Polyolefins

Polyethylene (PE) is one of the most widely used commercial polymer due to the excellent combination of low coefficient of friction, chemical stability and excellent moisture barrier properties [104]. The combination of a soft polymer matrix such as PE with nanosized rigid filler particles may provide new nanocomposite materials with largely improved modulus and strength. To improve the stiffness and rigidity of PE, CNTs can be used to make CNT/PE composites [104-107]. The mechanical properties of polyethylene (PE) reinforced by carbon nanotubes do not improve significantly because the weak polymer-CNT interfacial adhesion prevents efficient stress transfer from the polymer matrix to CNT [108-110]. The lack of functional groups and polarity of PE backbone results in incompatibility between PE and other materials such as glass fibres, clays, metals, pigments, fillers, and most polymers. A strategy for enhancing the compatibility between nanotubes and polyolefins consists in functionalising the sidewalls of CNT with polymers either by a 'grafting to' or a 'grafting from' approach. As discussed before, the "grafting from" approach involves the growth of polymers from CNT surfaces via in situ polymerisation of olefins initiated from chemical species immobilised on the CNT. As an example, Ziegler-Natta or metallocene catalysts for ethylene polymerisation can be immobilised on nanotubes to grow PE chains from their surface. However, covalent linkages or strong interactions between PE chains and nanotubes cannot be created during polymerisation [90, 111-113]. The "grafting to" technique involves the use of addition reactions between the polymer with reactive groups and the CNT surface. However, the synthesis of end-functionalized polyethylene (PE), which is necessary in the "grafting to" approach, is difficult [114]. Another promising route for a chemical modification of MWCNTs by PE is to use free radical initiators such as peroxides. The general mechanism of free radical grafting of vinyl compound from hydrocarbon chains detailed by Russell [115], Chung [116] and Moad [117] seems to express a widespread view. The grafting reaction starts with hydrogen abstraction by alkoxyl radicals generated from thermal decomposition of the peroxide. Then, the active species generated onto the hydrocarbon backbone react with unsaturated bonds located on the MWCNTs surface. This chemical modification is thus conceivable during reactive extrusion because the radicals' lifetimes (in the range of few milliseconds) are compatible with typical residence time in an extruder (around one minute).

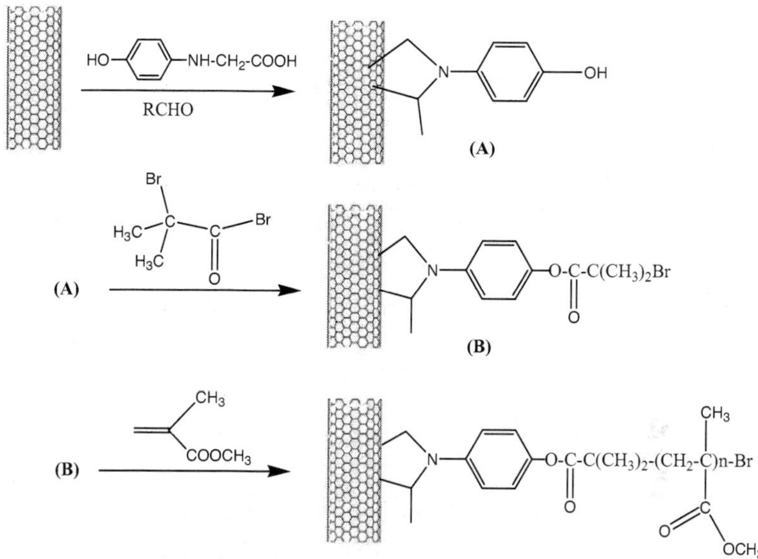

Figure 9. ATRP 'grafting from' modification approach. Reproduced from [92] with permission of ACS publications.

3.1. Radical grafting of polyethylene onto MWCNTs

The main drawback of the free radical grafting is the low selectivity of the radical center, specially at high temperatures (in the range of 150-200°C, required for extrusion of polyethylene), leading to side reactions such as coupling and chain scission [115, 118]. Moreover, performing this chemical modification by reactive processing brings in many constraints inherent to the processing (e.g. short reaction time, viscous dissipation and high temperature). For instance, the difference of viscosity between the monomer and the molten polymer could enhance these side reactions. So, to separate these physical influences from the chemical modification, the grafting reaction can be predicted with a model compound approach based on a radical grafting reaction between peroxide-derived alkoxyl radicals, and a low molar mass alkane representing characteristics moieties of PE.

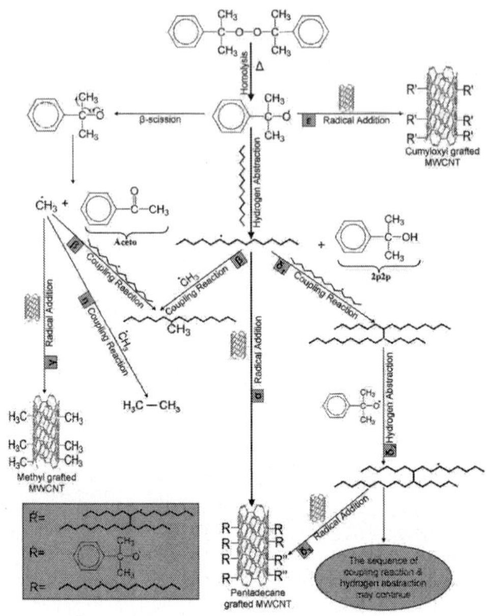

Figure 10. General reactive pathways of free radical grafting of pentadecane onto MWCNTs ; Reproduced from [119] with permission of Elsevier.

3.1.1. A model compound approach through the use of pentadecane

Covalent functionalization of pentadecane-decorated multiwalled carbon nanotubes (MWCNTs) has been studied as a model compound approach for the grafting of poly (ethylene-co-1-octene) onto MWCNTs by reactive extrusion [119]. It was accomplished through radical addition onto unsaturated bonds located on the MWCNTs' surface using dicumyl peroxide as hydrogen abstractor. Pentadecane ($C_{15}H_{32}$) has been resorted as model for polyethylene because high boiling points of long chain alkanes permit study under high temperature conditions, typically over 150°C. It also gives clues about low viscosity at 150°C, on top of that the formed products in the grafting experiment can hence be analysed more easily than in the polymer melt. Figure 10 sums up main reactive pathways of free radical grafting of pentadecane onto MWCNTs with dicumyl peroxide as initiator. The hydrogen abstraction reactions from alkyl hydrocarbon bonds was studied starting from the reaction of DCP-derived radicals with pentadecane.

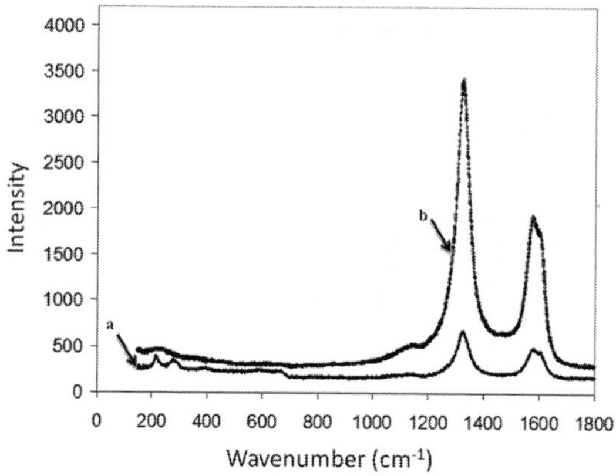

Figure 11. Raman spectra of: p-MWCNTs (a) and penta-g-MWCNTs (b).

However, the alkoxy radicals can undergo additional reactions including β-scission leading to the formation of methyl radicals [117]. These latter preferentially induce coupling reaction (Fig. 10, route b and h) or attack onto the sp2 carbon of the MWCNTs (Fig. 10, route g) whereas cumyloxyl radicals are more prone to hydrogen abstraction from pentadecane [120]. The formed pentadecyl radicals through hydrogen abstraction are able to react with MWCNTs by radical addition onto sp2 carbon of theMWCNTs (Fig.10, main route a) and with other radical species via the common radical coupling reactions (Fig. 10, routes d1 and b). According to the results of Johnston [121,122], based on a study of the crosslinking reaction of poly (ethylene-co-1-octene) in the presence of DCP at 160 °C, coupling reactions are four times more prone to happen than scission reactions so the authors assumed that pentadecyl radicals do not undergo scission reactions [119]. Direct evidence for covalent sidewall functionalization has been found by Raman spectroscopy [123-124]. G band is a characteristic feature of the graphitic layers and corresponds to the tangential vibration of the carbon atoms. The second characteristic mode is a typical sign for defective graphitic structures (D band). The A_D/A_G ratio, which was defined as the intensity ratio of the D-band to G-band of CNTs, directly indicates the structural changes in nanotubes. Some authors used D to G area ratios (A_D/A_G) rather than intensity [125] which is a better indicator. The relatively high area ratio of the G band relative to D band for penta-g-MWCNT (A_D/A_G = 1.51) in comparison with that of p-MWCNT (i.e. (A_D/A_G =1.20) could be designated as an indicator of grafting species. The ratio between the G band and D band is a good indicator of the changes in chemistry of CNTs. Interestingly, Raman spectra of p-MWCNTs and penta-g-MWCNTs (Figure11a and 11b, respectively) showed two main peaks around 1350cm^{-1} (D band) and 1586cm^{-1} (G band). The relatively high intensity of the G band relative to D band (I_G/I_D=1.25) for penta-g-MWCNT sample in comparison with that of p-MWCNT (i.e. I_G/I_D=0.95) was designated as an indicator of grafting species [119].

It is important to determine whether the results of a CNT surface modification process agree qualitatively with expectations, and equally important is the need for a quantitative assessment of the extent and nature of surface modifications. The course of the generated radical species and the extent of the grafting reaction in regards to the DCP concentration and temperature can be studied through gas chromatography and thermogravimetric analysis (TGA) [119, 126]. TGA permits measurement of the total weight fraction of surface modifiers introduced onto the surfaces of CNTs if the surface-modified CNTs are free of impurities. Indeed, it is well known that heating functionalized nanotubes in an inert atmosphere removes the organic moieties and restores the pristine nanotubes structure. TGA can indicate the degree of surface modification because the type and quantity of surface modifier is identified. It was found that the higher grafting density, as high as 1.46 mmol/g, was obtained at 150°C. At higher temperatures, the grafting density decreases because the β-scission reaction of cumyloxyl radical accelerates as the temperature increases, leading to the formation of methyl radicals. These latter preferentially react by combination whereas cumyloxyl radicals are more prone to hydrogen abstraction from pentadecane. At 150°C, for initiator concentration higher than 3wt%, the grafting density decreases from 1.464mmol/g to 0.371mmol/g upon increasing DCP concentration up to 5%. Thus, to get high grafting efficiency, one should opt for optimal initiator concentration, i.e. 3wt%, and choose the most favourable reaction temperature, i.e. 150°C. Incorporation of TEMPO as radical scavenger in the grafting reaction of pentadecane onto MWCNTs serves two purposes: firstly, it actively suppresses radical combination reactions and hence promotes pentadecyl radicals' addition to nanotubes (~16% increase in grafting density); and secondly, it effectively changes the polarity balance of the grafted species, making pentadecane and TEMPO functionalised nanotubes soluble in solvents such as THF and chloroform [126].

3.1.2. Synthesis of PE grafted carbon nanotubes via peroxide

To make full use of the strength of carbon nanotubes in a composite, it is important to have a high-stress transfer at the matrix-nanotube interface via strong chemical bonding, as discussed by Mylvaganam et al [127]. They have investigated the possible polyethylene-nanotube bonding with the aid of a quantum mechanics analysis with the polyethylene chains represented by alkyl segments, and the nanotubes modeled by nanotube segments with H atoms added to the dangling bonds of the perimeter carbons. Their study has predicted (i) covalent bond formation between alkyl radicals and carbon nanotubes is energetically favourable; and, (ii) this reaction may take place at multiple sites of nanotubes [126]. Hence one way to improve the load transfer of carbon nanotubes/PE composite via chemical bonds at the interface is to use free-radical generators such as peroxide or incorporate nanotubes by means of in situ polymerisation.

Figure 12 sums up main reactive pathways of free radical grafting of PE onto MWCNTs with dicumyl peroxide as initiator and TEMPO as radical scavenger.

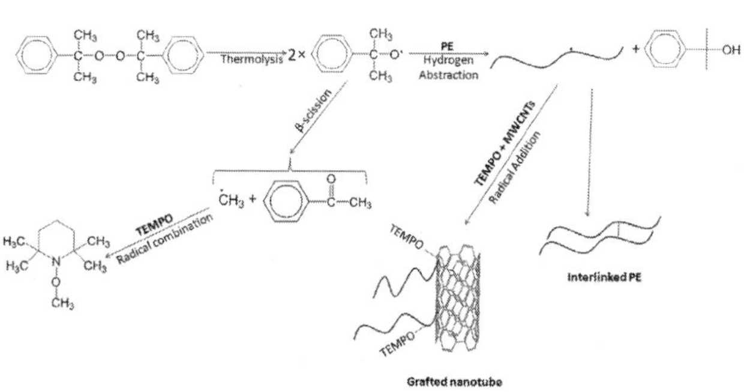

Figure 12. Reaction scheme for PE grafting onto MWCNTs with TEMPO as a radical scavenger. Reproduced from [128] with permission of John Wiley and Sons.

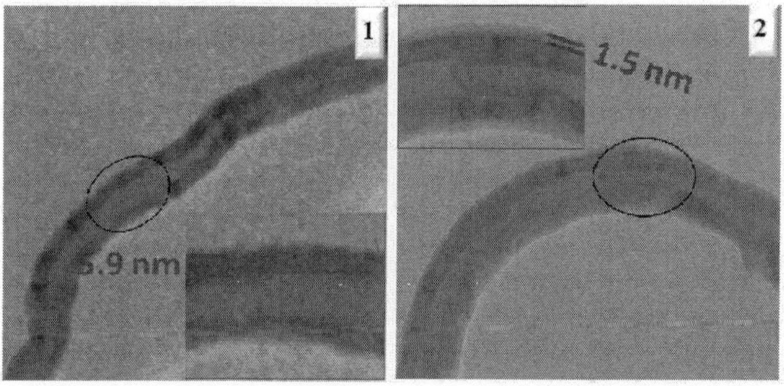

Figure 13. TEM images of p-MWCNTs (1) and PE-g-MWCNTs (2). Reproduced from [128] with permission of John Wiley and Sons.

The formed PE-based radicals are able to react with MWCNTs by radical addition onto sp^2 carbon of the MWCNTs (Figure 12) and with other radical species via the common radical coupling reactions. As discussed using a model compound approach (section 2.1.1), the presence of TEMPO radicals creates competitive combinations reactions (that are actually reversible reactions) which may favour the addition of PE-based radicals to MWCNTs (Figure 12). Before carrying out thermogravimetric analysis to gain a quantitative picture of the extent of nanotubes' functionalisation, the adsorbed (non-covalently attached) PE chains were removed from the grafted ones (covalently attached) by extensive washings with dichlorobenzene (DCB). PE-grafted onto MWCNTs are well known to degrade at 300-540°C, which are nearly the same temperatures as pure PE reactants and the weight of grafted PE is estimated to be in the range 20-24% depending on the experimental procedure [128]. The corresponding grafting densities, calculated using a specific surface area (SSA) of 225m^2/g for MWCNTs [115, 129] are varying from 1.1mg.m^{-2} to 1.4mg.m^{-2}. LDPE grafting density on nanotubes is 1.1mg.m^{-2} while incorporation of TEMPO raises the grafting density to 1.4mg.m^{-2} [128]. Then, it is usual to examine the morphological structures of p-MWCNTs and PE-grafted MWCNTs by transmission electron microscopy (TEM). In these experiments, a few drops of dilute solutions of PE-grafted nanotubes in hot DCB are initially deposited onto a carbon-coated copper grid and further observed in a dried state after evaporation of the solvent (Figure 13).

A "grafting to" approach based on a radical process can also involve a polymer with reactive end groups.

3.1.3. Synthesis of PE grafted carbon nanotubes via end functionalised PE

Recently, D'Agosto and Boisson [130-134] developed new strategies that rely on a one step *in situ* functionalisation reaction within an ethylene polymerisation reactor to introduce a variety of functional groups including alkoxyamine [130-132] and thiol [132,134] functions at the end of polyethylene chains. Di-polyethylenyl magnesium compound ($MgPE_2$) were prepared using a neodymium metallocene complex which catalysed polyethylene chain growth on magnesium compounds. $MgPE_2$ was in situ reacted with 2,2,6,6-tetramethylpiperidin-yl-1-oxy (TEMPO) radical or elemental sulphur to provide a macroalcoxyamine (PE-TEMPO) and polysulphur based product (PE-S_n-PE) respectively. PE-SH was obtained by simple reduction of PE-S_n-PE. According to these results, a strategy based on the use of those polyethylenes was investigated to generate radical-terminated chains formed either by thermal loss of a nitroxide (PE-TEMPO) or H-abstraction onto a thiol (PE-SH) and graft them onto CNTs (Figure 14) [128].

Indeed, Lou [87] showed an efficient attachment of poly(2-vinylpyridine) (P2VP) end-capped by TEMPO to CNT sidewalls by heating of TEMPO-terminated P2VP. Using the same strategy Liu [137] functionalized shortened CNTs with PS and poly[(*tert*-butyl acrylate)-*b*-styrene] and Wang [138] grafted poly(4-vinylpyridine-*b*-styrene) onto CNTs. In figure 14a, the homolytic cleavage of TEMPO-terminated PE leads to the formation of stable nitroxyl radicals and PE radicals. The reversible termination of the polyethylene chain is the key step for reducing the overall concentration of the radical chain end. The extremely low concentration of reactive chain ends is expected to minimize irreversible termination reactions, such as combination or disproportionation [135] (Figure 14a). Thiol-terminated polyethylene has been also grafted onto CNTs using a similar procedure. The thiol based compounds are widely used for controlling molar mass in free radical polymerizations via a chain transfer process. The chain transfer process displays two contiguous steps: transfer of the thiyl hydrogen to the growing polymer chain followed by re-initiation, whereby a thiyl radical adds to a monomeric double bond. In the presence of DCP-derived radicals and MWCNTs, thiyl radicals are formed and are expected to react by radical addition onto sp^2 carbon of the MWCNTs. For both samples PE-TEMPO and PE-SH, the weight loss observed by thermogravimetric analysis (TGA) has been increased to 36% and 34%, respectively despite their low molar masses (e.g. 1400g/mol and 980g/mol) in comparison with that of LDPE (weight loss = 20-24% ; Mw = 90000g/mol ; see section 2.1.2.). These results indicated that the use of short end-functionalized PE chains has permitted a significant increase of the grafting density

http://dx.doi.org/10.5772/50710

(e.g. 1.78 and 2.80μmol.m^{-2}, respectively). These values are approximately increased by two orders of magnitude in comparison with the LDPE grafting density (e.g. 0.012μmol.m^{-2}) suggesting that longer polymer chains cover a larger surface decreasing the grafting density, as previously described by Jerome et al [87,136] for the attachment of poly(2-vinylpyridine) (P2VP) and polystyrene (PS) onto MWCNTs. Indeed, they observed that PS grafting density decreases from 0.045μmol.m^{-2} to 0.01μmol.m^{-2} by increasing the molecular weight of PS-TEMPO from 3000g/mol to 30000g/mol [138]. The TEM observations (Figure 15) were consistent with the TGA results : the grafted polymer contents can be highered by using end-functionalized PE [129].

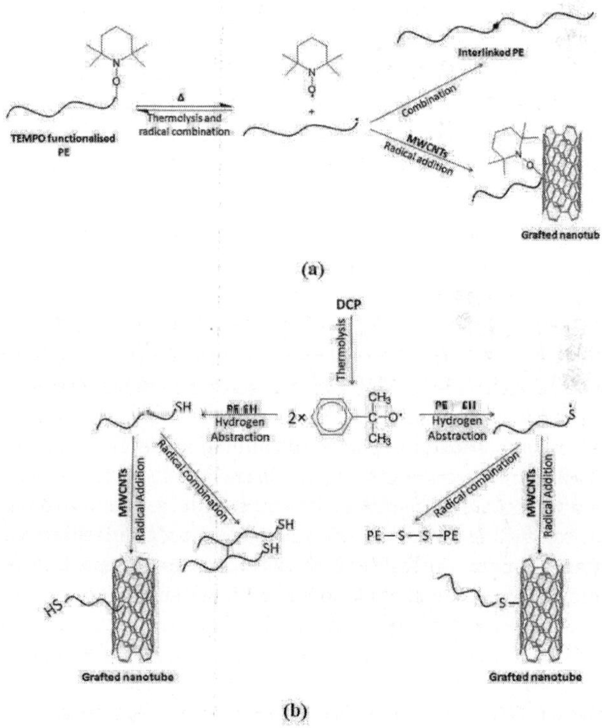

Figure 14. Reaction scheme for end functionalised PE grafting onto MWCNTs: (a) via PE-TEMPO; (b) via PE-SH. Reproduced from [128] with permission of John Wiley and Sons.

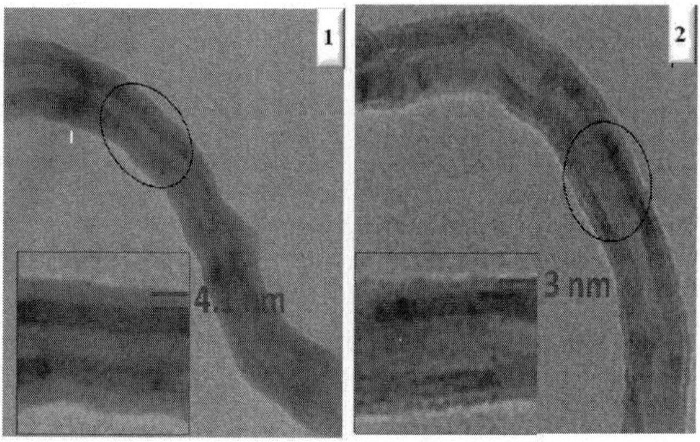

Figure 15. TEM images of $PE_{f\text{-}TEMPO}$-g-MWCNTs (1) and $PE_{f\text{-}SH}$-g-MWCNTs (2). Reproduced from [128] with permission of John Wiley and Sons.

3.2. Radical grafting of polypropylene on carbon nanotubes

Polypropylene (PP) is a widely used commercial polymer due to the excellent combination of mechanical resistance, chemical stability and excellent moisture barrier properties [104]. Although physical blending with CNTs is an economic way to modify polypropylene performance, compatibilizing agents are necessary for creating strong interface between filler particles and the polymer phase. Maleic anhydride grafted polypropylene (MA-*g*-PP) is often used as a compatibilizer which can improve the PP/CNTs composite properties by strong hydrogen bonding between hydroxyl groups located on the acidic-treated CNTs surface and anhydride groups of MA-g-PP [139, 140]. Recently, an original and simple method for promoting mobility sensitivity of carbon nanotubes (CNTs) to an external stress field in polypropylene (PP) matrix was developed [141]. In particular, an interfacial melt reaction initiated by free radicals were used as a tool to prepare PP/CNTs nanocomposites. The presence of tetrakis(phenylmethyl)thioperoxydi(carbothioamide) (TBzTD) increased the interfacial reaction between the PP chains and the CNTs. In addition, the grafted TBzTD to PP backbone could form a physical interaction with CNTs *via* a π–π interaction [141]. According to their previous results [119, 126] based on a study of the radical grafting of polyethylene derivatives onto MWCNTs, Farzi et al. investigated MWCNTs' sidewall functionalization by tetramethylpentadecane and PP in the presence of 1wt% DCP at 160°C.

3.2.2. *A model compound approach through the use of 2,6,10,14-tetramethylpentadecane*

Simlilarly to the pentadecane grafting procedure (see section 2.1.1.), 2,6,10,14-tetramethylpentadecane (TMP, $C_{19}H_{40}$) has been used as model for the grafting reaction of PP onto MWCNTs [129]. Thermolysis of dicumyl peroxide initiator performed in TMP and in presence of MWCNTs is depicted in Figure 16.

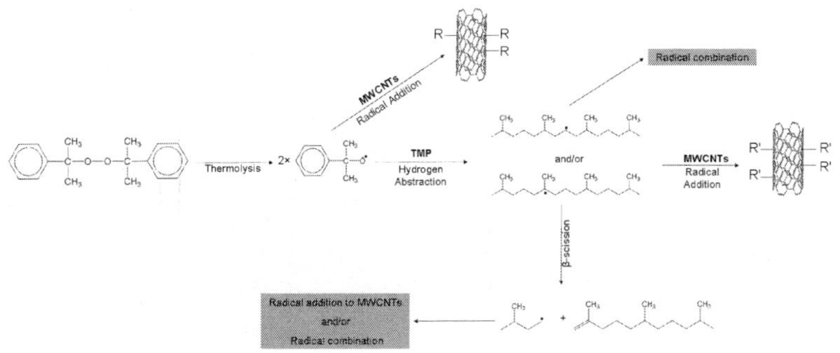

Figure 16. Reaction scheme for the addition of TMP onto CNT in the presence of DCP. Reproduced from [129] with permission of Elsevier.

As shown in Figure 16, the formed peroxide radicals are prone to hydrogen abstraction from hydrocarbon substrates and it is expected that the active species generated onto the hydrocarbon backbone react with unsaturated bonds located on the MWCNTs surface keeping in mind that side reactions such as chain scission for PP derivatives may occur at high temperatures (Figure16) [115, 118]. For experiments conducted in dichlorobenzene (DCB) as solvent at 160°C with 1.5wt% DCP, TMP grafting density was as high as 0.92 mg/m^2.

3.2.2. Synthesis of PP grafted carbon nanotubes via peroxide

Farzi et al. [129] have successfully grafted PP onto MWCNTs through a radical grafting reaction, carried out under similar experimental conditions to PE [128] and TMP [129] (1.5wt% DCP, 160°C) and using 1,2-dichlorobenzene (DCB) as solvent able to solubilize PP partially at elevated temperature. The corresponding PP-grafted nanotubes were analysed by TGA after a purification by soxhlet extraction in DCB. However, the authors were not able to obtain reproductible results with weight losses varying from 50% to 80% for the above-mentioned experiment. This behaviour was attributed to the purification procedure which did not permit to remove all the free PP chains. The authors have also speculated on the degradation behaviour of PP through the well-known β-scission reaction occurring in the presence of radical species therefore the authors were not able to give a PP grafting density. Then, the aforementioned PP coated MWCNTs have been dispersed within a commercially available PP matrix using a contra-rotating Haake Rheomixer and the amount of nanofiller in the final composites has been fixed to 3wt%. The evaluation of MWCNTs dispersion has been examined by using scanning electron microscopy (SEM) (Figure 17).

SEM images of the PP/PP-*g*-MWCNTs composites MWCNTs containing of 3wt% (Figure 17) demonstrated that there were still some areas where PP-*g*-MWCNTs were not found which was obviously connected with improper filler distribution. For a simple melt blend of PP with untreated MWCNTs, SEM images of the resulting material only showed clusters of a few tens micrometers of diameter evidencing a poor interfacial adhesion in the material

(Figure 18), as reported by Lee [140] for untreated MWCNT/PP composites MWCNTs containing of 2wt%.

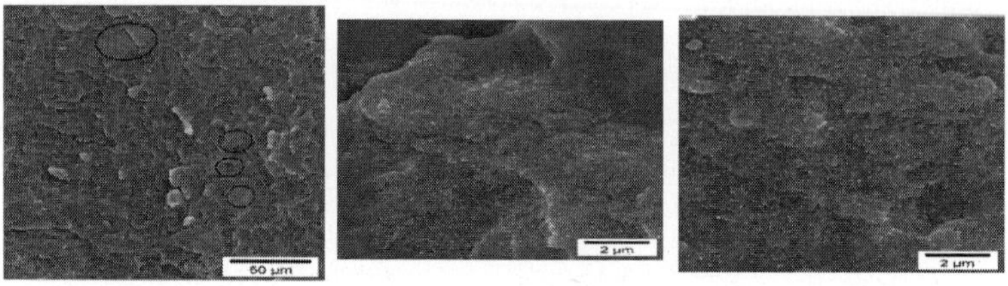

Figure 17. SEM micrographs of PP/PP-g-MWCNT composites with MWCNTs loading of 3wt%. Reproduced from [129] with permission of Elsevier.

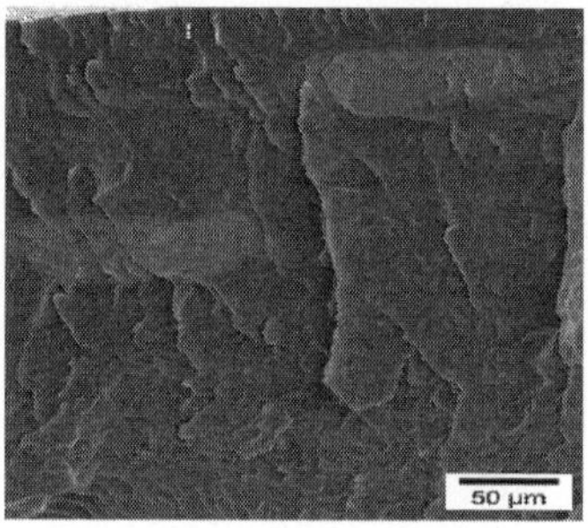

Figure 18. SEM micrographs of PP/MWCNTs composites with MWCNTs loading of 3wt%. Reproduced from [129] with permission of Elsevier.

It was concluded from these results that the grafting of PP onto MWCNTs provided a low steric barrier against the strong intermolecular Van der Waals interactions among nanotubes within the PP matrix.

In a similar approach, isotactic polyrpopylene (iPP) was successfully grafted onto multiwalled carbon nanotubes (MWCNTs) by direct macroradical addition by sonication in hot xylene with BPO as an initiator [142]. It was found that both iPP macromolecular radicals and small-molecular benzoic acid free radicals were grafted onto MWNTs. iPP-g-MWNTs dispersed more uniformly in iPP than pristine MWNTs.

3.3. Rheological behaviour of polyolefin based carbon nanotubes nanocomposites

It has been well known for a century that the addition of fillers, mostly carbon black, to rubber compounds has a strong impact on the viscoelastic properties of materials. In recent years, polymer nanocomposites have been developed as a new class of composites [143]. Ac-

http://dx.doi.org/10.5772/50710

tually from a rheological point of view, a direct consequence of incorporation of fillers in molten polymers is the significant change in the steady shear viscosity behavior and the viscoelastic properties. The level of filler dispersion is expected to play a major role in determining the filler effects on non linear responses of nanocomposites. Generally speaking, thermoplastic polymers filled with nano-particles show a solid-like behavior response which includes a non–terminal zone of relaxation, apparent yield stress and a shear-thinning dependence of viscosity on particle concentration, aspect ratio and dispersion.

Since the melt rheological properties of filled polymers are sensitive to the structure, concentration, particle size, shape and surface characteristics of the fillers, rheology offers original means to assess the state of the dispersion in nanocomposites and to investigate the influence of flow conditions upon nano-filler dispersion itself [144]. As discussed previously, one of the most important challenges in filled polymer developments and applications is to obtain a homogeneous dispersion of CNT in polymer matrix by overcoming the van der Waals interaction between elementary tubes. As a result, it can be expected that the rheological percolation, and subsequently the non-linearity effect, depend on nanotube dispersion and aspect ratio. As matter of fact, a great level of activity in the domain of the rheology of polymers filled with CNT is reported in the more recent scientific literature. The rheological behavior of melt thermoplastic polymer filled with NTC was reported to depend on nanotube dispersion, aspect ratio and alignment under flow. However, among the different studies on liquid systems filled with carbon nanotubes two types of relaxation mechanisms of CNT must be differentiated according to the matrix viscosity: Do the carbon nanotubes behave as Brownian particles? The Doi-Edwards theory for dilute regime (the nantotubes are free to rotate without any contacts) allows the rotary diffusivity D_{r0} of a rod (Length: L and diameter: d) in an isotropic suspension to be calculated by equation (1), in which k_B is the Boltzmann constant and η_m is the viscosity of the suspending medium

$$D_{r0}=\frac{3k_BT(\ln(L/d)-0.8)}{\pi\eta_mL^3} \quad (1)$$

In semi-dilute regime, the rotary diffusivity D_r can be written as equation (2), where *A* is a dimensionless constant whose value is generally large (A~1000).

$$D_r=AD_{r0}(\nu L^3)^{-2} \quad (2)$$

Consequently, the rotary diffusion of CNT varies according to matrix viscosity:

$$D_r\alpha 1/\eta_m \quad (3)$$

Actually, the rheological behavior of CNT suspension is observed close to the Doi-Edwards theory on the Brownian dynamics of rigid rods. However, it was observed that low shear deformations induced an aggregation mechanism, but these aggregates broke down at high shear, forming small aggregates with less entanglements [145]. The shear rheology of such carbon nanotube suspensions was reviewed by Hobbie [146] from the perspective of colloid and polymer science.

According to the Doi-Edwards theory, Marceau et al [147] have shown that the suspensions of CNT, at low concentration (ϕ=0.2%) and in low fluid matrix (η_m=5 Pa.s), behave as Brownian entities (D_r=5.0x10^{-5} s^{-1}). The diffusion time of these CNTs is then $\tau_r = 1/2D_r \approx 10^4 s$. If we imagine that these same CNT are dispersed in high viscous polymer matrix such as molten PP (η_m~1x10^3 Pa.s), their relaxation time will be then: $\tau_r \approx 2x10^6 s$. The order magnitude of the relaxation time is then one month! Consequently, the carbon nanotubes cannot be considered anymore as Brownian entities in most of the papers that have been addressed to the viscoelastic behavior of carbon nanotubes dispersed in high viscous molten polymers. The main challenge in such nanocomposite systems is to control the dispersion of the nanotubes in high viscous fluid in order to have the lowest percolation threshold regarding the electrical properties. For example, by improving the CNT dispersion using functionalized single wall nanotubes, Mitchell et al [148] observed that the percolation threshold dropped from 3wt% to 1.5% in PS nanocomposites.

Actually, the dispersion of CNT in polymer matrix is strongly difficult mainly due to the nanotube-nanotube interactions higher than the nanotube-polymer interaction. However, optimal dispersion of CNTs can be achieved in polar polymers such as polyamides, polyesters or polycarbonate. This optimal dispersion is generally measured, at least from a qualitative point of view, from the electrical and/or rheological percolation threshold. Nevertheless, the dispersion of CNT in polyolefin (PP, PE or copolymer of ethylene) is most of the time a real challenge due to unfavorable and low nanotube-polymer interactions. On the other hand, the fact that CNT have a high aspect ratio and are not Brownian in polymer matrix leads to the conclusion that the different works of the literature are difficult to compare. The samples, studied in rheology or electric conductivity, have generally undergone different processing conditions. As a result, CNT nanocomposites are totally out of isotropic dispersion and the isotropic equilibrium of CNTs can never be achieved. However, general trends in CNT nancomposite can be described from the open literature.

From a sample preparation point of view, dispersion of CNTs in polyolefins were generally prepared via conventional melt processing, i.e melt blending in batch mixer or in twin screw extruder). Marginal methods may also be used as for example solid-state shear pulverization [149] or in situ lubrication methods [150]. Numerous studies [151-156] have been devoted to the linear viscoelasticity of PP nanocomposites based on CNT dispersion.

All of these papers reported an increase in shear viscosity and storage and loss moduli of the nanocomposites with increasing the CNT concentration as shown in Figure 19.

Furthermore, a general rheological trend for nanocomposites studied in most of these papers is the appearance of a transition from a liquid-like behaviour to a solid-like behaviour, i.e. the apparition of a plateau (second plateau modulus, $G_o = \lim_{\omega \to 0} G'(\omega)$) of the storage modulus at low frequency which is obviously higher than the loss modulus. Obviously, it is admitted that the increase of the CNT concentration is driving this transition. Above this critical transition, generally associated with the percolation threshold, these nanocomposites show a solid-like behavior response, which includes a non-terminal zone of relaxation leading to apparent yield stress and a shear-thinning dependence on viscosity $|\eta^*(\omega)| \alpha \omega^{-1}$.

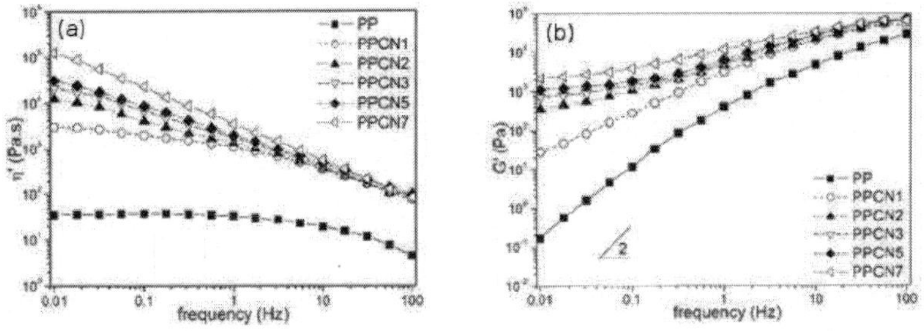

Figure 19. Variation versus frequency of the storage shear modulus G'(ω) and absolute complex viscosity η*(ω) at different concentrations of CNT (1 to 7%). Reproduced from [151] with permission of John Wiley and Sons.

This non-terminal frequency behavior is generally attributed to the formation of an interconnected nanotube network in the polymer matrix. Therefore, the solid-like behavior is associated nanotube–nanotube interactions which increase with the CNT content. Eventually, these interactions lead to percolation and the formation of an interconnected structure of nanotubes in the matrix. Due to the high aspect ratio of CNT (generally, L/d>150), the existence of this percolation threshold is expected at low concentration. For example, from Fig 19, the percolation threshold can be estimated to be less than 2% of CNT. This percolation threshold is generally observed in the range 0.5%-5% depending on CNT nature (aspect ratio, surface chemical modification) and on the processing methods for nanocomposite preparation.

If a lot of works have been devoted to the linear viscoelasticity of CNT nanocomposites whereas a few works have been reported on non-linear properties such as for example the melt flow instabilities. Interestingly, Palza et al [157] showed that carbon nanotubes modify the main characteristics of the spurt instability developed by the linear polyethylene. Furthermore, the sharkskin instability, developed in short chain branched polyethylene, is reduced at low amounts of MWCNT. Furthermore, the critical shear rate for the on-set of the spurt and the sharkskin instabilities decreases in the nanocomposites probably due to the physical interactions between the polymer and the nanofiller. Finally, at high shear rates, the gross melt fracture instability is completely erased in the nanocomposites based on the linear polymer whereas in short chain branched polyethylene the amplitude of this bulk distortion is rather moderated. Clearly, the carbon nanotubes have a drastic effect on the main flow instabilities observed in polyethylene. Consequently, the processing of CNT nanocomposites, i.e under non-linear deformation, is an open investigation domain.

4. Conclusion

The most frequent method for preparing polymer nanotubes composites has been mixing nanotubes and polymer in a suitable solvent and to evaporate the solvent to form composite film. But to increase the advantages at its best, one needs: (i) high interfacial area between nanotubes and polymer; and, (ii) strong interfacial interaction. Unfortunately this solvent technique does not help much in achieving these targets; and as a result a nanocomposite having properties much inferior to theoretical expectations are obtained. In order to obtain higher contact area between nanotubes and polymer, the issue of dispersion needs to be addressed. Uniform dispersion of these nanotubes produces immense internal interfacial area, which is the key to enhancement of properties of interest. On the other hand, modification of nanotubes surface through functionalisation is required for creating an effective interaction with the host matrix and to make nanotubes soluble and dispersible.

The idea of grafting PE or PP with the help of peroxide during extrusion is exciting. It was shown that cumlyoxly radical generated by thermolysis of DCP can abstract hydrogen from polyolefin chains, thus creating polyolefin macroradicals. Then, these macroradicals add to the unsaturated carbon bonds on the surface of the nanotubes. The upside of this strategy is that radicals have short lifetimes which make the procedure possible in an extruder where the residence time is generally low. On the contrary, the downside is the low selectivity of radicals leading to a competition between radical combination reactions and radical addition reactions. Alkanes can be used as model for PE to perform the grafting of PE onto nanotubes. The results were interesting however the degree of PE grafting remained lower than the model (weight loss by TGA = 22% as compared to a weight loss of 30% in case of pentadecane). LDPE grafting density on nanotubes was $1.1 mg.m^{-2}$ while incorporation of TEMPO raised the grafting density to $1.4 mg.m^{-2}$. End functionalised PE can also be used for PE grafting onto nanotubes with dichlorobenzene as solvent. As emphasized by TEM images, a layer of considerable thickness has been grafted around the periphery of the nanotubes.

In order to follow the same strategy for nanotubes functionalisation with PP, tetramethylpentadecane has been selected as a model compound for PP. It was successfully grafted onto carbon nanotubes with a grafting density of $0.92 mg/m^2$. However, the grafting of PP onto nanotubes did not permit to obtain reproductible results. SEM images of the PP-*g*-MWCNTs nanocomposites with filler loadings of 3wt% in PP matrix did not show a significant improvement in MWCNTs dispersion within the PP matrix although sizes of the aggregates were slightly reduced.

In addition, it can be expected that the rheological percolation, and subsequently the nonlinearity effect, depend on nanotube dispersion and aspect ratio. Low shear deformations induced an aggregation mechanism, but these aggregates broke down at high shear, forming small aggregates with less entanglements. In a high viscous polymer media, it was shown that carbon nanotubes could not be considered anymore as Brownian entities. A general rheological trend for CNTs-based nanocomposites is the appearance of a transition from a liquid-like behaviour to a solid-like behaviour increasing with the CNT content because it is associated to nanotube–nanotube interactions. Due to the high aspect ratio of CNT the percolation threshold can be expected to be less than 2% of CNT.

Author details

Emmanuel Beyou, Sohaib Akbar, Philippe Chaumont and Philippe Cassagnau

Université de Lyon, Université Lyon 1, CNRS UMR5223, Ingénierie des Matériaux Polymères: IMP@UCBL, 15 boulevard Latarget, F-69622 Villeurbanne, France

References

[1] Iijima, S. (1991). Helical microtubules of graphitic carbon. *Nature*, 354, 56-58.

[2] Baughman, R. H., Zakhidov, A.A., & de Heer, W. A. (2002). Carbon Nanotubes--the Route Toward Applications. *Science*, 297(5582), 787-792.

[3] Moniruzzaman, M., & Winey, K. I. (2006). Polymer Nanocomposites Containing Carbon Nanotubes. *Macromolecules*, 39(16), 5194-5205.

[4] Grossiord, N., Loos, J., Regev, O., & Koning, C. E. (2006). Toolbox for dispersing carbon nanotubes into polymers to get conductive nanocomposites. *Chemistry of Materials*.

[5] Ciardelli, F., Coiai, S., Passaglia, E., Pucci, A., & Ruggeri, G. (2008). Nanocomposites based on polyolefins and functional thermoplastic materials. *Polymer International*, 57(6), 805-836.

[6] Judeinstein, P., & Sanchez, C. (1996). Hybrid organic-inorganic materials:. *A land of multi-disciplinarity Journal of Materials Chemistry*, 6(4), 511-525.

[7] Connell, M.O. Carbon Nanotubes. *Properties and Applications; CRC Press, 2006.*, 360, 9780849327483.

[8] Gogotsi, Y. *Carbon NanomaterialsCRC Press, 2006.*, 344, 9780849393860.

[9] Spitalsky, Z., Tasis, D., Papagelis, K., & Galiotis, C. (2010). Carbon nanotube-polymer composites: Chemistry, processing, mechanical and electrical properties. *Progress in Polymer Science*, 35(3), 357-401.

[10] Osswald, S., Flahaut, E., & Gogotsi, Y. (2006). In situ Raman spectroscopy study of oxidation of double- and single-wall carbon nanotubes. *Chemistry of Materials*, 18(6), 1525-1533.

[11] Yang, Q., Wang, L., Xiang, W., Zhu, J., & Li, J. (2007). Grafting polymers onto carbon black surface by trapping polymer radicals. *Polymer*, 48(10), 2866-2873.

[12] Deng, X., Jia, G., Wang, H., Sun, H., Wang, X., Yang, S., Wang, T., & Liu, Y. (2007). Translocation and fate of multi-walled carbon nanotubes in vivo. *Carbon*, 45(7), 1419-1424.

[13] Shaffer, M. S. P., & Windle, A. H. (1999). Fabrication and Characterization of Carbon Nanotube/Poly(vinyl alcohol) Composites. *Advanced Materials*, 11(11), 937-941.

[14] Baskaran, D., Mays, J. W., & Bratcher, M.S. (2005). Noncovalent and nonspecific molecular interactions of polymers with multiwalled carbon nanotubes. *Chemistry of Materials*, 17(13), 3389-3397.

[15] Haggenmuller, R., Gommans, H. H., Rinzler, A. G., Fischer, J. E., & Winey, K. I. (2000). Aligned single-wall carbon nanotubes in composites by melt processing methods. *Chemical Physics. Letters*.

[16] Kang, Y. J., & Taton, T. A. (2003). Micelle-encapsulated carbon nanotubes: A route to nanotube composites. *Journal of the American Chemical Society*, 125(19), 5650-5651.

[17] Star, A., & Stoddart, J. F. (2002). Dispersion and solubilization of single-walled carbon nanotubes with a hyperbranched polymer. *Macromolecules*, 35(19), 7516-7520.

[18] Barraza, H. J., Pompeo, F., O'Rear, E. A., & Resasco, D. E. (2002). SWNT-filled thermoplastic and elastomeric composites prepared by miniemulsion polymerization. *Nano Letters*, 2(8), 797-802.

[19] Bonduel, D., Mainil, M., Alexandre, M., Monteverde, F., & Dubois, P. (2005). Supported coordination polymerization: a unique way to potent polyolefin carbon nanotube nanocomposites. *Chemical Communications*, 6, 781-783.

[20] Park, S., Yoon, W., Lee, B., Kim, J., Jung, Y. H., Do, Y., Paik, H. J., & Choi, I. S. (2006). Carbon Nanotubes as a Ligand in Cp2ZrCl2-Based Ethylene Polymerization. *Macromolecular Rapid Communication*, 27(1), 47-50.

[21] Liu, T. X., Phang, I. Y., Shen, L., Chow, S. Y., & Zhang, W. D. (2004). Morphology and mechanical properties of multiwalled carbon nanotubes reinforced nylon-6 composites. *Macromolecules*, 37(19), 7214-7222.

[22] Siochi, E. J., Working, D. C., Park, C., Lillehei, P. T., Rouse, J. H., & Topping, C. C. Melt processing of SWCNT-polyimide nanocomposite fibers. Composites Part B. (2004)., 35(5), 439-446.

[23] Potschke, P., Fornes, T. D., & Paul, D. R. (2002). Rheological behavior of multiwalled carbon nanotube/polycarbonate composites. *Polymer*, 43(11), 3247-3255.

[24] Li, C., Zhao, Q., Deng, H., Chen, C., Wang, K., Zhang, Q., Chena, F., & Fua, Q. (2011). Preparation, structure and properties of thermoplastic olefin nanocomposites containing functionalized carbon nanotubes. *Polymer international*, 60(11), 1629-1637.

[25] Zhang, Z. N., Zhang, J., Chen, P., Zhang, B. Q., He, J. S., & Hu, G. H. (2006). Enhanced interactions between multi-walled carbon nanotubes and polystyrene induced by melt mixing. *Carbon*, 44(4), 692-698.

[26] Shofner, M. L., Khabashesku, V. N., & Barrera, E. V. (2006). Chemistry of Materials. *Processing and mechanical properties of fluorinated single-wall carbon nanotube-polyethylene composites.*, 18(4), 906-913.

[27] Mahfuz, H., Adnan, A., Rangari, V. K., & Jeelani, S. (2005). Manufacturing and characterization of carbon nanotube/polyethylene composites. *International Journal of Nanoscience*, 4(1), 55-72.

[28] Ruan, S. L., Gao, P., & Yu, T. X. (2006). Ultra-strong gel-spun UHMWPE fibers reinforced using multiwalled carbon nanotubes. *Polymer*, 47(5), 1604-1611.

[29] Shofner, M. L., Khabashesku, V. N., & Barrera, E. V. (2006). *Chemistry of Materials*, 18, 906-913.

[30] Bahr, J. L., & Tour, J. M. (2001). Processing and mechanical properties of fluorinated single-wall carbon nanotube-polyethylene composites. *Chemistry of Materials*, 13(4), 3823-3824.

[31] Georgakilas, V., Kordatos, K., Prato, M., Guldi, D. M., Holzinger, M., & Hirsch, A. J. Organic functionalization of carbon nanotubes. . Journal of the American Chemical Society (2002). , 124(5), 760-761.

[32] Peng, H., Reverdy, P., Khabashesku, V. N., & Margrave, J. L. (2003). Sidewall functionalization of single-walled carbon nanotubes with organic peroxides. *Chemical Communications*, 362-363.

[33] Qin, S., Qin, D., Ford, W. T., Resasco, D. E., & Herrera, J. E. (2004). Functionalization of single-walled carbon nanotubes with polystyrene via grafting to and grafting from methods. *Macromolecules*.

[34] Riggs, J. E., Guo, Z., Carroll, D. L., & Sun-P, Y. (2000). Strong luminescence of solubilized carbon nanotubes. *Journal of the American Chemical Society*, 122(24), 5879-5884.

[35] Umek, P., Seo, J. W., Hernadi, K., Mrzel, A., Pechy, P., Mihailovic, D. D., & Forro, L. (2003). Addition of carbon radicals generated from organic peroxides to single wall carbon nanotubes. *Chemistry of Materials*, 15(25), 4751-4755.

[36] Liu, Y. K., Yao, Z. L., & Adronov, A. (2005). Functionalization of single-walled carbon nanotubes with well-defined polymers by radical coupling. *Macromolecules*, 38(4), 1172-1179.

[37] Shaffer, M. S. P., & Koziol, K. (2002). Polystyrene grafted multi-walled carbon nanotubes. *Chemical Communications*.

[38] Kong, H., Gao, C., & Yan, D. Y. (2004). Controlled functionalization of multiwalled carbon nanotubes by in situ atom transfer radical polymerization. *Journal of the American Chemical Society*, 126(2), 412-413.

[39] Baskaran, D., Mays, J. W., & Bratcher, M.S. (2004). Polymer-grafted multiwalled carbon nanotubes through surface-initiated polymerization. *Angewandte Chemie International Edition*.

[40] Breuer, O., & Sundararaj, U. (2004). Big returns from small fibers. *A review of polymer/ carbon nanotube composites. Polymer Composites*, 25(6), 630-645.

[41] Tiwari, I., Singh, K. P., & Singh, M. (2009). An insight review on the application of polymer-carbon nanotubes based composite material in sensor technology. *Russian Journal of General Chemistry*.

[42] Martínez- Hernández, A.L., Velasco-Santos, C., & Castaño, V.M. (2010). Carbon Nanotubes Composites: Processing, Grafting and Mechanical and Thermal Properties. *Current Nanoscience*, 6(1), 12-39.

[43] Jin, L., Bower, C. L., & Zhou, O. (1998). *Applied Physics Letters*, 73(9), 1197-1199.

[44] Coleman, J. N., Cadek, M., Blake, R., Nicolosi, V., Ryan, K. P., & Belton, C. (2004). High-performance nanotube-reinforced plastics: Understanding the mechanism of strength increase. *Advanced Functional Materials*, 14(8), 791-798.

[45] Cadek, M., Coleman, J. N., Ryan, K. P., Nicolosi, V., Bister, G., & Fonseca, A. (2004). Reinforcement of polymers with carbon nanotubes: The role of nanotube surface area. *Nano Letters*, 4(2), 353-356.

[46] Cadek, M., Coleman, J. N., Barron, V., Hedicke, K., & Blau, W. J. (2002). Morphological and mechanical properties of carbon-nanotube-reinforced semicrystalline and amorphous polymer composites. *Applied Physics Letters*, 81(27), 5123-5125.

[47] Andrews, R., Jacques, D., Minot, M., & Rantell, T. (2002). Fabrication of carbon multiwall nanotube/polymer composites by shear mixing. Macromolecular. *Materials and Engineering*, 287(6), 395-403.

[48] Breuer, O., & Sundararaj, U. (2004). Big returns from small fibers. *A review of polymer/ carbon nanotube composites. Polymer Composites*, 25(6), 630-645.

[49] Potschke, P., Bhattacharyya, A. R., Janke, A., & Goering, H. (2003). Melt mixing of polycarbonate/multi-wall carbon nanotube composites. *Composite Interfaces*, 10(4), 389-404.

[50] Andrews, R., Jacques, D., Qian, D. L., & Rantell, T. (2002). Multiwall carbon nanotubes: Synthesis and application. *Accounts of Chemical Research*, 35(12), 1008-1017.

[51] Tang, W., Santare, M. H., & Advani, S. G. (2003). Melt processing and mechanical property characterization of multi-walled carbon nanotube/high density polyethylene (MWNT/HDPE) composite films. *Carbon*, 41(14), 2779-2785.

[52] Jia, Z., Wang, Z., Xu, C., Liang, J, Wei, B., & Wu, D. (1999). Study on poly(methyl methacrylate)/carbon nanotube composites. *Materials Science and Engineering A*, 271(1), 395-400.

[53] Bonduel, D., Mainil, M., Alexandre, M., Monteverde, F., & Dubois, P. (2005). Supported coordination polymerization: a unique way to potent polyolefin carbon nanotube nanocomposites. *Chemical Communications*, 41(6), 781-783.

[54] Bryning, M. B., Milkie, D. E., Islam, M. F., Kikkawa, J. M., & Yodh, A. G. (2005). Thermal conductivity and interfacial resistance in single-wall carbon nanotube epoxy composites. *Applied Physics Letters*, 87(87), 161909-161911.

[55] Moniruzzaman, M., Du, F., Romero, N., & Winey, K. I. (2006). Increased flexural modulus and strength in SWNT/epoxy composites by a new fabrication method. *Polymer*, 47(1), 293-298.

[56] Zhu, J., Kim, J., Peng, H., Margrave, J. L., Khabashesku, V. N., & Barrera, E. V. (2003). Improving the dispersion and integration of single-walled carbon nanotubes in epoxy composites through functionalization. *Nano Letters*, 3(8), 1107-1113.

[57] Xia, H., Wang, Q., Li, K., & Hu, G. H. (2004). Preparation of polypropylene/carbon nanotube composite powder with a solid-state mechanochemical pulverization process. *Journal of Applied Polymer Science*, 93(1), 378-386.

[58] Kasimatis, K. G., Nowell, J. A., Dykes, L. M., Burghardt, W. R., Thillalyan, R., Brinson, L. C., Andrews, R., & Torkelson, J. M. (2005). Morphology, thermal, and rheological behavior of nylon 11/multi-walled carbon nanotube nanocomposites prepared by melt compounding. *Programme Making and Special Events Preprint*, 92, 255-256.

[59] Regev, O., El Kati, P. N. B., Loos, J., & Koning, C.E. (2004). Preparation of conductive nanotube-polymer composites using latex technology. *Advanced Materials*, 16(3), 248-251.

[60] Lee, J. U., Huh, J., Kim, K. H., Park, C., & Jo, W. H. (2007). Aqueous suspension of carbon nanotubes via non-covalent functionalization with oligothiophene-terminated poly(ethylene glycol). *Carbon*, 45(5), 1051-1057.

[61] Mcandrew, T., Roger, C., Bressand, E., & Laurent, P. W. (2008). *WO/2008/106572*.

[62] Vigolo, B., Penicaud, A., Coulon, C., Sauder, C., Pailler, R., Journet, C., Bernier, P., & Poulin, P. (2000). Macroscopic fibers and ribbons of oriented carbon nanotubes. *Science*, 290(5495), 1331-1334.

[63] Mamedov, A.A., Kotov, N.A., Prato, M., Guldi, D. M., Wicksted, J. P., & Hirsch, A. (2002). Molecular design of strong single-wall carbon nanotube/polyelectrolyte multilayer composites. *Nature Materials*, 1(3), 190-194.

[64] Calvert, P. (1999). Nanotube composites- A recipe for strength. *Nature*, 399(6733), 210-211.

[65] Hersam, M.C. (2008). Progress towards monodisperse single-walled carbon nanotubes. *Nature Nanotechnology*.

[66] Green, M.J. (2010). Analysis and measurement of carbon nanotube dispersions: nanodispersion versus macrodispersion. *Polymer Internantional*, 59(10), 1319-22.

[67] Kim, S. W., Kim, T., Kim, Y. S., Choi, H. S., Lim, H. J., Yang, S. J., & Park, C. R. (2012). Surface modifications for the effective dispersion of carbon nanotubes in solvents and polymers. *Carbon*, 50(8), 3-33.

[68] O'Connell, M.J., Boul, P., Ericson, L. M., Huffman, C., Wang, Y., & Haroz, E. (2001). Reversible water-solubilization of single-walled carbon nanotubes by polymer wrapping. *Chemical Physics Letters*, 342(3), 265-271.

[69] Star, A., Stoddart, J. F., Steuerman, D., Diehl, M., Boukai, A., & Wong, E. W. (2001). Preparation and properties of polymer-wrapped single-walled carbon nanotubes. *Angewandte Chemie International Edition*, 40(9), 1721-1725.

[70] Steuerman, D. W., Star, A., Narizzano, R., Choi, H., Ries, R. S., & Nicolini, C. Interactions between conjugated polymers and single-walled carbon nanotubes. Journal of Physical Chemistry B. (2002). , 106(12), 3124-3130.

[71] Star, A., Liu, Y., Grant, K., Ridvan, L., Stoddart, J. F., & Steuerman, D. W. (2003). Noncovalent side-wall functionalization of single-walled carbon nanotubes. *Macromolecules*, 36(3), 553-560.

[72] Star, A., & Stoddart, J. F. (2002). *Macromolecules*, 35(19), 7516-7520.

[73] Nativ-Roth, E., Levi-Kalisman, Y., Regev, O., & Yerushalmi-Rozen, R. (2002). On the route to compatibilization of carbon nanotubes. *Journal of Polymer Engineering*, 22(5), 353-368.

[74] Xia, H. S., Wang, Q., & Qiu, G. H. (2003). Polymer-encapsulated carbon nanotubes prepared through ultrasonically initiated in situ emulsion polymerization. *Chemistry of Materials*, 15(20), 3879-3886.

[75] Li, X., Qin, Y., Picraux, S. T., & Guo-X, Z. (2011). Noncovalent assembly of carbon nanotube-inorganic hybrids. *Journal of Material Chemistry*, 21(21), 25-47.

[76] Sahoo, N. G., Rana, S., Cho, J. W. L. S. H., & Chan, Li. (2010). Polymer nanocomposites based on functionalized carbon nanotubes. *Progress in Polymer Science*, 35(7), 837-867.

[77] Kitano, H., Tachimoto, K., & Anraku, Y. J. (2007). Functionalization of single-walled carbon nanotube by the covalent modification with polymer chains. *Journal of Colloid and Interface Science*, 306(1), 28-33.

[78] Park, H., Zhao, J., & Lu, J. P. (2006). Effects of sidewall functionalization on conducting properties of single wall carbon nanotubes. *Nano Letters*, 6(5), 916-9.

[79] Han, S. J., Kim, B., & Suh, K. D. (2007). Electrical properties of a composite film of poly(acrylonitrile) nanoparticles coated with carbon nanotubes. *Macromolecular Chemistry and Physics*, 208(4), 377-383.

[80] Zhou, C., Wang, S., Zhang, Y., Zhuang, Q., & Han, Z. (2008). situ preparation and continuous fiber spinning of poly(p-phenylene benzobisoxazole) composites with oligo-hydroxyamide-functionalized multi-walled carbon nanotubes. *Polymer*, 49(10), 2520-2530.

[81] Baskaran, D., Mays, J. W., & Bratcher, M.S. (2005). Polymer adsorption in the grafting reactions of hydroxyl terminal polymers with multi-walled carbon nanotubes. *Polymer*, 46(14), 5050-5057.

[82] Zhao, B., Hu, H., Yu, A., Perea, D., & Haddon, R. C. (2005). Synthesis and characterization of water soluble single-walled carbon nanotube graft copolymers. *Journal of American Chemical Society*, 127(22), 8197-8203.

[83] Zhang, N., Xie, J., Guers, M., & Varadan, V. K. *Chemical bonding of multiwalled carbon nanotubes to polydimethylsiloxanes and modification of the photoinitiator system for microstereolithography processing*.

[84] Qin, S., Qin, D., Ford, W. T., Resasco, D. E., & Herrera, J. E. (2004). Functionalization of Single-Walled Carbon Nanotubes with Polystyrene via Grafting to and Grafting from Methods. *Macromolecules*, 37(3), 752-757.

[85] Li, H., Cheng, F., Duft, A. M., & Adronov, A. (2005). Functionalization of single-walled carbon nanotubes with well-defined polystyrene by "click" coupling. *Journal of American Chemical Society*, 127(41), 14518-24.

[86] Huang, H. M., Liu, I. C., Chang, C. Y., Tsai, H. C., Hsu, C. H., & Tsiang, R. C. C. (2004). Preparing a polystyrene-functionalized multiple-walled carbon nanotubes via covalently linking acyl chloride functionalities with living polystyryllithium. *Journal of Polymer Science Part A: Polymer Chemistry*, 42(22), 5802-5810.

[87] Lou, X., Detrembleur, C., Pagnoulle, C., Sciannamea, V., & Jerome, R. (2004). Grafting of alkoxyamine end-capped (co)polymers onto multi-walled carbon nanotubes. *Polymer*, 45(18), 6097-6102.

[88] Viswanathan, G., Chakrapani, N., Yang, H., Wie, B., Chung, H., & Cho, K. (2003). Single-step in situ synthesis of polymer-grafted single-wall nanotube composites. *Journal of American Chemical Society*, 9258-9259.

[89] Liu, Y., Wu, D. C., Zhang, W. D., Jiang, X., He, C. B., & Chung, T. S. (2005). Polyethylenimine-grafted multiwalled carbon nanotubes for secure noncovalent immobilization and efficient delivery of DNA. *Angewandte Chemie-International Edition*, 44(30), 4782-4785.

[90] Bonduel, D., Mainil, M., Alexandre, M., Monteverde, F., & Dubois, P. (2005). Supported coordination polymerization: a unique way to potent polyolefin carbon nanotube nanocomposites. *Chemical Communication*, 6, 781-783.

[91] Park, J. J., Park, D. M., Youk, J. H., Yu, W. R., & Lee, J. (2010). Functionalization of multi-walled carbon nanotubes by free radical graft polymerization initiated from photoinduced surface groups. *Carbon*, 48(10), 2899-2905.

[92] Yao, Z., Braidy, N., Botton, G. A., & Adronov, A. (2003). Polymerization from the surface of single-walled carbon nanotubes- Preparation and characterization of nanocomposites. *Journal of American Chemical Society*, 125(51), 16015-16024.

[93] Kong, H., Gao, C., & Yan, D. (2004). Controlled functionalization of multiwalled carbon nanotubes by in situ atom transfer radical polymerization. *Journal of American Chemical Society*, 126(2), 412-413.

[94] Wang, M., Pramoda, K. P., & Goh, S. H. (2005). Enhancement of the mechanical properties of poly (styrene-co-acrylonitrile) with poly(methyl methacrylate)-grafted multiwalled carbon nanotubes. *Polymer*, 46(25), 11510-11516.

[95] Baskaran, D., Mays, J. W., & Bratcher, MS. (2004). Polymer-grafted multiwalled carbon nanotubes through surface-initiated polymerization. *Angewandte Chemie-International Edition*, 43(16), 2138-2142.

[96] Shanmugharah, A. M., Bae, J. H., Nayak, R. R., & Ryu, S. H. (2007). Preparation of poly(styrene-co-acrylonitrile)-grafted multiwalled carbon nanotubes via surface-initiated atom transfer radical polymerization. *Journal of Polymer Science Part A-Polymer Chemistry*, 45(3), 460-470.

[97] Cui, J., Wang, W. P., You, Y. Z., Liu, C., & Wang, P. (2004). Functionalization of multiwalled carbon nanotubes by reversible addition fragmentation chain-transfer polymerization. *Polymer*, 45(26), 8717-8721.

[98] Hong, C. Y., You, Y. Z., & Pan, C. Y. (2005). Synthesis of water-soluble multiwalled carbon nanotubes with grafted temperature-responsive shells by surface RAFT polymerization. Chemistry of Materials , 17(9), 2247-2254.

[99] Wang, G. J., Huang, S. Z., Wang, Y., Liu, L., Qiu, J., & Li, Y. (2007). Synthesis of water-soluble single-walled carbon nanotubes by RAFT polymerization. *Polymer*, 48(3), 728-733.

[100] Datsyuk, V., Guerret-Piecourt, C., Dagreou, S., Billon, L., Dupin, J. C., & Flahaut, E. (2005). Double walled carbon nanotube/polymer composites via in-situ nitroxide mediated polymerisation of amphiphilic block copolymers. *Carbon*, 43(4), 873-876.

[101] Zhao, X. D., Fan, X. H., Chen, X. F., Chai, C. P., & Zhou, Q. F. (2006). Surface modification of multiwalled carbon nanotubes via nitroxide-mediated radical polymerization. *Journal of Polymer Science Part A-Polymer Chemistry*, 44(15), 4656-4667.

[102] Dehonor, M., Masenelli-Varlot, K., Gonzalez-Montiel, A., Gauthier, C., Cavaille, J. Y., & Terrones, H. (2005). Nanotube brushes: Polystyrene grafted covalently on CNx nanotubes by nitroxide-mediated radical polymerization. *Chemical Communications*, 42, 5349-5351.

[103] Fan, D. Q., He, J. P., Tang, W., Xu, J. T., & Yang, Y. L. (2007). Synthesis of polymer grafted carbon nanotubes by nitroxide mediated radical polymerization in the presence of spin-labeled carbon nanotubes. *European Polymer Journal*, 43(1), 26-34.

[104] Kaminsky, W. (2008). Trends in polyolefin chemistry. *Macromolecular Chemistry and Physics*, 209(5), 459-466.

[105] Mokashi, V. V., Mokashi, V. V., Qian, D., & Liu, Y. J. A. (2007). study on the tensile response and fracture in carbon nanotube-based composites using molecular mechanics. *Composites science and technology*, 67(3-4), 530-540.

[106] Tang, W. Z., Santare, M. H., & Advani, S. G. (2003). Melt processing and mechanical property characterization of multi-walled carbon nanotube/high density polyethylene (MWNT/HDPE) composite films. *Carbon*, 41(14), 2779-2785.

[107] Yang, B. X., Pramoda, K. P., Xu, G. Q., & Goh, S. H. (2007). Mechanical reinforcement of polyethylene using polyethylene-grafted multiwalled carbon nanotubes. *Advanced Functional Materials*, 17(13), 2062-2069.

[108] Shofner, M. L., Khabashesku, V. N., & Barrera, E. V. (2006). Processing and mechanical properties of fluorinated single-wall carbon nanotube-polyethylene composites. *Chemistry of Materials*, 18(4), 906-913.

[109] Mahfuz, H., Adnan, A., Rangari, V. K., & Jeelani, S. (2005). Manufacturing and Characterization of Carbon Nanotube/Polyethylene Composites. *International Journal of Nanoscience*, 4(1), 55-72.

[110] Ruan, S., Gao, P., & Yu, T. X. (2006). Ultra-strong gel-spun UHMWPE fibers reinforced using multiwalled carbon nanotubes. *Polymer*, 47(5), 1604-1611.

[111] Tong, X., Liu, C., Cheng, H. M., Zhao, H., Yang, F., & Zhang, X. (2004). Surface Modification of Single-walled Carbon Nanotubes with Polyethylene via in situ Ziegler-Natta Polymerization. *Journal of Applied Polymer Science*, 92(6), 3697-3700.

[112] Funk, A., & Kaminsky, W. (2007). Polypropylene carbon nanotube composites by in situ polymerization. *Composites Science and Technology*, 67(5), 906-915.

[113] Bonduel, D., Bredeau, S., Alexandre, M., Monteverde, F., & Dubois, P. (2007). Supported metallocene catalysis as an efficient tool for the preparation of polyethylene/ carbon nanotube nanocomposites: effect of the catalytic system on the coating morphology. *Journal of Materials Chemistry*, 17(22), 2359-2366.

[114] Dong, J. Y., & Hu, Y. (2006). Design and synthesis of structurally well-defined functional polyolefins via transition metal-mediated olefin polymerization chemistry. *Coordination Chemistry Reviews*.

[115] Russell, K.E. (2002). Free radical graft polymerization and copolymerization at higher temperatures. *Progress in Polymer Science*, 27(6), 1007-1038.

[116] Chung, T.C. (2002). Synthesis of functional polyolefin copolymers with graft and block structures. *Progress in Polymer Science*, 27(1), 39-85.

[117] Moad, G. (1999). The synthesis of polyolefin graft copolymers by reactive extrusion. *Progress in Polymer Science*, 24(1), 81-142.

[118] Hettema, R., Van Tol, J., & Janssen, L. P. B. M. (1999). In-situ reactive blending of polyethylene and polypropylene in co-rotating and counter-rotating extruders. *Polymer Engineering and Science*, 39(9), 1628-1641.

[119] Akbar, S., Beyou, E., Cassagnau, P., Chaumont, P., & Farzi, G. (2009). Radical grafting of polyethylene onto MWCNTs: A model compound approach. *Polymer*, 50(12), 2535-2543.

[120] Badel, T., Beyou, E., Bounor-Legare, V., Chaumont, P., Flat, J. J., & Michel, A. (2007). Melt grafting of polymethyl methacrylate onto poly(ethylene-co-1-octene) by reactive extrusion: Model compound approach. *Journal of Polymer Science Part A-Polymer Chemistry*, 45(22), 5215-5226.

[121] Johnston, R.T. (2003). Monte Carlo simulation of the peroxide curing of ethylene elastomers. *Rubber Chemistry and Technology*, 76(1), 174-201.

[122] Johnston, R.T. (2003). Modelling peroxide crosslinking in polyolefins. *Sealing Technology*, 76(2), 6-9.

[123] Osswald, S., Havel, M., & Gogotsi, Y. (2007). Monitoring oxidation of multiwalled carbon nanotubes by Raman spectroscopy. *Journal of Raman Spectroscopy*, 38(6), 728-736.

[124] Pastine, S. J., Okawa, D., Kessler, B., Rolandi, M., Llorente, M., Zettl, A., & Frechet, J. M. J. (2008). A facile and patternable method for the surface modification of carbon nanotube forests using perfluoroarylazides. *Journal of the American Chemical Society*, 130(13), 4238-4239.

[125] Park, M.J., Lee, J. K., Lee, B.S., Lee-W, Y., Choi, I. S., & Lee-G, S. (2006). Covalent Modification of Multiwalled Carbon Nanotubes with Imidazolium-Based Ionic Liquids. *Effect of Anions on Solubility. Chemistry of Materials*, 18, 1546-51.

[126] Akbar, S., Beyou, E., Chaumont, P., & Mélis, F. (2010). Effect of a Nitroxyle-Based Radical Scavenger on Nanotube Functionalisation with Pentadecane. *A Model Compound Study for Polyethylene Grafting onto MWCNTs, Macromolecular Chemistry and Physics*, 211(22), 2396-2406.

[127] Mylvaganam, K., & Zhang, L. C. (2004). Nanotube functionalization and polymer grafting: An ab initio study. *Journal of Physical Chemistry B*, 108(39), 5217-5220.

[128] Akbar, S., Beyou, E., Chaumont, P., Mazzolini, J., Espinosa, E., D'Agosto, F., & Boisson, C. (2011). Synthesis of Polyethylene-Grafted Multiwalled Carbon Nanotubes via a Peroxide-Initiating Radical Coupling Reaction and by Using Well-Defined TEMPO and Thiol End-Functionalized Polyethylenes. Journal of Polymer Science Part A:Polymer Chemistry , 49(4), 957-965.

[129] Farzi, G., Akbar, S., Beyou, E., Cassagnau, P., & Mélis, F. (2009). Effect of radical grafting of tetramethylpentadecane and polypropylene on carbon nanotubes' dispersibility in various solvents and polypropylene matrix. *Polymer*, 50(25), 5901-5908.

[130] Lopez, R. G., Boisson, C., D'agosto, F., Spitz, R., Boisson, F., Bertin, D., & Tordo, P. (2004). Synthesis and characterization of macroalkoxyamines based on polyethylene. *Macromolecules*, 37(10), 3540-3542.

[131] Lopez, R. G., Boisson, C., D'agosto, F., Spitz, R., Boisson, F., Gigmes, D., & Bertin, D. (2007). Catalyzed chain growth of polyethylene on magnesium for the synthesis of macroalkoxyamines Application to the production of block copolymers using controlled radical polymerization . *Journal of Polymer Science, Part A: Polymer Chemistry*, 45(13), 2705-2718.

[132] Mazzolini, J., Espinosa, E., D'Agosto, F., & Boisson, C. (2010). Catalyzed chain growth (CCG) on a main group metal: an efficient tool to functionalize polyethylene. Polymer Chemistry , 1(6), 793-800.

[133] D'Agosto, F., & Boisson, C. (2010). A RAFT Analogue Olefin Polymerization Technique Using Coordination Chemistry. *Australian Journal of Chemistry*, 63(8), 1155-1158.

[134] Mazzolini, J., Mokthari, I., Briquet, R., Boyron, O., Delolme, F., Monteil, V., Gigmes, D., Bertin, D., D'Agosto, F., & Boisson, C. (2010). Thiol-End-Functionalized Polyethylenes. *Macromolecules*, 43(18), 7495-7503.

[135] Hawker, C. J., Bosman, A. W., & Harth, E. (2001). New polymer synthesis by nitroxide mediated living radical polymerizations. *Chemical Review*, 101(12), 3661-3688.

[136] Lou, X., Detrembleur, C., Sciannamea, V., Pagnoulle, C., & Jerome, R. (2004). Grafting of alkoxyamine end-capped (co)polymers onto multi-walled carbon nanotubes. *Polymer*, 45(18), 6097-6102.

[137] Liu, Y., Yao, Z., & Adronov, A. (2005). Functionalization of single-walled carbon nanotubes with well-defined polymers by radical coupling. *Macromolecules*, 38(4), 1172-1179.

[138] Wang, H. C., Li, Y., & Yang, M. (2007). Sensors for organic vapor detection based on composites of carbon nonotubes functionalized with polymers. *Sensors and Actuators B-Chemical*, 124(2), 360-367.

[139] Wang, Y., Wu, J., & Wei, F. (2003). A treatment method to give separated multi-walled carbon nanotubes with high purity, high crystallization and a large aspect ratio *Carbon*, 41(15), 2939-2948.

[140] Lee, S. H., Cho, E., Jeon, S. H., & Youn, J. R. (2007). Rheological and electrical properties of polypropylene composites containing functionalized multi-walled carbon nanotubes and compatibilizers. *Carbon*, 45(14), 2810-2822.

[141] Wang, Y., Wen, X., Wan, D., Zhang, Z., & Tang, T. (2012). Promoting the responsive ability of carbon nanotubes to an external stress field in a polypropylene matrix: A synergistic effect of the physical interaction and chemical linking. *Journal Material Chemistry*, 22, 3930-3938.

[142] Zheng, J., Zhu, Z., Qi, j., Zhou, Z., Li, P., & Peng, M. (2011). Preparation of isotactic polypropylene-grafted multiwalled carbon nanotubes (iPP-g-MWNTs) by macroradical addition in solution and the properties of iPP-g-MWNTs/iPP composites. *Journal Material Science*, 46, 648-658.

[143] Jancar, J., Douglas, J. F., Starr, F. W., Kumar, S. K., Cassagnau, P., Lesser, A. J., Sternstein, S. S., & Buehler, M. J. (2010). Current issues in research on structure-property relationships in polymer nanocomposites. *Polymer*, 51(15), 3321-3343.

[144] Cassagnau, P. (2008). Melt rheology of organoclay and fumed silica nanocomposites. *Polymer*.

[145] Moreira, L., Fulchiron, R., Seytre, G., Dubois, P., & Cassagnau, P. (2010). Aggregation of Carbon Nanotubes in Semidilute Suspension. *Macromolecules*, 43(3), 1467-1472.

[146] Hobbie, E.K. (2010). Shear rheology of carbon nanotube suspensions. *Rheologica Acta*.

[147] Marceau, S., Dubois, P., Fulchiron, R., & Cassagnau, P. (2009). Viscoelasticity of Brownian Carbon Nanotubes in PDMS Semidilute Regime. *Macromolecules*, 42(5), 1433-1438.

[148] Mitchell, C. A., Bahr, J. L., Arepalli, S., Tour, J. M., & Krishnamoorti, R. (2002). Dispersion of functionalized carbon nanotubes in polystyrene. *Macromolecules*, 35(23), 8825-8830.

[149] Pujari, S., Ramanathan, T., Kasimatis, K., Masuda, J., Andrews, R., Torkelson, J. M., Brinson, L. C., & Burghardt, W. R. (2009). Preparation and Characterization of Multiwalled Carbon Nanotube Dispersions in Polypropylene: Melts Mixing versus Soli-State Shear Pulverization. *Journal of Polymer Science, Part B: Polymer Physics*, 47(14), 1426-1436.

[150] Hong, J. S., Hong, I. G., Lim, H. T., Ahn, K. H., & Lee, S. J. (2012). In situ lubrication dispersion of Multi-walled carbon nanotubes in Polypropylene melts. *Macromolecular Material Engineering*, 297(3), 279-287.

[151] Wu, D., Sun, Y., Wu, L., & Zhang, M. (2008). Linear Viscoelasticity Properties and Crystallisation Behavior of Multi-Walled Carbon Nanotube/Polypropylene Composites. *Journal of Applied Polymer Science*, 108(3), 1506-1513.

[152] Wu, D., Sun, Y., & Zhang, M. (2009). Kinetics Study on Melt Compounding of Carbon Nanotube/Polypropylene Nanocomposites. *Journal of Polymer Science, Part B: Polymer Physics*, 47(6), 608-618.

[153] Jin, S. H., Kang, C. H., Yoon, K. H., Bang, D. S., & Park, Y. B. (2009). Effect of Compatibilizer on Morphology, Thermal and Rheological Properties of Polypropylene/Functionalized Multi-walled Carbon Nanotubes composite. *Journal of Applied Polymer Science*, 111(2), 1028-1033.

[154] Menzer, K., Hrausee, B., Boldt, R., Kretzschmar, B., Weidisch, R., & Potschke, P. (2011). Percolation Behaviour of multiwalled carbon nanotubes of altered length and primary agglomerate morphology in melt mixed isotactic polypropylene-based composites. *Composite Science and Technology*, 71(16), 1936-1943.

[155] Hemmati, M., Rahimi, G. H., Kaganj, A. B., Sepehri, S., & Rashidi, A. M. (2008). Rheological and Mechanical Characterization of Multi-walled Carbon Nanotubes/Poly-

propylene Nanocomposites. *Journal of Macromolecular Science, Part B: Physics*, 47(6), 1176-1187.

[156] Prashantha, K., Soulestin, J., Lacrampe, M. F., Krawczak, P., Dupin, G., & Claes, M. (2009). Masterbatch-based multi-walled carbon nanotube filled polypropylene nanocomposites: Assessment of rheological and mechanical properties. *Composite Science and Technology*, 69(11-12), 1756-1763.

[157] Palza, H., Reznik, B., Kappes, M., Hennrich, F., Naue, I. F. C., & Wilhelm, M. (2010). Characterization of melt flow instabilities in polyethylene/carbon nanotube. *Polymer*, 51(16), 3753-3761.

Chapter 6

Characterization and Morphology of Modified Multi-Walled Carbon Nanotubes Filled Thermoplastic Natural Rubber (TPNR) Composite

Mou'ad A. Tarawneh and Sahrim Hj. Ahmad

Additional information is available at the end of the chapter

http://dx.doi.org/10.5772/50726

1. Introduction

Carbon nanotubes describes a specific topic within solid-state physics, but is also of interest in other sciences like chemistry or biology. Actually the topic has floating boundaries, because we are at the molecule level. In the recent years carbon nanotubes have become more and more popular to the scientists. Initially, it was the spectacularly electronic properties, that were the basis for the great interest, but eventually other remarkable properties were also discovered.

The first CNTs were prepared by M. Endo in 1978, as part of his PhD studies at the University of Orleans in France. Although he produced very small diameter filaments (about 7 nm) using a vapour-growth technique, these fibers were not recognized as nanotubes and were not studied systematically. It was only after the discovery of fullerenes, C60, in 1985 that researchers started to explore carbon structures further. In 1991, when the Japanese electron microscopist Sumio Iijima [1] observed CNTs, the field really started to advance. He was studying the material deposited on the cathode during the arc-evaporation synthesis of fullerenes and came across CNTs. A short time later, Thomas Ebbesen and Pulickel Ajayan, from Iijima's lab, showed how nanotubes could be produced in bulk quantities by varying the arc-evaporation conditions. However, the standard arc-evaporation method only produced only multiwall nanotubes. After some research, it was found that the addition of metals such as cobalt to the graphite electrodes resulted in extremely fine single wall nanotubes.

The synthesis in 1993 of single-walled carbon nanotubes (SWNTs) was a major event in the development of CNTs. Although the discovery of CNTs was an accidental event, it opened the way for a flourishing research into the properties of CNTs in labs all over the world, with many scientists demonstrating promising physical, chemical, structural, and optical properties of CNTs.

CNTs exhibit a great range of remarkable properties, including unique mechanical and electrical characteristics. These remarkable high modulus and stiffness properties have led to the use of CNTs to reinforce polymers in the past few years. Both theoretical (e.g. molecular structural mechanics and tight-binding molecular dynamics) and experimental studies have shown SWCNTs to have extremely high elastic modulus (≈1 TPa) [2-3]. The tensile strength of SWCNTs estimated from molecular dynamics simulation is ≈150 MPa [4]. The experimental measurement of 150 MPa was found for the break strength of multi-walled carbon nanotubes (MWCNTs) [5].

The remarkable properties of CNTs offer the potential for improvement of the mechanical properties of polymers at very low concentrations. In practice, MWCNTs are preferred over SWCNTs as the reinforcing fillers for polymers due to their lower production cost. However, slippage between the shells of MWCNTs would undermine the capability of the fillers to bear the external applied load.

Mixed 1 wt.% MWCNTs with polystyrene (PS) in toluene via ultrasonication, achieved about 36–42% increase in the elastic modulus and a 25% increase in the tensile strength of the PS–MWCNT film compared to pure PS [6]. They found that nanotube fracture and pull-out are responsible for the failure of the composite. The fracture of MWCNTs in a PS matrix implies that certain load transfer from the PS to the nanotubes has taken place. However, the pullout of MWCNTs from the PS matrix indicating that the PS–nanotube interfacial strength is not strong enough to resist debonding of the fillers from the matrix. It is considered that some physical interactions exist at the PS–MWCNT interface, thereby enabling load transfer from the matrix to the fillers.

The additions of 0.25– 0.75 wt.% SWCNTs to polypropylene (PP) considerably its tensile strength and stiffness as well as storage modulus. The elongation at break reduces from 493 (PP) to 410% with the addition of 0.75 wt.% filler, corresponding to -17% reduction in ductility. At 1 wt.% SWNT, both stiffness and strength are significantly reduced due to the formation of aggregates [7].

The morphology and mechanical properties of the melt-compounded polyamide 6 (PA6)–MWNT nanocomposites were studied by [8]. The MWCNTs were purified by dissolving the catalyst in hydrochloric acid followed by refluxing in 2.6 Mnitric acids to increasing the carboxylic and hydroxyl groups. It was also found that with the addition of only 1 wt.% MWCNTs, the tensile modulus and the tensile strength are greatly improved by ≈115 and 120%, respectively compared to neat PA6. The tensile ductility drops slightly from 150 to 125%. They attributed the improvements of these mechanical properties to a better dispersion of MWCNTs in PA6 matrix, and to a strong interfacial adhesion between the nanofillers and PA6 matrix which leads to favorable stress transfer across the polymer to the MWCNTs.

The influence of SWNT and carbon nanofiber additions on the mechanical performances of silicone rubber was reported by [9]. They reported that SWCNTs are effective reinforcements for silicone rubber due to their large aspect ratio and low density. The initial modulus (measured by fitting a straight line to the data below 10% strain) tends to increase almost linearly with increasing filler content. The effect of SWCNT and carbon fiber additions on the tensile ductility of silicone rubber is shown that the strain to failure drops from 325 to 275% upon loading with 1 wt.% SWCNTs, corresponding to ≈15% reduction.

The carbon nanotube additions to polyurethane (PU) improve the mechanical properties such as increased modulus and yield stress, without loss of the ability to stretch the elastomer above 1000% before final failure; the addition of CNTs increases the modulus and strength of PU without degrading deformabilty. The elongation at break decreases very slightly with CNT loading up to 17 wt.%. At this filler loading, the nanocomposite still maintains a very high value of elongation at rupture, i.e. 1200% [10].

Theoretical prediction showed an extremely high thermal conductivity (6000 W/mK) of an isolated SWCNTs [11]. High thermal conductivity of the CNTs may provide the solution of thermal management for the advanced electronic devices with narrow line width. Revealed the thermal conductivity of epoxy-based composites reinforced with 1.0 wt.% SWCNTs increased over 125% reaching a value of ~0.5 W/mK [12]. The variation of thermal conductivity with the values of 35 and 2.3 W/mK for a densepacked mat and a sintered sample, respectively [13]. High thermal conductivity of 42 and ~18W/mK of the aligned and the random bucky paper mats, respectively. However, the thermal conductivity drops significantly by almost an order of magnitude when the aligned bucky paper mats were loaded with epoxy, the volume fraction of the aligned bucky paper composites is about 50%[14].

Developed an infiltration method to produce CNTs/epoxy composites and showed a 220% increase in thermal conductivity (~0.61 W/mK) at 2.3 wt.% SWCNT loading, and they found that the electrical resistance between SWCNT-polymer is more severe than that of SWCNT–SWCNT [15]. Prepared SWCNT and MWCNT films and reported the thermal conductivity of 1.64 and 1.51 W/mK, respectively; they concluded that the intra-tube spacing affects the thermal transfer more significantly than that of the nanotubes themselves [16].

The thermal conductivities of composites reinforced with 1.0 wt.% SWCNTs and 1.0 wt.% MWCNTs are 2.43 W/mK and 3.44 W/mK, respectively. Composites reinforced with the unpurified CNTs have higher thermal conductivity than that of the purified CNTs reinforced composite. This is attributed to the generation of defects on the CNT surface during acid treatment. Moreover, due to longer phonon propagation length, it is found that thermal conductivity increases with temperatures over the range from 25 to 55°C for both SWCNTs/Poly (methyl methacrylate) PMMA and MWCNTs/PMMA composites. However, the thermal conductivities of CNT films decrease with increasing temperature, which results from phonon scattering during transfer due to the presence of defects coupled with smaller phonon mean free path at higher temperature [17].

The differences in the composite manufacturing methods, powder-(MWCNTs and ball milled SWCNTs) or liquid- (chemically treated SWCNTs) based approach, can not account for the differences in the properties, since both methods were used for the SWNT-composites and resulted in similar thermal behaviour [18]. Thus, they concluded that in this case, there must be a very large interface resistance to the heat flow associated with poor phonon coupling between the stiff nanotubes and the (relatively) soft polymer matrix. In addition it is possible that the phonon vibrations in the SWCNTs are dampened by the matrix interaction, while in the MWCNTs the phonons can be carried in the inner walls without hindrance.

The precise sectioning of CNTs provides an effective way to shorten carbon nanotubes with controlled length and minimum sidewall damage [19]. For shortened nanotubes they found that they are easily dispersed into polymer matrices, which effectively improved the percolation. The minimum CNT sidewall damage and improved percolation in short SWCNT composites led to an obvious improvement of thermal conductivity. Hence, their research suggests an effective way to improve dispersion of CNTs into polymer matrices and also retain the perfect electronic structure of the CNTs, resulting in desired functional materials.

Accurate measurement of the thermal conductivity of composites and nanocomposites can be done using the transient hot-wire technique which is capable of measuring the thermal conductivity of solid materials in an absolute way. The enhancement in the thermal conductivity was measured as 27% in relation to the thermal conductivity of the epoxy-resin polymer, which is satisfactory taking into account the low volume fraction (28%) of the glass fibres used in the composite [20]. They reported that when 2% by weight C-MWNT were mixed with the epoxy-resin, the enhancement of thermal conductivity was 9% while using both glass fibres and C-MWNT the enhancement was 48%.

For sufficient enhancement of most of the nanocomposites' properties, the dispersion of the CNTs should be very fine in the polymer matrix, which means that the surface of interaction between the filler and the matrix should be optimised. However, this is difficult to achieve since their long length results in them becoming entangled. Moreover, their very large surface-to-volume ratio and strong van der Waals interactions keep them tied together, which in most cases leads to the formation of large agglomerates in polymer matrices. The interfacial adhesion between CNTs and the polymeric matrix is also crucial. In order to increase the interfacial adhesion between the polymer and the CNTs various routes of surface modification of the nanotubes have been considered. One is non-covalent functionalisation of molecules and the other is covalent functionalisation from the walls of the nanotubes. Non-covalent functionalisation is based on weak Van der Waals forces [21]. The advantage of non-covalent functionalisation is that the perfect structure of the nanotubes is not altered while the covalent attachment can greatly improve the load transfer to the matrix; however, it usually introduces structural defects on the nanotubes' surface.

Although both probe style and bath style ultrasonic systems can be used for dispersing CNTs, it is widely believed that the probe style ultrasonic systems work better for dispersing CNTs [22]. It is also widely known that adding a dispersing reagent (surfactant) into the solution will accelerate the dispersion effect.

The most common procedure used for covalent attachment of reactive groups is the treatment with inorganic acids. Usually the nanotubes are refluxed with a nitric acid solution or a mixture of nitric and sulfuric acid, sometimes concurrently with the application of high power sonication [23]. These oxidative treatments usually result in shortening of the CNTs' length and formation of surface reactive groups, such as hydroxyl, carbonyl and mainly carboxylic acid. Oxidation of the nanotubes starts at the tips and gradually moves towards the central part of the tube and the layers are removed successively [24].

The synthesized carbon nanotubes usually exist as agglomerates of the size of several hundred micrometers [25]. Such entanglements make it difficult to disperse nanotubes uniformly in a polymer matrix. To overcome the dispersion problem, it is necessary to tailor the chemical nature of the nanotube surface. One of the most straightforward methods for nanotube dispersion is direct mixing; however, it does not always yield a homogeneous distribution of nanotubes because of the lack of compatibility between the MWCNTs and polymer matrix. Solution processing has been a commonly used method in fabrication of the well-dispersed carbon nanotube composites. However, it is hard to achieve homogeneous dispersion of nanotubes in a polymer matrix because carbon nanotubes are insoluble and bundled.

Chemical functionalization of the MWCNTs surface increases the interfacial interaction between MWCNTs and the polymer matrix. This enhances the adhesion of the MWCNTs in various organic solvents and polymers, reduces the tendency to agglomerate, and improves dispersion. The improved interactions between MWCNTs and the polymer matrix govern the load-transfer from the polymer to the nanotubes and, hence, increase the reinforcement efficiency. Attachment of oxygen containing functional groups (i.e., carboxyl groups, carbonyl groups, hydroxyl groups, etc.) on the surface of the MWCNTs could be achieved by applying several chemical treatments.

The chemically functionalized MWCNTs can be easily mixed with the polymer matrix. Acid treatment of the nanotube is an especially well-known technique to remove catalytic impurities, generate functional groups on open ends or sidewalls of nanotubes, and facilitate good dispersion of MWCNTs in polymeric solutions or melts.

The emergence of thermoplastic elastomers (TPEs) is one of the most important developments in the area of polymer science and technology. TPEs are a new class of material that combines the properties of vulcanized rubber with the ease of processability of thermoplastics [26]. Thermoplastic elastomers can be prepared by blending thermoplastic and elastomers at a high shear rate. Thermoplastics, for example, polypropylene (PP), polyethylene (PE) and polystyrene (PS), and elastomers, such as ethylene propylene diene monomer (EPDM), natural rubber (NR) and butyl rubber (BR), are among the materials used in thermoplastic elastomer blends.

Blends of natural rubber (NR) and polypropylene (PP) have been widely reported by previous researchers [26]. According to them, polypropylene is the best choice for blending with natural rubber due to its high softening temperature (150°C) and low glass transition temperature (-60°C, is Tg for NR), which makes it versatile in a wide range of temperatures. Even though NR and PP are immiscible, their chemical structure is nearly the same. Thus, stable dispersion of NR and PP is possible. Incompatibility between NR and PP can be overcome by the introduction of a compatibiliser that can induce interactions during blending. Compatibility is important as it may affect the morphology, mechanical and thermal properties of the blends. Among the commonly used compatibilisers are dicumyl peroxide (DCP), m-phenylene bismaleimide (HVA-2) and liquid natural rubber (LNR). Apart from compatibility, mixing torque and curing are interrelated in determining the homogeneity of the TPNR blend.

Mechanical blending of PP and NR with the addition of LNR as a compatibiliser has been reported to be optimal at a temperature of 175-185°C and a rotor speed of 30-60rpm. The percentage of LNR used depends on the ratio of NR to PP. For a NR:PP ratio of 30:70 the best physical properties are obtained at 10% LNR [27]. The compatibiliser helps to induce the interaction between the rubber and plastic interphase and thereby increases the homogeneity of the blend.

MWCNTs/TPNR composites with different amounts of MWCNT were prepared and their thermal properties have been investigated by [28]. The higher thermal conductivity was achieved in the samples with 1 and 3wt% of MWCNTs compared to the pristine TPNR. Any sample with MWCNTs content higher than 3wt% caused the conductivity to decrease. In addition, the improvement of thermal diffusivity and specific heat was also achieved at the same percentage. DMA confirmed that the glass transition temperature (Tg) increased with the increase in the amount of MWCNTs.

The tensile strength, tensile modulus, and also the impact strength of TPNR/MWCNTs are improved significantly while sacrificing high elongation at break by incorporating MWCNTs. The reinforcing effect of MWCNTs was also confirmed by DMA where the addition of nanotubes has increased the storage modulus, the loss modulus, and also the glass transition temperature (Tg). Homogeneous dispersion of MWCNTs throughout the TPNR matrix and strong interfacial adhesion between MWCNTs and matrix as confirmed by SEM images are proposed to be responsible for the significant mechanical enhancement [29].

The reinforcing effect of two types of MWCNTs has also confirmed by dynamic mechanical analysis where the addition of nanotubes have increased in the storage modulus E', and the loss modulus E'', in the addition the glass transition temperature (Tg) increased with an increase in the amount of MWCNTs. The addition of MWCNTs in the TPNR matrix improved the mechanical properties. The tensile strength and elongation at break of MWCNTs 1 increased by 23%, and 29%, respectively. The Young's modulus had increased by increasing the content of MWCNTs. For MWCNTs 2 the optimum result of tensile strength and Young's modulus was recorded at 3% which increased 39%, and 30%, respectively. The laser flash technique was used to measure the thermal conductivity, thermal diffusivity and specific heat, from the results obtained. The high thermal conductivity was achieved at 1 wt% and 3 wt% of MWCNTs compared with TPNR after 3 wt% it decreased, also the improvement of thermal diffusivity and specific heat was achieved at the same percentage. The MWCNTs 1 and 2/TPNR nanocomposites were fabricated and the tensile and properties were measured [30].

In this chapter, the effect of multi-walled carbon nanotubes with and without acid treatment on the properties of thermoplastic natural rubber (TPNR) was investigated. Two types of MWCNTs were introduced into TPNR, which are untreated multi-walled carbon nanotubes (UTMWCNTs) (without acid treatment) and treated multi-walled carbon nanotubes (TMWCNTs) (with acid treatment). Using this method, MWCNTs are dispersed homogeneously in the TPNR matrix in an attempt to increase the properties of these nanocomposites. The effect of MWCNTs on the mechanical and thermal properties of TPNR nanocomposites is reported in this chapter.

2. Experiment Details

Polypropylene, with a density of 0.905 g cm-3, was supplied by Propilinas (M) Sdn. Bhd, natural rubber was supplied by Guthrie (M) Sdn. Bhd, and polypropylene (PP) with a density of 0.905 g/cm3 was supplied by Polipropilinas (M) Sdn. Bhd were used in this research. Maleic anhydride–grafted–polypropylene (MAPP) with a density of 0.95 g/cm3 was supplied from Aldrich Chemical Co., USA. Liquid natural rubber (LNR) was prepared by the photochemical degradation technique.

A Multi-walled carbon nanotubes (MWCNTs) were provided by Arkema (GraphistrengthTM C100). Table 1 shows the properties of multi-walled carbon nanotubes (MWCNTs).

MWCNTs	Purity	Length	Diameter	Manufactured
MWCNTs	"/90%	0.1-10 μm	10-15 nm.	Catalytic Chemical Vapour Deposition (CVD)

Table 1. Properties of multi-walled carbon nanotubes (MWCNTs).

2.1. Preparation of TPNR-Multi-Walled Carbon Nanotubes (MWCNTs) Composite

Mixing was performed by an internal mixer (Haake Rheomix 600P). The mixing temperature was 180°C, with a rotor speed of 100 rpm and 13 min mixing time. The indirect technique (IDT) was used to prepare nanocomposites, this involved mixing the MWCNTs with LNR separately, before it was melt blended with PP and NR in the internal mixer. TPNR nanocomposits were prepared by melt blending of PP, NR and LNR with MWCNTs in a ratio of 70 wt% PP, 20 wt% NR and 10wt% LNR as a compatibiliser and 1,3,5 and 7% MWCNTs.

2.2. Acid Treatment of MWCNTs

Two types of MWCNTs were introduced to the TPNR which is untreated MWCNTs (MWCNTs 1) and treated MWCNTs (MWCNTs 2), MWCNTs 2 were treated by immersing neat MWCNTs in a mixture of nitric and sulfuric acid with a molar ratio of 1:3, respectively. In a typical experiment, 1g of raw MWCNTs was added to 40ml of the acid mixture. Then, the oxidation reaction was carried out in a two-necked, round-bottomed glass flask equipped with reflux condenser, magnetic stirrer and thermometer. The reaction was carried out for 3 hours at 140°C. After that, this mixture was washed with distilled water on a sintered glass filter until the pH value was around 7, and was dried in a vacuum oven at 70°C for 24hours [31].

2.3. Characterizations

Fourier transform infrared (FTIR) spectroscopy analysis was carried out on the Perkin Elmer spectrum V-2000 spectrometer by the potassium bromide (KBr) method for MWCNTs. The samples were scanned between 700 to 4000 cm-1 wave number. Differences in the peaks as well as the new peaks of MWCNTs and MWCNTs after acid treatment were observed to identify any functional groups on the MWCNTs tubes surface.

The tensile properties were tested using a Testometric universal testing machine model M350-10CT with 5 kN load cell according to ASTM 412 standard procedure using test specimens of 1 mm thickness and a crosshead speed 50 mm min-1. At least five samples were tested for each composition, and the average value was reported.

The impact test was carried out using a Ray Ran Pendulum Impact System according to ASTM D 256-90b. The velocity and weight of the hammer were 3.5m/s and 0.898kg, respectively.

Dynamic mechanical analysis for determining the glass transition temperature, storage and loss modulus was carried out using DMA 8000 (PerkinElmer Instrument), operating in single cantilever mode from -100 to 150°C at a constant frequency of 1 Hz, with a heating rate of 5°C/min. The dimensions of the samples were 30 x 12.5 x 3 mm.

The thermal conductivity was measured by a laser flash method. Disk-type samples (12.7 mm in diameter and 1mm in thickness) were set in an electric furnace. Specific heat capacities were measured with a differential scanning calorimeter DSC. Thermal diffusivity (λ, Wm_1 K_1) was calculated from thermal diffusivity (α, m2 s_1), density (ϱ, g cm_3) and specific heat capacity (Cp, J g_1 K_1) at each temperature using the following:

$$\lambda = \alpha . \rho . C. \tag{1}$$

The reference used for the heat capacity calculation was a 12.7mm thick specimen of pyroceram. The reference sample was coated with a thin layer of graphite before the measurement was performed. The thermal conductivity of MWCNTs reinforced TPNR matrix composites of all volume fractions was studied from 30°C to 150°C. The morphology of the MWCNTs and the composite were examined using a scanning electron microscope (Philips XL 30). The samples were coated with a thin layer of gold to avoid electrostatic charging during examination.

3. Results and Discussion

3.1. Fourier-Transform Infrared Spectroscopy

The method used to functionalize the pristine MWCNTs in this study was the acid treatment method, which is described in section 2.2. Through this process, MWCNTs were oxidized and purified by eliminating impurities such as amorphous carbons, graphite particles, and metal catalysts [32]; the functional group of the surface of the CNTs are as shown in Figure 1.

http://dx.doi.org/10.5772/50726

The generation of chemical functional groups on MWCNTs was confirmed using Fourier transform infrared spectroscopy (FT-IR) spectra which were recorded between 400 cm_1 and 4000 cm_1.

The FT-IR spectra of pure MWCNTs and the surface treated MWCNTs are shown in Figure. 2 and Figure 3. The characteristic bands due to generated functional groups are observed in the spectrum of each chemically treated MWCNTs. In figure 2 we could not see any band compared with the treated MWCNTs. The acid treated MWCNTs shows new peaks in comparison with the FT-IR spectrum of the untreated MWCNTs, which lack the hydroxyl and carbonyl groups. The peaks around 1580 cm_1 are assigned to the O–H band in C-OH, and the peaks at 674 cm-1 are assigned to COOH, as shown in Figure 3. This demonstrates that hydroxyl and carbonyl groups have been introduced on the nanotube surface [33].

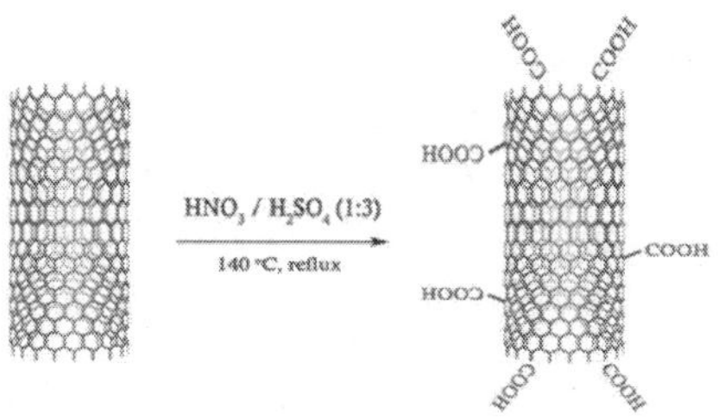

Figure 1. Example of chemical functionalization of carbon nanotubes.

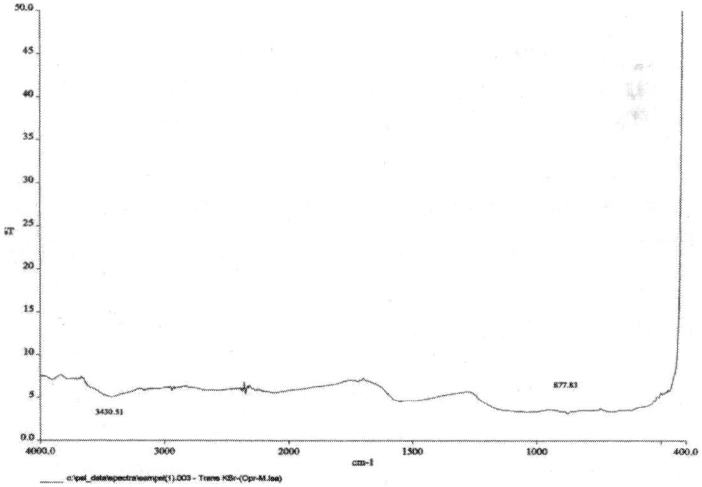

Figure 2. FTIR spectra of MWCNTs (before acid treatment).

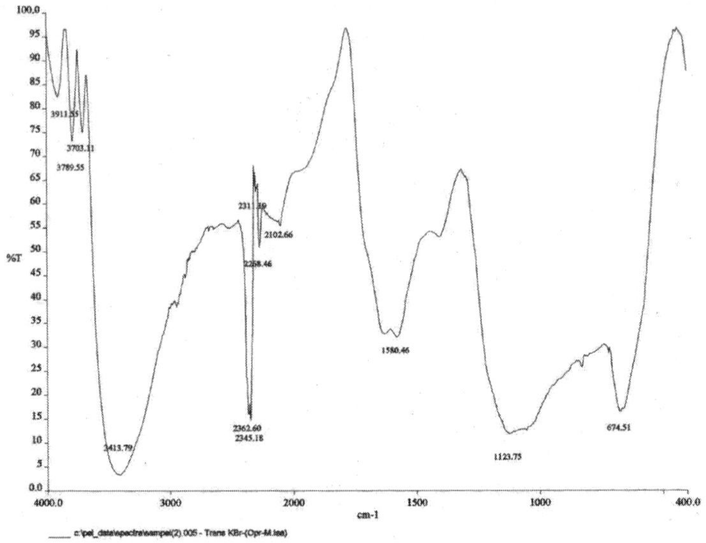

Figure 3. FTIR spectra of MWCNTs (after acid treatment).

3.2. Transmission Electron Microscopy (TEM)

TEM microphotographs of pure MWCNTs are shown in Figure 4 (A and B). The figure presents unmodified MWCNTs containing particles with diameters of 5–12 nm. The nanoparticles may be impurities from amorphous carbon and can be removed by acid treatment. According to the supplier, the unmodified MWCNT contains approximately 5% amorphous carbon. Figure 4 B demonstrates that most of the nanoparticles were deposited on the surface of the carbon nanotubes and some of them were dispersed throughout the solution used to view the MWCNTs by TEM.

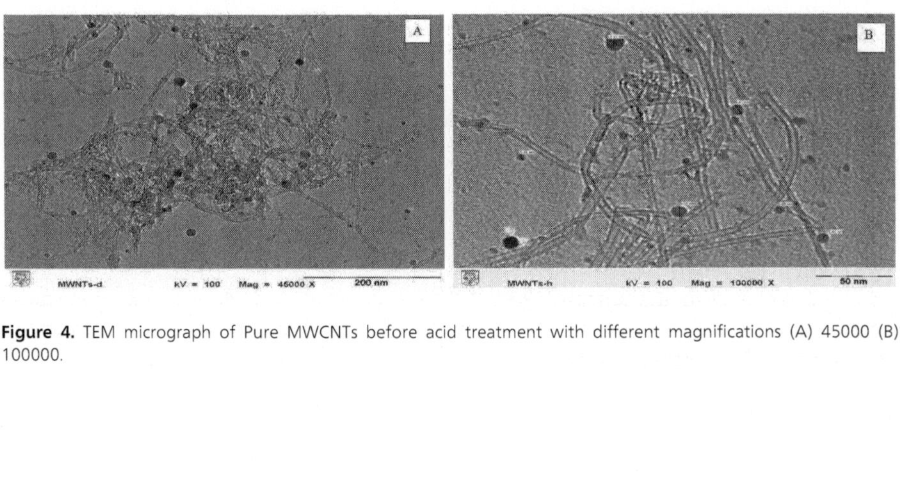

Figure 4. TEM micrograph of Pure MWCNTs before acid treatment with different magnifications (A) 45000 (B) 100000.

http://dx.doi.org/10.5772/50726

Figure 5 (A and B) displayed no nanoparticles in the acid-modified MWCNTs. The particles might have been removed during acid modification. This reveals that the acid-modified MWCNTs were straight and that some of them aggregated in bundles, which were dispersed well in the matrix. The length of the MWCNTs were reduced during acid modification, since the mixed acid corroded the MWCNTs. TEM microphotographs of the unmodified and acid-modified, curled and entangled MWCNTs demonstrate that the MWCNTs are straight.

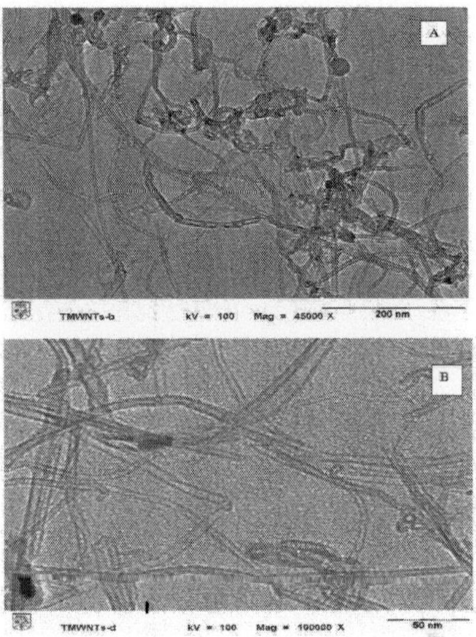

Figure 5. TEM micrograph of MWCNTs after acid treatment with different magnifications (A) 45000 (B) 100000.

3.3. Mechanical Properties

3.3.1. Tensile strength

The tensile strengths of TPNR reinforced with MWCNTs (with and without treatment) of different percentages (1%, 3%, 5% and 7%) are shown in Figure 6. Generally, both MWCNTs exhibited an increasing trend up to 3wt% content. Further increments in MWCNTs content decreased the tensile strength compared to the optimum filler loading.

From Figure 6, TPNR with UTMWCNTs and TMWCNTs have optimum results at 3 wt%, which, compared with TPNR, increased by 23% and 39%, respectively. The tensile strength increased radically as the amount of MWCNTs concentration increased. The mechanical performance, such as tensile properties, strongly depends on several factors such as the properties of the filler reinforcement and matrix, filler content, filler length, filler orientation, and processing method and condition. The improvement in the tensile strength may be caused by the good dispersion of MWCNTs in the TPNR matrix, which leads to a strong interaction between the TPNR matrix and MWCNTs. These well-dispersed MWCNTs may have the effect of physically crosslinking points, thus, increasing the tensile strength.

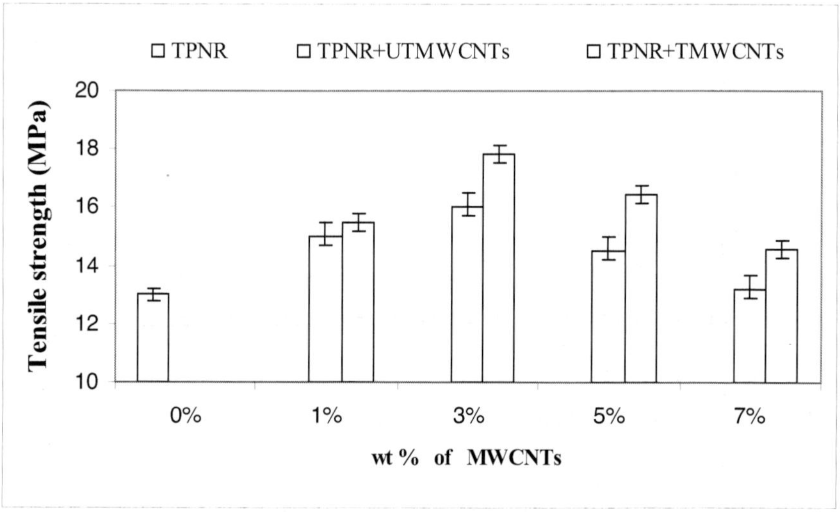

Figure 6. Tensile strength of TPNR reinforced with MWCNTs (with and without treatment).

A good interface between the CNTs and the TPNR is very important for a material to stand the stress. Under load, the matrix distributes the force to the CNTs, which carry most of the applied load. The order of these value is TPNR/TMWCNTs > TPNR/UTMWCNTs > TPNR. The better properties in tensile strength for the TPNR/TMWCNTs nanocomposites could be due to the improved dispersion of the MWCNTs, as well as the response to the opportunities offered by the acid treated MWCNTs. Furthermore, the MWCNTs after acid treatment contain many defects as well as acidic sites on CNTs, such as carboxylic acid, carbonyl and hydroxyl groups. These will greatly enhance the combination of CNTs in a polymer matrix, thus improving the mechanical strength of the nanocomposites [34]. When the content of MWCNTs is higher, the MWCNTs cannot disperse adequately in the TPNR matrix and agglomerate to form a big cluster. This is because of the huge surface energy of MWCNTs as well as the weak interfacial interaction between MWCNTs and TPNR, which leads to inhomogeneous dispersion in the polymer matrix and negative effects on the properties of the resulting composites that cause a decrease in the tensile strength [35].

http://dx.doi.org/10.5772/50726

3.3.2. Young's Modulus

Figure 7 shows the effect of filler content on the tensile modulus of TPNR reinforced by TMWNTs and UTMWCNTs. The same trend as for the tensile strength in Figure 6 was observed for the tensile modulus of TMWCNTs. Figure 6 clearly shows that the presence of MWCNTs has significantly improved the tensile modulus of the TPNR.

The remarkable increase of Young's modulus with TMWCNTs content shows a greater improvement than that seen in the tensile strength at high content, which indicates that the Young's modulus increases with an increase in the amount of the TMWCNTs. At 3 wt% of TMWCNTs the Young's modulus is increased by 34 % compared to TPNR. The Young's modulus of UTMWCNTs increased with the increase in the amount of UTMWCNTs. The maximum result was achieved at 3wt%, with an increase of about 22%, which was due to the good dispersion of nanotubes displaying perfect stress transfer [36].The improvement of modulus is due to the high modulus of MWCNTs [37]. The further addition of TMWCNTs and UTMWCNTs from 5 to 7 wt% increased the Young modulus dropped respectively.

As explained before, a reduction in performance occurred at higher filler contents for both types of MWCNTs, as depicted in Figure 7. Initially it increases with filler content and then decreases when exceeding the filler loading limit due to the diminishing interfacial filler-polymer adhesion. It is assumed that aggregates of nanotube ropes effectively reduce the aspect/ratio (length/diameter) of the reinforcement.

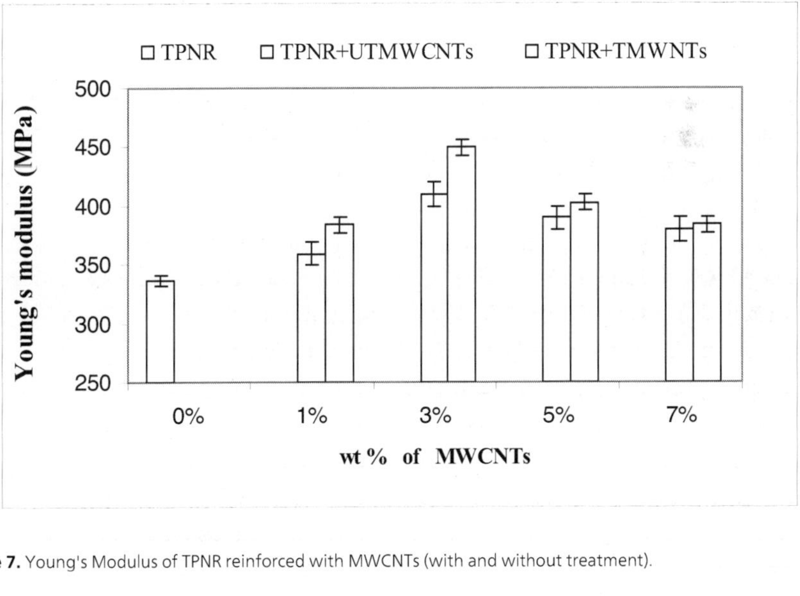

Figure 7. Young's Modulus of TPNR reinforced with MWCNTs (with and without treatment).

3.3.3 Elongation at Break

The elongation at the break of TPNR with TMWCNTs and UTMWCNTs is shown in Figure 8. For TMWCNTs and UTMWCNTs, the elongation at break decreased with the increase in the amount of MWCNTs, compared with TPNR.

It can be deduced that the reinforcing effect of MWCNTs is very marked. As the MWCNTs content in the TPNR increases, the stress level gradually increases, however, the strain of the nanocomposites decreased at the same time. This is because the MWCNTs included in the TPNR matrix behave like physical crosslinking points and restrict the movement of polymer chains. This indicates that, when the amount of CNTs incorporated into the rubber increase it tends to decrease the ductility and the material become stronger and tougher, however, at the same time, it is also

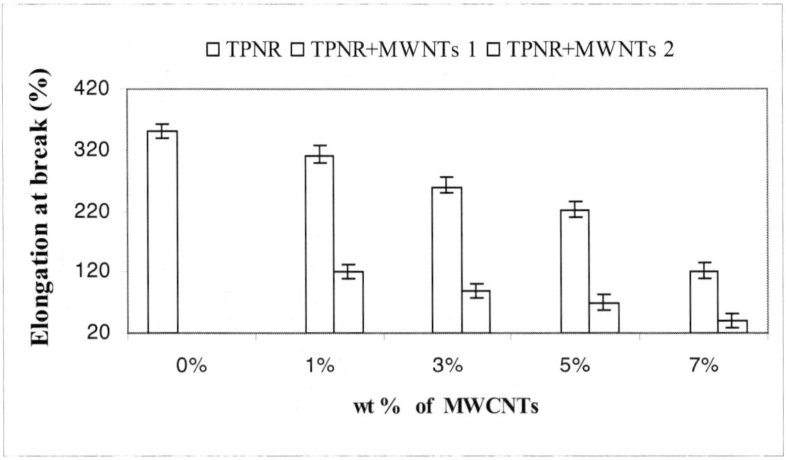

Figure 8. Elongation at Break of TPNR reinforced with MWCNTs (with and without treatment).

3.3.4 Impact Strength

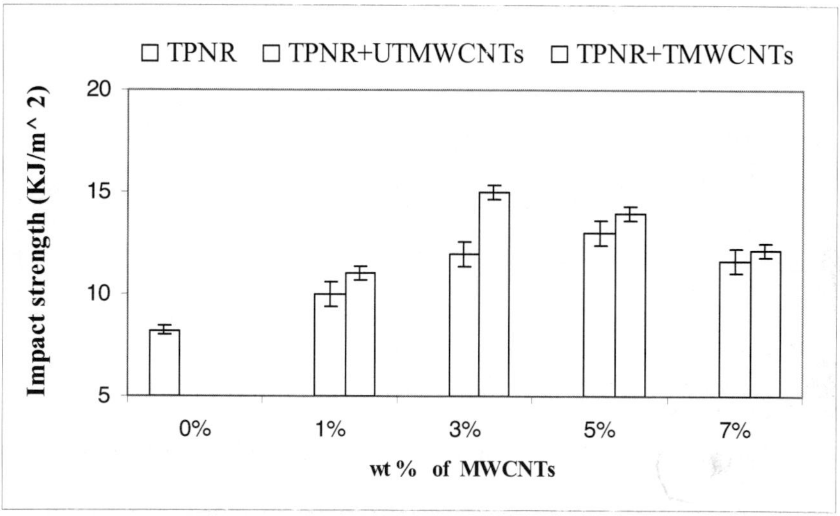

Figure 9. Impact Strength of TPNR reinforced with MWCNTs (with and without treatment).

The effect of filler loading on the impact strength of TPNR/TMWCNTs and TPNR/UTMWCNTs nanocomposites is given in Figure 9. It shows that incorporation of MWCNTs into TPNR considerably affects the impact strength of TPNR nanocomposites.

The results exhibited better impact strength for TMWCNTs and UTMWCNTs at 3 wt% with an increase of about 82% and 46%, respectively. This is due to the better dispersion of carbon nanotubes in the matrix, which generated a significant toughening effect on the TPNR/TMCWNTs nanocomposite compared with TPNR/UTMWCNTs nanocomposites. However, when the load is transferred to the physical network between the matrix and the filler, the debonding of the chain segments from the filler surface facilitates the relaxation of the matrix entanglement structure, leading to higher impact toughness.

The low impact energy was attributed to the filler content being more than 3wt%. This will reduce the ability of reinforced composites to absorb energy during fracture propagation. However, in the case of elastomer-toughened polymer, the presence of the elastomer basically produces stress redistribution in the composite, which causes micro cracking or crazing at many sites, thereby resulting in a more efficient energy dissipation mechanism [38].

Consequently, because of their higher surface energy and large aspect ratio, it will be difficult for the nanotubes to disperse in the TPNR when the TMWCNTs and UTMWCNTs content are higher. This will lead to less energy dissipating in the system due to the poor interfacial bonding and induces micro spaces between the filler and polymer matrix. This causes micro-cracks when impact occurs, which induces easy crack propagation. Therefore, the higher agglomeration of MWCNTs can cause the mechanical properties of the composites to deteriorate [39].

3.4. Thermal Properties

3.4.1. Glass Transition Temperature

The dynamic mechanical data shows that the glass transition temperature of the TPNR/UTMWCNTs and TPNR/TMWCNTs is affected by the addition of the different amounts of MWCNTs, as depicted in Figure 10.

From the figures, the Tg for the TPNR/TMWCNTs nanocomposites is higher than the corresponding temperature for the TPNR and TPNR/UTMWCNTs nanocomposites, usually the Tg of a polymeric matrix tends to increase with the addition of carbon nanotubes. The rise in Tg in any polymeric system is associated with a restriction in molecular motion, reduction in free volume and/or a higher degree of crosslinking (TPNR/TMWCNTs > TPNR/UTMWCNTs) due to the interactions between the polymer chains and the nanoparticles, and the reduction of macromolecular chain mobility.

With the high amount of MWCNTs (after 3wt %) of TMWCNTs and UTMWCNTs the Tg drops. This might be due to the phase separation/agglomeration of MWCNTs, this allows the macromolecules to move easily. When the content of MWCNTs is higher, the MWCNTs congregate, possibly because the intrinsic van der Waals forces occurs, which leads to bubbles and small aggregates. The conglomerations and matrix holes existing in the network of MWCNTs may perform as defects, which make the macromolecules move easily, and the Tg of the matrix is decreased.

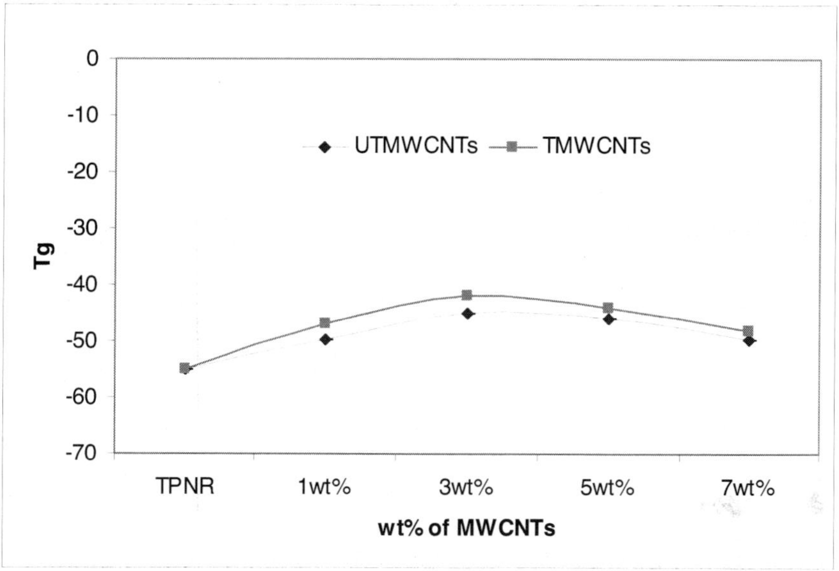

Figure 10. Glass Transition Temperature of TPNR reinforced with MWCNTs (with and without treatment).

3.4.2. Thermal Conductivity

To study the effect of MWCNTs filler on thermal conductivity, the temperature was varied from (30 – 150) °C. The carbon filler loading was from 1wt% to 7wt% for two types of carbon nanotubes (UTMWCNTs and TMWCNTs). Introducing MWCNTs to TPNR can significantly enhance the thermal conductivity of the TPNR matrix, as shown in Figure 11 and Figure 12.

As shown in figure 11 at 30°C the thermal conductivity of TPNR/TMWCNTs composites, Thermal conductivity increased at 3wt% compared to 1wt%, 5wt% and 7wt%, respectively, and for TPNR/UTMWCNTs, the thermal conductivity increased at 3wt%, as compared to TPNR at the same temperature as shown in figure 12. Thermal transport in the CNT composites includes phonon diffusion in the matrix and ballistic transportation in the filler.

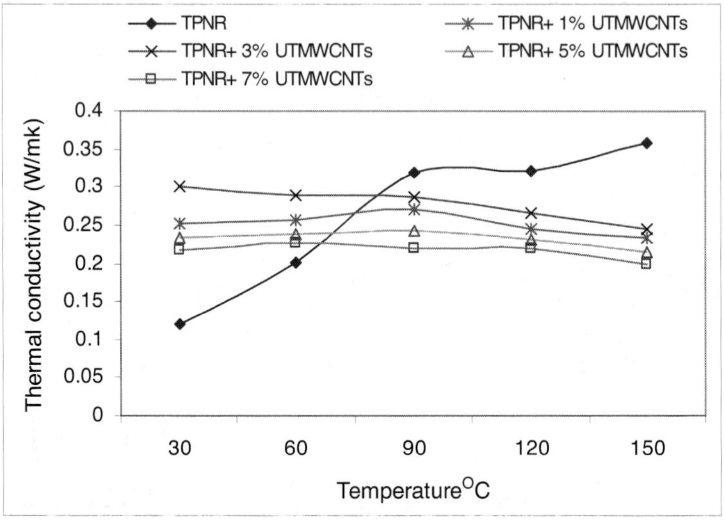

Figure 11. Thermal Conductivity of TPNR reinforced with UTMWCNTs.

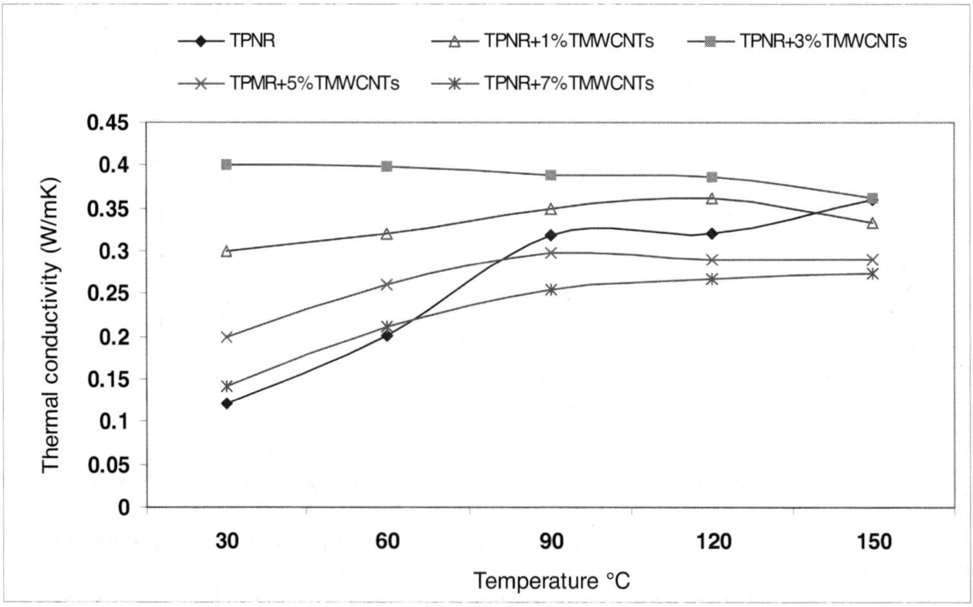

Figure 12. Thermal Conductivity of TPNR reinforced with TMWCNTs.

The improvement in thermal conductivity in TMWCNTs/TPNR may stem from the improved percolation because of the better dispersion and formation of a network [40]. The dispersion of 1wt% and 3wt% TMWCNTs is better than 5wt% and 7wt% in TPNR, at 5% and 7% the TMWCNTs agglomerated inside TPNR. Therefore, the large amounts of junctions among the carbon nanotubes form a single conducting path, which is believed to be the reason why the measured thermal conductivity is low. For the UTMWCNTs the conductivity at 3wt% and 1wt% is better than 5wt% and 7wt%, respectively.

The significant enhancement in the thermal conductivity of CNT nanocomposites is possibly attributed to the kinks or twists of UTMWCNTs. When the phonon travels along the nanotubes, if it meets the kinks or twists it would be blocked at those sites. The existence of such kinks or twists in CNTs would lead to a decrease in the effective aspect ratio of the nanotubes [41] when the amount of UTMWCNTs increases, and, thus, the thermal conductivity of UTMWCNTs-TPNR nanocomposites would be reduced. Therefore, the acid treatment of MWCNTs in TPNR could reduce these kinks or twists of TMWCNTs due to the good dispersion of MWCNTs in TPNR, causing the thermal conductivity of the nanocomposites to increase.

Two factors have been proposed to explain the significant enhancement of thermal conductivity with TMWCNTs (1) the rigid linkage between TMWCNTs and TPNR matrix with provides good interface compatibility which may reduce interface thermal resistance; (2) the good interface compatibility allows TMWCNTs to disperse well in the matrix, consequently, the results of the TEM indicate that TMWCNTs possess good dispersion and good compatibility in the TPNR matrix.

The formation of the UTMWCNTs bundles restrict the phonon transport in composites, which maybe be attributed to two reasons (1) the UTMWCNTs aggregation reduces the aspect ratio, consequently, decreasing the contact area between the UTMWCNTs and the TPNR matrix; (2) the UTMWCNTs bundles cause the phenomenon of reciprocal phonon vector, which acts like a heat reservoir and restricts heat flow diffusion.

The resistance to phonon movement from one nanotube to another through the junction will hinder phonon movement and, hence, limit the thermal conductivity. The low thermal conductivity could be partly due to the non-uniform diameter and size, as well as, the defects in and the nano-scale dimension of UTMWNTs. However, the numerous junctions between carbon nanotubes involved in forming a conductive path and the exceptionally low thermal conductance at the interface [42] are believed to be the main reason for the low thermal conductivity.

The effect of reducing the thermal conductivity is the transfer of phonons from nanotube to nanotube. This transition occurs by direct coupling between CNTs, in the case of the improper impregnated ropes, CNT-junctions and agglomerates, or via the matrix. In all these cases, the transition occurs via an interface and, thus, the coupling losses can be attributed to an intense phonon boundary scattering. At the same time the thermal conductivity decreases with the increase in temperature (if the temperature is near the melting point of the matrix). This indicates that the thermal conductivity of the composites is dominated by the interface thermal transport between the nanotube/matrix or nanotube/nanotube interface. Thus, it is believed that the decreased effective thermal conductivity of the studied composites could be due to the high interface thermal resistance across the nanotube/matrix or nanotube/nanotube interfaces.

As shown in Figure 11 and Figure 12, the thermal conductivity of TMWCNTs reinforced TPNR matrix composites for all volume fractions studied from 30°C to 150°C is better than UTMWCNTs. The effect of temperature on the thermal conductivity is clear from 30°C to 90°C, as shown in the figures. This is because of the opposing effect of temperature on the specific heat and thermal diffusivity. Eventually, at high temperatures, as the phonon mean free path is lowered, the thermal conductivity of the matrix approaches the lowest limit and the corresponding thermal resistivity approaches the highest limit.

3.4.3. Morphological Examination

The TEM can observe the morphology of UTMWCNTs/TPNR and TPNR/TMWCNTs nanocomposites, which indicates the dispersion abilities of MWCNTs in TPNR matrix before and after treatment of MWCNTs, which summarizes the TEM images of TPNR with 1wt%, 3wt% and 7wt% UTMWCNTs as shown in figure 12-14. Figure 13 shows the good dispersion of 3wt% of UTMWCNTs inside TPNR, and exhibits the better interfacial adhesion of UTMWCNTs and TPNR, Figure 14, 7wt% of UTMWCNTs, shows the poor dispersion and the large UTMWCNTs agglomerates of UTMWCNTs. This is because of the huge surface energy of MWCNTs, as well as, the weak interfacial interaction between UTMWCNTs and TPNR, which leads to inhomogeneous dispersion in the polymer matrix and negative effects on the properties of the resulting composites that causes a decrease in the tensile strength. This supports our results for thermal behavior, which due to the kinks or twists of CNTs can affect the thermal conductivity. Therefore, so when the phonon travels along the nanotube and the phonon meets the kinks or twists, it could be blocked at those sites. The existence of those kinks or twists in CNTs would result in a decrease in the effective aspect ratio of nanotubes at 7wt% UTMWCNTs because of agglomeration compared with 3wt% of MWCNTs due to the good dispersion. The homogenous dispersion of TMWCNTs in the composites is confirmed by TEM after acid treatment. Figure 15 shows the 3% TMCWNTs, which are very well dispersed in the matrix, there by suggesting a strong polymer nanotubes interfacial. Strong interfacial adhesion is essential for efficient stress transfer from the matrix to the nanotubes; this supports our observation that the higher efficiency of carbon nanotubes assists in enhancing the properties of TPNR. Low magnification was necessary to observe the poor dispersion of 7wt% of TMWCNTs in TPNR as depicted in Figure 17. The figure clearly shows a large number of unbroken carbon nanotubes but less than Figure 14, indicating a poor polymer/nanotube adhesion which is attributed to the reduction in the properties of TPNR/MWCNTs nanocompsites.

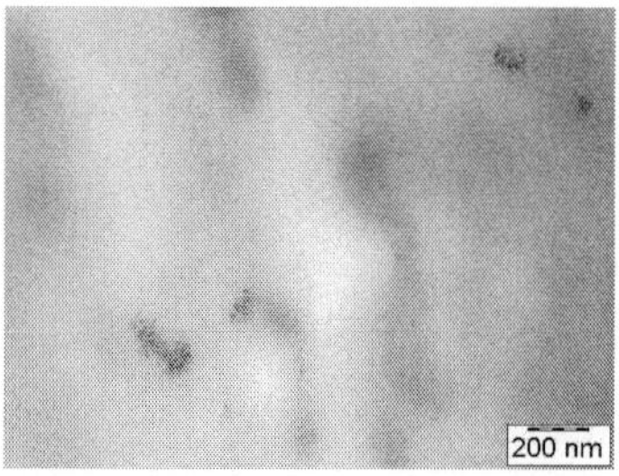

Figure 13. TPNR with 1wt% UTMWCNTs.

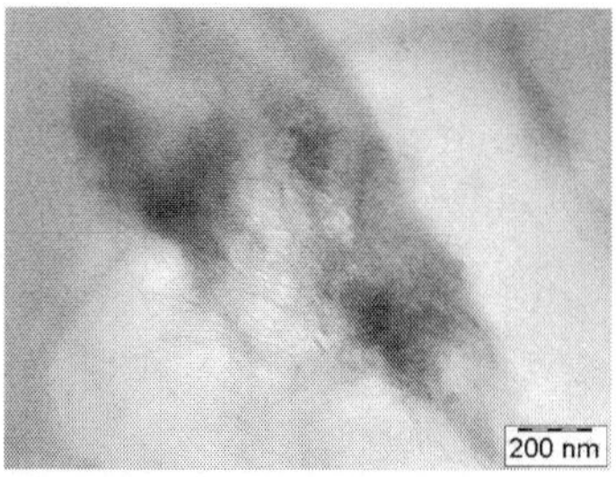

Figure 14. TPNR with 3wt% UTMWCNTs.

Figure 15. TPNR with 7wt% UTMWCNTs.

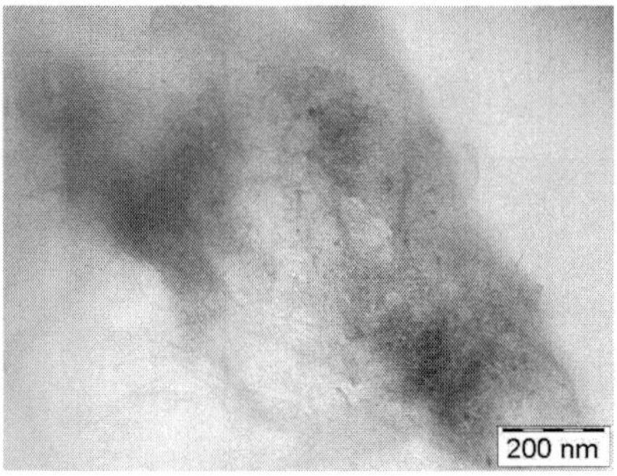

Figure 16. TPNR with 1wt% TMWCNTs.

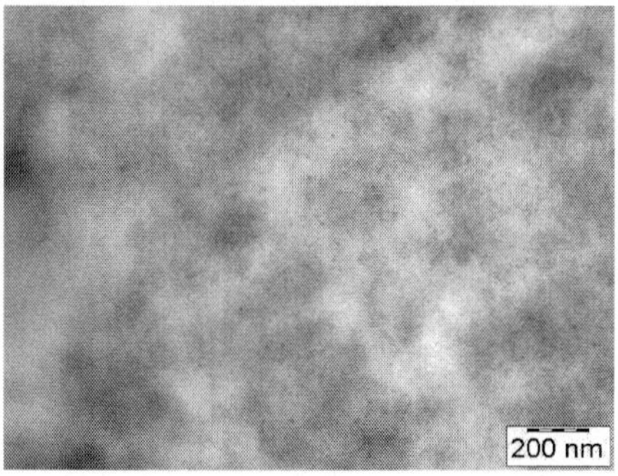

Figure 17. TPNR with 3wt% TMWCNTs.

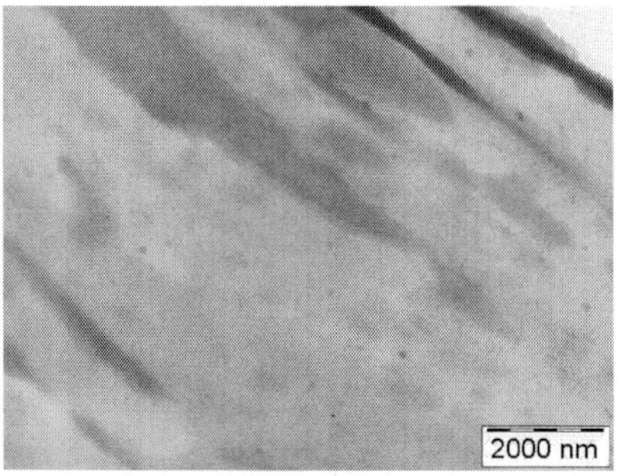

Figure 18. TPNR with 7wt% TMWCNTs.

4. Conclusion

Recently, it is believed that single-wall carbon nanotubes (SWCNTs), multi-walled carbon nanotubes (MWCNTs), coiled nanotubes and carbon nanofibers (CNFs) can be used as filler in the polymer matrix leading to composites with many enhanced properties, especially in mechanical properties. Furthermore, the inclusion of CNTs in a polymer holds the potential to improve the mechanical, electrical or thermal properties by orders of magnitude well above the performance possible with traditional fillers. In addition many researchers revealed that using functionalized MWCNTs or surface modification of MWCNTs as filler enhanced the properties of nanocomposites. This enhancement was probably suggested because of the homogenous dispersion and stronger interaction between the MWCNTs and the polymer matrix. After being treated with an acid, some functional groups were introduced onto the MWCNTs surface, which can form a physical interaction with the polymer chain. In this chapter, the effect of multi-walled carbon nanotubes with and without acid treatment on the properties of thermoplastic natural rubber (TPNR) was investigated. Two types of MWCNTs were introduced into TPNR, which are untreated UTMWCNTs (without acid treatment) and treated TMWCNTs (with acid treated MWCNTs). The acid treatment of MWCNTs removed catalytic impurities and generated functional groups such as hydroxyl, carbonyl and mainly carboxylic acid.

The results in this chapter show that the properties of MWCNTs can be improved by using this method. The TEM micrograph has shown that the effect of acid treatment has roughened the MWCNTs surface and also reduced the agglomeration. Various functional groups have been confirmed using FTIR. The TPNR nanocomposite was prepared using the melt blending method. MWCNTs are incorporated in the TPNR nanocomposite at different compositions which is 1, 3, 5 and 7 wt%. The addition of MWCNTs in the TPNR matrix improved the mechanical properties. At 3wt%, the tensile strength and Young's modulus of TPNR/UTMWCNTs increased 23% and 22%, respectively. For TPNR/TMWCNTs the optimum result of tensile strength and Young's modulus was recorded at 3% which increased 39% and 34%, respectively. In the addition the elongation of break decreased by increasing the amount of both types of MWCNTs.

The results exhibited better impact strength for UTMWCNT and TMWCNT at 3 wt% with an increase of almost 46 % and 82%, respectively. The reinforcing effect of two types of MWCNTs was also confirmed by dynamic mechanical analysis where the addition of MWCNTs have increased in the glass transition temperature (Tg) with an increase in the amount of MWCNTs (optimum at 3wt %) and it increased with the TMWCNTs more than the UTMWCNTs. Thermal conductivity improved with TMWCNTs compared to the UTMWCNTs. The homogeneous dispersion of two types of the MWNTs throughout the TPNR matrix and strong interfacial adhesion between the MWCNTs and the matrix as confirmed by the TEM images are proposed to be responsible for the significant mechanical enhancement.

Acknowledgements

The authors would like to thank the Malaysian Government and Universiti Kebangsaan Malaysia (UKM) under Science Fund Grant UKM-OUP-NBT-29-142/2011 and UKM-OUP-2012-135 for financial support.

Author details

Mou'ad A. Tarawneh* and Sahrim Hj. Ahmad

*Address all correspondence to: moaath20042002@yahoo.com

School of Applied Physics, Faculty of Science and Technology, Universiti Kebangsaan Malaysia, Malaysia

References

[1] Sakurada, I. (1985). Polyvinyl Alcohol Fibres. *International Fibre Science and Technology Series*, 6, Marcel Dekker, Inc., New York.

[2] Li, C. Y., & Chou, T. W. (2003). A structural mechanics approach for the analysis of carbon nanotubes. *Int. J. Solids Struct.*, 40, 2487-2499.

[3] Hernandez, E., Goze, C., Bernier, P., & Rubio, A. (1998). Elastic Properties of C and BxCyNz Composite Nanotubes. *Phys. Rev. Lett.*, 80, 4502-4505.

[4] Yakobson, B. I., Campbell, M. P., Brabec, C. J., & Bernholc, J. (1997). High strain rate fracture and C-chain unraveling in carbon nanotubes. *Comp. Mater. Sci*, 8, 341-348.

[5] Demcyzk, B. G., Wang, Y. M., Cumings, J., Hetman, M., Han, W., Zettl, A., & Ritchie, R. O. (2002). Direct mechanical measurement of the tensile strength and elastic modulus of multiwalled carbon nanotubes. *Mater. Sci. Eng. A*, 334, 173-178.

[6] Qian, E. C., Dickey, , Andrews, R., & Rantell, T. (2000). Load transfer and deformation mechanisms in carbon nanotube-polystyrene composites. *Appl. Phys. Lett*, 76, 2868-2870.

[7] Lopez, M. A., Valentine, L., Biagiotti, J., & Kenny, J. M. (2005). Thermal and mechanical properties of single-walled carbon nanotubes-polypropylene composites prepared by melt processing. *Carbon*, 43, 1499-1505.

[8] Liu, T., Phang, I. Y., Shen, L., Chow, S. Y., & Zhang, W. D. (2004). Morphology and mechanical properties of multi-walled carbon nanotubes reinforced nylon-6 nanocomposites. *Macromolecules*, 37, 7214-7222.

[9] Frogley, M. D., Ravich, D., & Wagner, H. D. (2003). Mechanical properties of carbon nanoparticle-reinforced elastomers. *Compos. Sci. Technol*, 63, 1647-1654.

[10] Koerner, H., Liu, W., Alexander, M., Mirau, P., Dowty, H., & Vaia, R. A. (2005). Deformation-morphology correlations in electrically conductive carbon nanotube- Thermoplastic polyurethane nanocomposites. *Polymer*, 46, 4405-4420.

[11] Berber, S., Kwon, Y. K., & Tomanck, D. (2000). Unusually High Thermal Conductivity of Carbon Nanotubes. *Phys. Rev. Lett*, 84, 4613-4616.

[12] Biercuk, M. J., Llaguno, M. C., Radosavljevic, M., Hyun, J. K., & Johnson, A. T. (2002). Carbon nanotube composites for thermal management. *Appl. Phys. Lett*, 80, 2767-2769.

[13] Hone, J. W., Piskoti, C., & Zettl, A. (1999). Thermal conductivity of single-walled carbon nanotubes. *Phys. Rev. B*, 59, 2514-2516.

[14] Gonnet, P., Liang, Z., Choi, E. S., Kadambala, R. S., Zhang, C., Brooks, J. S., Wang, B., & Kramer, L. (2006). Thermal conductivity of magnetically aligned carbon nanotube buckypapers and nanocomposites. *Current Applied Physics*, 6, 119-122.

[15] Fangming, D., Csaba, G., Takashi, K., John, E. F., & Karen, I. W. (2006). An infiltration method for preparing single-wall nanotube/epoxy composites with improved thermal conductivity. *J. Poly. Sci. Part B: Poly. Phys*, 44, 1513-1519.

[16] Sinha, S., Barjami, Iannacchione. G., Schwab, A., & Muench, G. (2005). Off-axis Thermal Properties of Carbon Nanotube Films. *J. Nanoparticle Res*, 7, 651-657.

[17] Wen-Tai, H., & Nyan-Hwa, T. (2008). Investigations on the thermal conductivity of composites reinforced with carbon nanotubes. *Diamond & Related Materials*, 17, 1577-1581.

[18] Moisala, Q. L., Kinloch, I. A., & Windle, A. H. (2006). Thermal and electrical conductivity of single- and multi-walled carbon nanotube-epoxy composites. *Composites Science and Technology*, 66, 1285-1288.

[19] Shiren, W., Richard, L., Ben, W., & Chuck, Z. (2009). Dispersion and thermal conductivity of carbon nanotube composites. *Carbon*, 47, 53-57.

[20] Assael, M. J., Antoniadis, K. D., & Tzetzis, D. (2008). The use of the transient hot-wire technique for measurement of the thermal conductivity of an epoxy-resin reinforced with glass fibres and/or carbon multi-walled nanotubes. *Composites Science and Technology*, 68, 3178-3183.

[21] Oconnell, M. J., Boul, P., Ericson, L. M., Huffman, C., Wang, Y., Haroz, E., Kuper, C., Tour, J., Ausman, K. D., & Smalley, R. D. (2001). Reversible water-solublization of single-walled carbon nanotubes by polymer wrapping. *Chem Phys Lett*, 342, 265-271.

[22] Niyogi, S., Hamon, M. A., Perea, D. E., Kang, C. B., Zhao, B., & Pal, S. K. (2003). Ultrasonic dispersion of single-walled carbon nanotubes. *J. Phys. Chem. B*, 107, 8799-8812.

[23] Viswanathan, G., Chakrapani, N., Yang, H., Wei, B., Chung, H., & Cho, K. (2003). Single-step in situ synthesis of polymer-grafted single-wallednanotube composites. *J Am Chem Soc*, 125, 9258-9259.

[24] Xie, X. L., Mai, Y. W., & Zhou, X. P. (2005). Dispersion and alignment of carbon nanotubes in polymer matrix: A review. *Mater Sci Eng*, 49, 89-112.

[25] Wang, Y., Wu, J., & Wei, F. (2003). A treatment method to give separated multiwalled carbon nanotubes with high purity, high crystallization and a large aspect ratio. *Carbon*, 41, 2939-2948.

[26] Abdullah, I., & Dahlan, M. (1998). Thermoplastic natural rubber blends. *Prog. Polym. Sci*, 23, 665-706.

[27] Abdullah, I., & Ahmad, S. (1992). Liquid NR as a compatibilizer in the blending of NR with PP. *Mater. Forum*, 16, 353-357.

[28] Tarawneh, Mou'ad A., Ahmad, Sahrim Hj., Rasid, Rozaidi, Yahya, S. Y., & Eh Noum, Se Yong. (2011). Thermal Behavior of a MWNT Reinforced Thermoplastic Natural Rubber Nanocomposite. *Journal of Reinforced Plastic and Composites*, 30(3), 216-221.

[29] Tarawneh, Mou'ad A., Ahmad, Sahrim Hj., Rasid, Rozaidi, Yahya, S. Y., & Eh Noum, Se Yong. (2011). Enhancement of the Mechanical Properties of Thermoplastic Natural Rubber Using Multi-walled Carbon Nanotubes. *Journal of Reinforced Plastic and Composites*, 30(4), 363-368.

[30] Tarawneh, Mou'ad A., & Ahmad, Sahrim Hj. Reinforced Thermoplastic Natural Rubber (TPNR) Composites with Different Types of Carbon Nanotubes (MWNTS), Book Name Carbon Nanotubes - Synthesis, Characterization, Applications (Chapter 21, Page 443-468). InTech. July 2011, 978-9-53307-497-9.

[31] Chang-Eui, H., Joong-Hee, L., Prashantha, K., & Suresh, G. A. (2007). *Composites Science and Technology*, 67, 1027-1034.

[32] Seung, H. L., Eunnari, C., So, H. J., & Jae, R. Y. (2007). Rheological and electrical properties of polypropylene composites containing functionalized multi-walled carbon nanotubes and compatibilizers. *Carbon*, 45, 2810-2822.

[33] Hirsch, A. (2002). Functionalization of single-walled carbon nanotubes. *Angew Chem Int Ed*, 41, 1853-1859.

[34] Sang, H. J., Young-Bin, P., & Kwan, H. Y. (2007). Rheological and mechanical properties of surface modified multi-walled carbon nanotube-filled PET composite. *Composites Science and Technology*, 67, 3434-3441.

[35] Potschke, P., Fornes, T. D., & Paul, D. R. (2002). Rheological behavior of multi-walled carbon nanotubes/polycarbonate composites. *Polymer*, 43, 3247-3255.

[36] Treacy, T. W. E., & Gibson, J. M. (1996). Exceptionally high Young's modulus observed for individual carbon nanotubes. *Nature*, 381, 678-680.

[37] Canche-Escamilla, G., Rodriguez-Laviada, J., Cauich-Cupul, J. I., Mendizabal, E., Puig, J. E., & Herrera-Franco, P. J. (2002). Flexural, impact and compressive properties of a rigid-thermoplastic matrix/cellulose fiber reinforced composites. *Compos. Part (A): Appl. Sci. & Manufact.*, 33, 539-549.

[38] Jianfeng, S., Weishi, H., Liping, W., Yizhe, H., & Mingxin, Y. (2007). The reinforcement role of different amino-functionalized multi-walled carbon nanotubes in epoxy nanocomposites. *Composites Science and Technology*, 67, 3041-3050.

[39] Kumar, S., Alam, M. A., & Murthy, J. Y. (2007). Effect of percolation on thermal transport in nanotube composites. *Appl Phys Lett*, 90, 104105-1-104105-3.

[40] Nan, C. W., Shi, Z., & Lin, Y. (2003). A simple model for thermal conductivity of carbon nanotube-based composites. *Chem Phys Lett.*, 375, 666-669.

[41] Yunsheng, X., Gunawidjaja, R., & Beckry, Abdel-Magid. (2006). Thermal behavior of single-walled carbon nanotube polymer-matrix composites. *Composites A*, 37(1), 114-121.

[42] Ramasamy, S., Shuqi, G., Toshiyuki, N., & Yutaka, K. (2007). Thermal conductivity in multi-walled carbon nanotubes/silica-based nanocomposites. *Scripta Materialia*, 56, 265-268.

Chapter 7

Mixtures Composed of Liquid Crystals and Nanoparticles

Vlad Popa-Nita, Valentin Barna, Robert Repnik and Samo Kralj

Additional information is available at the end of the chapter

http://dx.doi.org/10.5772/50896

1. Introduction

The past decade has witnessed an increased interest in the study of mixtures [1–3] of various soft materials and nanoparticles. A characteristic feature of a nanoparticle is that at least one of its dimensions is limited to between 1 and 100 nm. It is of interest to find combinations where each component introduces a qualitatively different behavior into the system. Such systems are expected to play an important role in the emerging field of nanotechnology and also in composites with extraordinary material properties.

In several cases various liquid crystalline phases [4] are chosen as a soft carrier matrix. Their main advantageous properties are as follows. LCs are optically anisotropic and transparent. Their structure can be readily controlled by the confining surfaces and by applying an external electric or magnetic field. LCs exhibit a rich pallet of different structures and phases that can display almost all physical phenomena. In addition the chemistry of LCs is relatively well developed, which can mean the synthesis of LC molecules with the desired properties. As a result of these properties, even pure LC systems have found several applications, in particular in the electro-optics industry.

In our study we will confine our interest to the nematic LC phase formed by rod-like anisotropic molecules. The molecules tend to be parallel, at least locally. In bulk equilibrium nematic phase LC molecules are on the average aligned homogeneously along a single symmetry breaking direction, while translational ordering is absent. In thermotropic LCs nematic ordering is reached from the isotropic (ordinary liquid) phase by lowering the temperature via a weakly first order phase transition. Reversely, in lyotropic LCs the nematic ordering could be obtained via a first order phase transition by increasing packing density of LC molecules .

Various NPs are added to LC matrices in order to introduce additional quality into the system. It has been shown that in such mixtures one can obtain dramatically enhanced [2] or even new material properties [5] (e.g., multiferroics), which is of particular interest for composite materials with exceptional properties. Because the LC phases are reached via continuous symmetry-breaking phase transitions, the presence of NPs can stabilize the LC domain structures and consequently give rise to topological defects [3]. These can strongly interact with NPs, yielding different patterns that depend on the conditions at the LC-molecule-NP interface.

In several studies one uses as NPs carbon nanotubes (CNTs) [6–12]. Most of CNTs extraordinary properties of potential use in various applications could be realized in relatively well aligned samples. Recently it has been shown that liquid crystal alignment could trigger spontaneous ordering of CNTs with remarkably high degree of ordering [9–12]. CNTs orient parallel to average direction of liquid crystal (LC) alignment with an orientational order parameter between 0.6 to 0.9 [13–16]. Both, thermotropic [13, 14, 16] and lyotropic nematic LC phases [15] have been successfully applied as aligning solvents.

The theoretical study of the collective behavior of anisotropic nanoparticles dispersed in isotropic solvents or in liquid crystals is based on the observation that they can be consider essentially as rigid-rod polymers with a large aspect ratio [17]. The Onsager's theory for the electrostatic repulsion of long rigid rods has been used to investigate the phase behavior of SWNTs dispersed into organic and aqueous solvents [18]. In a good solvent, when the van der Waals attractive interaction between CNTs is overcome by strong repulsive interrods potential, the ordered phases of CNTs can form at room temperature. On the contrary, if the solvent is not good, the van der Waals attractive interactions between the rods are strong and as a result, only extremely dilute solutions of SWNTs are thermodynamically stable and no liquid crystal phases form at room temperature. The liquid crystallinity of CNTs with and without van der Waals interactions has been analyzed by using the density functional theory [19]. In the presence of van der Waals interaction, the nematic as well as the columnar phases occur in the temperature-packing fraction phase diagram in a wide range of very high temperatures. In the absence of van der Waals interaction, with an increase of packing fraction, the system undergoes an isotropic-nematic phase transition via a biphasic region. The isotropic-nematic packing fraction decreases with the increase of the aspect ratio of CNTs. To describe the dispersion of SWNTs in superacids, the Onsager theory for rigid rods was extended to include the length polydispersity and solvent mediated attraction and repulsion [20]. The main conclusion of these theoretical models is that to obtain liquid crystal phases of CNTs at room temperature the strong van der Waals interaction between them must be screened out. This requires a good solvent with an ability to disperse CNTs down to the level of individual tube.

In the previous papers [6–8] we have presented a phenomenological theory for predicting the alignment of length monodisperse CNTs dispersions in thermotropic nematic liquid crystals. We combined the Landau-de Gennes free energy for thermotropic ordering of the liquid crystal solvent and the Doi free energy for the lyotropic nematic ordering of CNTs caused by excluded-volume interactions between them. In the first paper [6], the interaction between CNTs and liquid crystal molecules is thought to be sufficiently weak to not cause any director field deformations in the nematic host fluid. The principal results of this first study could be summarized as follows. (i) The coupling between the CNTs and a LC seems to be dominated by an anisotropic surface tension not by any deformation of the director field because the rods are thin on the scale of the extrapolation length. This means that CNTs dispersed in NLCs are in the weak-anchoring limit. (ii) The first order nematic-isotropic phase transition of CNTs dispersed in a LC disappears for a strong enough coupling to the nematic host fluid. A tricritical point can be defined that within the Landau-de Gennes model exhibits universal characteristics if expressed in the right dimensionless variables. (iii) Although in the weak-anchoring limit, the coupling between the CNTs and the LC host is so strong that in practice one should expect CNTs always to be in the strong-coupling limit, i.e., above the tricritical point. This means CNTs in LCs are always strongly paranematic. (iv)The degree of alignment of CNTs in NLCs can be tuned by varying the CNT concentration or the temperature.

The phase and structural behavior of a mixture of CNT and LC using a mean field-type phenomenological model in the strong anchoring regime was presented in the second paper [7]. We have considered cases where the nematic director field is either nonsingular or where topological defects are present in the LC medium. The effective field experienced by CNTs yields pretransitional ordering below the critical point. Above the critical point, a gradual variation of orientational order of CNTs appears. In practice one should expect CNTs always to be in the strong-coupling limit, i.e., far above the critical point. This means CNTs in the nematic phase of LC are always strongly paranematic. The model predicts an increase of nematic-isotropic phase transition temperature of LC with the volume fraction of CNTs as well as the presence of a triple point in the phase diagram. For realistic values of the coupling constant, the degree of ordering of CNTs is enslaved by the properties of the host nematic fluid.

The comparison of the results for weak anchoring and strong anchoring regimes, respectively was presented in the third paper [8]. In both anchoring cases, the first-order nematic - isotropic phase transition of CNTs dispersed in the nematic phase transforms into a continuous transition for a strong enough coupling to the nematic host fluid. The corresponding critical value of the coupling parameter increases with increasing temperature being larger in the strong anchoring limit case. The numerical estimate of the coupling constants in the two anchoring regimes indicates that the coupling is so strong that CNTs are far above the critical point, meaning that the nematic-isotropic phase transition is a continuous one. In both cases, we have plotted the phase diagram of the homogeneous mixture for the same value of the coupling parameter. In both cases, three regions of the phase diagram could be distinguished and correspondingly the existence of triple points are shown. We mention that in both anchoring cases, the nematic-isotropic phase transition temperature of LC increases with the volume fraction of CNTs, a well-known experimental result.

Usually, after the acid treatment and ultrasonication (to enhance the dispersion and stability of the CNTs suspensions), depending on the temperature of water bath and time of ultrasonication, the obtained CNTs have different lengths and diameters [14, 21]. The influence of length bidispersity on the phase diagram and alignment of CNTs is the subject of the present paper. As the first step we extend our mesoscopic model [6–8] and consider length bidispersity of CNTs. We mention that the effect of bidispersity of the long rigid rods has been discussed in the framework of Onsager theory [22–25]. We shall refer to their results in the last section of the paper.

The plan of the paper is as follows. In the first part of the paper we focus on mixtures of nematic LCs and isotropic NPs. A simple phenomenological model is used which is sufficient to identify key mechanisms which might trigger phase separation. Conditions for efficient trapping of NPs to cores of topological defects is discussed. In the second part we confine our interest to mixtures of nematic LCs and carbon nanotubes. Using a simple model we take into account length dispersity of CNTs and analyze corresponding phase diagrams. In the last section we summarize the main results.

2. Binary nematogen - nonnematogen mixtures

We first consider a mixture of nematic LC and isotropic NPs using a relatively simple phenomenological model. We identify key phase separation triggering mechanisms. We also discuss conditions enabling efficient trapping of NPs within cores of topological defects or strongly localized elastic distortions. If these trapping sites are relatively uniformly spatially distributed they might prevent phase separation.

2.1. Free energy

We use semiphenomenological model within which the volume concentration of isotropic NPs is given by the conserved parameter ϕ. The orientational order of LC molecules is described by the symmetric and traceless tensor order parameter [4] $Q = \sum_{i=1}^{3} \lambda_i \vec{e}_i \otimes \vec{e}_i$, where λ_i and $\vec{e}_i$ stand for its eigenvalues and corresponding eigenvectors, respectively. In the case of uniaxial ordering Q is commonly expressed in terms of the nematic director field $\vec{n}$ and the uniaxial orientational order parameter S as [4]

$$Q = S\left(\vec{n} \otimes \vec{n} - \frac{1}{3}I\right). \tag{1}$$

Here I stands for the identity tensor. The unit vector $\vec{n}$ points along the local uniaxial ordering direction, where states $\pm\vec{n}$ are equivalent (the so called head-to-tail invariance). The extent of fluctuations is determined by S, where $S = 1$ and $S = 0$ reflect rigid alignment along $\vec{n}$ and isotropic liquid ordering, respectively. If strong distortions are present biaxial states could be locally entered. Degree of biaxiality is assessed via parameter [26]

$$\beta^2 = 1 - \frac{6(trQ^3)^2}{(trQ^2)^3}, \tag{2}$$

ranging in the interval $[0, 1]$. Uniaxial configurations correspond to $\beta^2 = 0$ and an ordering with the maximum degree of biaxiality is signaled by $\beta^2 = 1$.

The free energy F of a mixture is expressed as

$$F = \int \left(f_m + f_c + f_e + f_i \delta(\vec{r} - \vec{r}_i)\right) d^3\vec{r}. \tag{3}$$

The quantity δ stands for the delta measure, $\vec{r}_i$ locates NP-LC interfaces and the integral runs over the LC volume. The role of different contributions in Eq.(3) is as follows.

The mixing term f_m describes the isotropic mixing of the two components. Within the Flory theory [27] it is expressed as

$$f_m = \frac{k_B T}{v_{lc}}(1-\phi)\ln(1-\phi) + \frac{k_B T}{v_{np}}\phi\ln\phi + \chi\phi(1-\phi). \tag{4}$$

Here k_B is the Boltzmann constant, T is the absolute temperature, and χ stands for the Flory-Huggins parameter [27]. The volume of a LC molecule and of a nanoparticle is given by v_{lc} and v_{np}, respectively,

The condensation contribution f_c enforces orientational LC ordering below a critical temperature T_{NI}. Is is expressed as [4]

$$f_c = \frac{3a(T-T^*)}{2}Q_{ij}Q_{ij} - \frac{9B}{2}Q_{ij}Q_{jk}Q_{kj} + \frac{9C}{4}(Q_{ij}Q_{ij})^2, \tag{5}$$

where summation over repeated indices is assumed. The quantities a, B, C, are material constants and T^* denotes the spinodal temperature limit of the isotropic phase of the pure LC. This condensation free energy term describes a weakly first order nematic-isotropic phase transition. At $T = T_{NI} = T^* + B^2/(4aC)$, the two phases, nematic ($S^{(NI)} = B/(2C)$) and isotropic ($S = 0$) coexist in equilibrium.

The deviations from homogeneous nematic ordering are penalized by the elastic term

$$f_e = \frac{L}{2}Q_{jk,i}Q_{jk,i}, \tag{6}$$

which is expressed within a single elastic constant approximation. Here L is the representative bare nematic elastic constant.

The conditions at the NP-LC interface are determined by the term f_i. We express it as

$$f_i = -we_kQ_{kj}e_j, \tag{7}$$

where $w > 0$ is the anchoring strength favoring the nematic ordering at an interface and $\overrightarrow{e}$ stands for the local surface normal.

2.2. Phase separation tendency

We proceed by identifying key mechanisms favoring phase separation tendency in a mixture of NPs and nematic liquid crystal. We describe global LC orientational ordering with a spatially averaged order parameter $\overline{S}$ and volume concentration of nanoparticles $\overline{\phi}$. Here the over-bar $\overline{(..)}$ denotes the spatial average. Relative presence of NPs and LC molecules in the mixture is therefore given by $\overline{\phi} = N_{np}v_{np}/V$ and $1-\overline{\phi}$, respectively. Here N_{np} counts the number of NPs and V stands for the volume of the sample.

The resulting average free energy density is expressed as $\overline{f} = \overline{f}_m + \overline{f}_c + \overline{f}_e + \overline{f}_i$, where

$$\overline{f}_m \sim \frac{k_B T}{v_{lc}}(1-\overline{\phi})\ln(1-\overline{\phi}) + \frac{k_B T}{v_{np}}\overline{\phi}\ln\overline{\phi} + \chi\overline{\phi}(1-\overline{\phi}), \tag{8}$$

$$\overline{f}_c \sim (1-\overline{\phi})\left(a(T-T^*)\overline{S}^2 - B\overline{S}^3 + C\overline{S}^4\right), \tag{9}$$

$$\overline{f}_e \sim (1-\overline{\phi})\, L\overline{S}^2/\overline{\xi_d}^2, \tag{10}$$

$$\overline{f}_i \sim -(1-\overline{\phi})\,\overline{\phi}w\overline{S}. \tag{11}$$

The factor $(1-\overline{\phi})$ present in terms $\overline{f}_c$ and $\overline{f}_e$ accounts for the part of the volume not taken up by LC. Furthermore, the factor $(1-\overline{\phi})\,\overline{\phi}$ in $\overline{f}_i$ accounts for absence of this term if $\overline{\phi}=0$ or $\overline{\phi}=1$. Note that in general $T^* = T_*(\phi)$. Simple binary modeling [28] suggests $T^* = T_0 - \lambda\overline{\phi}$, where T_0 and λ are positive material constants. It accounts for weaker interactions among LC molecules due to presence NPs. In general NPs could introduce spatially nonhomogeneous orientational ordering of LC molecules which is taken into account by $\overline{f}_e$. On average degree of elastic distortions in $\overrightarrow{n}$ is approximated by the average domain length $\overline{\xi_d}$.

From the expression for $\overline{f}$ one can extract the effective Flory-Huggins [27] parameter

$$\chi_{eff} = \chi + a\lambda\overline{S}^2 - w\overline{S}. \tag{12}$$

The phase transition takes place if χ_{eff} exceeds a threshold value χ_c. In typical LCs it holds [29] $\chi << a_0\lambda$ and $\chi < \chi_c$. Henceforth we limit our attention to such cases. Consequently, in the isotropic phase (where $\overline{S}=0$) homogeneous mixtures are established. On entering orientational ordered phase different scenaria can be realized. We first consider cases where the wetting interaction between LCs and NPs is negligible weak (i.e., $w \sim 0$). In this case phase separation is very likely. It is triggered providing $\chi + a\lambda\overline{S}^2 \sim a\lambda\overline{S}^2 > \chi_c$. However, strong enough surface wetting interaction could suppress phase separation providing $\chi + a\lambda\overline{S}^2 - w\overline{S} < \chi_c$.

Next, we consider cases where there exist localized regions in LC ordering exhibiting strong local distortions. Therefore, in some parts $\overline{\xi_d}$ entering the expression for $\overline{f}_e$ is relatively small. From Eq.(10) we infer that local free energy penalties could be reduced if they are occupied by NPs (i.e., $\overline{\phi} \sim 1$). Therefore, the structure of expression for $\overline{f}$ suggests that NPs tend to assemble at local elastic distortions in order to reduce the total free energy penalty of the system.

2.3. Interaction between NPs and topological defects

In this section we investigate in more detail interaction between NPs and localized elastic distortions. We estimate general conditions for which this interaction is attractive. For this purpose we study a specific example where we enforce a topological defect within a cell. We add NPs exhibiting different surface constraints and determine conditions for which attractive interaction is enabled.

We consider LC ordering within a cylindrical plane-parallel cell of thickness h and radius R. The cell is schematically depicted in Fig.1.

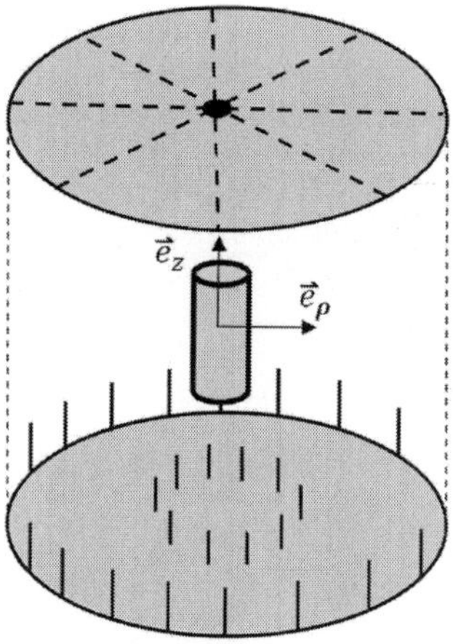

Figure 1. Schematic presentation of the hybrid plan-parallel cell hosting a nanoparticle at its symmetry axis. The diameter and the height of cylindrically shaped NP is in simulation set to be equal to the biaxial correlation length ξ_b. Furthermore, we set $h = 2R = 10\xi_b$, where h describes the height and R the radius of the cell.

We use the cylindrical coordinate system determined by unit vectors $\{\vec{e}_\rho, \vec{e}_\varphi, \vec{e}_z\}$. Here $\vec{e}_\rho$ is the radial unit vector, $\vec{e}_z$ points along the z-coordinate, while $\vec{e}_\varphi = \vec{e}_z \times \vec{e}_\rho$. We enforce a topological defect (boojum) [30] by imposing strong hybrid anchoring conditions at the confining plates. At the top plate ($z = h$) strong uniaxial radial anchoring is set imposed, i.e., we enforce $Q(z = h) = Q_{rad} \equiv S_{eq}\left(\vec{e}_\rho \otimes \vec{e}_\rho - \frac{1}{3}I\right)$. Here S_{eq} stands for the equilibrium nematic order parameter. At the bottom plate we impose strong homeotropic anchoring, i.e. $Q(z = h) = Q_{\text{hom}} \equiv S_{eq}\left(\vec{e}_z \otimes \vec{e}_z - \frac{1}{3}I\right)$. At the lateral wall we set free boundary conditions. The resulting equilibrium equations were solved numerically, where technical details are given in [30] and [31].

The boojum structure is well characterized by its biaxial spatial profile shown in Fig. 2.

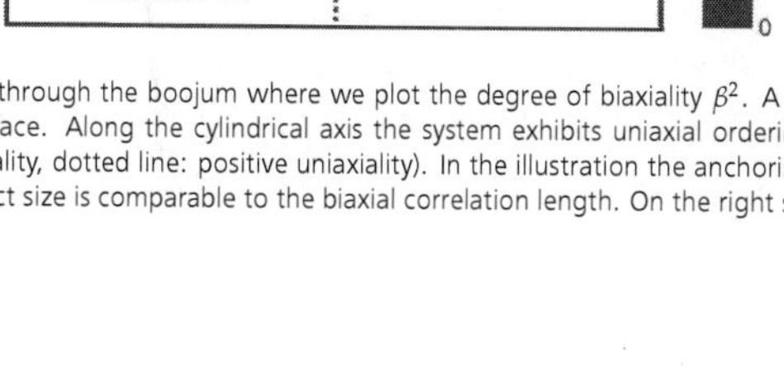

Figure 2. The cross-section through the boojum where we plot the degree of biaxiality β^2. A biaxial shell joins the isotropic finger tip and the upper surface. Along the cylindrical axis the system exhibits uniaxial ordering due to topological reasons (dashed line: negative uniaxiality, dotted line: positive uniaxiality). In the illustration the anchoring strength at the top plane is finite. The characteristic defect size is comparable to the biaxial correlation length. On the right side of the figure the grayscale bar of $1 - \beta^2$ is shown.

We plot $\beta^2(\rho, z)$ dependence (see Eq.(2)) in the plane through the defect core. Note that the system exhibits cylindrical symmetry. The boojum is characterized by a finger-like structure, where the finger tip is melted by topologically reasons. The boojum core structure is analyzed in detail in [30] and here we briefly summarize its main characteristics. The defect core is dominated by a *finger* protruding into the cell's interior along its symmetry axis. The center of the *finger*, residing at the cylinder axis, is negatively uniaxial ($S < 0$). It ends in a melted (isotropic) point ($S = 0$) to which we refer as the *finger tip*. It is placed roughly at the distance $\xi_f \sim \xi_b$ from the surface, where ξ_b stands for the biaxial correlation length. Below the *finger tip* the nematic configuration is positively uniaxial ($S > 0$) at the symmetry axis. The *finger* is enclosed within a biaxial shell exhibiting maximal biaxiality [26, 30] which joins the *finger* tip with the upper surface. By topological reasons the nematic order parameter melts only at the *finger tip* for realistic anchoring strengths. Note that in Fig. 2 the biaxial profile is plotted for a more realistic finite anchoring strength for which the finger-like profile is well pronounced.

At the cylinder axis we place a cylindrically shaped NP. The height and diameter of the NPs is set equal to ξ_b. Our interest is to estimate impact of NP surface treatment on interaction with its surrounding. For this purpose we impose three qualitatively different strong boundary conditions at the NP surface which is determined by the position vector $\overrightarrow{r}_i$: i) $Q(\overrightarrow{r}_i) = 0$, ii) $Q(\overrightarrow{r}_i) = Q_{rad}$, iii) $Q(\overrightarrow{r}_i) = Q_{\mathrm{hom}}$. These conditions locally enforce melting, radial, and spatially homogeneous configuration, respectively. We calculate the LC free energy within the cell as a function of the NP position along the z axis. Note that three qualitatively different areas exist within the cell. These are: i) region surrounding the melted point at the boojum *finger tip*, ii) prevailing radial ordering at $z \sim h$, and iii) the homogeneous ordering along $\overrightarrow{e}_z$ at $z \sim 0$. We vary the position of NP along the z-axis, and for each position we calculate the free energy of the system. In Fig. 3 we plot the total free energy as a function of z-coordinate for three different surface treatments.

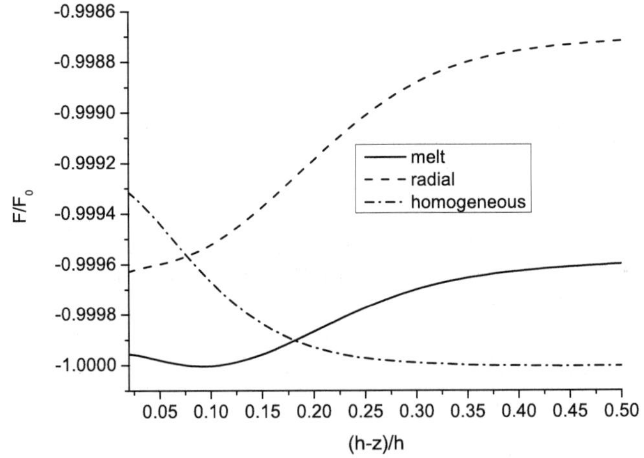

Figure 3. The free energy F of the system as a function of the nanoparticle position along the symmetry axis. The free energy is scaled with respect to the minimum of free energy F_0 calculated for the melted boundary condition. One sees that for the i) melted ii) radial and iii) homogeneous boundary condition the free energy exhibits minimum at i) the finger tip ii) $z = h$, iii) $z = 0$, respectively.

http://dx.doi.org/10.5772/50896

We see that NP enforcing i) melting, ii) radial, iii) homogeneous configuration tends to migrate towards the i) *finger tip*, ii) top plate, iii) bottom plate, respectively. Simulations shows that NPs tend to migrate towards regions which exhibit a similar local structure with respect to the conditions at the NP surface. Therefore, to assemble efficiently NPs at a defect core it is essential to make the surface coating such that the effective NP configuration resembles the defect core structure.

We next assume that NPs environment resembles a defect core structure and therefore tend to be trapped within the core of the defect. In the following we estimate the free energy gain if the NP is trapped within the core. For illustration we consider a line defect (disclination) of length h. Note that lattices of topological defects can be stabilized either by inherent LC property (e.g. chirality [3]) or imposed geometrically by imposing frustrating boundary conditions [32]. The corresponding condensation free energy penalty ΔF_c for introducing the defect line inside an orientational ordered medium is roughly given by

$$\Delta F_c \sim a(T_{NI} - T)\overline{S}^2 \pi \xi^2 h. \tag{13}$$

Here T_{NI} refers to I-N phase transition temperature and $T < T_{NI}$. The core average radius is roughly given by the relevant (uniaxial or biaxial) order parameter correlation length ξ. Within the core the LC order is either essentially melted (i.e. $S \sim 0$) or strongly biaxial [33].

We next assume that NPs are added to the LC medium and that they collect at the disclination line. If NPs do not apparently disrupt the defect core structure then the condensation free energy penalty is decreased due to the reduced volume occupied by the energetically costly essentially isotropic (or strongly biaxial) phase [34]. One refers to these effect as the *Defect Core Replacement mechanism* [3]. The resulting decrease in ΔF_c penalty reads

$$\Delta F_c() \sim a(T_{NI} - T)\overline{S}^2 \left(\pi \xi^2 h - N_{np}^{(def)} v_{np} \right), \tag{14}$$

where $N_{np}^{(def)}$ counts number of NPs trapped within the core.

3. Dispersions of carbon nanotubes

We next consider mixtures of nematic LCs and carbon nanotubes with length bidispersity. We furthermore assume homogeneous uniaxial orientational alignment of all components along a single symmetry breaking direction.

3.1. Free energy of three component mixtures

The mixture is characterized by the volume fractions of the three components:

$$\Phi_i = \frac{N_i v_i}{\sum\limits_{i=1}^{3} N_j v_j} \quad \text{with} \quad \sum_{i=1}^{3} \Phi_i = 1, \tag{15}$$

where N_i is the number of molecules of component i ($i = 1$ defines the liquid crystal with molecules of length $L_{01} = 3$ nm and diameter $D_1 = 0.5$ nm, $i = 2$ the CNTs of length $L_{02} = 400$ nm and diameter $D = 2$ nm, and $i = 3$ the CNTs of length $L_{03} = 800$ nm and diameter $D = 2$ nm) and v_i is the volume of a particle of component i. For latter convenience we introduce scaled CNT lengths $L_i = L_{0i}/D$ ($i = 2, 3$).

The degree of alignment of every component of the mixture is characterized by the scalar order parameter S_i [4]. The corresponding isotropic liquid of the component i is characterized by $S_i = 0$ while a perfectly oriented nematic phase would correspond to $S_i = 1$.

The free energy per unit volume of the mixture is expressed as

$$f = f_{CNT} + f_{LC} + f_C, \tag{16}$$

where f_{CNT} describes contribution of the two CNT components dispersed in the LC fluid, f_{LC} represents the free energy density of nematic liquid crystal order, while f_C takes into account the coupling between LC molecules and CNTs, respectively.

The free energy density of CNTs is given by [6, 7]

$$\frac{f_{CNT}}{k_B T} = \sum_{i=1}^{3} \frac{\Phi_i}{v_i} \ln \Phi_i + \sum_{i=2}^{3} \frac{L_i \Phi_i^2}{6 v_i} \left[\left(\frac{3}{L_i \Phi_i} - 1 \right) S_i^2 - \frac{2}{3} S_i^3 + S_i^4 \right] - \frac{\gamma_{23}}{k_B T} \Phi_2 \Phi_3 S_2 S_3. \tag{17}$$

The first sum represents the entropic isotropic contribution due to mixing of the two CNT components and LC neglecting their orientational degree of ordering [27]. The second sum describes a first order orientational phase transition of the i-species of CNT from the isotropic phase with $S_i = 0$, to the nematic phase with $S_i = (1 + \sqrt{9 - 24/L_i \Phi_i})/4$. The first order nematic-isotropic phase transition takes place at $\Phi_i^{(NI)} = 2.7/L_i$ and $S_i^{(NI)} = 1/3$. It is obtained starting from the Onsager theory [35] and using the Smoluchovsky equation [36, 37]. The model neglects the van der Waals attractions between CNTs which are responsible for their tendency to form bundles. The last term in Eq. (17) represents the interaction energy between the different CNTs species, where the interaction parameter γ_{23} is given by

$$\gamma_{23} = 8 k_B T / \pi D^3 \approx 10^6 \text{ N/m}^2.$$

This expression for γ_{23} is obtained using the same Doi procedure [36, 37] starting with the Onsager theory for a bidisperse hard rods system.

The second term in Eq. (16) is the Landau-de Gennes free energy density [4] which describes the weakly first-order nematic-isotropic phase transition of thermotropic LC

$$f_{LC} = \Phi_1 [a(T - T^*) S_1^2 - B S_1^3 + C S_1^4]. \tag{18}$$

For representative LC material we chose pentylcyanobiphenyl (5CB), for which $T^* = 307.55$ K, $a \approx 5.2 \cdot 10^4$ J/m^{3}K, $B \approx 5.3 \cdot 10^5$ J/m^3, $C \approx 9.7 \cdot 10^5$ J/m^3 [38]. This choice yields $S_1^{(NI)} = 0.275$ and $T_{NI} = 308.95$ K.

The third term in Eq.(16) represents the coupling between the liquid crystal molecules and the two CNT species. The resulting coupling term structure in both anchoring limits (weak and strong) was estimated in [6, 7]. The two limits are defined by the ratio DW/K, where W is the anchoring energy of a LC-nanotube interface, K is the average Frank nematic elastic constant. For typical values of $D = 2$ nm, $K \approx 10^{-11}$ N, $W \approx 10^{-6}$ N/m, $DW/K << 1$. Consequently, only the weak-anchoring limit needs to be considered, as already concluded by Lynch and Patrick [13]. The corresponding free energy density of coupling is approximately given by [6]:

$$f_C = -\gamma_{12}\Phi_2 S_1 S_2 \left(1 - \frac{1}{2}S_2\right) - \gamma_{13}\Phi_3 S_1 S_3 \left(1 - \frac{1}{2}S_3\right). \tag{19}$$

The terms in brackets ensures that $S_i \to 1$ when $\gamma_{12}(\gamma_{13}) \to \infty$ as it should. The coupling parameters γ_{12} and γ_{13} depend only on the anchoring energy of CNTs at the LC molecules surface and the diameter of CNTs [6] (no on their lengths). Therefore

$$\gamma_{12} = \gamma_{13} = \gamma_1 = 4W/3D \approx 10^3 \text{ N/m}^2$$

and the coupling free energy can be written as

$$f_C = -\gamma_1 S_1 \left[\Phi_2 S_2 \left(1 - \frac{1}{2}S_2\right) + \Phi_3 S_3 \left(1 - \frac{1}{2}S_3\right)\right]. \tag{20}$$

The free energy per unit volume of a monodisperse system (one species of CNTs of the diameter D and length L_{02} ($L_2 = L_{02}/D$, the volume v_2 and the volume fraction Φ) dispersed in a LC (with the molecular volume v_1 and the volume fraction $1 - \Phi$) is given by

$$\begin{aligned} f &= k_B T \left[\frac{1-\Phi}{v_1}\ln(1-\Phi) + \frac{\Phi}{v_2}\ln\Phi\right] \\ &+ k_B T \frac{L_2\Phi^2}{6v_2}\left[\left(\frac{3}{L_2\Phi_2} - 1\right)S_2^2 - \frac{2}{3}S_2^3 + S_2^4\right] \\ &+ (1-\Phi)[a(T-T^*)S_1^2 - BS_1^3 + CS_1^4] - \gamma_1\Phi S_1 S_2\left(1 - \frac{1}{2}S_2\right). \end{aligned} \tag{21}$$

3.2. Equilibrium equations

In the bidisperse case, the equilibrium values of the order parameters are obtained by minimization of the free energy density (Eqs. (16), (17), (18), and (19))with respect to S_1, S_2, and S_3, respectively. From the corresponding equations we find the equilibrium values of the order parameters in the nematic phase (S_{in}) and paranematic phase (S_{ip}), respectively. Once the minimization procedure has been solved the volume fractions of the coexisting phases are found by solving the equilibrium conditions:

$$\mu_2(S_{1n}, S_{2n}, S_{3n}, \Phi_{2n}, \Phi_{3n}) = \mu_2(S_{1p}, S_{2p}, S_{3p}, \Phi_{2p}, \Phi_{3p}),$$
$$\mu_3(S_{1n}, S_{2n}, S_{3n}, \Phi_{2n}, \Phi_{3n}) = \mu_3(S_{1p}, S_{2p}, S_{3p}, \Phi_{2p}, \Phi_{3p}),$$
$$g(S_{1n}, S_{2n}, S_{3n}, \Phi_{2n}, \Phi_{3n}) = g(S_{1p}, S_{2p}, S_{3p}, \Phi_{2p}, \Phi_{3p}), \quad (22)$$

where the chemical potential of the two species of CNTs μ_2, μ_3, and the grand potential g are defined by the equations

$$\mu_2 = \frac{\partial f}{\partial \Phi_2}; \ \mu_3 = \frac{\partial f}{\partial \Phi_3}; \ g = f - \mu_2\Phi_2 - \mu_3\Phi_3. \quad (23)$$

There are three equilibrium equations with four variables: Φ_{2n}, Φ_{3n}, Φ_{2p}, and Φ_{2p}. We take Φ_{3p} as freely variable and calculate the other three Φ_{2n}, Φ_{3n}, Φ_{2p} from the coexistence equations (22).

In the monodisperse case, the equilibrium values of the order parameters (S_{1p}, S_{2p}, S_{1n}, and S_{2n})are obtained minimizing the free energy (21) with respect to S_1 and S_2 and the equilibrium conditions are given by

$$\mu(S_{1n}, S_{2n}, \Phi_n) = \mu(S_{1p}, S_{2p}, \Phi_p),$$
$$g(S_{1n}, S_{2n}, \Phi_n) = g(S_{1p}, S_{2p}, \Phi_p), \quad (24)$$

where the chemical potential of CNTs and the grand potential are defined as: $\mu = \partial f / \partial \Phi$, and $g = f - \mu\Phi$, respectively.

3.3. Phase behavior

In the first part of this section we present the phase behavior of monodisperse CNTs immersed in LC as a function of T, Φ and γ_1, while in the last part the analyze of the phase behavior of the bidisperse system is analyzed as a function of T, Φ_2, Φ_3 and γ_1.

3.3.1. *Monodisperse CNTs*

The coupling term between CNTs and the LC molecules in (21) induces two different region in the phase diagram separated by a critical line $\gamma_1^{(c)}(T)$. i) For $\gamma_1 < \gamma_1^{(c)}$, the CNTs exhibit a first order (discontinuous) phase transition between a paranematic phase (a phase with a low degree of orientational order) and a nematic phase (the *subcritical* region). ii) On the contrary, for $\gamma_1 > \gamma_1^{(c)}$, CNTs display gradual variation of S_2 with Φ (the *supercritical* regime).

The critical line $\gamma_1^{(c)}(T)$ is obtained by solving the equations $\partial f / \partial S_1 = \partial f / \partial S_2 = \partial^2 f / \partial S_2^2 = \partial^3 f / \partial S_2^3 = 0$. They yield at the critical point universal values for the order parameter $S_2^{(c)} = 1/6$ and volume fraction $\Phi^{(c)} = 18/7L_2$. The $\gamma_1^{(c)}(T)$ dependence for the two species of CNTs is presented in Figure 4.

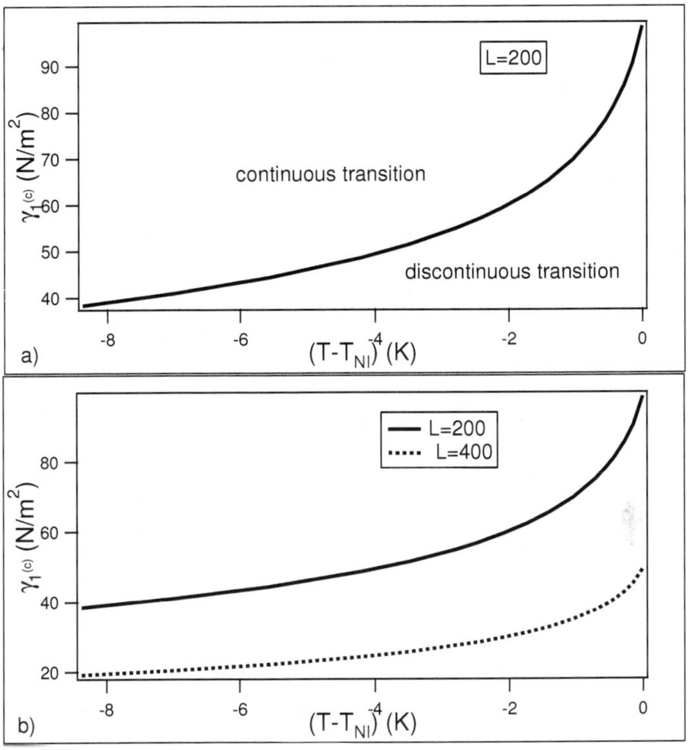

Figure 4. The critical value of the coupling parameter $\gamma_1^{(c)}$ as function of temperature calculated for both species of CNTs in the nematic LC phase.

On decreasing the temperature the external field felt by the CNTs is increasing due to increasing value of S_1 and the nematic-isotropic phase transition of CNTs becomes gradual for lower values of interaction parameter $\gamma_1^{(c)}$. Furthermore, the critical value of γ_1 decreases with increasing the length of CNTs. Therefore, the continuity of paranematic-nematic phase transition is favored by longer CNTs.

The phase diagram for the monodisperse system in the *subcritical* regime, for $L_2 = 200$ and $L_2 = 400$, respectively is shown in Figure 5.

Depending on temperature, we distinguish two regions in the phase diagram: i) for $T < T_{NI}$, the LC is in the nematic phase and the CNTs exhibit a first order phase transition with increasing Φ from a paranematic to a nematic phase. With decreasing temperature, the order parameter jump $(S_{2n} - S_{2p})$ as well as the difference in the volume fractions $(\Phi_n - \Phi_p)$ become lower and they cancel at some temperature lower for shorter CNTs. This is due to the fact that the value of $\gamma_1 = 40.33$ N/m^2 considered, corresponds to a critical temperature $T - T_{NI} = -7.31$ K for shorter CNTs ($L_2 = 200$) and $T - T_{NI} = -0.45$ K for longer CNTs ($L_2 = 400$), respectively. We emphasize that the longer CNTs become ordered at lower

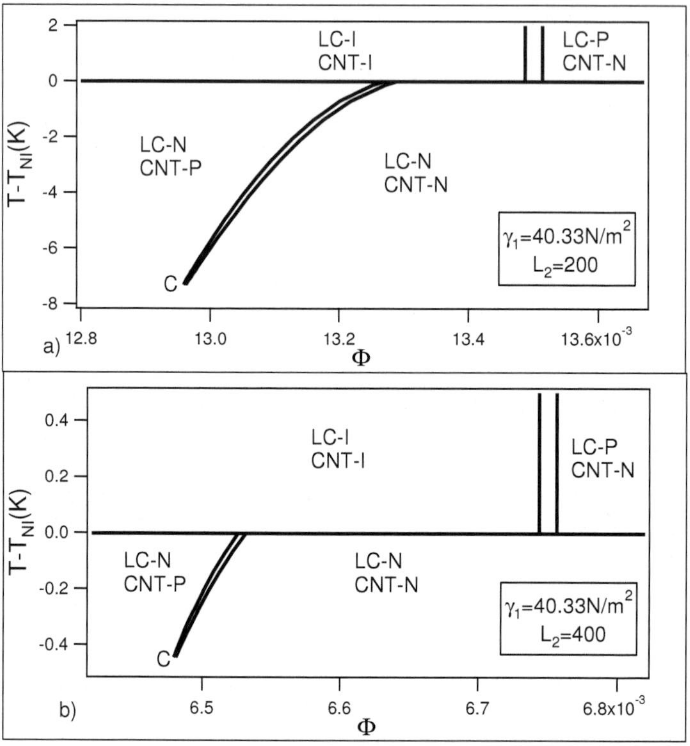

Figure 5. The (T, Φ) phase diagram of a monodisperse system in the *subcritical* regime. I means isotropic, P-paranematic, and N-nematic.

volume fraction. ii) for $T > T_{NI}$, the LC is in the isotropic phase ($S_1 = 0$) and the volume fraction gap of CNTs at the transition does not depend on temperature (the Flory horn [27]). Again the longer CNTs become aligned at lower volume fractions. It is important to note also the influence of CNTs on the LC alignment in this region. Above T_{NI}, the transition of CNTs from isotropic to nematic induces the transition of LC from isotropic to a paranematic phase (with a very very small degree of ordering for this value of the coupling constant). This problem of the influence of LC properties by the CNTs is not elucidated neither theoretically, nor experimentally yet and will be a subject of a future study.

To see in more detail the orientational order developed in the system, we have plotted in Figures 6 and 7 the order parameter variation as a function of the volume fraction Φ of the CNTs at a fixed temperature.

In Figure 6a, the temperature corresponds to a *subcritical* regime ($\gamma_1 < \gamma_1^{(c)}$) (see Figure 4), and the transition of CNTs is a discontinuous one with a jump in the order parameter. On the contrary, the temperature in Figure 6b corresponds to a *supercritical* regime because for this temperature $\gamma_1 > \gamma_1^{(c)}$ (see Figure 4). As a consequence, the CNTs phase transition is

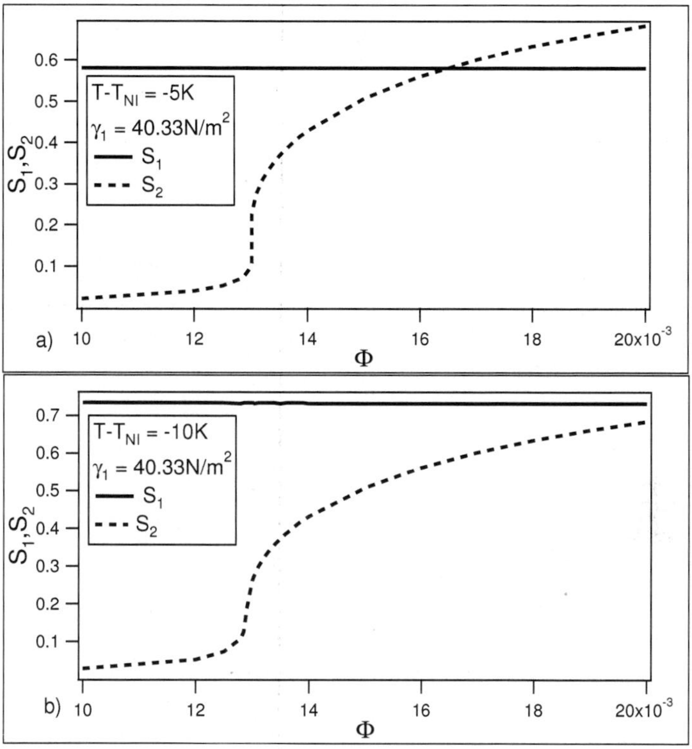

Figure 6. The order parameter variations for two temperatures lower than T_{NI} in the *subcritical* regime.

continuous, the order parameter is continuous at the transition. In both figures, the order parameter of the liquid crystal is constant. Therefore the CNTs are enslaved by LC. The degree of ordering of CNTs is present even at small volume fractions, so that the CNTs are in the paranematic phase.

In Figure 7, the temperature is greater than nematic-isotropic phase transition temperature, so that the LC is in the isotropic phase (even if the degree of ordering exists it is very small and can not be seen on the figure). For low volume fraction, the CNTs are in the isotropic phase and becomes nematic by a first order phase transition at some value of Φ.

The phase diagram for the monodisperse system in the *supercritical* regime, for $L_2 = 200$ and $L_2 = 400$, respectively is shown in Figure 8.

For a more realistic value of the coupling constant $\gamma_1 >> \gamma_1^{(c)}$, in the nematic phase of LC, there is only a gradual variation of the order parameter of CNTs with the volume fraction and temperature. Above T_{NI}, the transition isotropic-nematic of CNTs is first order and also an induced first order isotropic-paranematic phase transition takes place in LC. With increasing temperature, the transitions takes place at a larger volume fractions. For longer CNTs, the volume fractions at the transition are lower.

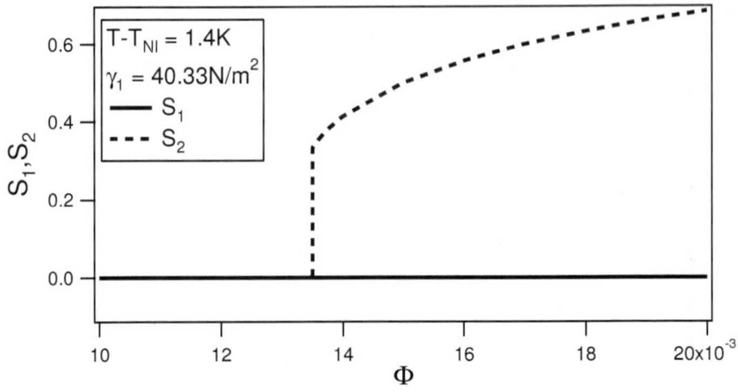

Figure 7. The order parameter variations for a temperature larger than T_{NI} in the *subcritical* regime.

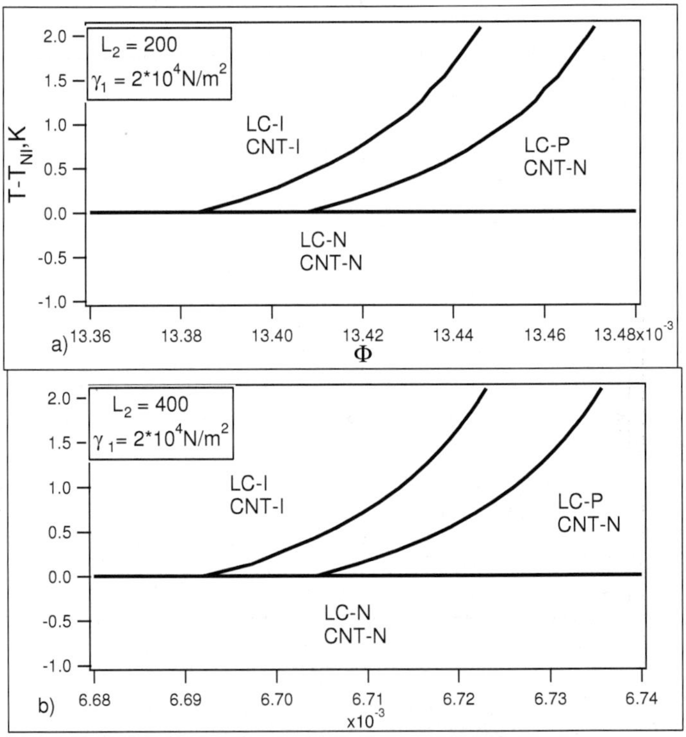

Figure 8. The (T, Φ) phase diagram of a monodisperse system in the *supercritical* regime. Signification of I, P, and N as in Figure 5.

3.3.2. *Bidisperse CNTs*

The (Φ_3, Φ_2) phase diagram of the bidisperse CNTs suspension in the LC in the *subcritical regime* is shown in Figure 9. In Figure 6a, the LC is in the nematic phase ($T - T_{NI} = -0.42$K), while in Figure 9b, the LC is in the isotropic phase ($T - T_{NI} = 1.4$K). Thick lines indicates phase boundary, while the thin lines connects the coexisting pairs (Φ_{2p}, Φ_{3p}) and (Φ_{2n}, Φ_{3n}).

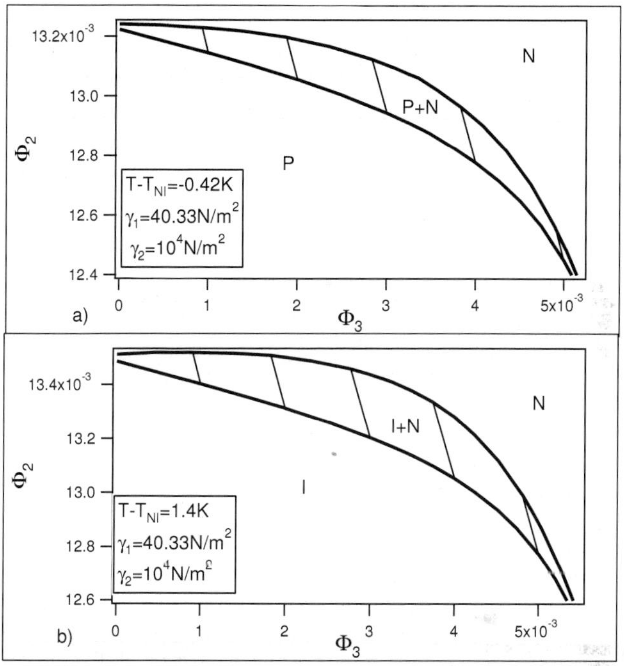

Figure 9. The (Φ_3, Φ_2) phase diagram in the *subcritical* regime for the bidisperse system, in nematic and isotropic phase of LC, respectively. I-isotropic,P-paranematic, and N-nematic.

The Figure 6 reveals two qualitatively different regimes, to which we refer as *decoupled* and *coupled* regime, respectively. In the *decoupled* regime defined by the conditions $\Phi_3 \to 0$ and $\Phi_2 \to 0$, respectively, the system exhibits monodisperse-type behavior. In the limit of very low volume fraction of the longer CNTs ($\Phi_3 \to 0$), the system is monodisperse containing only one species of CNTs of length $L_{02} = 400$ nm. In the limit of very low volume fraction of the shorter CNTs ($\Phi_2 \to 0$), the system is monodisperse containing only one species of CNTs of length $L_{03} = 800$ nm. In these two subregions, due to very small values of Φ_3 and Φ_2, respectively, the coupling term between CNTs species (the γ_{23} term in Eq. (17)) is relatively small and the species are independent. On the contrary, in the *coupled region* (intermediate region in Figure 6), the interaction term in Eq. (17) becomes important and the two CNTs species influence each other. In this region, the volume fraction of the longer CNTs increases in the nematic phase, while that of the shorter CNTs decreases.

In the case of a more realistic value of the coupling parameter between the LC molecules and CNTs $\gamma_1 = 2 * 10^3\ \mathrm{N/m^2} >> \gamma_1^{(c)}$ (*supercritical* regime, the Φ_3, Φ_2) phase diagram for the bidisperse CNTs in the isotropic phase of LC ($T - T_{NI} = 1.4$K is plotted in Figure 10.

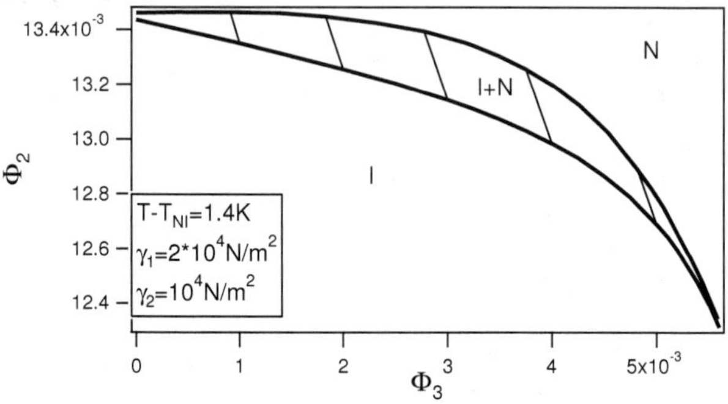

Figure 10. The (Φ_3, Φ_2) phase diagram in the *supercritical* regime for the bidisperse system, in the isotropic phase of LC, respectively. I-isotropic, and N-nematic.

The phase diagram is similar to that of Figure 9 showing again the existence of the *decoupled* and *coupled* regions that we have discussed previously.

4. Conclusions

In the paper we analyze phase behavior of mixtures consisting of LC soft carrier matrix and immersed NPs. A relatively simple phenomenological modeling is used where we focus on isotropic and nematic LC ordering.

In the first part we consider isotropic NPs. We derive the effective free energy of the mixture from which we extract the effective Flory-Huggins parameter. Its structure reveals that on entering nematic ordering phase separation is very probable. However, it could be suppressed by LC-NP interfacial contribution providing that it promotes nematic ordering. From the structure of the average elastic free energy term we also conclude that NPs have in general tendency to assemble at localized sites exhibiting relatively strong elastic distortions. We proceed by studying interaction between a nanoparticle and a topological defect, which is a typical representative of localized strong elastic distortions. It is of interest to identify conditions for which NPs could be effectively trapped to tunable localized distortions. For example, in such a way phase separation could be prevented. In addition, localized elastic distortions could be exploited controlled positional trapping of immersed NPs.

As a model system we use a cylindrical hybrid cell possessing the boojum topological defect. We consider different surface treatment of the nanoparticle and analyze where it is placed in order to minimize total free energy of the system. We find out that a nanoparticle is attracted to a region the structure of which is compatible with configuration enforced by the nanoparticle. Therefore, one could trap NPs to topological defects if its surface enforces configuration resembling the defect core structure. We further show that condensation penalty of forming defects could be in this case significantly reduced due to the *Defect Core Replacement mechanism* [3, 34].

In the second part of the paper we study interaction between nematic LC ordering and CNTs. We extend the mesoscopic model [6, 7] to include length bidispersity of CNTs dispersed in LC in the weak anchoring limit of the coupling between LC molecules and CNTs (in this limit, the coupling is dominated by the anisotropy of the surface tension not by the deformation of the director field). The main conclusions of our study can be summarized as follows:

1. Depending on the coupling between the LC molecules and CNTs (the value of the coupling constant γ_1), two different regimes can be defined: i) if $\gamma_1 < \gamma_1^{(c)}$, (the *subcritical* regime), the nematic-isotropic phase transition of CNTs dispersed in LC is first order and ii) if $\gamma_1 > \gamma_1^{(c)}$ (the *supercritical* regime), the transition is continuous. $\gamma_1^{(c)}$ is the critical value of the coupling parameter depending on the temperature (Figure 1). In both regimes, the isotropic phase of CNTs transforms into a phase with a small degree of ordering, a paranematic phase. Above the critical point this degree of orientational order is strongly increased.
2. The CNTs species are enslaved by the LC (the nematic LC order parameter depends only on the temperature not on the volume fractions of the two species).
3. The longer CNTs are driven into the nematic phase ($\Phi_{3n} - \Phi_{3p} > 0$ in Figures 6 and 7).
4. The longer CNTs induces a larger volume fractions differences $\Phi_{2n} - \Phi_{2p} > 0$ for the shorter CNTs.
5. In the nematic phase, the longer CNTs are more ordered than the shorter ones ($S_{3n} - S_{2n} > 0$).

We emphasize that the last three conclusions are similar with those obtained using the Onsager theory of nematic-isotropic phase transition of the hard rods [22–25], while the first two are specific to the dispersion of CNTs into LC.

Finally, we point out the mesoscopic model for length bidisperse CNTs dispersed in LC presented here is only a first step in considering the important influence of polydispersity on the ordering of CNTs in nematic fluids. Together with considering the attraction interaction between CNTs, this subject will be study in a future work.

Acknowledgments

V.P.-N. thanks to T. J. Sluckin for useful discussion, gratefully acknowledge the hospitality of l'Ecole Normale Supérieure de Lyon and the funding from CNRS. V.B. acknowledge support from the Romanian National Authority for Scientific Research, CNCS - UEFISCDI, project number PN-II-ID-PCE-2011-3-1007.

Author details

Vlad Popa-Nita[1,*], Valentin Barna[1], Robert Repnik[2] and Samo Kralj[2,3]

* Address all correspondence to: v.popanita@gmail.com

1 Faculty of Physics, University of Bucharest, Bucharest, Romania
2 Laboratory Physics of Complex Systems, Faculty of Natural Sciences and Mathematics, University of Maribor, Maribor, Slovenia
3 Jožef Stefan Institute, Ljubljana, Slovenia

References

[1] A.C. Balazs, T. Emrick, T.P. Russell, Science 314, 1107 (2006).

[2] F. Li, O. Buchnev, C. Cheon, A. Glushchenko, V. Reshetnyak, Y. Reznikov, T. J. Sluckin, and J. L. West, Phys. Rev. Lett. 97, 147801 (2006).

[3] B. Rozic, V. Tzitzios, E. Karatairi, U. Tkalec, G. Nounesis, Z. Kutnjak, G. Cordoyiannis, R. Rosso, E.G. Virga, I. Musevic, S. Kralj, Eur. Phys. J. E 34, 17 (2011).

[4] P.G. de Gennes and J. Prost, The Physics of Liquid Crystals, Oxford University Press, Oxford (1993).

[5] B. Rozic, M. Jagodic, S. Gyergyek, M. Drofenik, S. Kralj, G. Lahajnar, Z. Jaglicic, Z. Kutnjak, Ferroelectrics 410, 37 (2011).

[6] P. van der Schoot, V. Popa-Nita, and S. Kralj, J. Phys. Chem. B 112, 4512 (2008).

[7] V. Popa-Nita and S. Kralj, J. Chem. Phys. 132, 024902 (2010).

[8] V. Popa-Nita, M. Cvetko and S. Kralj, in *Electronic Properties of Carbon Nanotubes*, (Edited by Jose Mauricio Marulanda (Intech, 2011).

[9] C. Zakri, Liquid Crystals Today 16, 1 (2011).

[10] S. Zhang and S. Kumar, Small 4, 1270 (2008).

[11] J. P. F. Lagerwall and G. Scalia, J. Mater. Chem. 18, 2890 (2008).

[12] M. Rahman and W. Lee, J. Phys. D: Appl. Phys. 42, 063001 (2009).

[13] M. D. Lynch and D. L. Patrick, Nano Lett. 2, 1197 (2002).

[14] I. Dierking, G. Scalia, and P. Morales, J. of Appl. Phys. 97, 044309 (2005).

[15] J. Lagerwall, G. Scalia, M. Haluska, U. Dettlaff-Weglikowska, S. Roth, and F. Giesselmann, Adv. Mater. 19, 359 (2007).

[16] N. Lebovka, T. Dadakove, L. Lysetskiy, O. Melezhyk, G. Puchkovska, T. Gavrilko, J. baran, M. Drodz, J. Molecular Structure 887, 135 (2008).

[17] M. J. Green, N. Behabtu, M. Pasquali, W. W. Adams, Polymer 50, 4979 (2009).

[18] Y. Sabba and E. L. Thomas, Macromolecules 37, 4815 (2004).

[19] A. M. Somoza, C. Sagui, and C. Roland, Phys. Rev. B63, 081403-1 (2001).

[20] M. J. Green, A. N. G. Parra-Vasquez, N. Behabtu, M. Pasquali, J. Chem. Phys. 131, 084901 (2009).

[21] W. Song and A. H. Windle, Macromolecules 38, 6181 (2005).

[22] H. N. W. Lekkerkerker, Ph. Coulon and R. Van der Haegen, J. Chem. Phys. 80, 3427 (1984).

[23] T. Odijk, Macromolecules 19, 2313 (1986).

[24] G. J. Vroege and H. N. W. Lekkerkerker, Rep. Prog. Phys. 55, 1241 (1992).

[25] H. N. W. Lekkerkerker and G. J. Vroege, J. Philos. Trans. R. Soc. London, Ser A, 344, 419 (1993).

[26] P. Kaiser, W. Wiese, and S. Hess, J. Non-Equilib. Thermodyn. 17, 153 (1992).

[27] P. J. Flory, Proc. R. Soc. A 243, 73 (1956).

[28] S. Kralj, Z. Bradac, and V. Popa-Nita, J. Phys. Condens. Matter 20, 244112 (2008).

[29] V. J. Anderson, E. M. Terentjev, S. P. Meeker, J. Crain, and W. C. K. Poon, Eur. Phys. J E 4, 11 (2011).

[30] S. Kralj, R. Rosso, and E.G. Virga, Phys. Rev. E 78, 031701 (2008).

[31] S. Kralj, R. Rosso, and E.G. Virga, Phys. Rev. E 81, 021702 (2010).

[32] D. Coursault , J. Grand , B. Zappone , H. Ayeb , G. Levi, N. Felidj , and E. Lacaze, Adv. Mater. 24, 1461 (2012).

[33] S. Kralj, E.G. Virga, S. Zumer, Phys.Rev.E 60, 1858 (1999).

[34] H. Kikuchi, M. Yokota, Y. Hisakado, H. Yang, and T. Kajiyama, Nat. Mater. 1, 64 (2002).

[35] L. Onsager, Ann. N. Y. Acad. Sci. 51, 727 (1949).

[36] M. Doi, J. Polym. Sci., Part B: Polym. Phys. 19, 229 (1981); N. Kuzuu and M. Doi, J. Phys. Soc. Jpn. 52, 3486 (1983).

[37] M. Doi and S. F. Edwards, *Theory of Polymer Dynamics* (Clarendon, Oxford, 1989).

[38] P. Oswald and P. Pieranski, *Nematic and Cholesteric Liquid crystals; concepts and physical properties illustrated by experiments* (Taylor and Francis Group, CRC Press, in Liquid Crystals Book Series, Boca Raton, 2005).

Chapter 8

Toward Greener Chemistry Methods for Preparation of Hybrid Polymer Materials Based on Carbon Nanotubes

Carlos Alberto Ávila-Orta, Pablo González-Morones,
Carlos José Espinoza-González,
Juan Guillermo Martínez-Colunga,
María Guadalupe Neira-Velázquez,
Aidé Sáenz-Galindo and Lluvia Itzel López-López

Additional information is available at the end of the chapter

http://dx.doi.org/10.5772/51257

1. Introduction

Recent technological advances and the need for materials with new functionalities and better performance have generated an enormous demand for novel materials. Nanostructures such as carbon nanotubes (CNTs) possess outstanding mechanical, electrical, thermal and chemical properties which make them ideal for a wide variety of current or future applications [1], especially for the preparation of multifunctional *hybrid polymer materials*.

The incorporation of CNTs to polymer matrices have demonstrated to improve the mechanical, electrical, thermal and morphological properties of the produced nanocomposites [2]; however, the full exploitation of CNTs has been severely limited due to difficulties associated with dispersion of entangled CNTs during processing, and their poor interfacial interaction with the polymer matrix. Therefore, significant efforts have been directed toward improving the dispersion of CNTs by means of surface modification either by non-covalent functionalization or covalent functionalization [3].

Most strategies designed to functionalize CNTs involve the use of strong acids as reagents and organic solvents as reaction media, which can become environmental pollution and health hazard problems. Nowadays, the global environmental trends are seeking greener chemistry methods to prepare materials, thus, there is plenty of room for developing environmentally-friendly chemistry methods to functionalize CNTs.

"Green" chemistry is based on the use of a set of principles that reduces or eliminates the use of hazardous reagents and solvents in the design, preparation and application of materials [4]. In this context, functionalization of CNTs using microwaves, plasma, and ultrasound waves are strategies very promising for greener production of hybrid polymer materials, due to shorter reaction times, reduced energy consumption, and better yields.

The focus of this chapter will be on the microwaves, ultrasound and plasma assisted functionalization of CNTs as greener chemistry methods to produce hybrid polymer materials. After a brief overview on preparation of hybrid polymer materials containing CNTs, we will present the physical principles, mechanisms and processing conditions involved in the functionalization of CNTs for each of these *"Green"* chemistry methods, and then present our point of view on challenges and opportunities in both the immediate and long-term future.

2. Hybrid polymer materials

In polymer science, we can define a *hybrid polymer material* as a combination of two or more materials mixed at the nanometer level, or sometimes at the molecular level (0.1 – 100 nm) in a predetermined structural configuration, covering a specific engineering purpose. The term *hybrid material* is used to distinguish them from the conventionally known *composites* that are referred as simple mixtures of two or more materials at micro-scale level (> 1 μm).

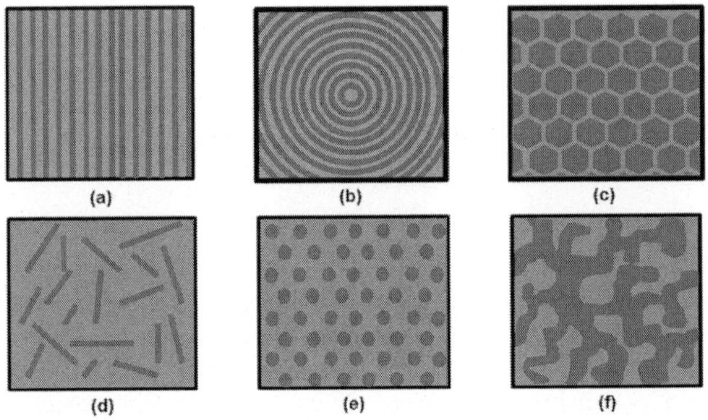

Figure 1. Some examples of structural configurations of hybrids of the composite type: (a) sandwich, (b) concentric cylindrical shells, (c) honeycomb, (d) chopped fibers, (e) particulate, and (f) amorphous blend.

An ideal hybrid polymer material requires an accurate molecular design or structural control of its components in order to obtain synergistic properties. As structural configuration of components moves away from its ideal configuration, the material properties will range from an arithmetic average value (average of the properties of each component) to below of that arithmetic value [5]. Thus, the shape and structural configuration of the components in

http://dx.doi.org/10.5772/51257

a hybrid polymer material play a key role in determining its properties. Figure 1 shows a scheme of hybrids materials composed by two components, in which one of them is arranged so that synergistic properties can be achieved.

The hybrid polymer materials can be classified depending of the nature of interactions between their components. In particular, when structural materials in the form of particles, flakes or fibers are incorporated into polymer matrices, this type of hybrid polymer materials can be classified in (i) *class I* hybrid materials, which show weak interactions between the two components, such as van der Waals, hydrogen bonding or weak electrostatic interactions, and (ii) *class II* hybrid materials, which show covalent interactions between both components such that there is no tendency for the components to separate at their interfaces when the hybrid material is loaded [6].

Hybrid polymer materials containing CNTs have attracted considerable attention due to the unique atomic structure, high surface area-to-volume ratio and excellent electronic, mechanical and thermal properties of carbon nanotubes. Although the incorporation of CNTs to polymer matrices have significantly improved the mechanical, electrical and morphological properties of polymers, there is plenty of room for controlling the structural configuration of the hybrid polymer material, thus, different efforts have been focused in the preparation methods.

3. Polymer-CNTs hybrid materials

3.1. Structural configuration

Since the first ever materials based on polymer-CNTs were reported in 1994 by Ajayan *et al.* [7], several processing methods have been developed for fabricating polymer-CNTs hybrid materials. These methods mainly include solution mixing, *in-situ* polymerization, and melt blending [8].

Because the unique mechanical properties of CNTs, such as the high modulus, tensile strength and strain to fracture, there have been numerous efforts to obtain hybrid materials with improved mechanical properties [2]. Within the structural configurations for this specific application, the "*chopped fibers*" configuration, as seen in Figure 1(d), has been the most desired.

On the other hand, for other unique properties of CNTs such as high electrical and thermal conductivity, the obtaining of multiphase polymer amorphous blends, as seen in Figure 1(f), offers a much higher potential for the development of conductive composites containing CNTs. The selective localization of the CNTs either in one of the blend phases or at the interface of an immiscible co-continuous blend can form an ordered network of conductive phase, creating the so-called segregated systems [9]. In such systems, considerably lower value of percolation threshold compared to "*chopped fibers*" structural configuration can be achieved.

The building of polymer-CNTs hybrid materials with desired structural configurations is potentially promising to develop advanced hybrid materials; however, the full exploitation of properties of CNTs by means the manufacturing of those desired structural configurations has been severely limited, because difficulties associated with dispersion of the entangled CNTs during processing and their poor interfacial interaction with some polymer matrices.

3.2. Chemical and physical functionalization of CNTs

The efficient exploitation of the unique properties associated with CNTs depends on its uniform and stable dispersion in the host polymer matrix, as well as the nature of the interfacial interactions with the polymer. Thus, obtaining of polymer-CNTs hybrid materials with desired properties has represented a great challenge, because CNTs exhibit strong inter-tube van der Waals' forces of attraction that impede its uniform and stable dispersion in the matrix, in addition to certain properties of the polymer matrix like wetting, polarity, crystallinity, melt viscosity, among others [2, 10].

Surface modification of CNTs has been one of the most used strategies in order to improve its affinity with the polymer matrix, and therefore to achieve a better uniform dispersion. These methods have been conveniently divided into chemical functionalization and physical functionalization [3, 11].

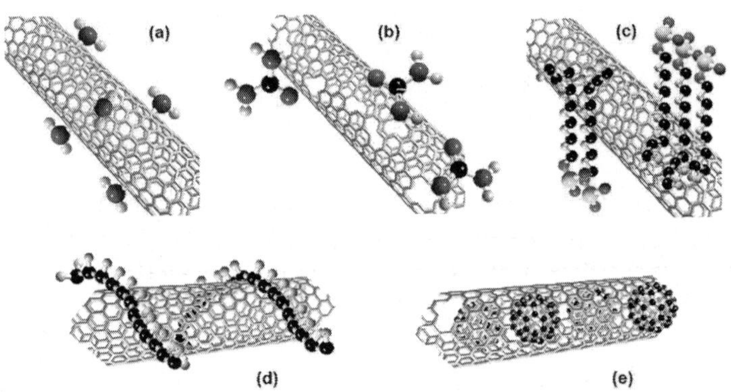

Figure 2. Strategies for chemical and physical functionalization of CNTs: a) covalent sidewall functionalization, b) covalent defect sidewall functionalization, c) non-covalent adsorption of surfactants, d) wrapping of polymers, and e) endohedral functionalization (case for C_{60}).

Chemical functionalization method is based on the covalent linkage of functional groups such as –COOH or –OH on the surface of CNTs. These methods can be also divided in sidewall functionalization and defect functionalization (see Figure 2). The reaction mechanisms that take place at their sidewall include fluorination and derivate reactions, hydrogenation, cycloaddition, and radical (R•) attachment; whilst the reaction mechanisms by amidation, esterification, thiolation, silanization, and polymer grafting (*grafting to* and *grafting from*) takes advantages of chemical transformation of defect sites on CNTs.

Physical functionalization method is based in the formation of non-covalent interactions between molecules and CNTs. These methods include the wrapping of polymer around the CNTs, the physical adsorption of surfactants and the endohedral method (see Figure 2). In the latter, molecules are stored in the inner cavity of CNTs through the capillary effect, where the insertion often takes place at defect sites localized at the ends or on the sidewalls.

In particular, the covalent functionalization of CNTs has been one of the most preferred methods since it allows an efficient interaction between polymer-CNT interface through the functional moieties of the CNTs surface and the available functional groups of the polymer. However, these methods involve rough acid treatment conditions during functionalization which damage the nanotube framework and decrease the electrical conductivity of the hybrid material. In addition, the use acids and organic solvents as the reaction media represent problems of environmental pollution and health hazard.

In this context, the global trend of seeking for "*Green*" chemistry methods is demanding to researchers in the field to develop environment-friendly methods to functionalize CNTs.

4. Greener production of polymer-CNTs hybrid materials

4.1. "*Green*" chemistry: definition and principles

Diverse definitions of "*Green*" chemistry can be found in the literature. According to EPA (Environment Protection Agency) "*Green*" chemistry philosophy speaks of chemicals and chemical processes designed to reduce or eliminate negative environmental impacts, where the use and production of these chemicals may involve reduced waste products, non-toxic components, and improved efficiency. Anastas and Warner [12], who are considered the founders of this field that born in 1990s, define "*Green*" chemistry as the utilization of a set of principles that reduce or eliminates the use or generation of hazardous substances in the design, manufacture and application of chemical products.

The 12 Principles of "*Green*" chemistry (defined by Anastas and Warner) help us think about how to prevent pollution when creating new chemicals and materials:

1. *Prevention*. It is better to prevent waste to treat or clean up waste after it has been created.
2. *Atom Economy*. Synthetic methods should be designed to maximize the incorporation of all materials used in the process into the final product.
3. *Less Hazardous Chemical Synthesis*. Synthetic methods should be designed to use and generate substances that possess little or no toxicity to people or the environment.
4. *Designing Safer Chemicals*. Chemical products should be designed to affect their desired function while minimizing their toxicity
5. *Safer Solvents and Auxiliaries*. The use of auxiliary substances should be made unnecessary whenever possible and innocuous when used.

6. *Design for Energy Efficiency*. Energy requirements of chemical processes should be recognized for their environmental and economic impacts and should be minimized. If possible, synthetic methods should be conducted at ambient temperature and pressure.
7. *Use of Renewable Feedstocks*. A raw material or feedstock should be renewable rather that depleting whenever technically and economically practicable.
8. *Reduce Derivatives*. Unnecessary derivatization like use of blocking group, protection/de-protection, and temporary modification of physical/chemical processes, should be minimized or avoided if possible, because such steps require additional reagents and can generate waste.
9. *Catalysis*. Catalytic reagents should be superior to stoichiometric reagents.
10. *Design for Degradation*. Chemical products should be designed so that at the end of their function they break down into innocuous degradation products and do not persist in the environment.
11. *Real-Time Analysis for Pollution Prevention*. Analytical methodologies need to be further developed to allow for real-time, in-process monitoring and control prior to the formations of hazardous substances.
12. *Inherently Safer Chemistry for Accident Prevention*. Substances and the form of a substance used in a chemical process should be chosen to minimize the potential for chemical accidents, including releases, explosions, and fires.

"*Green*" chemistry is a highly effective approach to pollution prevention since it applies innovative scientific solutions to real-world environmental situations. The preparation of polymer-CNTs hybrid materials can be considered as "*Green*" as more of those principles are applied to the design, production and processing of hybrid materials.

4.2. Greener processing technologies

4.2.1. Microwaves

4.2.1.1. Background and physical principles

Microwaves are electromagnetic waves with wavelengths ranging from 1 mm to 1 m and frequencies between 0.3 GHz and 300 GHz, respectively. 0.915 GHz is preferably used for industrial/commercial microwave ovens and 2.45 GHz is mostly used for household microwave ovens. Since the first ever report of a microwave-assisted organic synthesis in the 80s, it is being further developed and extended to polymer science, in particular in the field of microwave-assisted polymer synthesis and polymer nanocomposites [13].

In polymer chemistry, microwave-assisted reactions present a dramatic increasing in reaction speed and significant improvements in yield compared with conventional heating. These advantages are attributed to instantaneous and direct heating of the reactants, which lead to reduction in reaction time, energy savings and low operating costs. The principles 2, 5, 6 and 11 of '*Green*' chemistry describe these strengths.

http://dx.doi.org/10.5772/51257

How does microwave irradiation lead chemical reactions? When a dielectric material (i.e. molecules containing polar groups in their chemical structure) is placed under microwave irradiation, the dipolar molecules will tend to align their dipole moment along the field intensity vector. As the field intensity vector varies sinusoidally with time, the polar molecules re-align with the electro-magnetic field and generate both translational and rotational motions of the dipoles. These movements generate heat because the internal friction, so a portion of the electromagnetic field is converted in thermal energy.

The power absorbed per unit, P (V/m^3) is expressed as [14]:

$$P = 2\pi f \varepsilon_0 \varepsilon_r^{''} \mid E \mid^2 \tag{1}$$

where f is the microwave frequency (GHz), ε_0 the permittivity of free space (ε_0 = 8.86x10^{-12} F/m), $\varepsilon_r^{''}$ the dielectric loss factor and E (V/m) is the magnitude of the internal field.

The dielectric loss factor is a measurement of the efficiency with which microwave energy is converted into heat, and depends on the dielectric conductivity σ and on the microwave frequency f according to

$$\varepsilon_r^{''} = \sigma / 2\pi f \tag{2}$$

The degree of energy coupling in the reaction system is expressed by the dissipation factor D, which is defined by the loss tangent tan δ

$$D = \tan\delta = \varepsilon_r^{''} / \varepsilon_r^{'} \tag{3}$$

where $\varepsilon_r^{'}$ is the relative dielectric constant and describes the ability of molecules to be polarized by the electric field. Thus, the dissipation factor defines the ability of a medium at a given frequency and temperature to convert electromagnetic energy into heat.

Therefore, the absorbed microwave energy into dielectric material produces the molecular friction, which leads the rapid heating of the reaction medium and the subsequent chemical reactions. The dramatic rate enhancements of these reactions have been explained by means of very well-known Arrhenius law:

$$k = A\exp[-E_a / RT] \tag{4}$$

Some authors have suggested that, the microwave dielectric heating increases the temperature of the medium in a way that cannot be achieved by conventional heating (superheating), so the rate enhancements are considered essentially a result of thermal effects, although the exact temperature reaction has been difficult to determine experimentally [15]. Other authors, however, suggest that the microwave energy produces an increase in molecular vibrations which could affect anyway the pre-exponential factor A, and also produce an alteration in the exponential factor by affecting the activation energy [16, 17].

After 50 years of research, microwave chemistry is still a research field in expansion and also seems it as green technology; however, some questions regarding microwave heating mechanisms remain unsolved. The microwave-assisted production of polymer-CNTs hybrid materials is a recent field of research, in which additional questions have emerged. Beyond to give an overview on microwaves-assisted preparation of hybrid materials, this section is addressed under one of those questions: how could microwave energy be controlled to prepare more efficiently these hybrid materials? As discussed below, the answer to this question is still not understood.

4.2.1.2. Carbon nanotubes-microwaves interaction

Carbon nanotubes have demonstrated to act as highly efficient absorbers of microwave energy, producing heating, outgassing and light emission [18]. Over the past few years, the investigation on microwave heating mechanisms in CNTs has been a focus of interest. It has been proposed that the microwave irradiation might cause heating by two plausible mechanisms [19]: (i) Joule heating and (ii) vibrational heating.

The mechanism of *Joule heating* postulate that the electric field component of the microwave induces the motion of the electrons in electrically conductive impurities present at as-synthesised CNTs such as metallic catalysts, leading a localised superheating at the site of impurities which increase the temperature of CNTs. In addition, another suggested potential source of localized superheating has been the generation of gas plasma from absorbed gases (particularly H_2) in CNTs, introduced during the synthesis phase or via atmospheric absorption.

The sources of superheating in the *Joule heating* mechanism are focus of discussion. It has been argued that the nano-sized magnetic particles should be impacted minimally by microwave irradiation at low frequencies and therefore, plays no significant role in the microwave energy absorption. Paton *et al.* work [18], among others, demonstrated that even with the removal of iron and other catalytic particles, the CNTs still present microwave heating. On the other hand, regarding to gas plasma, it is still unclear if the plasma is directly generated by microwave irradiation or by other superheating effect. Moreover, it is doubtful that plasma be generated under presence of solvents, since their conductivity is higher than air.

Paton *et al.* [18] hypothesized that *Joule heating* mechanism in CNTs can be explained by the motion of free electrons distributed on the surface of the CNTs, induced by the electric field component of the electromagnetic field. This theory was supported by the measurements of DC conductivity of as-synthesised, heat and acid treated CNTs. The microwave energy absorption was significantly increased as the crystallinity and electrical conductivity of CNTs were improved.

Regarding to *vibrational heating* mechanism, Ye [20] described the heating of non-bounded CNTs in terms of non-linear dynamics of a vibrating nanotube. CNTs subjected to microwaves undergo superheating due to transverse vibrations attributed to parametric resonance, similarly to forced longitudinal vibrations of a stretched elastic string. Ye found that

CNTs present a resonance frequency between 2.0 – 2.5 GHz, which is in the region of the frequency of microwaves of the most operating systems used in this field (2.45 GHz). However, the intensity of vibration modes might be attenuated by the presence of impurities, a viscous environment, and highly entangled CNTs.

Both *Joule heating* and *vibration heating* mechanisms help to explain the different obtained results of microwave energy absorption, in the presence of solvents or dry conditions. However, the need of an in-depth understanding of the microwave heating mechanisms is more tangible as microwave-CNTs systems become more complex.

4.2.1.3. Microwaves-assisted functionalization of CNTs

Within the standard procedures to chemically functionalize CNTs is firstly the purification phase. The most common techniques include acid reflux, oxidation and filtration, where most of them involve long processing times or multiple stages, the use of large acid volumes and some cases the structural damage of CNTs [3]. Microwave-assisted purification of CNTs has emerged as promising technique for effective purification of CNTs with minimal damages and significant reduction of the processing times and use of harmful reactants [19].

Purification has been attributed to generation of highly localized temperatures within metallic particles which burst any amorphous carbon coating. During purification phase in conventional techniques, the use of aggressive treatments facilitates the creation of defect sites on sidewall of CNTs, in order to graft desired functional groups; however, in microwave-assisted processes the energy absorbed by CNTs leads the activation of vacancy sites on surface and the subsequent reaction with active functional groups of molecules. At the same time, the microwave irradiation can supply enough energy to reorient any "damaged" sp^3 carbon bonds into sp^2 hybridization, thus leading an increase in CNTs quality.

As described previously, the main goal of functionalization of CNTs in the preparation of hybrid materials is to improve their dispersion degree and interaction with the polymer matrix. Thus, the challenge of microwave-assisted functionalization is to achieve a desired degree of functionalization on CNTs surface, whilst avoiding damages of the structure that could compromise the properties of the final product. A review on recent works in microwave-assisted functionalization of CNTs was published by Ling and Deokar [21], and it is not our intent to duplicate that effort here. Rather, we focus on some *"Green"* key issues that might improve the preparation of hybrid polymer materials through control of microwave energy absorption.

In this context, microwave-assisted functionalization under solvent-free conditions is a promising approach for large-scale functionalization of CNTs and paves the way to greener chemistry, because in the absence of solvents, the CNTs and reagents absorb the microwave energy more directly and so takes full advantage of the strong microwave absorption of such components. In addition, the solvent-free conditions open the possibility to all proposed microwave heating mechanisms, and therefore increase the absorption of microwave energy.

The use of solvent-free conditions involves the use of bulk CNTs, so dealing with the entangled CNTs results more complicated. Although some works have carried out using solvent-free conditions during microwave irradiation [22-25], a pre-dispersion stage of CNTs with ultrasound in solvent systems is still used. Recently, Ávila-Orta *et al.* [26, 27] developed a method of dispersion of nanostructures in gas phase assisted by ultrasound, enhancing the dispersion of bulk CNTs under solvent-free conditions.

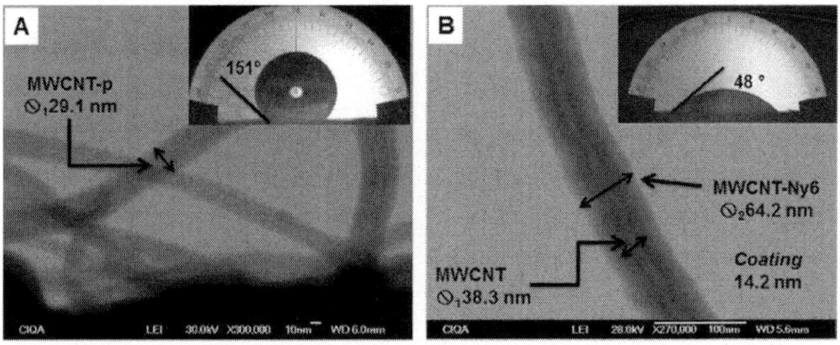

Figure 3. STEM images of MWCNTs. a) pristine MWCNTs (MWCNT-p), and b) functionalized MWCNTs with Nylon (MWCNT-Ny6). [28].

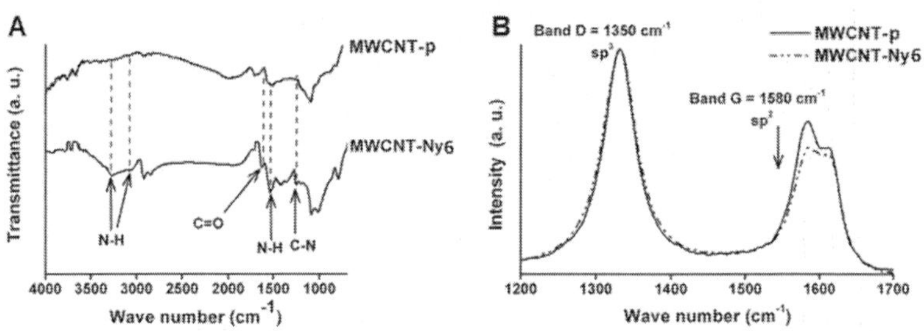

Figure 4. Evidence of the formation and grafting of Nylon-6 on surface of MWCNTs: a) FTIR spectrum, and b) RAMAN spectrum. [28].

González-Morones [28] used the dispersion method developed by Ávila-Orta *et al.* in order to functionalize multi-walled carbon nanotubes (MWCNTs) with Nylon through "*grafting from*" strategy, using ε-caprolactam and aminocaproic acid as monomers. The MWCNTs were previously dispersed into a recipient containing air and then blended with ε-caprolactam powder. The blend was treated for 30 min using a multimodal microwave oven (2.45 GHz) at 250 °C and microwave power of 600 W. Figure 3 shows a STEM image of functionalized carbon nanotubes, in which the average thickness of the polymeric coating was 14.2 nm. The contact angle measurements for pristine and functionalized MWCNTs are also showed in the Figure 3. The reduction in hydrophobic character of MWCNTs-p represented by a decreasing in their contact angle (from 151° to 48°) suggests the presence of a hydrophilic coating. Furthermore, the FTIR spectrum for functionalized MWCNTs shows the presence of the characteristic functional groups of Nylon-6 (Figure 4a), which demonstrates the formation of Nylon-6 by hydrolytic polymerization; whilst the RAMAN spectrum shows a decreasing in the G band intensity (sp^2), suggesting that Nylon-6 are grafted on the surface of MWCNTs through actives sites created during microwave irradiation.

Although the pre-dispersion stage of CNTs in gas phase assisted by ultrasound reduce the consumption of solvent (*"Green"* principle # 5), after the functionalization by microwaves, it is still necessary the use of organic solvents to eliminate the residual monomer (*"Green"* principle # 8). Thus, in order to boost the advantages of this pre-dispersion phase, the efforts should focus on pathways to increase the conversion of reagents (*"Green"* principles # 2, 6 and 9).

4.2.1.4. Preparation of polymer-CNTs hybrid materials

Some efforts have been performed in the preparation of hybrid materials under solvent-free conditions. Virtanen *et al*. developed a hybrid material with a structural configuration sandwich-like (similar to Figure 1a), composed by two polymer plates (extremes) and a film made up from functionalized CNTs (center) which were joined by microwave irradiation [29]. Lin *et al*. obtained hybrid material from microwave-assisted cured process of epoxy resin containing vertically aligned CNTs [30].

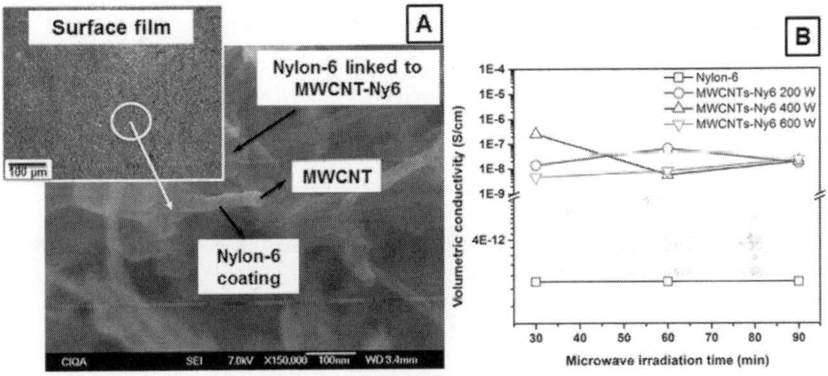

Figure 5. Nylon-6/MWCNTs hybrid material obtained by *in-situ* polymerization assisted by microwaves. a) STEM images of a film made from a hybrid material obtained at microwave power of 600 W, and b) conductivities values of hybrid materials as function of microwave power. [32].

Because a special interest is placed on one-step processes for preparation of those materials, the combined process of in-situ functionalization of CNTs by *"grafting from"* and in-situ bulk polymerization by microwave irradiation becomes a very attractive approach. In recent years, Dr. Ávila-Orta's group has focused on preparation of hybrid materials with electrical conductivity properties; so a structural configuration with an interconnection between CNTs is mostly desired. The combined process described above has demonstrated to be a good approach to enhance this structural configuration.

In this context, Yañez-Macías *et al.* [31, 32] prepared Nylon-6/MWCNTs hybrid material films with high electrical conductivities (values ranged from 10^{-9} to 10^{-7} S/cm) using this combined process. In that work, the MWCNTs were also previously pre-dispersed using the method developed by Ávila-Orta *et al.* The influence of microwave power on polymerization was studied for 200, 400 and 600 W. Figure 5 shows STEM images of a film made from Nylon-6/MWCNTs hybrid material obtained after 90 min of reaction at 600 W. The image shows as the MWCNTs are interconnected and coated by Nylon-6.

From Yañez-Macías *et al.* work, the microwave power intensity demonstrated to play a crucial role in the hydrolytic polymerization. As microwave power intensity increases the yield of Nylon-6 increases, however, at higher microwave power intensity degradation mechanism occurs. These results show that efficient production of polymer-based CNTs in solvent-free conditions can be boosted through control of microwave energy applied to bulk medium.

4.2.1.5. Future perspectives

Microwave irradiation under solvent-free conditions in combination with a pre-dispersion stage of CNTs in gas phase represents a promising approach to large-scale greener production of polymer-based CNTs hybrid materials. The pre-dispersion stage of CNTs allows increasing the efficiency in the microwave energy absorption and the available surface to their functionalization. However, although great efforts have been developed for in-situ preparation of polymer-CNTs hybrids, it is still required to improve the yield.

The control in the yield of functionalization and polymerization reactions can be performed through an in-depth understanding of the mechanisms of microwave heating and kinetic reactions studies. Since the increasing of microwave power intensity increases the temperature of medium reaction, after further research, an optimum microwave energy supply can be found as function of microwave power intensity. In addition, the use of mono-modal microwave ovens can improve the efficiency in the microwave energy absorption, because the microwaves are only concentered in a reaction volume and are not dispersed around chamber volume like multi-modal microwave ovens.

4.2.2. Ultrasound

4.2.2.1. Background and physical principles of sonochemistry

Ultrasound (US) is defined as sound that is beyond human listening range (i.e. 16 Hz to 18 kHz.). In its upper limit, ultrasound is not well defined but is generally considered to 5 MHz for gases and 500 MHz for liquids and solids, and is also subdivided according to applications of interest. The range of 20 to 100 KHz (although in certain cases up to 1 MHz) is designated as the region of high power ultrasound (*sonochemistry*), while the frequencies above 1 MHz are known as high frequency or ultrasound diagnostics (e.g. the imaging technique using echolocation, as SONAR system to detect or US in the health care).

Since the first report on the chemical effects of high power ultrasound in 1927, when Loomis and Richards [33] studied the hydrolysis of dimethyl sulfate and iodine as a catalyst; the studies on chemical effects of ultrasound have further extended to several areas such as organic and organometallic chemistry, materials science, food, and pharmaceutical, among others [34].

The use of ultrasound to accelerate chemical reactions has proven to be a particularly important tool for meeting the "*Green*" Chemistry goals of minimization of waste, reduction of energy and time requirements ("*Green*" principles # 6, 8 and 11). Thus, nowadays the applications of ultrasonic irradiation are playing an increasing role in chemical processes, especially in cases where classical methods require drastic conditions or prolonged reaction times [35].

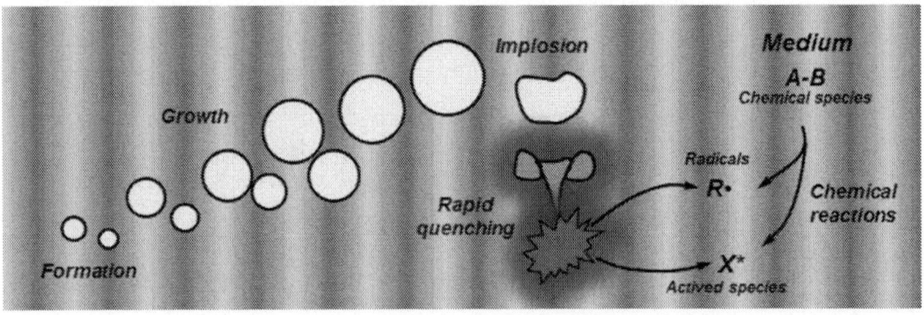

Figure 6. Chemical effects of the high power ultrasound derived from bubble collapse.

The chemical effects of ultrasound in liquids systems are derived from the formation, growth and implosion of small bubbles that appears when the liquid is irradiated by ultrasound waves, phenomenon called "acoustic cavitation" [36]. During bubble collapse, the conversion of kinetic energy of the liquid into thermal energy generates high temperatures (1000 – 10,000 K, most often in the range 4500 to 5500 K) and pressure conditions (~ 500 atm), which lead the formation of free radicals and active species as a result of the heating of the bubble content (Figure 6). On the other hand, the surrounding liquid quenches these portions of the medium in less than 10^{-6} seconds. Thus, the high local temperatures and pressures, combined with extraordinarily rapid cooling, provide a unique means for driving chemical reactions under extreme conditions [34].

A combination between the capability of ultrasonic irradiation to induce chemical reactions and also to achieve a full dispersion of nanostructures in different systems represents a synergistic approach to produce polymer-carbon nanotubes hybrid materials, because surface modification and dispersion of CNTs might take place at the same time; however, unlike microwaves and plasma technologies, there have been very few efforts for exploring it. In this section, we discuss some keys issues associated with the functionalization of CNTs, in order to foster the use of ultrasonic irradiation as greener method for preparation of hybrid materials.

4.2.2.2. Mechanisms for ultrasound activation

The studies on sonochemistry have demonstrated that the ultrasonic irradiation differs from traditional energy sources (such as heat, light or ionizing radiation), so it has been used as a source of alternating activation to assist chemical processes, such as in synthetic methods for obtaining organic molecules and macromolecules and inorganic [37, 38], extraction of natural and synthetic products [39], and medicine [40]. The enormous local temperatures, pressures as well as the heating and cooling rates generated during bubble collapse provide an unusual mechanism for generating high-energy chemistry. However, despite chemical effects of ultrasound have been studied for many years, the mechanisms underlying these effects are too complex and not well-understood.

It has established that during bubble cavitation, three sites for chemical reactions can be identified [41]: i) the interior of the bubble, ii) the interface region at around the bubble surface, and iii) the liquid region outside the interface region (Figure 6). In the interior of a bubble, volatile solute is evaporated and dissociated due to extreme high temperature, where depending of the nature of the system, different free radicals and excited species are generated. Those chemical species with a relatively long lifetime can diffuse out of the interface region and chemically react with solutes or the bulk medium; whilst in the interface region, in addition to high temperatures due to the thermal conduction from the heated interior of a bubble, the presence of relatively short lifetime species such as OH• and O• can lead more interesting chemical reactions.

When a solid material is present in a cavitation medium, the high speed of the liquid jet generated during bubble collapse produces a violent impact on solid surface, in which some material can be removed (e.g. ultrasonic cleaning processes). On the other hand, from a chemical point of view, the shock waves emitted by the pulsating bubbles and the liquid flow around the bubble enhance a mass transfer toward the solid surface during bubble collapse, so the free radicals and the active species generated are available to induce different chemical reactions on solid surface.

4.2.2.3. Ultrasound-assisted functionalization of CNTs

One of the most common methods to functionalize CNTs is acid treatment at elevated temperatures. In this process, functional groups such as hydroxyl (-OH), carboxyl (-COOH), and carbonyl (-C=O) can be introduced into a carbon nanotube network through their physical defects and sites with imperfections. In particular, electron-spin-resonance (ESR) studies on the acid-oxidized CNTs demonstrated that sites with unpaired electrons are generated on CNTs surface, and are significantly increased when acid-treatment functionalization is assisted by high power ultrasound. Moreover, Cabello-Alvarado *et al.* [42] reported the ultrasound-assisted functionalization oxidation of MWCNTs using H_2SO_4/HNO_3. The MWCNTs were subjected to ultrasonic radiation by 8 hours at 60 °C, obtaining similar results to those reported using high temperatures H_2SO_4/HNO_3 mixture [43].

Ultrasound-assisted acid-treatment functionalization is an ideal alternative for reducing reaction conditions and increase rates of reaction, but the use of strong acids as reagents does not contribute largely to "*Green*" chemistry. However, the high speed of the liquid jet generated during bubble collapse can be strong enough to disperse the CNTs agglomerates and also rupture some covalent carbon-carbon bonds of CNTs, and so, generates those sites with unpaired electrons or "active sites" and subsequently induces desired chemical reactions with the surrounding molecules (Figure 7). As proof of this, in 2006, Chen and Tao reported the functionalization of SWCNTs with polymethyl methacrylate by *"grafting from"* [44]. In that work, SWCNTs were irradiated by ultrasound in methyl methacrylate monomer and polymer grafted CNTs were obtained by *in-situ* sonochemically initiated radical polymerization.

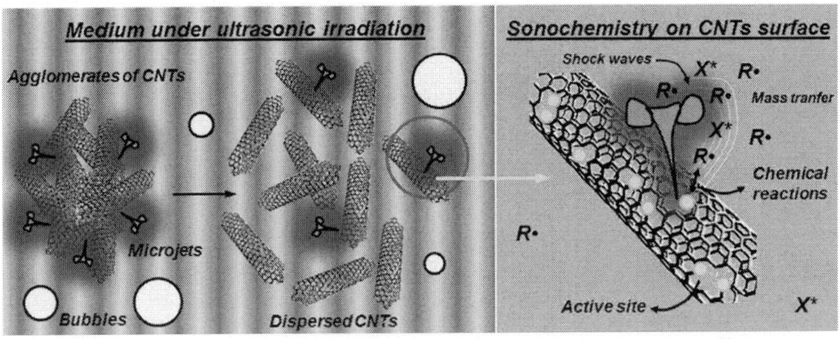

Figure 7. Scheme of the activation mechanism of the CNTs surface treated with ultrasound.

However, the damages of the sp^2-carbon network derived from rupture of carbon-carbon bonds trends to reduce both mechanical and electrical properties of CNTs. Therefore, further research on new pathways to preserve such properties is required. In this context, recently Gebhardt *et al.* [45] developed a novel covalent sidewall functionalization method of CNTs that allowing preserves the integrity of the entire σ-framework of SWCNTs in contrast to classical oxidation. The reductive carboxylation of SWCNTs under ultrasonic treatment resulted in a highly versatile reaction with respect to electronic type selectivity, since functionalization occurs preferentially on semiconducting CNTs. Also, the degree of functionalization can be controlled thought handling of external variables such as pressure.

The emerging on new pathways on ultrasound-assisted functionalization methods could displace to conventional acid treatment methods of CNTs and therefore, opening the possibility for more efficient and greener chemistry methods.

4.2.2.4. Preparation of polymer-CNTs hybrid materials by sonochemistry

The preparation of polymer-CNTs hybrid materials assisted by ultrasound is a feasibility approach since ultrasound has influence on the dispersion of the CNTs and activation of their surface, thereby facilitating interaction between the polymer and the CNTs. In addition, the sonochemical activation can lead polymerization reactions, so offers more attractive features such as low reaction temperatures and short reaction times compared with conventional methods.

Thus, the obtaining of polymer-CNTs hybrid materials by *in-situ* bulk polymerization assisted by ultrasound represents a viable method to exploit all these features: i) a full dispersion of CNTs can be obtained in monomer solution, at same time that ii) the effects of bubble collapse activates the surface of CNTs and lead the *in-situ* functionalization with monomer molecules, in which also iii) the polymerization is started sonochemically. Park *et al.* reported the preparation of poly(methyl methacrylate) (PMMA)-MWCNTs nanocomposites with AIBN and different content of MWCNTs [46]. The molecular weight of polymer matrix increased as MWCNTs content was increased due to generation of initiator radicals on CNTs surface. Kim *et al.* prepared polystyrene-MWCNTs nanocomposites without any added initiator, in which an electrical percolation threshold less than 1 wt% was obtained [47].

Although the role of CNTs as effective initiators and control agents for radical polymerizations have been recently demonstrated by Gilbert *et al.* [48], the ultrasound irradiation is more widely used as an alternative method of dispersion, but not as a source of sonochemical activation, leaving large areas of opportunities to explore.

Ultrasound technology is also applied in dispersion and preparation of nanocomposites in melt blending; however the concepts of sonochemistry and surface activation of CNTs that address this section might not be applied to such systems, due to cavitation phenomenon is dramatically suppressed and the chemical effects might be dominated by mechanochemical phenomenon different from cavitation [49].

4.2.2.5. *Future perspectives*

Ultrasound is a viable source of "*Green*" activation in the context of "*Green*" chemistry, since it has demonstrated to promote low reaction temperatures, faster reaction rates and higher yields in functionalization and polymerization processes. Notably, each of these chemical processes require a source of activation that efficiently furnishes the energy necessary to activate the carbon-carbon bonds in the CNTs network, so in order to perform the "*activation*" of the surface of the CNTs, further research on mechanism of interaction between ultrasound-CNTs has to be addressed taking into account factors such as ultrasound power, environmental solvents, temperature and being one of the most important the sonication time. In particular, the use of environmental solvents could be an interesting factor due the solvent is crucial in the process of bubble cavitation.

4.2.3. *Plasma*

Plasma, the fourth state of the matter, is generated when atoms and molecules are exposed to high sources of energy such as those produced by direct current (DC), alternating current (AC), microwaves and radio frequency (RF). The absorbed energy by such atoms and molecules induces particle collision processes which generate electrons, photons, and excited atoms and molecules. Because of the unique active species present in the plasma, this state of the matter is used in the synthesis [50] and surface modification of CNTs [51].

Regarding to preparation of polymer-CNTs hybrid materials, unlike microwaves and ultrasound technologies described earlier, plasma is mainly used for functionalization of CNTs in order to later use them in the preparation of hybrid materials by *in-situ* polymerization [52, 53] and melt blending [54]. As described before, the surface modification of CNTs leads to improve the dispersion and interactions between CNTs and polymer matrix and represent a critical issue to enhance properties of polymer-CNTs hybrid materials; thus, this section is focused

http://dx.doi.org/10.5772/51257

on the role of plasma-CNTs interactions in the functionalization of CNTs by polymer grafting, in order to exploit more efficiently the unique properties of CNTs (Green principles # 2, 6).

4.2.3.1. Plasma-assisted functionalization of CNTs

The functionalization mechanisms of CNTs by plasma are carried out by both *etching* and *coating* processes [55]. A physical etching process is presented when ions bombard to CNTs leading erosion on CNTs surface and therefore vacancies or "active sites", whilst a chemical etching process is induced by surface reactions between reactive ions and CNTs. Chemical processes are limited to non-inert plasma gases, whilst both inert and reactive gases can produce the physical effects. On the other hand, the coating process results from physical or chemical deposition of the active species present at plasma environment, in which the chemical deposition is induced through the active sites generated from the etching process. A scheme of these mechanisms of functionalization of CNTs is presented in Figure 8.

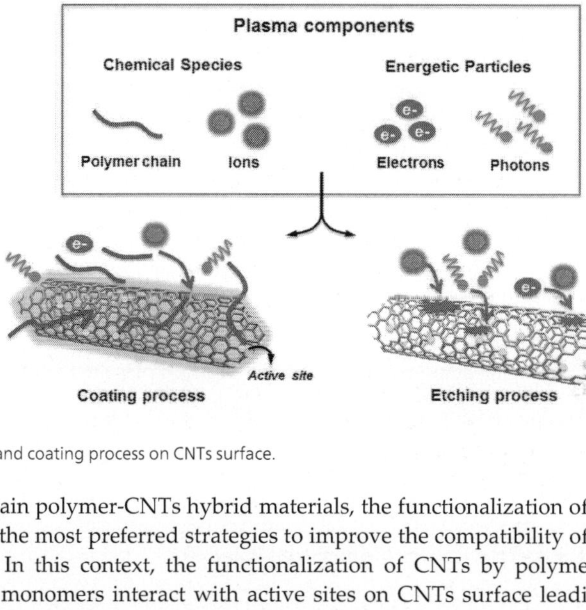

Figure 8. Etching and coating process on CNTs surface.

In order to obtain polymer-CNTs hybrid materials, the functionalization of CNTs with polymers is one of the most preferred strategies to improve the compatibility of CNTs with polymer matrices. In this context, the functionalization of CNTs by polymer grafting occurs when ionized monomers interact with active sites on CNTs surface leading the growth of polymer chains (grafting from/etching process), and also when polymer chains present into plasma environment are physically or chemically deposited on CNTs surface (grafting to/coating process). However, both etching and coating process can occur simultaneously, thus the surface coating can retard the etching process whilst etching can remove the surface coating depending of the plasma gas chemistry.

The surface modification of CNTs based on plasma polymerization presents some advantages compared to wet chemical processes: i) surface modification without altering CNTs bulk properties ("*Green*" principles # 2, 6), ii) large amount of reagents and extreme temperatures are avoided ("*Green*" principles #3, 5 and 6), and iii) a product with no or very low amount of residual monomer can be obtained ("*Green*" principles # 2 and 8).

Despite these "*Green*" advantages, there are few reports about functionalization of CNTs based on plasma polymerization. Chen *et al.* reported the functionalization of CNTs using the monomers acetaldehyde and ethylenediamine [51]; Shi *et al.* reported the plasma deposition of polypyrrole on CNTs surface [56]; Ávila-Orta *et al.* modified MWCNTs using ethylene glycol as monomer [57]; and more recently, Chen *et al.* reported the preparation of MWCNTs grafted with polyacrylonitrile [58], and Rich *et al.* reported the surface modification of MWCNTs using methyl methacrylate and allylamine as monomers [59].

Because of the structural and chemical character of the polymer coating play an important role on interaction between CNTs and polymer matrix, the structural and chemical nature of the polymer coating obtained by plasma can be controlled through processing parameters. Recently, González-Morones [28] studied the effect of power excitation on chemical nature of the polymer deposited on CNTs surface by plasma polymerization of acrylic acid. In that work, firstly the CNTs were pre-dispersed using the method developed by Ávila-Orta *et al.* [27], then the CNTs were exposed to acrylic acid plasma. It was observed that at low power excitation (20 W) the CNTs surface is partially coated by polyacrylic acid and –COOH groups. At power excitation of 40 W, the polyacrylic acid and -COOH groups are mostly removed since the etching process is favored. Whilst at power excitation of 100 W, the CNTs are partially coated by the polymer and functional groups, thus suggesting a competition between etching and coating process. Figure 9a shows FTIR spectra of the functionalized CNTs at different power excitation. For each case, the solubility in water of the functionalized CNTs is showed in a photograph. A STEM image of the polymer coating on the CNT surface obtained at 20 W is also showed in Figure 9b.

Within the actions that can be executed in order to increase the efficiency of the plasma polymerization process and achieving some of the "*Green*" principles are the following:

1. *Frequency and power excitation:* An increase in value of these factors will result in an increase in both the level of ionization of the species and polymer deposition rates. ("*Green*" principles # 2, 6 and 8).

2. *Monomer flow rate/pressure:* Low flow rates and pressure lead an increase in the level of ionization of the species. ("*Green*" principles # 2 and 8).

3. *Geometry of the plasma reactor:* Reactors with cylindrical geometry (without corners) distribute homogeneously the generated plasma. ("*Green*" principles # 6, 8 and 12).

4. *Temperature of substrate:* Very low temperatures promote a higher insertion of monomer molecules on CNTs surface with minimal molecular modifications. ("*Green*" principles # 2, 6, 8).

5. *Treatment time:* This factor will depend on desired final structure and morphology. Long treatment times could produce thick polymer coatings or a rugous surface. The later, due to competition between the etching and coating process. ("*Green*" principle # 2).

Another interesting modification to plasma polymerization process includes the pre-treatment of CNTs with an inert of non-inert gas in order to induce the etching process and the saturation of the CNTs surface with active sites. Subsequently, the active CNTs are subjected to monomer plasma so a better polymer coating can be achieved [58].

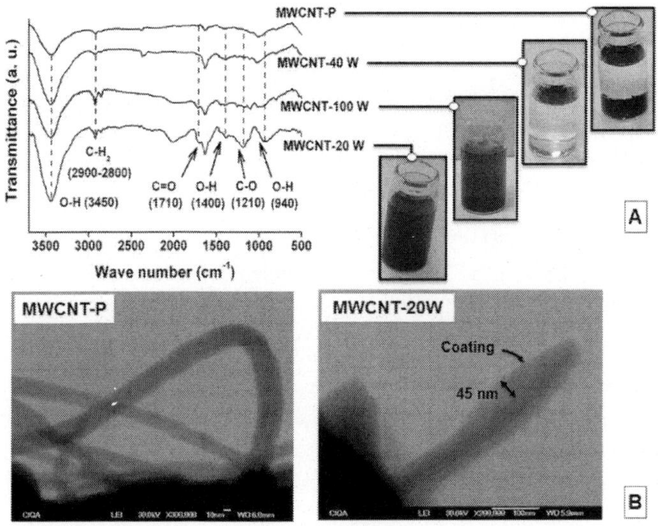

Figure 9. Functionalized CNTs with polyacrylic acid by plasma polymerization. a) FTIR spectra and tests of solubility in water. b) STEM images of pristine MWCNTs (left) and MWCNTs functionalized at 20W (right). [28].

Plasma polymerization is a complex process which makes difficult to achieve an efficient functionalization of CNTs. Due to high number of involved factors, one or two factors have been only studied.

4.2.3.2. Future perspectives

Plasma-assisted functionalization has demonstrated to be a successful method for creating environmentally friendly polymer coatings on CNTs surface. The functionalization of CNTs with desired structural and chemical characteristics can be performed by means of control of the involved processing conditions; however, there is a need for a complete understanding of the interactions between plasma-CNTs which allow controlling successfully the etching and coating process.

A better understanding of the plasma-CNTs interactions can be enhanced if the efficiency of the plasma-CNTs interactions is improved. The stages of pre-dispersion and pre-activation of CNTs should be added previous to plasma polymerization, which allow increasing the surface area exposed to plasma as well as the interactions with the active species, respectively. Further research on such issues could launch to plasma polymerization process to large-scale applications.

5. Summary and final remarks

The properties of CNTs and its application in the preparation of polymer hybrid materials have been very active research fields over the last decade. The preparation of polymer-CNTs hybrid materials faces considerable research challenges because their performance depends on the dispersion of CNTs in the polymer matrix and interfacial interactions between the CNTs and the polymer. In order to achieve the maximum performance, the surface modification of CNTs through several functionalization methods is one of the most used strategies to achieve this goal. Nowadays, the global environmental trends provide convincing reasons for exploring greener chemistry methods for functionalization of CNTs, which represent additional research challenges.

Microwave, ultrasound and plasma technology have emerged as promising green approaches since they have demonstrated to reduce the energy consumption, shorter reaction times and increase yields. These greener methods offer a range of energies that are not available from other conventional sources such as thermal, so leave open the possibility to explore more efficient functionalization routes, taking into account the 12 principles of "*Green*" chemistry as a framework.

Despite the progress described within this chapter, there are still considerable research challenges within this field that remain to be addressed. The successful functionalization of CNTs depends on availability of active sites on CNTs surface; however, the mechanisms that generate these sites remain unclear. Therefore, an in-depth understanding of the mechanisms of interaction between microwave/ultrasound/plasma-CNTs is required. For satisfying this need, further studies on correlation between the irradiation conditions and the level of the functionalization are suggested, in order to establish the most efficient and greener conditions for each particular system.

In addition, the incorporation of a dispersion stage of CNTs in gas phase previous to irradiation is highly recommended, because the surface area exposed to irradiation is increased and therefore the interactions with the matter are highly promoted.

As a general statement, the "*Green*" chemistry methods based on microwave, ultrasound and plasma energy can be easily incorporated to emerging fields of nanotechnology, in particular into strategies for preparation polymer-CNTs hybrid materials. However, there are still a number of important research challenges to study.

Acknowledgements

The authors acknowledge the financial support from project 132699 funded by CONACyT, México.

Author details

Carlos Alberto Ávila-Orta[1*], Pablo González-Morones[1], Carlos José Espinoza-González[1], Juan Guillermo Martínez-Colunga[2], María Guadalupe Neira-Velázquez[3], Aidé Sáenz-Galindo[4] and Lluvia Itzel López-López[4]

*Address all correspondence to: cavila@ciqa.mx

1 Department of Advanced Materials, Research Center for Applied Chemistry

2 Department of Plastics Transformation Processing, Research Center for Applied Chemistry

3 Department of Polymer Synthesis, Research Center for Applied Chemistry

4 Department of Organic Chemistry, Autonomous University of Coahuila, Saltillo, México

References

[1] Jorio, A., Dresselhaus, G., & Dresselhaus, M. S. (2008). *Carbon Nanotubes: Advanced Topics in the Synthesis, Structure, Properties and Applications*, Berlin, Springer.

[2] Bose, S., Khare, R. A., & Moldenaers, P. (2010). Assessing the Strengths and Weaknesses of Various Types of Pre-treatments of Carbon Nanotubes on the Properties of Polymer/Carbon Nanotubes Composites: A Critical Review. *Polymer*, 51(5), 975-993.

[3] Jeon, I. Y., Chang, D. W., Kumar, N. A., & Baek, J. B. (2011). Functionalization of Carbon Nanotubes. In: Yellampalli S, (ed). Carbon Nanotubes- Polymer Nanocomposites Rijeka InTech Available from http://www.intechopen.com/books/carbon-nanotubes-polymer-nanocomposites/functionalization-of-carbon-nanotubes(accessed 25 May 2012), 91-110.

[4] Dahl, J. A., Maddux, B. L. S., & Hutchison, J. E. (2007). Toward Greener Nanosynthesis. *Chemical Reviews*, 107(6), 2228-2269.

[5] Ashby, M. F., & Bréchet, Y. J. M. (2003). Designing Hybrid Materials. *Acta Materialia*, 51(19), 5801-5821.

[6] Kickelbick, G. (2007). *Hybrid Materials*, Weinheim, Wiley-VCH.

[7] Ajayan, P. M., Stephan, O., Colliex, C., & Trauth, D. (1994). Aligned Carbon Nanotube Arrays Formed by Cutting a Polymer Resin-Nanotube Composite. *Science*, 265(5176), 1212-1214.

[8] Spitalsky, Z., Tasis, D., Papagelis, K., & Galiotis, C. (2010). Carbon Nanotube-Polymer Composites: Chemistry, Processing, Mechanical and Electrical Properties. *Progress in Polymer Science*, 35(3), 357-401.

[9] Mamunya, Y. Carbon Nanotubes as Conductive Filler in Segregated Polymer Composites- Electrical Properties. In: Yellampalli S, (ed). Carbon Nanotubes- Polymer Nanocomposites Rijeka InTech (2011). Available from http://www.intechopen.com/books/carbon-nanotubes-polymer-nanocomposites/carbon-nanotubes-as-conductive-filler-in-segregated-polymer-composites-electrical-properties(accessed 25 May 2012), 173-196.

[10] Rahmat, M., & Hubert, P. (2011). Carbon Nanotube-Polymer Interactions in Nanocomposites: A Review. *Composites Science and Technology*, 72(1), 72-84.

[11] Chen, L., Xie, H., & Yu, W. Functionalization Methods of Carbon Nanotubes and Its Applications. In: Marulanda JM, (ed). Carbon Nanotubes Applications on Electron Devices Rijeka InTech (2011). Available from http://www.intechopen.com/books/carbon-nanotubes-applications-on-electron-devices/functionalization-methods-of-carbon-nanotubes-and-its-applications(accessed 25 May 2012)

[12] Anastas, P. T., & Warner, J. C. (1998). *Green Chemistry. Theory and Practice*, New York, Oxford University Press.

[13] Kempe, K., Becer, C. R., & Schubert, U. S. (2011). Microwave-Assisted Polymerizations: Recent Status and Future Perspectives. *Macromolecules*, 44(15), 5825-5842.

[14] Sutton, W. H. (1989). Microwave Processing of Ceramic Materials. *American Ceramic Society Bulletin*, 68(2), 376-386.

[15] Kappe, C. O. (2004). *Controlled Microwave Heating in Moderm Organic Synthesis* (Angewandte Chemie International Edition), 43(46), 6250-6284.

[16] Jacob, J., Chia, L. H. L., & Boey, F. Y. C. (1995). Thermal and Non-Thermal Interaction of Microwave Radiation with Materials. *Journal of Materials Science*, 30(21), 5321-5327.

[17] Binner, J. G. P., Hassine, N. A., & Cross, T. E. (1995). The Possible Role of the Pre-exponential Factor in Explaining the Increased Reaction Rates Observed During the Microwave Synthesis of Titanium Carbide. *Journal of Materials Science*, 30(21), 5389-5393.

[18] Paton, K. R., & Windle, A. H. (2008). Efficient Microwave Energy Absorption by Carbon Nanotubes. *Carbon*, 46(14), 1935-1941.

[19] Mac, Kenzie. K., Dunens, O., & Harris, A. T. (2009). A Review of Carbon Nanotube Purification by Microwave Assisted Acid Digestion. *Separation and Purification Technology*, 66(2), 209-222.

[20] Ye, Z. (2005). Mechanism and the effect of microwave-carbon nanotube interaction. *PhD thesis*, University of North Texas.

[21] Ling, Y. C., & Deokar, A. (2011). Microwave-Assisted Preparation of Carbon Nanotubes with Versatile Functionality. In: Marulanda JM, (ed). Carbon Nanotubes Applications on Electron Devices Rijeka InTech Available from http://www.intechopen.com/books/carbon-nanotubes-applications-on-electron-devices/microwave-assisted-preparation-of-carbon-nanotubes-with-versatile-functionality(accessed 15 May 2012), 127-142.

[22] Brunetti, F. G., Herrero, M. A., Muñoz, JdM., Giordani, S., Díaz-Ortiz, A., Filippone, S., et al. (2007). Reversible Microwave-Assisted Cycloaddition of Aziridines to Carbon Nanotubes. *Journal of American Chemical Society*, 129(47), 14580-14581.

[23] Brunetti, F. G., Herrero, M. A., Muñoz, JdM., Díaz-Ortiz, A., Alfonsi, J., Meneghetti, M., et al. (2008). Microwave-Induced Multiple Functionalization of Carbon Nanotubes. *Journal of American Chemical Society*, 130(25), 8094-8100.

[24] Economopoulos, S. P., Pagona, G., Yudasaka, M., Iijima, S., & Tagmatarchis, N. (2009). Solvent-Free Microwave-Assisted Bingel Reaction in Carbon Nanohorns. *Journal of Materials Chemistry*, 19, 7326-7331.

[25] Rubio, N., Herrero, M. A., Meneghetti, M., Díaz-Ortiz, Á., Schiavon, M., Prato, M., et al. (2009). Efficient Functionalization of Carbon Nanohorns via Microwave Irradiation. *Journal of Materials Chemistry*, 19, 4407-4413.

[26] González-Morones, P., Yañez-Macias, R., Navarro-Rodríguez, D., Ávila-Orta, C., & Quintanilla, M. L. (2011). Study of Dispersion of Carbon Nanotubes in Gas Phase by Ultrasound and its Effect on Plasma Surface Modification. *Ide@s CONCYTEG*, 6, 727-738.

[27] Ávila-Orta, C., Neira-Velázquez, M., Borgas-Ramos, J., Valdéz-Garza, J., González-Morones, P., & Espinoza-González, C. (2010). CIQA, assignee Process of desagglomeration, fragmentation and size reduction of agglomerated nanoparticles in gas phase assisted by ultrasound. México patent MX/a/2010/014326 (20 December 2010)

[28] González-Morones, P. (2011). Nylon-6/MWCNTs nanocomposites: Functionalization and hybridization. *PhD thesis*, Research Center for Applied Chemistry.

[29] Virtanen, J., Tilli, M., & Keinanen, . (2006). Novel hyibride materials and related methods and devices. U.S.A. patent WO 2006040398 (2 July 2009)

[30] Lin, W., Moon, K. S., & Wong, C. P. (2009). A Combined Process of In-Situ Functionalization and Microwave Treatment to Achieve Ultrasmall Thermal Expansion of Aligned Carbon Nanotube-Polymer Nanocomposites: Toward Applications as Thermal Interface Materials. *Advanced Materials*, 21(23), 2421-2424.

[31] Yañez-Macías, R., González-Morones, P., Ávila-Orta, C., Torres-Rincón, S., Valdéz-Garza, J., Rosales-Jasso, A., et al. (2011, August). Polymer nanohybrids with high

electrical conductivities. Cancún, México. *IMRC meeting, MRS proceedings*, 14-19, Cambridge University Press, 2012.

[32] Yañez-Macías, R. (2011). Polymerization of Nylon-6 on surface of MWCNTs by microwave irradiation. *Master thesis*, Research Center for Applied Chemistry.

[33] Richards, W. T., & Loomis, A. L. (1927). The Chemical Effects of High Frequency Sound Waves. I. A Preliminary Study. *Journal of the American Chemical Society*, 49(12), 3086-3089.

[34] Suslick, K. S. (1989). The Chemical Effects of Ultrasound. *Journal of the American Chemical Society*, 111(6), 2342-2344.

[35] Pizzuti, L., Franco, M. S. F., Flores, A. F. C., Quina, F. H., & Pereira, C. M. P. Recent Advances in the Ultrasound-Assisted Synthesis of Azoles. In: Kidwai M, (ed). Green Chemistry- Environmentally Benign Approaches Rijeka InTech (2012). Available from http://www.intechopen.com/books/green-chemistry-environmentally-benign-approaches/recent-advances-in-the-ultrasound-assisted-synthesis-of-azoles(accessed 25 May 2012), 81-102.

[36] Peters, D. (1996). Ultrasound in Materials Chemistry. *Journal of Materials Chemistry*, 6, 1605-1618.

[37] Ovalle, R. J. (2009). Synthesis and study of aromatic and aliphatic amides obtained through green chemistry. *Master thesis*, Autonomous University of Coahuila.

[38] Sánchez, E. B. N. (2010). Obtaining such anilia aromatic amides using green methods with potential pharmacological application. *Master thesis*, Autonomous University of Coahuila.

[39] Keith, L. H., Gron, L. U., & Young, J. L. (2007). Green Analytical Methodologies. *Chemical Reviews*, 107(6), 2695-2708.

[40] Qu, W., Ma, H., Jia, J., He, R., Luo, L., & Pan, Z. (2012). Enzymolysis Kinetics and Activities of ACE Inhibitory Peptides from Wheat Germ Protein Prepared with SFP Ultrasound-Assisted Processing. *Ultrasonics Sonochemistry*, 19(5), 1021-1026.

[41] Mason, T. J. (1999). *Sonochemistry*, Oxford University Press.

[42] Cabello, C., Sáenz, A., López, L., & Ávila, C. (2011). Surface Modification of (MWCNTs) with Ultrasound H_2SO_4/HNO_3 . *Afinidad*, 68(555), 370-374.

[43] Aravind, S. S. J., Baskar, P., Baby, T. T., Sabareesh, R. K., Das, S., & Ramaprabhu, S. (2011). Investigation of Structural Stability, Dispersion, Viscosity, and Conductive Heat Transfer Properties of Functionalized Carbon Nanotube Based Nanofluids. *Journal of Physical Chemistry C*, 115(34), 16737-16744.

[44] Chen, W., & Tao, X. (2006, 27 Nov- 1 Dec). Ultrasound-induced functionalization and solubilization of carbon nanotubes for potential nanotextiles applications. Boston, MA. *MRS Fall Meeting: MRS Proceedings*, Cambridge University Press, 0920-S0902-0902.

[45] Gebhardt, B., Hof, F., Backes, C., Muller, M., Plocke, T., Maultzsch, J., et al. (2011). Selective Polycarboxylation of Semiconducting Single-Walled Carbon Nanotubes by Reductive Sidewall Functionalization. *Journal of American Chemical Society*, 133(48), 19459-19473.

[46] Park, C., Ounaies, Z., Watson, K. A., Crooks, R. E., Smith, J., Lowther, S. E., et al. (2002). Dispersion of Single Wall Carbon Nanotubes by In Situ Polymerization under Sonication. *Chemical Physics Letters*.

[47] Kim, S. T., Choi, H. J., & Hong, S. M. (2007). Bulk Polymerized Polystyrene in the Presence of Multiwalled Carbon Nanotubes. *Colloid and Poymer Science*, 285(5), 593-598.

[48] Gilbert, A. C. C., Derail, C., Bounia, N. E. E., & Billon, L. (2012). Unexpected Behaviour of Multi-Walled Carbon Nanotubes During "In-Situ" Polymerization Process: When Carbon Nanotubes Act as Initiators and Control Agents for Radical Polymerization. *Polymer Chemistry*, 3, 415-420.

[49] Ávila-Orta, C., Espinoza-González, C., Martínez-Colunga, J. G., Bueno-Baqués, D., Maffezzoli, A., & Lionetto, F. (2012). An Overview of Progress and Current Challenges in Ultrasonic Treatments of Polymer Melts. *Advances in Polymer Technology*, Accepted for publication (13 July 2012).

[50] Lee, S. J., Baik, H. K., Yoo, J. E., & Han, J. H. (2002). Large Scale Synthesis of Carbon Nanotubes by Plasma Rotating Arc Discharge Technique. *Diamond and Related Materials*, 11(3-6), 914-917.

[51] Chen, Q., Dai, L., Gao, M., Huang, S., & Mau, A. (2001). Plasma Activation of Carbon Nanotubes for Chemical Modification. *Journal of Physical Chemistry B*, 105(3), 618-622.

[52] Cruz-Delgado, V. J., España-Sánchez, B. L., Ávila-Orta, C. A., & Medellín-Rodríguez, F. J. (2012). Nanocomposites Based on Plasma-Polymerized Carbon Nanotubes and Nylon-6. *Polymer Journal*, doi:10.1038/pj.2012.49.

[53] Lee, J. I., Yang, S. B., & Jung, H. T. (2009). Carbon Nanotubes–Polypropylene Nanocomposites for Electrostatic Discharge Applications. *Macromolecules*, 42(21), 8328-8334.

[54] Ruelle, B., Peeterbroeck, S., Bittencourt, C., Gorrasi, G., Patimo, G., Hecq, M., et al. (2012). Semi-Crystalline Polymer/Carbon Nanotube Nanocomposites: Effect of Nanotube Surface-Functionalization and Polymer Coating on Electrical and Thermal Properties. *Reactive & Functional Polymers*, 72(6), 383-392.

[55] Hou, Z., Cai, B., Liu, H., & Xu, D. (2008). Ar, O2, CHF3, and SF6 Plasma Treatments of Screen-Printed Carbon Nanotube Films for Electrode Applications. *Carbon*, 46(3), 405-413.

[56] Shi, D., Lian, J., He, P., Wang, L. M., Ooij, W., Jv, Schulz. M., et al. (2002). Plasma Deosition of Ultrathin Polymer Films on Carbon Nanotubes. *Applied Physics Letters*, 81(27), 5216-5218.

[57] Ávila-Orta, C. A., Cruz-Delgado, V. J., Neira-Velázquez, M. G., Hernández-Hernández, E., Méndez-Padilla, M. G., & Medellín-Rodríguez, F. J. (2009). Surface Modification of Carbon Nanotubes with Ethylene Glycol Plasma. *Carbon*, 47(8), 1916-1921.

[58] Chen, I. H., Wang, C. C., & Chen, C. Y. (2010). Preparation of Carbon Nanotube (CNT) Composites by Polymer Functionalized CNT under Plasma Treatment. *Plasma Processes and Polymers*, 7(1), 59-63.

[59] AbouRich, S., Yedji, M., Amadou, J., Terwagne, G., Felten, A., Avril, L., et al. (2012). Polymer Coatings to Functionalize Carbon Nanotubes. *Physica E*, 44(6), 1012-1020.

Chapter 9

Carbon Nanotubes and Their Composites

Veena Choudhary, B.P. Singh and R.B. Mathur

Additional information is available at the end of the chapter

http://dx.doi.org/10.5772/52897

1. Introduction

Carbon nanotubes (CNTs), a fascinating material with outstanding properties has inspired the scientist, engineer and technologist for wide range of potential applications in many areas [1]. Since all these properties are concerned directly to the atomic structure of nanotubes, it is quite necessary to have a thorough understanding of the phenomenon to control nanotube size, the number of shells (walls), the helicities and the structure during growth. The full potential of nanotubes for applications will not be realized until the structure of nanotubes during their growth is optimized and well controlled. For utilization of CNTs properties in real world applications, like composite preparation, it is desired to obtain high quality and in bulk quantity using growth methods that are simple, efficient and inexpensive. Significant work has been carried out in this field and various methods have been studied to synthesize CNTs by several researchers.

2. Properties

Carbon nanotubes are endowed with exceptionally high material properties, very close to their theoretical limits, such as electrical and thermal conductivity, strength, stiffness, toughness and low density.

2.1. Mechanical properties of CNTs

The strength of C-C bond gives a large interest in mechanical properties of nanotubes. Theoretically, these should be stiffer than any other known substance. Young's modulus of the single walled carbon nanotubes (SWCNTs) can be as high as 2.8-3.6 TPa and 1.7-2.4 TPa for multiwalled carbon nanotubes (MWCNTs) [2]which is approximately 10 times higher than steel, the strongest metallic alloy known. Experimental values of Young's modulus for SWCNTs are reported as high as to 1470 GPa and 950 GPa [3, 4] for MWCNTs, nearly 5 times of steel. There are no direct mechanical testing experiments that can be done on individual nanotubes (nanoscopic specimens) to determine directly their axial strength. However, the indirect experiments like AFM provide a brief view of the mechanical properties as well as scanning probe techniques that can manipulate individual nanotubes, have provide some basic answers to the mechanical behavior of the nanotubes [5]. The analysis performed on several MWCNTs gave average Young's modulus values of 1.8 TPa, which is higher than the in – plane modulus for single crystal graphite. So the high stiffness and strength combined with low density implies that nanotubes could serve as ideal reinforcement in composite materials and provide them great potential in applications such as aerospace and other military applications.

2.2. Electrical properties of CNTs

The nanometer dimensions of CNTs, together with the unique electronic structure of a graphene sheet, make the electronic properties of these one-dimensional (1D) structures extraordinary. The one dimensional structure of CNTs helps them in making a good electric conductor. In a 3D conductor the possibility of scattering of electrons is large as these can scatter at any angle. Especially notable is the fact that SWCNTs can be metallic or semiconducting depending on their structure and their band gap can vary from zero to about 2 eV, whereas MWCNTs are zero-gap metals. Thus, some nanotubes have conductivities higher than that of copper, while others behave more like silicon. Theoretically, metallic nanotubes having electrical conductivity of 10^5 to 10^6 S/m can carry an electric current density of 4×10^9 A/cm2 which is more than 1000 times greater than copper metal and hence can be used as fine electron gun for low weight displays. Due to the large diameter of MWCNTs, their transport properties approaches those of turbostatic graphite. Theoretical study also shows that in case of MWCNTs the overall behavior is determined by the electronic properties of the external shell. Conductivities of individual MWCNTs have been reported to range between 20 and 2×10^7 S/m [6], depending on the helicities of the outermost shells or the presence of defects [7].The electronic properties of larger diameter MWCNTs approach those of graphite. Nanotubes have been shown to be superconducting at low temperatures. As probably CNTs are not perfect at ends and end defects like pentagons or heptagons are found to modify the electronic properties of these nanosystems drastically. There is great interest in the possibility of constructing nanoscale electronic devices from nanotubes and some progress is being made in this area. SWCNTs have been recently used to form conducting and semiconducting layers (source, drain and gate electrodes) in thin films transistors. So the high electrical conductivity of CNTs makes them an excellent additive to impart electrical conductivity in otherwise insulating polymers. Their high aspect ratio means that a very low loading is needed to form a connecting network in a polymer compared to make them conducting.

2.3. Thermal properties of CNTs

CNTs are expected to be very good thermal conductors along the tube, but good insulators laterally to the tube axis. Experiments on individual tubes are extremely difficult but measurements show that a SWCNT has a room-temperature thermal conductivity along its axis of about 3500 W m^{-1} K^{-1} and MWCNTs have a peak value of ~ 3000 W m^{-1} K^{-1} at 320 K; compare this to copper, a metal well-known for its good thermal conductivity, which transmits 385 W m^{-1} K^{-1} [8]. Although for bulk MWCNTs foils, thermal conductivity limits to 20 W m −1 K−1 suggesting that thermally opaque junctions between tubes severely limit the large scale diffusion of phonons. The thermal conductivity of CNTs across axis (in the radial direction) is about 1.52 W m−1 K−1, which is about as thermally conductive as soil. Both SWCNT and MWCNT materials and composites are being actively studied for thermal management applications, either as "heat pipes" or as an alternative to metallic addition to low thermal conductive materials. In case of composites, the important limiting factors are quality of dispersion and interphase thermal barriers.

3. Synthesis of CNTs

A variety of synthesis methods now exist to produce carbon nanotubes. The three main production methods used for synthesis of CNTs are d.c. arc discharge, laser ablation and chemical vapor deposition (CVD).

3.1. d.c. arc discharge technique

The carbon arc discharge method, initially used for producing C60 fullerenes, is the most common and perhaps easiest way to produce carbon nanotubes as it is rather simple to undertake. In this method two carbon rods placed end to end, separated by approximately 1mm, in an enclosure that is usually filled with inert gas (helium, argon) at low pressure (between 50 and 700 mbar) as shown in Figure 1. Recent investigations have shown that it is also possible to create nanotubes with the arc method in liquid nitrogen [9]. A direct current of 50 to 100 A driven by approximately 20 V creates a high temperature (~4000K) discharge between the two electrodes. The discharge vaporizes one of the carbon rods (anode) and forms a small rod shaped deposit on the other rod (cathode). Large-scale synthesis of MWCNTs by a variant of the standard arc-discharge technique was reported by Ebbesen and Ajayan [10]. A potential of 18 V dc was applied between two thin graphite rods in helium atmosphere. At helium pressure of ~500 Torr, the yield of nanotubes was maximal of 75% relative to the starting graphitic material. The TEM analysis revealed that the samples consisted of nanotubes of two or more concentric carbon shells. The nanotubes had diameters between 2 and 20 nm, and lengths of several micrometers. The tube tips were usually capped with pentagons.

If SWCNT are preferable, the anode has to be filled with metal catalyst, such as Fe, Co, Ni, Y or Mo. Experimental results show that the width and diameter distribution depends on the composition of the catalyst, the growth temperature and the various other growth conditions. If both electrodes are graphite, the main product will be MWCNTs. Typical sizes for MWCNTs are an inner diameter of 1-3nm and an outer diameter of approximately 10nm. Because no catalyst is involved in this process, there is no need for a heavy acidic purification step. This means MWCNT can be synthesized with a low amount of defects.

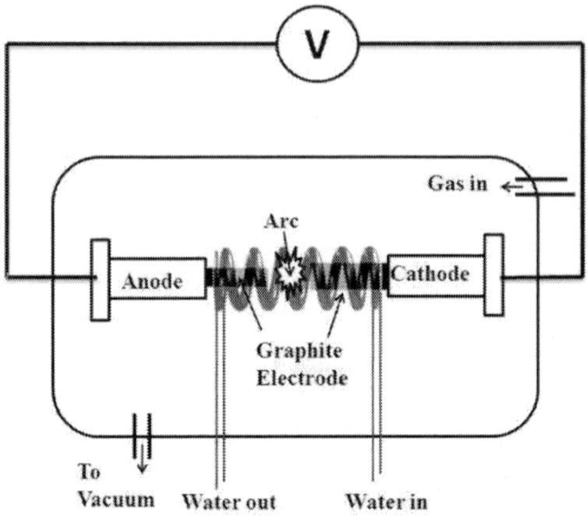

Figure 1. Schematic diagram of dc-arc discharge set-up

In most of the studies, SWCNTs are synthesized using the dc-arc discharge process by filling the catalyst powder into a hole drilled in a graphite elctrode act as an anode and arcing takes place between this anode and a pure graphite based cathode in optimized chamber conditions. In one of the study by Mathur et al. [11] SWCNTs and MWCNTs were synthesized simultaneously in a single experiment selectively. In their experiment, However, instead of filling the catalyst powder into a hole drilled in a graphite electrode; they prepared a catalyst/graphite composite electrode. Coke powder, catalyst powder, natural graphite powder and binder pitch were thoroughly mixed together in a ball mill in appropriate proportions and molded into green blocks using conventional compression molding technique. A mixture of Ni and Co powders was used as catalyst. The green blocks were heated to 1200° C in an inert atmosphere to yield carbonized blocks with varying compositions of coke, natural graphite powder, Ni and Co. These electrodes were used as the anodes in the arcing process and a high density graphite block was used as the cathode. A uniform gap of 1– 2 mm was maintained between the electrodes during the arcing process with the help of a stepper motor for a stable arc-discharge (dc voltage 20–25 V, current 100–120 A, 600 torr helium).The SWCNTs yield was found to be doubled in this case.

3.1.1. Characteristics of CNTs produced by d.c. arc discharge technique

Arc discharge is a technique that produces a mixture of components and requires separating nanotubes from the soot and the catalytic metals present in the crude product. In this technique both SWCNT and MWCNT can be produced and it has been described by several researchers.

The scanning electron microscope (SEM) and transmission electron microscope (TEM) are generally used to observe the physical appearance of any carbon based soot. Similarly, Mathur et al. [11] used SEM and TEM for the observation of SWCNT and MWCNT produced from the arc discharge technique as shown in Figure 2. In this technique, the carbon material deposits on the chamber and cathode. The arcing process resulted in the formation of web-like deposits on the inner walls of the arc chamber. A typical SEM micrograph of such deposit (Figure 2a) revealed the presence of SWNT bundles along with the amorphous carbon and catalyst particles. Rod-like microstructures aligned preferentially along the length of the cathode were also found at the tip of the cathode as shown in Figure 2b. The inset in Figure 2b shows the presence of graphitized carbon and sharp needle-like nanostructure when these rods are powdered. Upon detailed electron microscopic examination, these needles exhibited the MWCNT structure with an outer diameter of 20–25 nm (Figure 2c).

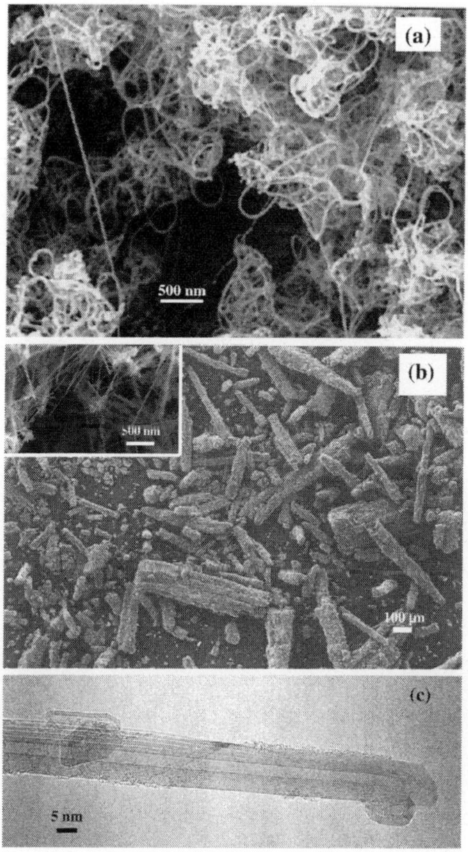

Figure 2. (a) SEM micrograph of the chamber deposit showing the presence of long and flexible carbon nanotubes. (b) SEM micrograph of the cathode deposit showing the presence of rod-like microstructures. The inset figure shows the presence of needle-like nanostructures present within each microstructure. (c) TEM micrograph of a single needle-like nanostructure (Reprinted with permission from Elsevier (11))

The nature of the soot can be identified using Raman spectrometer and generally used for confirmation of the quality of CNTs. The nature of these two deposits obtained using this arc discharge process was confirmed from their respective Raman spectrum (Figure 3). The Raman spectrum of the chamber deposit showed the characteristic radial breathing and tangential bands at 165–183 and 1591 cm^{-1}, respectively. The strong G-band at 1580 cm^{-1} in the Raman spectrum of the cathode deposit and its TEM image depicted in Figure 2c, confirmed that the cathode deposit predominantly contained MWCNTs. The prominent D-band around 1350 cm^{-1}seen in both spectrum is attributed to the presence of disordered carbonaceous material present in the as-prepared deposits. In their study, Mathur et al. [11] show that SWCNTs deposit on the arc chamber and MWCNTs on cathode deposit.

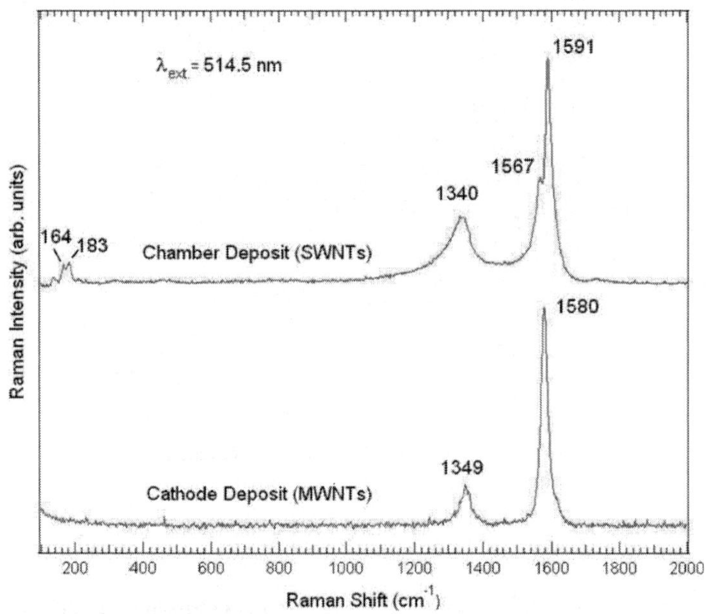

Figure 3. Room temperature Raman spectrum of the chamber and cathode deposit (Reprinted with permission from Elsevier (11))

3.2. Chemical vapor deposition

Pyrolysis of organometallic precursors such as metallocenes (e.g. ferrocene) in a furnace provides a straight forward procedure to prepare CNT by CVD technique. Different hydrocarbons, catalyst and inert gas combinations have been used by several researchers in the past for the growth of CNT by CVD technique. In one of the study by Mathur et al. [12], CNTs were grown inside the quartz reactor by thermal decomposition of hydrocarbons, e.g. toluene in presence of iron catalyst obtained by the decomposition of organometallic like ferrocene. The furnace provided a constant temperature zone of 18 cm in the centre as shown in Figure 4. The reaction zone was maintained at 750°C. Once the temperature was reached, the solution containing a mixture of ferrocene and toluene in particular proportion (0.077 g ferrocene in 1 ml toluene) was injected in the reactor at a point where the temperature was 200°C. Argon was also fed along with the charge as a carrier gas and its flow rate was adjusted so that the maximum amount of precursor is consumed inside the desired zone.

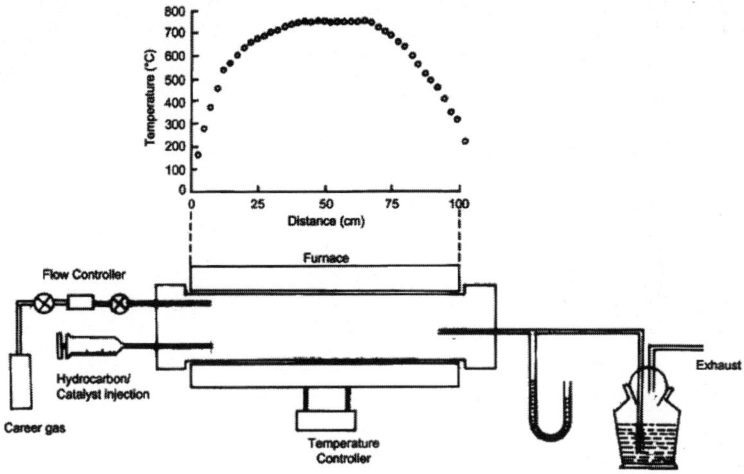

Figure 4. Schematic diagram of the CVD reactor along with the temperature profile (Reprinted with permission from Elsevier (12))

3.2.1. *Characteristics of CNTs Produced by CVD Technique*

CNTs are produced in the form of big bundles using CVD technique. The physical appearance of the as produced CNTs is shown in Figure 5a and Figure 5b for SEM and TEM respectively. Figure 5a shows a big CNT bundle of length >300μm and the inset image of Figure 5a shows very good quality of uniform CNTs. Figure 5b shows the TEM image of as produced CNTs confirming the presence of MWCNT with metallic catalytic impurites either on the tip of the tube or in the cavity of of CNTs (inset of Figure 5b).

Further confirmation of the quality and type of CNTs can be obtained using Raman spectrometer as shown in Figure 6. This shows the tangential band at 1580 cm^{-1} (G band) of high intensity and the disorder-induced band at 1352 cm^{-1} (D band) as a perfect MWCNT nature [13]. The ratio of intensity of G to D band gives the information regarding the quality of the CNTs. The high value of intensity ratio of G/D band confirms the better quality of CNTs.

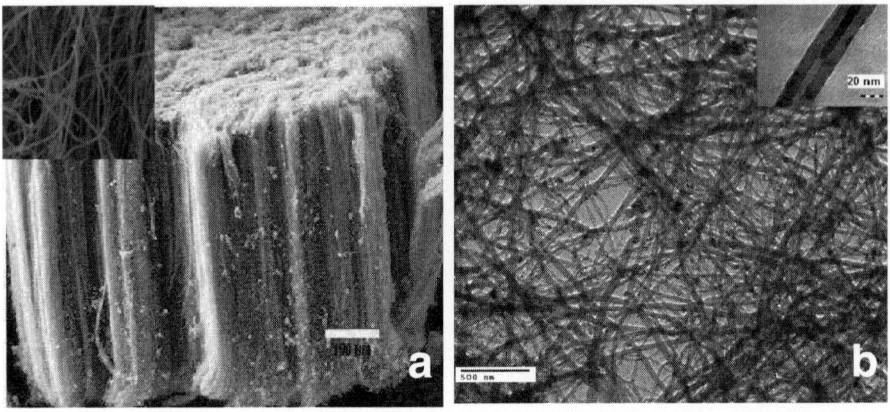

Figure 5. (a) SEM image of aligned CNT bundle synthesized by CVD technique.The inset figure shows the very good quality of uniform CNTs (b) TEM image of as grown MWCNT and inset image shows the MWCNTs with encapsulated metallic impurities

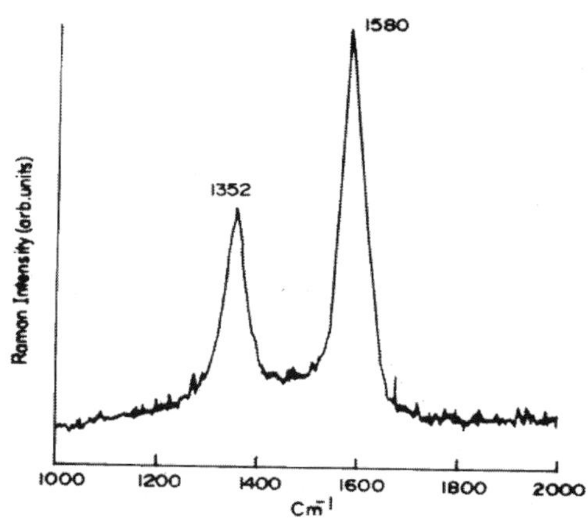

Figure 6. Raman spectrum of CVD-grown MWCNTs.

3.3. Laser ablation

In the laser ablation process, a pulsed laser vaporizes a graphite target containing small amounts of a metal catalyst [14] as shown in Figure 7. The target is placed in a furnace at roughly 1200°C in an inert atmosphere. The nanotubes develop on the cooler surface of the

http://dx.doi.org/10.5772/52897

reactor, as the vaporized carbon condenses. The yield of nanotube synthesis by this process is roughly 70%.

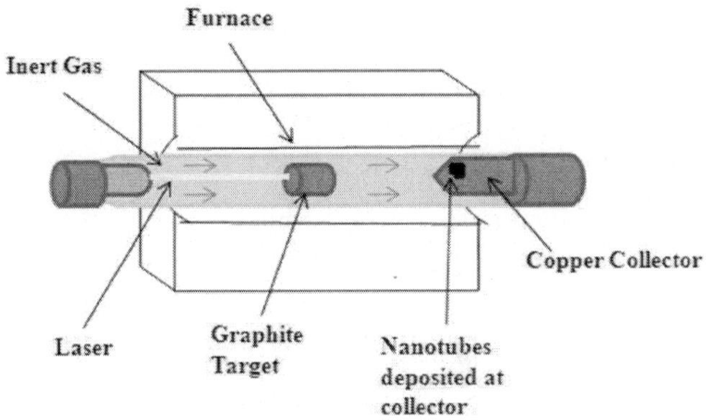

Figure 7. Schematic diagram of Laser ablation set-up for CNT synthesis

3.3.1. Characteristics of CNTs produced by laser ablation technique

The laser-ablation prepared samples usually contain >70% nearly endless, highly tangled ropes of SWCNTs along with nanoscale impurities. The SWCNTs formed in this case are bundled together by van der Waals forces. Laser vaporisation results in a higher yield for SWCNT synthesis and a narrower size distribution than SWCNTs produced by arc-discharge [15]. The nanotubes generated by the laser ablation and arc discharge technique are relatively impure, with presence of unwanted carbonaceous impurities and not operated at higher scale; therefore, the overall production costs are high.

Compared to other methods for synthesis of CNTs, more parameters, including temperature, feeding gases, flow rate, catalyst components and heating rate are accessible to control the growth process in CVD. By changing the growth conditions, we can control the properties of the produced CNTs such as length, orientation and diameter to some extent. It has been observed that the gas phase processes produces CNTs with fewer impurities and are most amenable to large scale processing. So the gas phase techniques such as CVD, for nanotube growth offer the greatest potential for scaling up nanotube production for processing of composites.

4. Purification

During CNTs synthesis, impurities in the form of catalyst particles, amorphous carbon and non tubular fullerenes are also produced. The most of the production methods involve the use of catalysts which are normally transition metals (Fe, Co, Ni or Y), these remains in the resulting nanotubes as spherical or cylindrical particles after experiments. Through careful control of process parameters one could minimize the formation of amorphous carbon particles, so that the main impurities in CNTs are the remaining catalyst particles. However as most of these catalytic particles may either hide in internal cavity or stick firmly to the walls of CNTs, it is almost impossible to get rid of these effectively without damaging the nanotubes. Several purification methods have been tried to overcome these impurities. In one of the study by Mathur et al. [11], SWCNT soot prepared by dc arc discharge process was purified by removing various forms of impurities, such as amorphous carbon, graphitic nanoshells and catalyst particles present in the chamber deposit by applying a judicious combination of wet and dry chemical methods (acid treatment and oxidation). In this process, initially SWCNT soot were oxidized at 350 °C for 6h in air which remove the amorphous carbon followed by refluxing in HCl for the removal of metallic impurities like Ni and Co and again oxidation at 550°C for 30 min for the removal of graphitic nanoshell.The final product gives 97% purified SWCNT. The MWCNTs produced by CVD technique contains mainly ~10% metallic impruites which can be removed by heating it in the inert atmosphere at 2500°C in graphitization furnace. This process gives >99% pure MWCNTs and also helps in annealing out the defects in the tubes. This graphitization process at high temperature can also be useful for removal of impurities in the arc discharge produced SWCNTs soots with the combination of other purification steps.

5. Nanocomposites

Because of the high strength and stiffness of CNTs, they are ideal candidates for structural applications. For example, they may be used as reinforcements in high strength, low weight and high performance composites. Presently there is a great interest in exploiting the exciting properties of these CNTs by incorporating them into some form of polymer matrix.

5.1. Composite fabrication techniques

A large number of techniques have been used for the fabrication of CNT-polymer nanocomposites based on the type of polymer used.

5.1.1. Solvent casting

The solution casting is most valuable technique to form CNTs/polymer nanocomposites. However, its use is restricted to polymers that are soluble. Solvent casting facilitates nanotube dispersion and involves preparing a suspension of CNTs in the desirable polymer solution via energetic agitation (magnetic stirring or sonication) and then allowing the solvent to evaporate to produce CNT-polymer nanocomposites. A lot of study is available in open literature for the formation of CNT nanocomposites by this method [16-18]. Mathur et al. [18] cast the solutions of the MWCNT/polystyrene (PS)/toluene and MWCNT/ polymethyl methacrylate (PMMA)/toluene suspensions after sonication into a petry dish to produce nanotubes composites with enhanced electrical and mechanical properties. Benoit et al. [19]

obtained electrically conductive nanocomposites by dispersing CNT and PMMA in toluene, followed by the drop casting on substrate. The choice of solvent is generally made based on the solubility of the polymer. The solvent selection for nanotube dispersion also had a significant influence on the properties of the nanocomposites and studied by Lau and co-workers [20].Their results demonstrates that, contrary to the general belief that small traces of CNTs alone will serve to strengthen the epoxy composites, the choice of the solvent used in the dispersion of CNTs also can have a significant impact. The change trend of the mechanical properties was found to be related to the boiling point of respective solvent used. In the samples observed in their study, only acetone-dispersed nanocomposites displayed improvements in flexural strength over the pure epoxy, while ethanol and DMF used in CNTs dispersion actually countered the benefits of CNTs in the resulting nanocomposites.It is reasonable that, easier the solvent can evaporate, less solvent will remain to affect the curing reaction. Their results of thermogravimetric analysis (TGA) proved the existence of residual solvent in the resulting nanocomposites. Further evidence of the solvent influence was obtained by Fourier transform infrared (FTIR) spectra, which displayed the difference in the molecular structure of the final nanocomposites depending on the solvent used. The solvent influence is attributed to the different amount of unreacted epoxide groups and the extent of cure reaction in the manufacturing process. The presence of residual solvent may alter the reaction mechanism by restricting the nucleophile–electrophile interaction between the hardener and epoxy, henceforth, affect the cross-linking density and thus degrade the transport properties [21]and mechanical properties of the cured structures. The residual solvent may absorb some heat energy from the composite systems in the pre-cured process, causing a change in local temperature. Nanocomposites with other thermoplastic materials with enhanced properties have been fabricated by solvent casting [16-18, 22]. The limitation of this method is that during slow process of solvent evaporation, nanotubes may tend to agglomerate, that leads to inhomogeneous nanotube distribution in polymer matrix. The evaporation time can be decreased by dropping the nanotube/polymer suspension on a hot substrate (drop casting) [19]or by putting suspension on a rotating substrate (spin-casting) [23]. Du et al. [24] developed a versatile coagulation method to avoid agglomeration of CNTs in PMMA-CNT nanocompositses that involves pouring a nanotube/polymer suspension into an excess of solvent. The precipitating polymer chains entrap the CNT, thereby preventing the CNT from bundling.

5.1.2. Melt mixing method

The alternative and second most commonly used method is melt mixing, which is mostly used for thermoplastics and most compatible with current industrial practices. This technique makes use of the fact that thermoplastic polymers softens when heated. Melt mixing uses elevated temperatures to make substrate less viscous and high shear forces to disrupt the nanotubes bundle. Samples of different shapes can then be fabricated by techniques such as compression molding, injection molding or extrusion. Andrews and co-workers [25] formed composites of commercial polymers such as high impact polystyrene, polypropylene and acrylonitrile–butadiene–styrene (ABS) with MWCNT by melt processing. Initially these polymers were blended in a high shear mixer with nanotubes at high loading level to form master batches that were thereafter diluted with pure polymer to form lower mass fraction samples. Compression molding was used to form composite films. A similar combination of shear mixing and compression molding is studied by many other groups discussed elsewhere [16]. Also Meincke et al. [5] mixed polyamide-6, ABS and CVD-MWCNT in a twin screw extruder at 260°C and used injection molding to make nanocomposites. Tang et al. [26] used both compression and twin-screw extrusion to form CNT/polyethylene composites. Although melt-processing technique has advantages of speed and simplicity, it is not much effective in breaking of agglomeration of CNTs and their dispersion. Bhattacharyya et al. [27] made 1 wt% CNT/polypropylene (PP) nanocomposites by melt mixing, but found that melt mixing alone did not provide uniform nanotube dispersion. Niu et el. [28] studied both methods to prepare polyvinylidene fluoride (PVDF)-CNT nanocomposites to study electrical properties and found it better in composites formed by solution casting.

5.1.3. In-situ polymerization

In addition to solvent casting and melt mixing the other method which combines nanotubes with high molecular weight polymers is in-situ polymerization starting with CNTs and monomers. In-situ polymerization has advantages over other composite fabrication methods. A stronger interface can be obtained because it is easier to get intimate interactions between the polymer and nanotube during the growth stage than afterwards [29, 30].The most common in situ polymerization methods involve epoxy in which the monomer resins and hardeners are combined with CNTs prior to polymerizing [31]. Pande and coworkers [32] performed the in-situ polymerization of MWCNT/ PMMA composites for the enhancement in flexural strength and modulus of composites. Li et al. [33] reported the fabrication and characterization of CNT/ polyaniline (PANI) composites. Xiao and Zhou [34]deposited polypyrrolre (PP) and poly(3-methylthiophene) (PMet) on the surface of MWCNTs by in situ polymerization. Saini et al. [35] reported fabrication process of highly conducting polyaniline (PANI)–(MWCNT) nanocomposites by in-situ polymerization. This material was used in polystyrene for the fabrication of MWCNT-PANI-PS blend for microwave absorption [36]. Moniruzzaman [17] reported many other studies of in-situ polymerization of CNTs with different polymers. Generally, in situ polymerization can be used for the fabrication of almost any polymer composites containing CNT that can be non-covalently or covalently bound to polymer matrix. This technique enables the grafting of large polymer molecules onto the walls of CNT. This technique is particularly important for the preparation of insoluble and thermally unstable polymers, which cannot be processed by solution or melt processing.

Some studies have been also carried out using combined methods, such as solvent casting in conjunction with sonication, followed by melt mixing. Haggenmueller et al. [37] observed considerable nanotube dispersion in CNT-polymer nanocomposites using combination of solvent casting and melt mixing. Pande et al. [32, 38] also prepared MWCNT bulk composites with PMMA and PS using a two-step method of solvent casting followed by compression molding and obtained better electrical and mechanical properties. Singh et al. [39] also prepared MWCNT-LDPE composites using solvent casting followed by compression moulding and obtained better electrical conductivity. Jindal et al. [29, 40] prepared

MWCNT-polycarbonate composite using solvent casting followed by compression moulding for the enhancement in the impact properties.

The other less commonly known methods for CNT- polymer nanocomposites formation are twin screw pulverization [41], latex fabrication [42], coagulation spinning [43] and electrophoretic deposition [44].

6. Challenges in MWCNT polymer composites fabrication and possible solutions

Although these fabrication methods helped to enhance the properties of CNT reinforced composites over neat polymer but there are several key challenges that hinders the excellent CNT properties to be fruitful in polymer composite formation.

6.1. Dispersion

Disperion of nanoscale filler in a matrix is the key challenge for the formation of nanocomposite. Dispersion involves separation and then stabilization of CNTs in a medium. The methods described above for the nanocomposites fabrication require CNTs to be well dispersed either in solvent or in polymer for maximizing their contact surface area with polymer matrix. As CNTs have diameters on nanoscale the entanglement during growth and the substantial van der Waals interaction between them forces to agglomerate into bundles. The ability of bundle formation of CNTs with its inert chemical structure makes these high aspect ratio fibers dissolving in common solvents to form solution quite impossible. The SEM of MWCNTs synthesized by CVD technique seems to be highly entangled and the dimensions of nanotube bundles is hundreds of micrometer. This shows several thousands of MWCNTs in one bundle as shown in Figure 5a. These bundles exhibits inferior mechanical and electrical properties as compared to individual nanotube because of slippage of nanotubes inside bundles and lower aspect ratio as compared to individual nanotube. The aggregated bundles tend to act as defect sites which adversely affect mechanical and electrical properties of nanocomposites. Effective separation requires the overcoming of the inter-tube van der Waal attraction, which is anomalously strong in CNT case. To achieve large fractions of individual CNT several methods have been employed. The most effective methods are by attaching several functional sites on the surface of CNTs through some chemical treatment or by surrounding the nanotubes with dispersing agents such as surfactant. Thereafter the difficulty of dispersion can be overcome by mechanical/physical means such as ultrasonication, high shear mixing or melt blending. Another obstacle in dispersing the CNTs is the presence of various impurities including amorphous carbon, spherical fullerenes and other metal catalyst particles. These impurities are responsible for the poor properties of CNTs reinforced composites [45].

6.2. Adhesion between CNTs and polymer

The second key challenge is in creating a good interface between nanotubes and the polymer matrix. From the research on microfiber based polymer composites over the past few decades, it is well established that the structure and properties of filler-matrix interface plays a major role in determining the structural integrity and mechanical performance of composite materials. CNTs have atomically smooth non-reactive surfaces and as such there is a lack of interfacial bonding between the CNT and the polymer chains that limits load transfer. Hence the benefits of high mechanical properties of CNTs are not utilized properly. The first experimental study focusing on interfacial interaction between MWCNT and polymer was carried out by Cooper et al. [46]. They investigated the detachment of MWCNTs from an epoxy matrix using a pullout test for individual MWCNT and observed the interfacial shear stress varied from 35-376 MPa. This variation is attributed to difference in structure and morphology of CNTs.

There are three main mechanisms for load transfer from matrix to filler. The first is weak van der Waal interaction between filler and polymer. Using small size filler and close contact at the interface can increase it. The large specific surface area of CNTs is advantageous for bonding with matrix in a composite, but is a major cause for agglomeration of CNTs. Therefore, uniformally dispersed individual nanotubes in matrix is helpful. The second mechanism of load transfer is micromechanical interlocking which is difficult in CNTs nanocomposites due to their atomically smooth surface. Although local non uniformity along length of CNTs i.e. varying diameter and bends due to non hexagonal defects contributes to this micromechanical interlocking. This interlocking can increase by using long CNTs to block the movement of polymer chains. The contribution of this mechanism may reach saturation at low CNT content. The third and best mechanism for better adhesion and hence load transfer between CNTs and polymer is covalent or ionic bonding between them. The chemical bonding between CNTs and polymer can be created and enhanced by the surface treatment such as oxidation of CNTs with acids or other chemicals. This mode of mechanism have much importance as it provides strong interaction between polymer and CNT and hence efficiently transfers the load from polymer matrix to nanotubes necessary for enhanced mechanical response in high-performance polymers.

6.3. Chemical functionalization of CNTs

The best route to achieve individual CNT to ensure better dispersion is chemical modifications of CNT surface. The chemical functionalization involves the attachment of chemical bonds to CNT surface or on end caps. Nanotube functionalization typically starts with oxidatative conditions, commonly by refluxing in nitric or sulfuric acid or combination of both to attach carboxylic acid moieties to the defect sites. The end caps of nanotubes have extra strain energy because of their high degree of curvature with pentagons and heptagonal carbon atoms are most vulnerable to reaction with acid. The side walls also containing defects like pentagon-heptagon pairs, sp^3 hybridized defects and vacancies in nanotube lattice and are easily supplemented by oxidative damage and can be stabilized by formation of functional groups mainly carboxylic acid and hydroxide group. These acid moieties and hydroxide groups can be further replaced to more reactive groups like –COCl or –$CORNH_2$. The addition of these functional groups on CNTs possesses intermolecular repulsion between functional groups on surface that overcomes the otherwise weak van der Waal attraction between CNTs. It is also vital to stabilize the dispersion to prevent reagglomeration of the CNTs. Chemical functionalization can prevent reagglomeration of CNTs also. Sen et al. [47] carried out chemical functionalization to form ester functionalized CNTs and found that it is an effective approach to exfoliate the CNTs bundles and improve their processibility with polymer matrix. Georgakilas et al. [48] observed that CNT covalently functionalized with pyrrolidone by 1,3-dipolar cycloaddition of azomethine ylides show a solubility of 50 mg/mL in chloroform, even without sonication whereas the pristine CNT is completely insoluble in this solvent. Liang et al. [49] performed reductive alkylation of CNTs using lithium and alkyl halides in liquid ammonia for sidewall functionalization of CNTs and observed their extensive debundling by inspection of HRTEM images. Kinloch et al. [50] studied the rheological behavior of oxidized CNTs and found that the composites filled with functionalized CNTs had better dispersion. It has been observed by the researchers that amine modified CNTs is very important for the enhancement in the mechanical properties with epoxy. Garg et al. [31] shows the reaction mechanism for the formation of acid functionalized and amine functionalized CNT and their interaction with the epoxy resin as Figure 8a and b respectively. Two different types of functional groups were attached on the CNT surface. In the first case MWCNTs were refluxed for 48 h in HNO_3 (400 ml, 60% concentration) to achieve reasonable surface oxidation of the tubes. The mixture was then filtered and the residue (treated material) was washed several times with distilled water till washings were neutral to pH paper. The treated MWCNT were dried in oven before use. In a second step the oxidized nanotubes were dispersed in benzene by stirring, and then refluxed with excess $SOCl_2$ along with a few drops of DMF used as catalyst for chlorination of MWCNT surfaces. After the acyl chlorination, $SOCl_2$ and DMF were removed through repeatedly washing by tetrahydrofuran (THF). 100 ml of triethylene tetra-amine (TETA) was added to react with acyl chlorinated MWCNT at 100 °C for 24 h reflux until no HCl gas evolved. After cooling to room temperature, MWCNTs were washed with deionized water 5 times to remove excess TETA. Finally, the black solid was dried at room temperature overnight in vacuum and named as amine modified CNTs. These functionalized CNTs were characterized by FTIR, TGA and HRTEM and clearly showed the presence of these types of functional group. Gojny et al. [51]achieved surface modified MWCNTs by refluxing of oxidized MWCNTs with multifunctional amines and observed from TEM images that these were completely covered by epoxy matrix that confirmed the bonding between them. Sinnott [52]has provided an in depth review of chemical functionalization of CNTs where the chemical bonds are used to tailor the interactions between nanotubes and polymers or solvent. The chemical functionalization of CNTs has also been accomplished through irradiation with electrons or ions [53]. In this manner one may hope to improve the binding of CNTs by interdigitation of active sites on its sides into polymer matrix.

Covalent bond also benefits phonon transferring between nanotubes and polymer matrix, which is a key factor for improving thermal conductivity of the nanocomposites. To ensure the adhesion between polymer and nanotubes various surfactant and chemical modification procedures have been adopted to modify the surface of otherwise inert surface of CNTs that provides bonding sites to the polymer matrix.

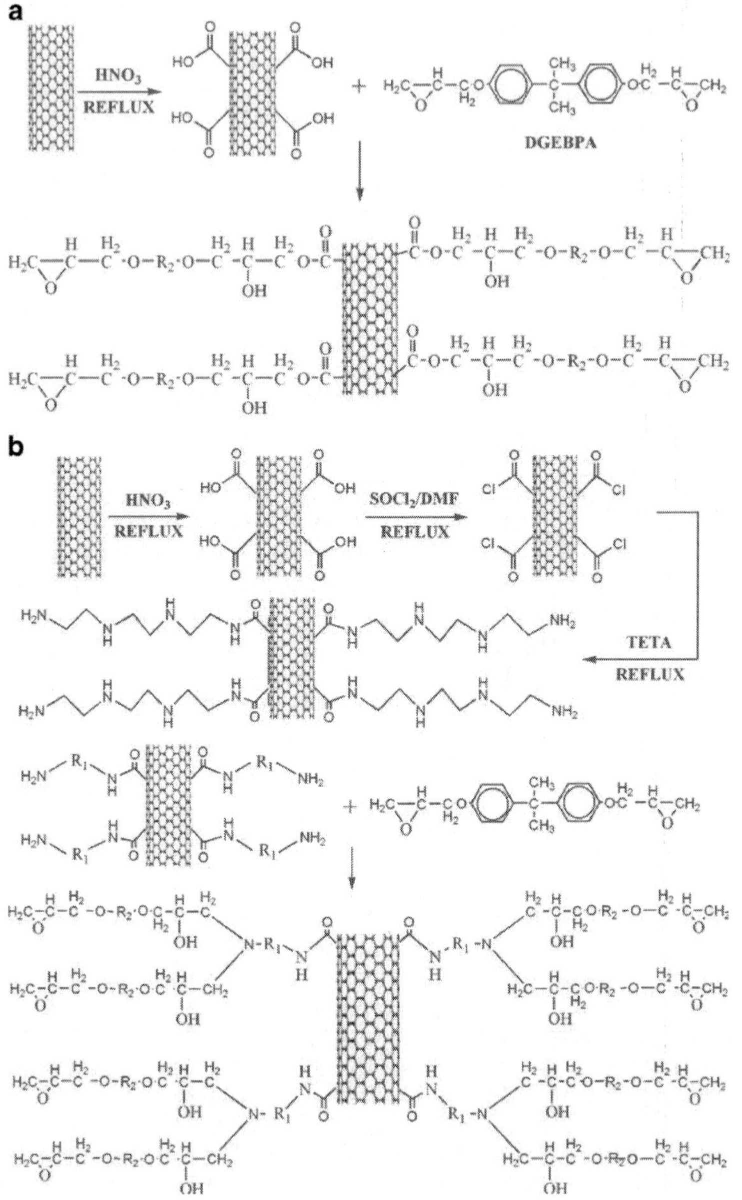

Figure 8. Reaction mechanism for (a) acid modified-MWCNT (b) amine modified-MWCNT with epoxy resin (Reprinted with permission from Springer(31))

So the surface modification of CNTs is the crucial factor that decides the effective dispersion and improves the interactions between CNTs and matrix. However there are certain drawbacks of using chemically functionalized CNTs. Chemical functionalization normally employs harsh techniques resulting in tube fragmentation and also disrupts the bonding between graphene sheets and thereby reduces the properties of CNTs. Studies revealed that different chemical treatments may decrease the maximum buckling force of nanotubes by 15% [16]. Also the chemical functionalized CNTs significantly decrease the electrical conductivity of CNTs nanocomposites due to unbalance polarization effect, shortening of length and physical structure defects during acidic treatment [54]. But it is still necessary for increased dispersion and strengthens the interfacial bonding of CNTs with polymer matrix that is more important in structural applications.

The solubility or dispersion of CNTs in certain specified solvents or polymers can also increase by non covalent association which is more fragile. The non-ionic surfactant such as sodium dodecylbenzene sulfonate (SDS) or polyoxyethylene-8-lauryl (PoEL) has two segments. The hydrophobic segment of surfactant shows strong interactions with carbon of CNTs via van der Waal force and the hydrophilic segment shows hydrogen bonding with solvent or polymer used for dispersion. Islam et al. [55] reported that ~ 65% CNT bundles exfoliated into individual nanotubes even with a very low of 20 mg (CNT)/ml of water containing SDS as surfactant. Barrau et al. [56] used palmitic acid as surfactant to disperse CNTs into epoxy resin and observed that electrical percolation threshold decreases indicating better CNT dispersion. Gong et al. [57] added PoEL as surfactant in CNT/epoxy composite to assist the dispersion of CNTs. The improvement in dispersion in chitosan with nitric acid treated CNTs was also reported by Ozarkar et al. [58] and the stability of dispersion prepared by using functionalized CNTs was observed to be better. However, CNTs treated with different surfactants are wrapped in it and hence contacts between CNTs decreases thereby the transport properties (electrical and thermal conductivities) of CNTs/polymer nanocomposites are adversely affected.

6.4. Dispersion of high loading of CNTs in polymer matrix

Dispersion of high loading of CNTs in any polymer is very difficult due to the formation of agglomerates by the conventional techniques. To maximize the improvment in properties, higher loading of CNTs is preferred [59]. However, polymer composites synthesized by using the conventional methods generally have low CNT contents. It has been observed that beyond 1 wt.-% of loading, CNTs tend to agglomerate [60] resulting in poor mechanical properties of the composites. It is therefore important to develop a technique to incorporate higher CNT loading in the polymer matrices without sacrificing their mechanical properties. Recently, several methods have been developed for fabricating CNT/polymer composites with high CNT loadings. One such technique is mechanical densification technique where vertically aligned CNTs were densified by the capillary induced wetting with epoxy resin [61]. By this technique dimensions of sample preparation are limited. In another technique, a filtration system was used to impregnate the epoxy resin into CNT bucky paper [62, 63]. However, it was very difficult to completely impregnate the bucky paper with epoxy resin.

Recently, Feng et al., [64] reported a mixed curing assisted layer by layer method to synthesize MWCNT/epoxy composite film with a high CNT loading from ~15 to ~36 wt.-%. The mixed-curing-agent consists of two types of agents, one of which is responsible for the partial initial curing at room temperature to avoid agglomeration of the CNTs, and the other for complete curing of epoxy resin at high temperature to synthesize epoxy composite films with good CNT dispersion. In another study by Feng et al. [65] upto ~39.1 wt. % SWCNT-epoxy composites were fabricated using same mixed curing layer by layer method and their mechanical properties were enhanced significantly. Bradford et al. [66] reported a method to quickly produce macroscopic CNT composites with a high volume fraction upto 27% of millimeter long, well aligned CNTs. Specifically, they used the novel method, shear pressing, to process tall, vertically aligned CNT arrays into dense aligned CNT preforms, which are subsequently processed into composites. In another study by Ogasawara et al. [67] aligned MWCNT/epoxy composites were processed using a hot-melt prepreg method. Vertically aligned ultra-long CNT arrays (forest) were converted to horizontally aligned CNT sheets by pulling them out. An aligned CNT/epoxy prepreg was fabricated using hot-melting with B-stage cured epoxy resin film. The final composites contains 21.4 vol% of MWCNTs.

6.5. Alignment of CNTs in polymer matrix

The other key challenge is to understand the effect of nanotube alignment on nanocomposites properties because the nanotubes have asymmetric structure and properties. Like other one-dimensional fiber fillers CNTs displays highest properties in the oriented reinforced direction and the mechanical, electrical, magnetic and optical performance of its composites are linked directly to their alignment in the matrix. So to take the full advantage of excellent properties of CNTs these should be aligned in a particular direction. For example, the alignment of CNT increases the elastic modulus and electrical conductivity of nanocomposites along the nanotube alignment direction.

Several methods like application of electric field during composite formation and carbon arc discharge [68], composite slicing [69], film rubbing [70], chemical vapor deposition [71, 72], mechanical stretching of CNT-polymer composites [73] and magnetic orientation [74] have been reported for aligning nanotubes in composites. Electrospinning is also an effective method for the alignment of CNTs in polymer matrix.

7. Properties of the nanocomposites

7.1. Mechanical properties of MWCNTs polymer nanocomposites

Different thermoplastic and thermoset polymer matrices have been tried to realize the superior mechanical properties of CNTs for development of light weight strong material. NASA scientists are considering CNT-polymer composite for space elevator. To date, a volume of literature is available on the improvement of mechanical performance of polymers with addition of CNTs. The first study for formation of CNT-polymer composites was carried out by Ajayan et al. [69]. CNTs were aligned within the epoxy matrix by the shear forces induced by cutting with a diamond knife, however no quantitative mechanical measurements were made. The first true study for tensile and compression properties of CNT polymer composites was carried by Schadler et al. [75] with epoxy. On addition of 5-wt% MWCNTs the tensile modulus increased from 3.1 GPa to 3.71 GPa. and compression modulus increased from 3.63 to 4.5 GPa. However, no significant increases in toughness values were observed. Bai et al. [76] observed doubling of Young's modulus from 1.2 to 2.4 GPa and significant increase in strength from 30 to 41 MPa on addition of 1 wt.% MWCNTs. Also excellent matrix–nanotube adhesion was confirmed by the observation of nanotube breakage during fracture surface studies. Zhou et al. [77] reported steady increase of flexural modulus in CNT-epoxy composite with higher CNT weight percentage and found an improvement of 11.7% in modulus with 0.4 wt% loading of CNTs and 28% enhancement in flexural strength with 0.3 wt% loading. Garg et al. [31] reported an increment of 155 % in flexural strength of epoxy with addition of merely 0.3% amine functionalized MWCNTs and an increment of 38% in flexural modulus. Mathur et al. [13] reported an increment of 158% in flexural strength of phenolic with addition of 5 vol% of MWCNT. Colemann et al. [16] reviewed the mechanical properties of a large number of CNT reinforced polymer (thermoplastic and thermosetting) composites fabricated by various methods and reported enhancement in mechanical properties. Du et al. [78] studied the experimental results for mechanical performance of CNTs nanocomposites carried out by different research groups and observed that the gains are modest and far below the simplest theoretical estimates. Haggenmueller [79] applied the Halpen Tsai composite theory to CNT nanocomposites and observed that the experimental elastic modulus is smaller than predicted by more than one order. It is attributed to the lack of perfect load transfer from nanotubes to matrix due to non uniform dispersion and small interfacial interaction. Although chemical functionalization of CNTs has sorted out those problems to an extent yet the best results have to be achieved. Also aspect ratio is other source of uncertainty in mechanical properties. Defects on the CNT surface also expected to influence the mechanical properties significantly. The methods of handling nanotubes, including acid treatments and sonication for long time are known to shorten nanotubes results in decreasing aspect ratio and are detrimental to mechanical properties. The mechanical properties of CNT based composites increased upto a certain loading of CNTs and beyond which it starts decreasing. This may be because of increase in viscosity of polymers at higher CNTs loading and also cause some surface of CNTs not to be completely covered by polymers matrix due to the large specific surface area of CNTs.Therefore, few studies have been carried out to disperse high loading of CNTs in polymer matrices with improved mechanical properties. Bradford et al. [66] reported 400 MPa (tensile strength) and 22.3 GPa (tensile modulus) for 27vol% of MWCNT–epoxy composites. Feng et al. [80] also reported 183% and 408% improvement in tensile strength and tensile modulus respectively at 39.1 wt% SWCNTs loading compared with those of the neat epoxy. Ogasawara et al. [67] found 50.6 GPa and 183 MPa modulus and ultimate tensile strength respectively of CNT (21.4 vol.%)/epoxy composite. These values were 19 and 2.9 times those of the epoxy resin respectively.

Andrews et al. [80] prepared aligned CNT/pitch composites and found the significant improvement in the electrical and mechanical properties especially due to orientation of CNTs.

Du et al. [78] compared the mechanical performance of randomly oriented and aligned CNTs polymer composites. Their study revealed that in aligned CNT polymer nanocomposites tensile strength and modulus even reached to 3600 MPa and 80 GPA respectively which is much higher than the general value of 100 MPa and 6 GPa in case of randomly oriented CNT polymer nanocomposites. They also observed that the mechanical properties are always higher for aligned CNTs composites with higher loading while the case is different for isotropic CNT polymer composites.

7.2. Electrical properties of MWCNTs polymer nanocomposites

CNTs because of their extraordinary electrical conductivity are also excellent additive to impart electrical conductivity to polymer. Many experimental results shows that the conductive CNT composites can be constructed at low loading of CNTs due to low percolation threshold originated from the high aspect ratio and conductivity of CNTs [70, 78]. Figure 9 shows the general trend of electrical conductivity of CNT- polymer nanocomposites. It can be found from almost all the experimental results and also obvious from figure that CNT nanocomposites exhibit a typical percolation behavior and CNT reinforcement to polymers can increase the conductivity of resulting composites to several order of magnitude or even some times higher than ten orders of magnitude.

According to percolation theory the conductivity follow the following power law close to threshold percolation.

$$\sigma = \sigma o(p - p_o)^t \qquad \text{for} \quad p > p_o$$

where σ is the composite conductivity, σ_o is a constant, p the weight fraction of nanotubes, p_o is the percolation threshold and t the critical exponent [81]. Theoretical and experimental results have shown that percolation laws are applicable to CNT-based composites and that the enhanced maximum conductivity and percolation can be achieved with significantly lower filler concentrations than with other carbon and other conductive fillers. Depending on the polymer matrix, the processing technology and the type of nanotubes used, recent experimental studies have achieved percolation thresholds between 0.0021 to 9.5% by weight and critical exponents varying from 0.9 to 7.6 [78].

Sandler et al. [82] observed the percolation threshold of CNTs/epoxy nanocomposites between 0.0225 and 0.04 wt %. They further observed very low percolation threshold at 0.0025 wt% for aligned CNT- epoxy composites [83]. The current voltage behavior measurements exhibited non-ohmic behavior, which is most likely due to tunneling conduction mechanism. The main mechanism of conduction between adjacent nanotubes is probably electron hopping when their separation distance is small. At concentration greater than percolation threshold, conducting paths are formed through the whole nanocomposites, because the distance between the conductive CNT filler (individual or bundles) is small enough to allow efficient electron hopping.

The electrical conductivity of CNT/polymer composites also effected by dispersion and aspect ratio of CNTs and was studied by Barrau et al. [56]. They used palmitic acid as surfactant to improve the nanotube dispersion and reduced the threshold concentration from 0.18 to 0.08 wt%. To study the effect of aspect ratio on electrical conductivity of CNT nanocomposites Bai et al. [84] pretreated MWCNTs to alter their aspect ratios before preparing epoxy/MWCNTs composites and found that the threshold concentration varied from 0.5 to > 4 wt % with decreasing aspect ratio. The effect of alignment of CNTs in polymer composites was also studied. Du et al. [24] found some contradictory results with respect to alignment of rod like fillers and observed the lowest percolation threshold and maximum conductivity with their random orientation. They found that the electrical conductivity of 2 wt% CNT/PMMA nanocomposites decrease significantly (from ~10^{-4} to ~10^{-10} S/cm) when CNTs were highly aligned. In contrast Choi et al. [85] observed that the nanotube alignment increased the conductivity of a 3 wt% CNT/epoxy composites from ~10^{-7} to ~10^{-6}S/cm. In most of the cases the CNT nanocomposites with isotropic nanotubes orientation have greater electrical conductivity than the nanocomposites with highly aligned CNTs especially at lower CNT loadings. By alignment of CNTs in polymers, the percolation pathway is destroyed as aligned CNTs seldomly intersects each other. At higher CNTs loading the conductivity is more in case of aligned CNTs as compared to randomly oriented CNTs.

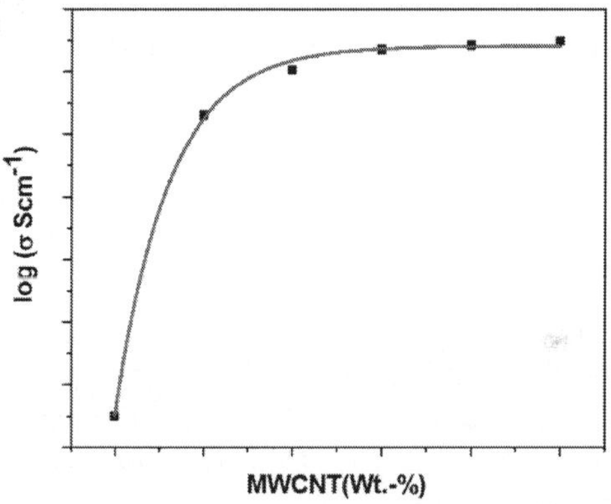

Figure 9. General trend of electrical conductivity of CNT polymer composites

The study carried out by different researchers also revealed that the composites with thermoplastic polymers have higher conductivity as compared to that of thermosetting polymers above percolation threshold. Transport properties in CNT-PMMA composites have been reported by Stephan et al. [86] and Benoit et al. [19] where low percolation threshold of 0.5 wt% and 0.33 wt% respectively were obtained. Singhai et al. [87] found that increase in number of defects lead to a decrease in conductivity. However Lau et al. [88] concluded that fuctionalization of CNTs can enhance the electrical conductivity of MWCNTs. The reason attributed to this phenomenon is electron transfer from the carbon atoms on MWCNTs to functionalized groups attached to the surface favorably promoting conductivity. The study carried out by Grimes et al. [89] revealed that the electrical response of as fabricated MWCNTs is significantly influenced by the presence of residual catalyst metal particles.

7.3. EMI shielding proerties of MWCNTs polymer nanocomposites

The electrical conductivity of CNT reinforced polymer composites makes them a very suitable candidate to be employed for electromagnetic interference (EMI) shielding. EMI is the process by which disruptive electromagnetic energy is transmitted from one electronic device to another via radiation or conduction. As we all know that the electromagnetic waves produced from some electronic instrument have an adverse effect on the performance of the other equipments present nearby causing data loss, introduction of noise, degradation of picture quality etc. The common example is the appearance of noise in television signal when a telephone or mobile rings. Also recent reports of deterious effects of electromagnetic radiations on electro medical devices have caused concern among health care providers. The overlapping of signals transmitted in air traffic system with signals from other electronic equipments became cause of several accidents in past. Also mobile phones and passing taxi radios have been known to interfere with anti-skid braking system (ABS), airbags and other electronic equipments causing drivers to lose control. In today's scenario where rapid communication is required, there is an increase in electromagnetic radiations within the spectrum in which the wireless, cordless and satellite system operates. So it a strong desire to shield electronics equipments from the undesired signals. Problems with EMI can be minimized or sometime eliminated by ensuring that all electronic equipments are operated with a good housing to keep away unwanted radio frequency from entering or leaving. The shielding effectiveness (SE) of the shielding material is its ability to attenuate the propagation of electromagnetic waves through it and measured in decibels (dB) given by

$$SE(\text{dB}) = -10\log(P_t / P_0),$$

where P_t and P_0 are, respectively, the transmitted and incident electromagnetic power. A SE of 10 dB means 90% of signal is blocked and 20 dB means 99% of signal is blocked.

One of the important criterion for a material to be used for EMI shielding material is that it should be electrically conducting. Because of their high electrical conductivity metals have been used for past several years as EMI shielding materials. But the shortcomings of metals like heavy weight, physical rigidity and corrosion restricts their use. The most notable substance that could overcome these shortcomings is the CNT-polymer composites. As discussed in previous sections these are electrically conductive, having low density, corrosion resistant and can be molded in any form. Due to easy processing and good flexibility, CNT–polymer composites have been employed for application as promising EMI shielding materials. The SE of the CNT-polymer composites depends on various factors like,type of CNTs (either SWCNT or MWCNT), aspect ratio of CNTs, quality of CNTs, thickness and electrical conductivity of the shielding material. Several studies have been reported on EMI shielding properties of randomly oriented CNT based polymer composites. Mathur and co-workers [18] have prepared MWNT-PMMA and MWNT-PS composites and observed 18dB and 17dB SE respectively with 10-wt % MWCNT loading. Singh et al. [90] reported a SE of 51 dB by using MWCNT grown carbon fibre fabric based epoxy composites with improved mechanical properties [91].The effect of the length (aspect ratio) of CNTs on EMI SE of composites was also studied by few researchers. Huang et al. [92] reported EMI SE of 18 dB with 15 wt.-% small CNTs and 23-28 dB with 15 wt% long CNTs in X band (8-12.4 GHz). Li et al. [93] also observed that SE with long length CNTs is more as compared to small length CNTs at the same 15wt % loading composites. The residual catalyst metal particle in the cavity of CNTs also effects the SE of the composites.

There are few additional advantages of using MWCNTs as EMI shielding material. The EMI SE also depends on the source of origin of electromagnetic waves. Electrically conducting material can effectively shield the electromagnetic waves generated from an electric source, whereas magnetic materials effectively shield the electromagnetic waves generated from a magnetic source. The MWCNTs exhibits electrical properties because of presence of pi electrons and magnetic properties because of the presence of catalytic iron particles in tubes. Also one common problem experienced with commonly used composite materials for EMI shielding is build up of heat in the substance being shielded. The possible solution for this is to add thermal conducting material. Composites with MWCNTs can easily overcome this problem as it has high thermal conductivity.

7.4. Thermal properties of MWCNTs polymer nanocomposites

As discussed above that the CNTs have thermal conductivity as high as 6600W/mK predicted for SWCNTs [94] at room temperature and have experimental value 3000W/mK for isolated MWCNT. So it is quite expected that the reinforcement of CNTs can significantly enhance the thermal properties of CNT-polymer nanocomposites. The improvement in thermal transport properties of CNT polymer composites leads their applications for usage as printed circuit boards, connectors, thermal interface materials, heat sinks.

8. Conclusion

Synthesis of high quality and reproducible CNTs is still remain a very importnat issue. Chemical vapor deposition has been found an efficient process for the synthseis of bulk quantity of CNTs. The CNT-polymer composites have been developed with improved mechanical properties but for actual structural applications, these have to compete with the existing carbon fibre based composites. Dispersion of high loading of CNTs and their alignment in any polymer matrix without sacrificing their mechanical properties is still a challenge for using CNTs in high performance composites for specific applications such as as automobile, defence, aerospace, sports etc. CNT- carbon fibres-polymer multiscale composites could be an alternative route for further improvement in the mechanical properties of the composites over commercially available CF-polymer composites. Till then electrical properties of CNT polymer composites provides exciting possibility as antistatic and electromagnetic interference shielding material.

Author details

Veena Choudhary[1], B.P. Singh[2] and R.B. Mathur[2]

1 Centre for Polymer Science and Engineering, Indian Institute of Technology Delhi, India

2 Physics and Engineering of Carbon, Division of Materials Physics and Engineering, CSIR-National Physical Laboratory, New Delhi, India

References

[1] Ajayan PM, Zhou OZ. Applications of carbon nanotubes. Carbon Nanotubes2001. p. 391-425.

[2] Lourie O, Wagner HD. Evaluation of Young's modulus of carbon nanotubes by micro-Raman spectroscopy. Journal of Materials Research 1998;13:2418-22.

[3] Yu MF, Files BS, Arepalli S, Ruoff RS. Tensile loading of ropes of single wall carbon nanotubes and their mechanical properties. Phys Rev Lett 2000;84:5552-5.

[4] Yu MF, Lourie O, Dyer MJ, Moloni K, Kelly TF, Ruoff RS. Strength and breaking mechanism of multiwalled carbon nanotubes under tensile load. Science 2000;287:637-40.

[5] Meincke O, Kaempfer D, Weickmann H, Friedrich C, Vathauer M, Warth H. Mechanical properties and electrical conductivity of carbon-nanotube filled polyamide-6 and its blends with acrylonitrile/butadiene/styrene. Polymer 2004;45:739-48.

[6] Ebbesen TW, Lezec HJ, Hiura H, Bennett JW, Ghaemi HF, Thio T. Electrical conductivity of individual carbon nanotubes. Nature 1996;382:54-6.

[7] Lee JO, Park C, Kim JJ, Kim J, Park JW, Yoo KH. Formation of low-resistance ohmic contacts between carbon nanotube and metal electrodes by a rapid thermal annealing method. Journal of Physics D-Applied Physics 2000;33:1953-6.

[8] Hone J. Dekker Encyclopedia of Nanoscience and Nanotechnology 2004.

[9] Jung SH, Kim MR, Jeong SH, Kim SU, Lee OJ, Lee KH, et al. High-yield synthesis of multi-walled carbon nanotubes by arc discharge in liquid nitrogen. Applied Physics a-Materials Science & Processing 2003;76:285-6.

[10] Ebbesen TW, Ajayan PM. LARGE-SCALE SYNTHESIS OF CARBON NANOTUBES. Nature 1992;358:220-2.

[11] Mathur RB, Seth S, Lal C, Rao R, Singh BP, Dhami TL, et al. Co-synthesis, purification and characterization of single- and multi-walled carbon nanotubes using the electric arc method. Carbon 2007;45:132-40.

[12] Mathur RB, Chatterjee S, Singh BP. Growth of carbon nanotubes on carbon fibre substrates to produce hybrid/phenolic composites with improved mechanical properties. Composites Science and Technology 2008;68:1608-15.

[13] Mathur RB, Singh BP, Dhami TL, Kalra Y, Lal N, Rao R, et al. Influence of Carbon Nanotube Dispersion on the Mechanical Properties of Phenolic Resin Composites. Polymer Composites 31:321-7.

[14] Guo T, Nikolaev P, Thess A, Colbert DT, Smalley RE. CATALYTIC GROWTH OF SINGLE-WALLED NANOTUBES BY LASER VAPORIZATION. Chem Phys Lett 1995;243:49-54.

[15] Thess A, Lee R, Nikolaev P, Dai HJ, Petit P, Robert J, et al. Crystalline ropes of metallic carbon nanotubes. Science 1996;273:483-7.

[16] Coleman JN, Khan U, Blau WJ, Gun'ko YK. Small but strong: A review of the mechanical properties of carbon nanotube-polymer composites. Carbon 2006;44:1624-52.

[17] Moniruzzaman M, Winey KI. Polymer nanocomposites containing carbon nanotubes. Macromolecules 2006;39:5194-205.

[18] Mathur RB, Pande S, Singh BP, Dhami TL. Electrical and mechanical properties of multi-walled carbon nanotubes reinforced PMMA and PS composites. Polymer Composites 2008;29:717-27.

[19] Benoit JM, Corraze B, Lefrant S, Blau WJ, Bernier P, Chauvet O. Transport properties of PMMA-carbon nanotubes composites. Synth Met 2001;121:1215-6.

[20] Lau KT, Lu M, Lam CK, Cheung HY, Sheng FL, Li HL. Thermal and mechanical properties of single-walled carbon nanotube bundle-reinforced epoxy nanocomposites: the role of solvent for nanotube dispersion. Composites Science and Technology 2005;65:719-25.

[21] Bryning MB, Milkie DE, Islam MF, Kikkawa JM, Yodh AG. Thermal conductivity and interfacial resistance in single-wall carbon nanotube epoxy composites. Applied Physics Letters 2005;87.

[22] Singh BP, Singh D, Mathur RB, Dhami TL. Influence of Surface Modified MWCNTs on the Mechanical, Electrical and Thermal Properties of Polyimide Nanocomposites. Nanoscale Research Letters 2008;3:444-53.

[23] de la Chapelle ML, Stephan C, Nguyen TP, Lefrant S, Journet C, Bernier P, et al. Raman characterization of singlewalled carbon nanotubes and PMMA-nanotubes composites. Synth Met 1999;103:2510-2.

[24] Du FM, Fischer JE, Winey KI. Coagulation method for preparing single-walled carbon nanotube/poly(methyl methacrylate) composites and their modulus, electrical conductivity, and thermal stability. Journal of Polymer Science Part B-Polymer Physics 2003;41:3333-8.

[25] Andrews R, Jacques D, Qian DL, Rantell T. Multiwall carbon nanotubes: Synthesis and application. Accounts of Chemical Research 2002;35:1008-17.

[26] Tang WZ, Santare MH, Advani SG. Melt processing and mechanical property characterization of multi-walled carbon nanotube/high density polyethylene (MWNT/HDPE) composite films. Carbon 2003;41:2779-85.

[27] Bhattacharyya AR, Sreekumar TV, Liu T, Kumar S, Ericson LM, Hauge RH, et al. Crystallization and orientation studies in polypropylene/single wall carbon nanotube composite. Polymer 2003;44:2373-7.

[28] Niu C, Ngaw L, Fischer A, Hoch R, Fischer AB, Ngam L. Piezoelectric high damping material for vibration suppression in vehicle, has carbon nano tube as piezoelectric electroconductive particle for packing. Hyperion Catalysis Int Inc.

[29] Jia ZJ, Wang ZY, Xu CL, Liang J, Wei BQ, Wu DH, et al. Study on poly(methyl methacrylate)/carbon nanotube composites. Materials Science and Engineering a-Structural Materials Properties Microstructure and Processing 1999;271:395-400.

[30] Velasco-Santos C, Martinez-Hernandez AL, Fisher FT, Ruoff R, Castano VM. Improvement of thermal and mechanical properties of carbon nanotube composites through chemical functionalization. Chemistry of Materials 2003;15:4470-5.

[31] Garg P, Singh BP, Kumar G, Gupta T, Pandey I, Seth RK, et al. Effect of dispersion conditions on the mechanical properties of multi-walled carbon nanotubes based epoxy resin composites. Journal of Polymer Research 2011;18:1397-407.

[32] Pande S, Mathur RB, Singh BP, Dhami TL. Synthesis and Characterization of Multiwalled Carbon Nanotubes-Polymethyl Methacrylate Composites Prepared by In Situ Polymerization Method. Polymer Composites 2009;30:1312-7.

[33] Li XH, Wu B, Huang JE, Zhang J, Liu ZF, Li HI. Fabrication and characterization of well-dispersed single-walled carbon nanotube/polyaniline composites. Carbon 2003;41:1670-3.

[34] Xiao QF, Zhou X. The study of multiwalled carbon nanotube deposited with conducting polymer for supercapacitor. Electrochimica Acta 2003;48:575-80.

[35] Saini P, Choudhary V, Singh BP, Mathur RB, Dhawan SK. Polyaniline-MWCNT nanocomposites for microwave absorption and EMI shielding. Materials Chemistry and Physics 2009;113:919-26.

[36] Saini P, Choudhary V, Singh BP, Mathur RB, Dhawan SK. Enhanced microwave absorption behavior of polyaniline-CNT/polystyrene blend in 12.4-18.0 GHz range. Synth Met;161:1522-6.

[37] Haggenmueller R, Gommans HH, Rinzler AG, Fischer JE, Winey KI. Aligned single-wall carbon nanotubes in composites by melt processing methods. Chem Phys Lett 2000;330:219-25.

[38] Pande S, Singh BP, Mathur RB, Dhami TL, Saini P, Dhawan SK. Improved Electromagnetic Interference Shielding Properties of MWCNT-PMMA Composites Using Layered Structures. Nanoscale Research Letters 2009;4:327-34.

[39] Singh BP, Prabha, Saini P, Gupta T, Garg P, Kumar G, et al. Designing of multiwalled carbon nanotubes reinforced low density polyethylene nanocomposites for suppression of electromagnetic radiation. J Nanopart Res 2011;13:7065-74.

[40] Jindal P, Pande S, Sharma P, Mangla V, Chaudhury A, Patel D, et al. High strain rate behavior of multi-walled carbon nanotubes–polycarbonate composites. Composites Part B: Engineering.

[41] Xia HS, Wang Q, Li KS, Hu GH. Preparation of polypropylene/carbon nanotube composite powder with a solid-state mechanochemical pulverization process. Journal of Applied Polymer Science 2004;93:378-86.

[42] Regev O, ElKati PNB, Loos J, Koning CE. Preparation of conductive nanotube-polymer composites using latex technology. Advanced Materials 2004;16:248-+.

[43] Vigolo B, Penicaud A, Coulon C, Sauder C, Pailler R, Journet C, et al. Macroscopic fibers and ribbons of oriented carbon nanotubes. Science 2000;290:1331-4.

[44] Dhand C, Arya SK, Singh SP, Singh BP, Datta M, Malhotra BD. Preparation of polyaniline/multiwalled carbon nanotube composite by novel electrophoretic route. Carbon 2008;46:1727-35.

[45] Montoro LA, Rosolen JM. A multi-step treatment to effective purification of single-walled carbon nanotubes. Carbon 2006;44:3293-301.

[46] Cooper CA, Cohen SR, Barber AH, Wagner HD. Detachment of nanotubes from a polymer matrix. Applied Physics Letters 2002;81:3873-5.

[47] Sen R, Zhao B, Perea D, Itkis ME, Hu H, Love J, et al. Preparation of single-walled carbon nanotube reinforced polystyrene and polyurethane nanofibers and membranes by electrospinning. Nano Letters 2004;4:459-64.

[48] Georgakilas V, Kordatos K, Prato M, Guldi DM, Holzinger M, Hirsch A. Organic functionalization of carbon nanotubes. Journal of the American Chemical Society 2002;124:760-1.

[49] Liang F, Sadana AK, Peera A, Chattopadhyay J, Gu ZN, Hauge RH, et al. A convenient route to functionalized carbon nanotubes. Nano Letters 2004;4:1257-60.

[50] Kinloch IA, Roberts SA, Windle AH. A rheological study of concentrated aqueous nanotube dispersions. Polymer 2002;43:7483-91.

[51] Gojny FH, Nastalczyk J, Roslaniec Z, Schulte K. Surface modified multi-walled carbon nanotubes in CNT/epoxy-composites. Chem Phys Lett 2003;370:820-4.

[52] Sinnott SB. Chemical functionalization of carbon nanotubes. Journal of Nanoscience and Nanotechnology 2002;2:113-23.

[53] Crespi VH, Chopra NG, Cohen ML, Zettl A, Radmilovic V. Site-selective radiation damage of collapsed carbon nanotubes. Applied Physics Letters 1998;73:2435-7.

[54] Sulong AB. MN, Sahari J., Ramli R., Deros BM., Park J. Electrical Conductivity Behaviour of Chemical Functionalized MWCNTs Epoxy Nanocomposites. European Journal of Scientific Research 2009;29:13-21.

[55] Islam MF, Rojas E, Bergey DM, Johnson AT, Yodh AG. High weight fraction surfactant solubilization of single-wall carbon nanotubes in water. Nano Letters 2003;3:269-73.

[56] Barrau S, Demont P, Perez E, Peigney A, Laurent C, Lacabanne C. Effect of palmitic acid on the electrical conductivity of carbon nanotubes-epoxy resin composites. Macromolecules 2003;36:9678-80.

[57] Gong XY, Liu J, Baskaran S, Voise RD, Young JS. Surfactant-assisted processing of carbon nanotube/polymer composites. Chemistry of Materials 2000;12:1049-52.

[58] Ozarkar S, Jassal M, Agrawal AK. Improved dispersion of carbon nanotubes in chitosan. Fibers and Polymers 2008;9:410-5.

[59] Park JG, Louis J, Cheng QF, Bao JW, Smithyman J, Liang R, et al. Electromagnetic interference shielding properties of carbon nanotube buckypaper composites. Nanotechnology 2009;20.

[60] Yang K, Gu MY, Guo YP, Pan XF, Mu GH. Effects of carbon nanotube functionalization on the mechanical and thermal properties of epoxy composites. Carbon 2009;47:1723-37.

[61] Wardle BL, Saito DS, Garcia EJ, Hart AJ, de Villoria RG, Verploegen EA. Fabrication and characterization of ultrahigh-volume-fraction aligned carbon nanotube-polymer composites. Advanced Materials 2008;20:2707-+.

[62] Gou JH. Single-walled nanotube bucky paper and nanocomposite. Polymer International 2006;55:1283-8.

[63] Wang Z, Liang ZY, Wang B, Zhang C, Kramer L. Processing and property investigation of single-walled carbon nanotube (SWNT) buckypaper/epoxy resin matrix nanocomposites. Composites Part a-Applied Science and Manufacturing 2004;35:1225-32.

[64] Feng QP, Yang JP, Fu SY, Mai YW. Synthesis of carbon nanotube/epoxy composite films with a high nanotube loading by a mixed-curing-agent assisted layer-by-layer method and their electrical conductivity. Carbon 2010;48:2057-62.

[65] Feng QP, Shen XJ, Yang JP, Fu SY, Mai YW, Friedrich K. Synthesis of epoxy composites with high carbon nanotube loading and effects of tubular and wavy morphology on composite strength and modulus. Polymer 2011;52:6037-45.

[66] Bradford PD, Wang X, Zhao HB, Maria JP, Jia QX, Zhu YT. A novel approach to fabricate high volume fraction nanocomposites with long aligned carbon nanotubes. Composites Science and Technology 2010;70:1980-5.

[67] Ogasawara T, Moon SY, Inoue Y, Shimamura Y. Mechanical properties of aligned multi-walled carbon nanotube/epoxy composites processed using a hot-melt prepreg method. Composites Science and Technology 2011;71:1826-33.

[68] Wang XK, Lin XW, Dravid VP, Ketterson JB, Chang RPH. GROWTH AND CHARACTERIZATION OF BUCKYBUNDLES. Applied Physics Letters 1993;62:1881-3.

[69] Ajayan PM, Stephan O, Colliex C, Trauth D. ALIGNED CARBON NANOTUBE ARRAYS FORMED BY CUTTING A POLYMER RESIN-NANOTUBE COMPOSITE. Science 1994;265:1212-4.

[70] Deheer WA, Bacsa WS, Chatelain A, Gerfin T, Humphreybaker R, Forro L, et al. ALIGNED CARBON NANOTUBE FILMS - PRODUCTION AND OPTICAL AND ELECTRONIC-PROPERTIES. Science 1995;268:845-7.

[71] Fan SS, Chapline MG, Franklin NR, Tombler TW, Cassell AM, Dai HJ. Self-oriented regular arrays of carbon nanotubes and their field emission properties. Science 1999;283:512-4.

[72] Li WZ, Xie SS, Qian LX, Chang BH, Zou BS, Zhou WY, et al. Large-scale synthesis of aligned carbon nanotubes. Science 1996;274:1701-3.

[73] Jin L, Bower C, Zhou O. Alignment of carbon nanotubes in a polymer matrix by mechanical stretching. Applied Physics Letters 1998;73:1197-9.

[74] Smith BW, Benes Z, Luzzi DE, Fischer JE, Walters DA, Casavant MJ, et al. Structural anisotropy of magnetically aligned single wall carbon nanotube films. Applied Physics Letters 2000;77:663-5.

[75] Schadler LS, Giannaris SC, Ajayan PM. Load transfer in carbon nanotube epoxy composites. Applied Physics Letters 1998;73:3842-4.

[76] Bai J. Evidence of the reinforcement role of chemical vapour deposition multi-walled carbon nanotubes in a polymer matrix. Carbon 2003;41:1325-8.

[77] Zhou YX, Pervin F, Lewis L, Jeelani S. Experimental study on the thermal and mechanical properties of multi-walled carbon nanotube-reinforced epoxy. Materials Science and Engineering a-Structural Materials Properties Microstructure and Processing 2007;452:657-64.

[78] Du JH, Bai J, Cheng HM. The present status and key problems of carbon nanotube based polymer composites. Express Polymer Letters 2007;1:253-73.

[79] Haggenmueller R, Zhou W, Fischer JE, Winey KI. Production and characterization of polymer nanocomposites with highly aligned single-walled carbon nanotubes. Journal of Nanoscience and Nanotechnology 2003;3:105-10.

[80] Andrews R, Jacques D, Rao AM, Rantell T, Derbyshire F, Chen Y, et al. Nanotube composite carbon fibers. Applied Physics Letters 1999;75:1329-31.

[81] D. Stauffer AA. Introduction to percolation theory. Taylor & Francis; 1992.

[82] Sandler J, Shaffer MSP, Prasse T, Bauhofer W, Schulte K, Windle AH. Development of a dispersion process for carbon nanotubes in an epoxy matrix and the resulting electrical properties. Polymer 1999;40:5967-71.

[83] Sandler JKW, Kirk JE, Kinloch IA, Shaffer MSP, Windle AH. Ultra-low electrical percolation threshold in carbon-nanotube-epoxy composites. Polymer 2003;44:5893-9.

[84] Bai JB, Allaoui A. Effect of the length and the aggregate size of MWNTs on the improvement efficiency of the mechanical and electrical properties of nanocomposites - experimental investigation. Composites Part a-Applied Science and Manufacturing 2003;34:689-94.

[85] Choi ES, Brooks JS, Eaton DL, Al-Haik MS, Hussaini MY, Garmestani H, et al. Enhancement of thermal and electrical properties of carbon nanotube polymer composites by magnetic field processing. Journal of Applied Physics 2003;94:6034-9.

[86] Stephan C, Nguyen TP, Lahr B, Blau W, Lefrant S, Chauvet O. Raman spectroscopy and conductivity measurements on polymer-multiwalled carbon nanotubes composites. Journal of Materials Research 2002;17:396-400.

[87] Singjai P, Changsarn S, Thongtem S. Electrical resistivity of bulk multi-walled carbon nanotubes synthesized by an infusion chemical vapor deposition method. Materials Science and Engineering: A 2007;443:42-6.

[88] Lau CH, Cervini R, Clarke SR, Markovic MG, Matisons JG, Hawkins SC, et al. The effect of functionalization on structure and electrical conductivity of multi-walled carbon nanotubes. J Nanopart Res 2008;10:77-88.

[89] Grimes CA, Dickey EC, Mungle C, Ong KG, Qian D. Effect of purification of the electrical conductivity and complex permittivity of multiwall carbon nanotubes. Journal of Applied Physics 2001;90:4134-7.

[90] Singh BP, Choudhary V, Saini P, Mathur RB. Designing of epoxy composites reinforced with carbon nanotubes grown carbon fiber fabric for improved electromagnetic interference shielding. AIP Adv 2012;2:6.

[91] Mathur RB, Singh, B.P,Tiwari, P.K., Gupta, T.K.Choudhary, V. Enhancement in the thermomechanical properties of carbon fibre-carbon nanotubes-epoxy hybrid composites. International Journal of Nanotechnology 2012;9:1040-1049.

[92] Huang Y, Li N, Ma Y, Feng D, Li F, He X, et al. The influence of single-walled carbon nanotube structure on the electromagnetic interference shielding efficiency of its epoxy composites. Carbon 2007;45:1614-21.

[93] Li N, Huang Y, Du F, He XB, Lin X, Gao HJ, et al. Electromagnetic interference (EMI) shielding of single-walled carbon nanotube epoxy composites. Nano Letters 2006;6:1141-5.

[94] Berber S, Kwon YK, Tomanek D. Unusually high thermal conductivity of carbon nanotubes. Phys Rev Lett 2000;84:4613-6.

Chapter 10

Kinetics of Growing Centimeter Long Carbon Nanotube Arrays

Wondong Cho, Mark Schulz and Vesselin Shanov

Additional information is available at the end of the chapter

http://dx.doi.org/10.5772/50837

1. Introduction

Carbon nanotubes (CNTs) are fascinating materials with outstanding mechanical, optical, thermal, and electrical properties [1-4]. CNTs also have a huge aspect ratio and a large surface area to volume ratio. Because of their unique properties, vertically aligned centimeter long CNT arrays have generated great interest for environmental sensors, biosensors, spinning CNT into yarn, super-capacitors, and super-hydrophobic materials for self-cleaning surfaces [5-11]. Yun et al. studied a needle-type biosensor based on CNTs to detect dopamine. Their results showed advantages of using CNT biosensors for detecting neurotransmitters [11]. Most of the envisioned applications require CNTs with high quality, a long length, and well aligned vertical orientation. Although many researchers have studied the synthesis of vertically aligned CNT arrays, the CNT growth mechanism still needs to be better understood. In addition, CNT lengths are typically limited to a few millimeters because the catalyst lifetime is usually less than one hour [12- 16]. Many groups have studied the kinetics of CNT growth trying to improve CNT properties. Different observation methods [17-22] were used to determine the effect of the catalyst, buffer layers, carbon precursor, and deposition conditions on nanotube growth. One of the suggested growth mechanisms postulates several steps [23]. First, the carbon source dissociates on the surface of the substrate. Next, the carbon atoms diffuse to the molten catalyst islands and dissolve. The metal-carbon solution formed reaches a supersaturated state. Finally, the carbon nanotubes start to grow from the carbon- catalyst solution. *In situ* observation of CNTs during their nucleation and growth is a useful method to understand the growth mechanism, which might help to overcome the limitation of the short length of nanotubes, and to control array growth and quality. Various remarkable approaches of *in situ* observation have been performed to affirm the growth mechanism of vertically aligned CNTs and also to obtain kinetics data such as growth rate and activation energy [12, 24-27]. Puretzky *et al.* studied the growth kinetics of CNT arrays using *in situ* time-lapse photography and laser irradiation under diffusion-limited growth conditions [28]. *In situ* transmission electron microscopy (TEM) was used by Kim *et al.* to study the dynamic growth behavior of CNT arrays [29]. Additionally, a pseudo *in situ* monitoring method was used to investigate the kinetics of CNT array growth by creating marks on the side of the CNT array during the growth. Using this method, several groups demonstrated root growth for their catalyst systems. However, their studies were limited to short lengths and also required *ex-situ* observation with SEM to obtain the growth length as a function of time [25, 27, 30, 31]. Most reported methods are designed to operate and monitor the growth length with time for relatively short (a few millimeters long) CNT arrays and also do not provide kinetics data for growing centimeter long CNT arrays.

Recently, we have developed new catalyst systems which are able to grow over 1 centimeter long vertically aligned CNT arrays [6]. In this paper we examined the growth mechanism and kinetics of centimeter long carbon nanotube arrays using a real-time photography technique and the effect of growth temperature and growth time on morphology of CNTs.

2. Experimental Method

Fig. 1 shows the schematic of the chemical vapor deposition (CVD) system used for centimeter long CNT array growth. The reactor consists of a 2 inch quartz tube placed inside a high temperature furnace (Barnstead International, type F79400 tube furnace) and four mass-flow controllers (MFC) which control the flow rate of the gas reactants such as hydrogen, ethylene, water vapor and argon. A water bubbler is also installed to provide water vapor using argon carrier gas. A window on the side of the reactor is used to acquire real-time images of CNT arrays and to record data with a digital camera (Olympus E510) controlled by a computer.

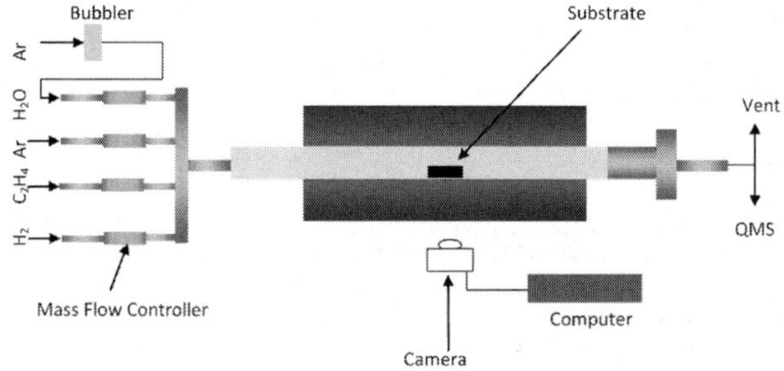

Figure 1. Schematic of the CVD system for direct observation of the centimeter long carbon nanotube arrays during their growth. The top view of the reactor is shown.

The substrates were parts of 4 inch silicon wafers (100) with SiO_2 (500 nm) on the top. The buffer and catalyst layers based on Al_2O_3 (15 nm thick)/Fe (1 nm thick) were deposited on the

http://dx.doi.org/10.5772/50837

wafers using e-beam evaporation. After the deposition, the substrates were annealed for several hours at 400 ºC in Air. All the experiments were performed using the following optimized recipe for centimeter long CNT arrays: 560 mmHg of argon, 60 mmHg of hydrogen, and 140 mmHg of ethylene as a carbon precursor. The water concentration in the reactor was near 900 ppm measured by a quadrupole mass spectrometer (QMS). The total pressure was kept at one atmosphere during the growth and the temperature varied from 690 ºC to 840 ºC. Real-time images of the CNT array growth were recorded from the moment that ethylene was introduced into the reactor. The images were used to study the growth mechanism and kinetics of the CNT growth. Scanning electron microscopy (SEM, Phillips XL30 ESEM), high resolution transmission electron microscopy (HRTEM, JOEL 2000 FX) and Micro-Raman spectroscopy (Renishaw inVia Reflex Micro-Raman) were employed to characterize the CNT morphology.

3. Results and discussion

3.1. Growth evolution by real-time photography

Real-time photography was used to study the growth mechanism and kinetics of centimeter long CNT array growth. The digital camera provided clear images showing details related to the dynamic changes of the array shape during the growth. This was achieved by controlling the intervals for taking pictures from a few seconds to several hours depending on the experimental conditions.

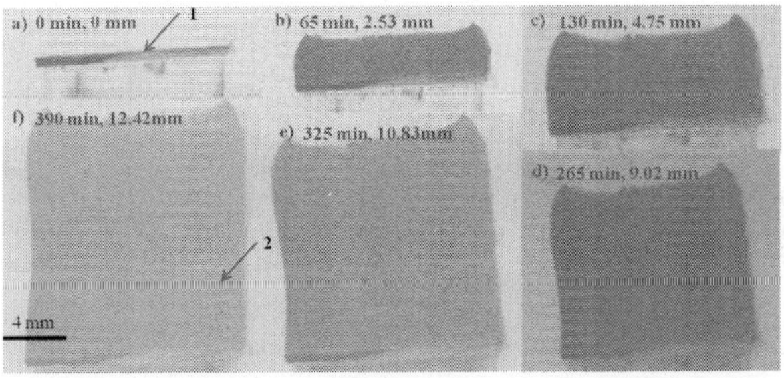

Figure 2. Real-time images of the centimeter long CNT array growth evolution with time during CVD at 780 ºC: (a) Side image of the substrate at zero growth time, (b) to (f) Images of CNT arrays grown for different times.

Fig. 2 illustrates sequential images of vertically aligned centimeter long CNT arrays grown at 780 ºC. As can be seen from Fig. 2, it is easy to distinguish the substrate from the CNT array. Arrow 1 in Fig. 2a points to a side view of the substrate. Arrow 2 in Fig. 2f shows the side view of the CNT array. The growth length can be obtained as a function of the deposition time from the images. Changes in the array shape can also be observed during the entire growth time. In Fig. 2f, the growth length was 12.47 mm and the catalyst lifetime was 450 min. This experiment was repeated several times at the same deposition conditions and the results were reproducible. Hence, Fig. 2 demonstrates the real-time images which allow measuring the growth length and observing the CNT array growth mechanism. This approach gives reproducible results for studying the kinetics of CNT array growth.

Movies composed of multiple images were created to investigate the morphological changes and growth mechanism of CNT arrays during their synthesis (Supporting material 1).

3.2. Growth mode of centimeter long CNT arrays

A carbon source interruption method combined with real-time photography was used to determine the CNT array growth mechanism (root vs. tip growth). Fig. 3 shows real-time images of CNT arrays obtained for different periods of growth employing 5 minute interruption of the ethylene supply. In this experiment, the ethylene flow was first stopped after 70 minutes growth while the rest of reactant gases such as Ar, H_2 and water vapor were continuously supplied with same partial pressure. Next, the ethylene was resupplied after the 5 min break. Fig. 3b, c and d show real-time images of CNT arrays before and after the 5 minute interruption of the carbon precursor. The growth length in Fig. 3b was 2.89 mm. The length did not change after the ethylene interruption (Fig. 3c). CNTs started to grow again after ethylene was resupplied (Fig. 3d). As can be seen in Fig. 3e and f, the first layer which grew before the 5 min ethylene interruption detached from the second layer grown after the 5 min interruption. The separation may be caused by water etching the interface between the root of the CNTs and the substrate during the interruption. Even though the first layer detached, the second layer continued to grow from the bottom. Fig. 3f shows that the growth length of the second layer was 2.85 mm. The calculated growth rate was 38.5 μm/min before and after the interruption and until the growth stopped.

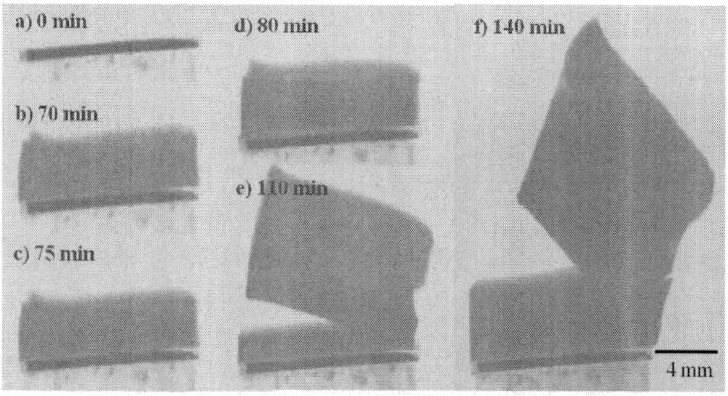

Figure 3. Real-time images of CNT array grown for different periods of time at a deposition temperature of 780 °C with 5 minutes interruption of ethylene: (a) Side image of the substrate at zero growth time, (b) to (f) Pictures of CNT arrays grown for different times.

It was obvious from the data that the interruption didn't affect the growth rate during the CNT synthesis. The images also reveal that the growth pathway of the centimeter long CNT array is "root growth" (Supporting materials 2).

The top surface of the centimeter long CNT array was studied by Energy Dispersive X-ray Spectroscopy (EDS) to determine if metal catalyst moved to the tips of the CNTs. No trace of iron catalyst was detected on the top of the CNT array. Thermogravimetric analysis (TGA) performed at a heating rate of 10 ºC/min in air showed that the change of CNT array weight was negligible at temperatures below 550 ºC. The combustion started slightly below 700 ºC and was completed at 750 ºC. The amount of residual matter was extremely small and was not measureable after the completion of combustion above 750 ºC. These results implied that the CNT array was almost "catalyst free".

3.3. Kinetics of centimeter long CNT array growth

The change in the growth length as a function of deposition time at different temperatures was investigated using real-time photography. Fig. 4 displays the dependence of growth length on the growth time in the range of 730 ºC to 840 ºC. The rest of the CVD experimental conditions were kept constant. As shown in Fig. 4a, the growth length increases linearly with the growth time within the entire temperature range of 730 ºC to 840 ºC. The data show that the growth rate remained constant until the catalyst was deactivated. Diffusion of the carbon precursor through the CNT forest apparently did not limit the growth rate even in the case of 12 mm long CNT arrays. It reveals the carbon source was able to reach the surface of catalyst particles without significant resistance. These results show that the growth rate followed a kinetic controlled mode. Zhu et al. reported that their catalyst system was controlled by a gas diffusion process and their growth rate decreased gradually with increasing the length of CNT array. They reported a diffusion controlled mode for a similar range of growth temperatures [31] used in the present paper.

In the current experiments, growth termination occurred abruptly for all experimental deposition temperatures. After growth termination, the array length remained constant with time. The reason for such abrupt catalyst deactivation is not clear and it could not be interpreted using the suggested mechanisms in the literature such as Oswald ripening, forming stable iron carbides, and depletion of the catalyst [32-34]. It is hypothesized that catalyst deactivation occurs due to several complex reasons. One reason is that amorphous carbon is built up during the CVD process and covers the catalyst active site at the surface of the substrate which results in passivation of the catalyst.

Fig. 4b shows the effect of temperature on the final growth length. The slope of the curve increases gradually as the CVD temperature increases from 730 ºC to 780 ºC, and then decreases. The longest CNT length of 12.42mm was achieved at 780 ºC. It was observed that the longest centimeter long CNT array was obtained neither at the highest growth rate nor at the longest catalyst lifetime. Analyzing Fig. 4 provides a better understanding of the catalyst lifetime at different temperatures. Fig. 4c shows that the catalyst lifetime decreases linearly with the increase of temperature in the range of 730 ºC to 840 ºC. In this experiment, the longest catalyst lifetime was 895 min at 730 ºC. The catalyst lifetime decreases dramatically down to a few minutes as the temperature approaches to 840 ºC. The results in this paper demonstrate that the catalyst lifetime and the final growth length are considerably longer than reported in the literature [12-16]. Water vapor usually plays an important role in enhancing the growth rate and also prolonging the catalyst lifetime [35] since water vapor may clean the surface of the catalyst particles by removing the amorphous carbon.

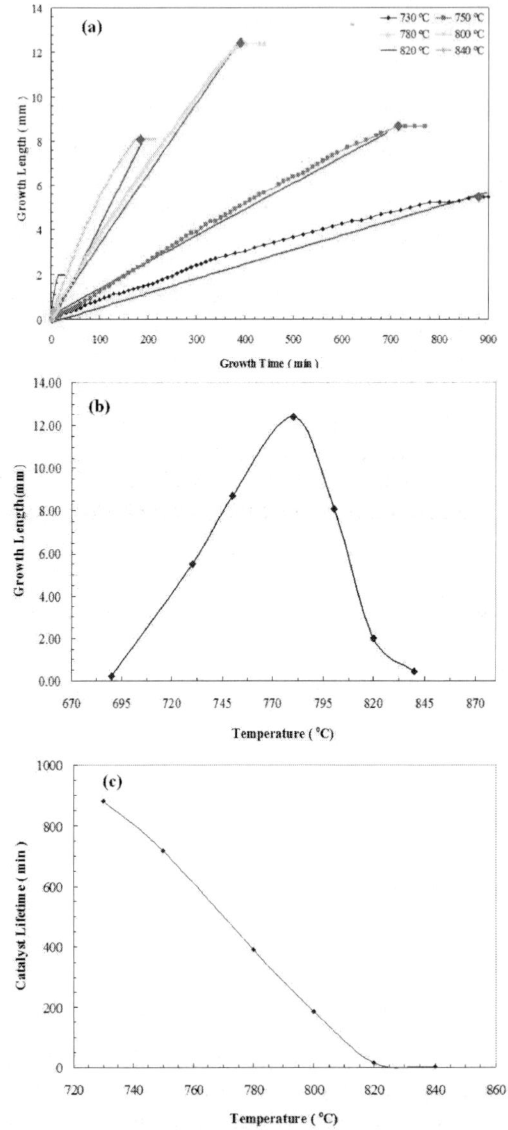

Figure 4. Experimental results from the kinetics study of centimeter long CNT array growth: (a) Growth length as a function of the growth time. The solid red lines indicate the fitting of the analytical growth equation to the experimental results. (b) Plot of the final growth length after termination vs. deposition temperature. (c) Plot of the catalyst lifetime as a function of the temperature.

Fig. 5 shows the Arrhenius plot of the CNT growth rate as a function of the temperature. The growth rate was calculated using the final growth length divided by the catalyst lifetime. The highest growth rate achieved was 193 μm/min at 840 ºC. The present results reveal that the growth rate is proportional to the concentration of ethylene, so first order reaction is assumed. The activation energy E_a was calculated based on the Arrhenius equation $k = k_0 \exp(-E_a/RT)$ where, k_0 is the frequency factor, T is the growth temperature (K), *k* is the reaction rate constant (in this case *k is* growth rate constant), and R is the ideal gas constant. The activation energy calculated from Arrhenius equation and Fig. 5 was about 248 kJ/mol. Similarly, Zhu et al. obtained activation energy of 201.2 kJ/mol for a $Fe/Al_2O_3/Si$ substrate using ethylene as a carbon source [27]. Li et al also reported activation energy of 158 kJ/mol for the temperature interval of 730 ºC to 780 ºC using ethylene [24]. Both groups concluded that the growth rate was not affected by the diffusion of the carbon source to the catalysts because the CNT length increased linearly with time. In the current paper, the same trend is observed although the catalyst lifetime and CNT growth length were much longer.

Based on the kinetics data, it was concluded that the centimeter long CNT array growth does not fit the diffusion controlled mode and is more reasonable to be considered as a kinetically controlled process. This conclusion is based on the specific process conditions and size of the substrate used here. Diffusion may play a role in growth for other experimental conditions.

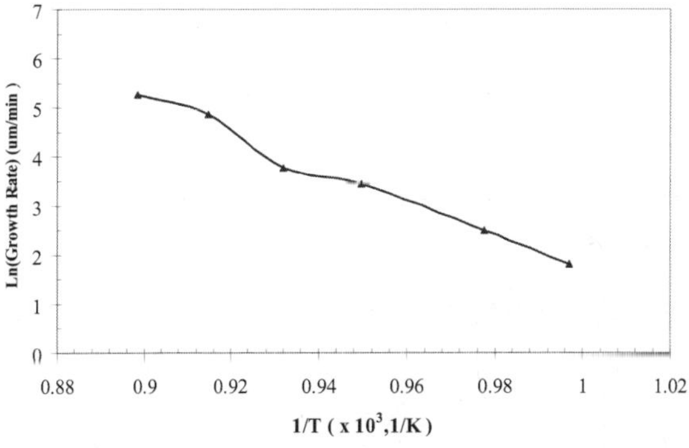

Figure 5. Arrhenius plot of centimeter long CNT array growth rate as a function of 1/temperature.

Zhu et al. adopted the silicon oxidation model for a diffusion controlled mode of CNT growth described with the equation: $h_0 = 0.5(A^2 + 4Bt)^{1/2} - 0.5A$. Futaba et al. assumed that the deactivation of catalyst was analogous to radioactive decay to model their super growth of CNT via the equation: $H(t) = \beta\tau_0 (1 - e^{-t/\tau 0})$. They demonstrated a good fit of the experimental data to the model equations. However, Futaba et al. could not extrapolate it for a longer growth time in order to obtain the final growth length and to predict the abrupt growth termination. Hence, the models described above cannot be adopted for the centimeter long CNT array growth. Based on the experimental data in this paper, it was attempted to derive a growth model using an analytical method that can be reasonably applied to the growth of centimeter long CNT arrays.

From Fig. 4, the growth length can be expressed by a linear equation:

$$GL_T(t) = GL_{Tf} - r_g(t_{Tf} - t) \tag{1}$$

where, $GL_T(t)$ is the array length at a certain growth time t and temperature T, GL_{Tf} is the final growth length of the centimeter long CNT array after termination at temperature T, r_g is the growth rate and t_{Tf} is the catalyst lifetime. The final growth length GL_{Tf} can be obtained by the following equation depending on the deposition temperature:

In case of temperatures below 780 ºC, the final CNT length

$$GL_{Tf} = 0.1367T(^{\circ}C) - 94.1 \tag{2}$$

Above deposition temperatures of 780 ºC, the final CNT length is

$$GL_{Tf} = -0.216T(^{\circ}C) + 181 \tag{3}$$

The catalyst lifetime t_{Tf} can also be expressed by a linear equation from Fig. 6b:

$$t_{Tf} = -10.08T(^{\circ}C) + 8253.7. \tag{4}$$

The solid red lines displayed in Fig. 4a are plotted based on the equation (1). As shown in Fig. 4a, the plot fits the experimental data well. The analytical model can also predict the final growth length, growth rate, and catalyst lifetime for a certain CVD temperature.

3.4. Morphology of centimeter long CNTs

Fig. 6a, b and c display SEM images of centimeter long CNT arrays obtained at different magnifications. At low magnification of 1000x (Fig. 6a) the image shows appearance of vertically aligned CNTs. Fig. 6b and c are taken at higher magnifications and reveal individual CNTs. The images indicate that despite the long growth time and centimeter length, the tubes grew vertically without any interruption until the catalyst activity was terminated.

HRTEM was used to study changes in the structure and diameter distribution of individual CNTs at different growth temperatures. The images shown in Fig. 7 reveal well defined multi-wall CNTs without metal catalyst incorporated into the tubes. At low temperatures amorphous carbon deposited on the walls was also observed.

Fig. 7e and f show the distribution of number of walls and average tube diameter of each wall at different growth temperatures. The diameter of CNTs is in the range of 5 to 9 nm and is independent on the number of walls.

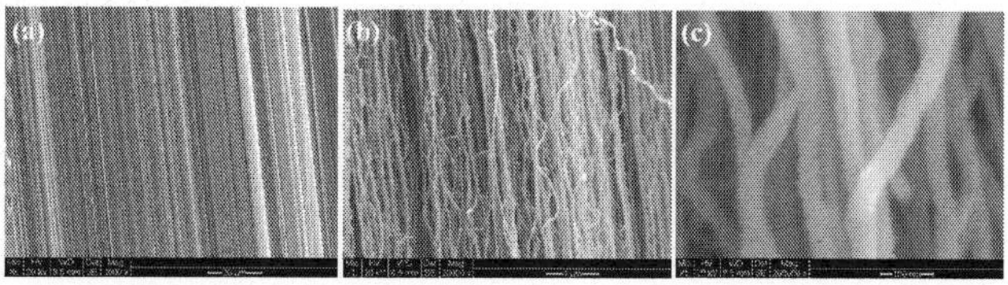

Figure 6. SEM images showing side view of a centimeter long CNT array at different magnifications. The magnification increases from (a) to (b) and (c).

This trend was observed to be similar for every growth temperature used. We found that the diameter distribution does not depend on the growth temperature. The distribution of the number of walls also revealed broad range from 1 to 10 walls for each growth temperature in Fig. 7e. Most of the tubes possess 2 to 5 walls at each growth temperature. We observed that the single wall CNTs were produced at 800 ºC and the double wall CNTs were yielded above 750 ºC. In our experiments, the distribution of the walls was not substantially affected by the growth temperature. Hence, we noted that the growth temperature is not an important factor affecting the structure of CNTs obtained in our growth process.

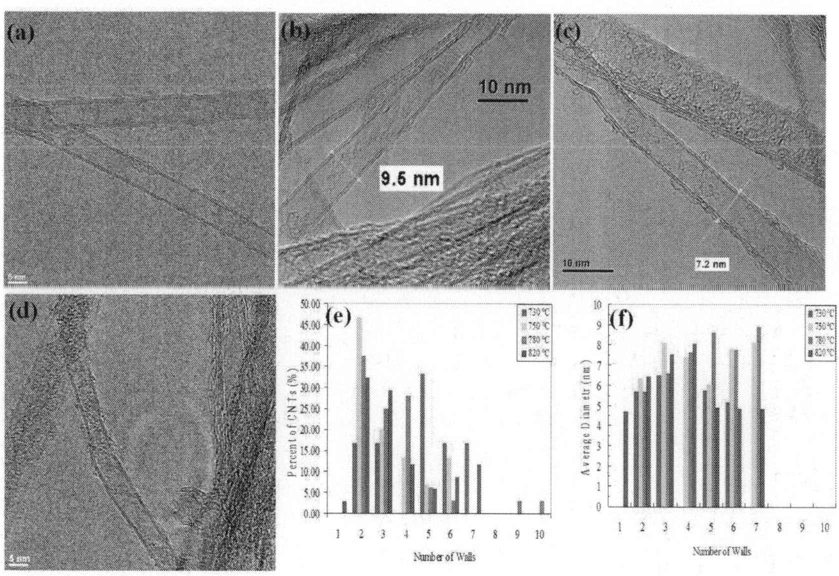

Figure 7. HRTEM images and distribution of the CNTs walls at different growth temperatures: (a) 730 °C, (b) 750 °C, (c) 780 °C, (d) 820 °C, (e) distribution of number of walls, (f) distribution of the average tube diameter of each wall for different multiwall CNTs.

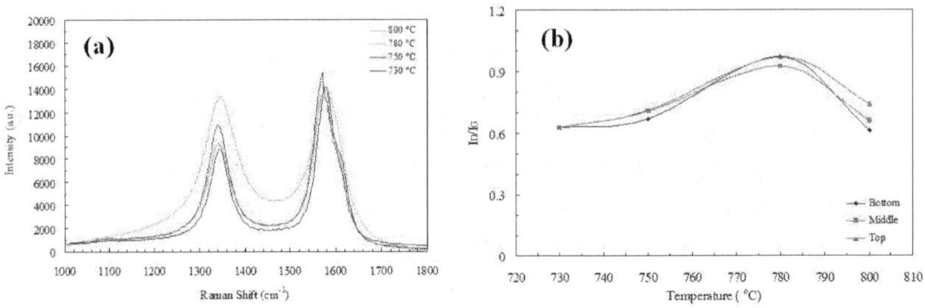

Figure 8. Raman spectra of CNTs taken from the middle position of the CNTs: (a) Spectra at various growth temperatures, (b) I_D/I_G peak intensity ratios as a function of growth temperature.

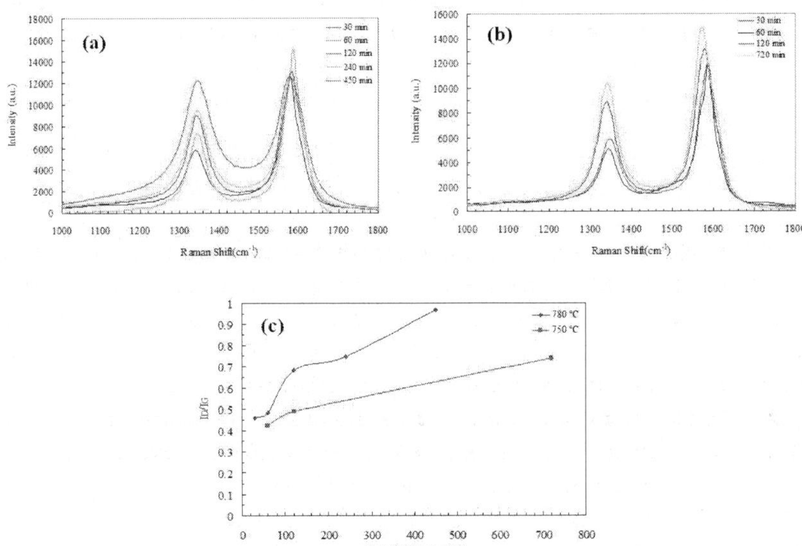

Figure 9. Raman spectra of CNTs synthesized at different temperature and growth time: (a) spectra at 780 °C, (b) spectra at 750 °C, (c) I_D/I_G peak intensity ratios vs. different growth time.

Micro-Raman spectroscopy (Renishaw inVia Reflex Micro-Raman) was used to investigate the effect of temperature and growth time on the quality of CNT arrays using a 514 nm excitation wavelength. Fig. 8 shows Raman shift and I_D/I_G ratio of the CNTs at different growth temperatures for three points along the array length. The spectra show distinguished D band peak (near 1350 cm^{-1}) which indicates presence of defects, disordered and amorphous carbon. A pronounced G band peak (at 1580 cm^{-1}) originating from graphitization of CNTs is also displayed [36-40]. As shown in Fig. 8a, the intensity of the D band increased with in-

http://dx.doi.org/10.5772/50837

creasing of the growth temperature up to 780 ºC. Fig. 8b displays the intensity ratio I_D/I_G as a function of the growth temperature from three different height positions of CNTs. The intensity ratio I_D/I_G was similar at bottom, middle and top positions for each growth temperature and rose from 0.62 to 1 as the temperature increases with a maximum at 780 ºC. Fig. 9 shows Raman data for CNT arrays obtained at different growth time and temperatures. The intensity of D peak increases linearly with rising the growth temperature as shown on Fig. 9a and b. Fig. 9c illustrates that the I_D/I_G ratio also increases near linearly with extending the growth time, which indicates that the quality of CNTs deteriorates when the tubes reside longer in the growth zone. At 750 ºC, I_D/I_G increased from 0.422 to 0.704 when prolonging the growth time. The highest value 0.968 of I_D/I_G was obtained at 780 ºC. Fig. 9c indicates that I_D/I_G values at 780 ºC are greater than those obtained at 750 ºC.

As the growth time increases, the gap between the two plots in Fig 9c broadens. Thus, the quality of CNTs decreases faster with time at higher temperatures. The reason for this is the accumulation of the amorphous carbon at high temperatures. These results are supported by the presented HRTEM images.

4. Conclusions

Real-time photography was used to record the growth of centimeter long CNT arrays during the CVD process. The kinetics of growing vertically aligned CNTs was studied based on the photographic images. Furthermore, we found that the CNT arrays grew by the root growth mechanism which was proved by the carbon source interruption method and real-time photography. The length of the CNT arrays increased linearly with growth time for all the tested temperatures followed by an abrupt growth termination. The catalyst lifetime decreased linearly with rising the deposition temperature and varied from a few minutes up to several hundred minutes depended on the growth conditions. We found out that the formation of centimeter long CNT array could not be described by diffusion controlled or exponentially decaying growth. This study suggests that the growth in this case is governed by a kinetically controlled mode within the temperature interval from 730 ºC to 840 ºC. The calculated activation energy is 248 kJ/mol. An analytical model for centimeter long CNT array synthesis was proposed which can predict the growth rate, final CNT length, and the catalyst lifetime. The obtained data indicated that the wall and diameter distribution of CNTs is independent on the growth temperature. The quality of CNTs deteriorates with increasing of the growth time and temperature. We found out that the amount of amorphous carbon on the CNTs depends on the residence time of the tubes in the CVD growth zone and on the deposition temperature. Longer residence time and higher deposition temperature accumulates greater amount of amorphous carbon.

Acknowledgements

The financial support from NSF through grant CMMI-07272500 and from NCA&T through DURIP-ONR is highly acknowledged. We also would like to thank Jay Yocis who helped to set up the real-time photography system and Dr. John Robertson from Cambridge University who suggested real time photography of studying kinetics of CNT arrays.

Appendix A. Supplementary data

Movies showing the centimeter long CNT array growth inside of the CVD reactor, including the root growth are available online.

Author details

Wondong Cho[1,2], Mark Schulz[2,3] and Vesselin Shanov[1,2*]

*Address all correspondence to: vesselin.shanov@uc.edu

1 Chemical and Materials Engineering, University of Cincinnati, Cincinnati, Ohio USA

2 Nanoworld Laboratory, Rh414, University of Cincinnati, Cincinnati, Ohio USA

3 Mechanical Engineering, University of Cincinnati, Cincinnati, Ohio USA

References

[1] Puretzky, A. A., Geohegan, D. B., & Eres, G. (2008). Real-time imaging of vertically aligned carbon nanotube array growth kinetics. *Nanotechnology*, 19, 055605.

[2] Amama, P. B., Pint, C. L., Mc Jilton, L., Kim, S. M., Stach, E. A., Murray, P. T., Hauge, R. H., & Maruyama, B. (2008). Role of Water in Super Growth of Single-Walled Carbon Nanotube Carpets. *Nano Letters*, 9, 44-49.

[3] Baughman, R. H., Zakhidov, A., & de Heer, W. A. (2002). Carbon nanotubes-the route toward applications. *Science*, 297, 787-792.

[4] Bronikowski, M. (2007). Longer nanotubes at lower temperatres: the influence of effective activation energies on carbon nanotube growth by thermal chemical vapor deposition. *J. Phys. Chem. C*, 111, 17705-17712.

[5] Gommes, C., Pirard, J. P., & Blacher, S. (2004). Influence of the operating conditions on the production rate of multi-walled carbon nanotubes in a CVD reactor. *Carbon*, 42, 1473-1482.

[6] Vix-Guterl, C., Couzi, M., & Delhaes, P. (2004). Surface charactrizations of carbon mutiwall nanotubes:comparison surface active sites and raman spectroscopy. *J. Phys. Chem. B*, 108, 19361-19367.

[7] Kim, D. H., Lee, H. R., & Jang, H. S. (2003). Dynamic growth rate behavior of a carbon nanotube forest characterized by in situ optical growth monitoring. *Nano Letters*, 3, 863-865.

[8] Du, C., Yeh, J., & Pan, N. (2005). High power density supercapacitors using locally aligned carbon nanotube electrodes. *Nanotechnology*, 16, 350-353.

[9] Einarsson, E., & Maruyama, Y. M. S. (2008). Growth dynamics of vertically aligned single-walled carbon nanotubes from in situ measurements. *Carbon* , 46, 923-930.

[10] Mckee, G. S. B., , J. S. F., & Vecchio, K. S. (2008). Length and the oxidation kinetics of chemical vapor deposition generated multiwalled carbon nanotubes. *J. Phys. Chem. C*, 112, 10108-10113.

[11] Guang-yong, Xiong. , Wang, D. Z., & Ren, Z. F. (2006). Aligned millimeter-long carbon nanotube arrays grown on single crystal magnesia. *Carbon* , 44, 969-973.

[12] Nii, H., Sumiyama, Y. , & Kunishige, A. (2008). Influence of diameter on the Raman of multi-walled carbon nanotubes. *Applied Physics Express*, 1, 064005.

[13] Hata, K. (2004). Water-assisted highly efficient synthesis of impurity-free single-walled carbon nanotubes. *Science*, 306, 1362-1364.

[14] Iijima, S. (1991). Helical microtubules of graphitic carbon. *Nature*, 354, 56-58.

[15] Ishikawa, T. (2006). Overview of trends in advanced composite research and applications in Japan. *Adv. Compos. Mater.*, 15, 3-37.

[16] Iwasaki, T., Zhong, G. F., Aikawa, T., Yoshida, T., & Kawarada, H. (2005). *J. Phys. Chem. B*, 109, 19556.

[17] Benit, J. M., , J. P. B., & Lefrant, S. (2002). Low frequency raman studies of muti wall carbon nanotubes: experiments and theory. Physical review B. Condensed matter and materials physics 073417 , 66, 1-4.

[18] Jiang, K., Li, Q., & Fan, S. (2002). Spinning continuous carbon nanotube yarns. *Nature*, 419, 801.

[19] Hasegawa, K., , S. N., & Yamaguchi, Y. (2008). Growth window and possible mechanism of millimeter-thick single-walled carbon nanotube forests. *Journal of Nanoscience and Nanotechnology*, 8, 6123.

[20] Lau, K. T., , C. G., & Hui, D. (2006). A critical review on nanotube and nanotube/nanoclay related polymer composite materials. *Composites, Part B*, 37, 425 .

[21] Zhu, L., Xu, J. , & Wong, C. P. (2007). The growth of carbon nanotubes stacks in the kinetics-controlled regime. *Carbon*, 45, 344-348.

[22] Li, Y., Kinloch, I. A., & Windle, A. H. (2004). Direct spinning of carbon nanotube fibers from chemical vapor deposition synthesis. *Science*, 304, 276-278.

[23] Lingbo, Zhu. D. W. H., & Ching-Ping, Wong. (2006). Monitoring Carbon Nanotube growth by Formation of nanotube stacks and investigation of the diffusion-control kinetics. *J.Phys. Chem. B*, 110, 5445-5449.

[24] Liu, K., Jiang, K. L., Feng, C., Chen, Z., & Fan, S. S. (2005). *Carbon*, 43, 2850.

[25] Matthews, M. J., Pimenta, M. A., & Endo, M. (1999). Origin of disperive effects of the raman D band in carbon materials. *Physical review B*, 59, R6585.

[26] Meshot, E. R., & Hart, A. J. (2008). Abrupt self-termination of vertically aligned carbon nanotube growth. *Applied Physics Letters*, 92, 113107-113103.

[27] Noda, S., Hasegawa, K., Sugime, H., Kakehi, K., Zhang, Z., Maruyama, S., & Yamaguchu, Y. (2007). *Jpn. J. Appl. Phys*, 46, L399.

[28] Oleg, V., & Yazyev, A. P. (2008). Effect of metal elements in catalytic growth of carbon nanotubes. *Physical review letters*, 100, 156102.

[29] Li, Qingwen. X. , Xiefei, Z., & Zhu, Yuntian T. (2006). Sustain growth of ultralong carbon nanotube arrays for fiber spinning. *Advanced Materials*, 18, 3160-3163.

[30] Brukh, R., & Mitra, S. (2006). Mechanism of carbon nanotube growth by CVD. *Chemical Physics Letters*, 424, 126-132.

[31] Xiang, R., , Z. Y., & Maruyama, S. (2008). Growth deceleration of vertically aligned carbon nanotube arrays: catalyst deactivation or feedstock diffusion controlled? *J. Phys. Chem. C*, 112, 4892-4896.

[32] Maruyama, S., Einarsson, E., & Edamura, T. (2005). Growth process of vertically aligned single-walled carbon nanotubes. *Chemical Physics Letters*, 403, 320-323.

[33] Pal, S. K., , S. T., & Ajayan, P. M. (2008). Time and temperature dependence of mutiwalled carbon nanotube growth on inconel 600. Nanotechnology , 19, 045610.

[34] Shim, J. S., Yun, Y. H., Cho, W., Shanov, V., Schulz, M. J., & Ahn, C. H. (2010). Self-Aligned Nanogaps on Multilayer Electrodes for Fluidic and Magnetic Assembly of Carbon Nanotubes. *Langmuir*, 26, 11642-11647.

[35] Stadermann, M., Sherlock, S. P., In, J. B., Fornasiero, F., Park, H. G., Artyukhin, A. B., Wang, Y., De Yoreo, J. J., Grigoropoulos, C. P., Bakajin, O., Chernov, A. A., & Noy, A. (2009). Mechanism and Kinetics of Growth Termination in Controlled Chemical Vapor Deposition Growth of Multiwall Carbon Nanotube Arrays. *Nano Letters*, 9, 738-744.

[36] Shanov, V., , W. C., Schulz, M., & Malik, N. (2008). Advances in synthesis and application of carbon nanotube materials. *Materials science and technology*, 2253.

[37] Zhao, X., Saito, R., & Ando, Y. (2002). Characteristic raman spectra of mutiwalled carbon nanotubes. *Physica B*, 323, 265-266.

[38] Yun, Y. H., , A. B., Shanov, V. N., & Schulz, M. J. (2006). A nanotube composite microelectrode for monitoring dopamine levels using cyclic voltammetry and differential pulse voltammetry. 220, 53 -60.

[39] Yun, Y., Shanov, V., Tu, Y., Subramaniam, S., & Schulz, M. J. (2006). Growth Mechanism of Long Aligned Multiwall Carbon Nanotube Arrays by Water-Assisted Chemical Vapor Deposition. *The journal of physical chemistry B*, 110, 23920-23925.

[40] Zhong, G., Iwasaki, T., Robertson, J., & Kawarada, H. (2007). Growth Kinetics of 0.5 cm Vertically Aligned Single-Walled Carbon Nanotubes. *J. Phys. Chem. B*, 111, 1907 -1910.

Chapter 11

Carbon Nanotubes from Unconventional Resources: Part A: Entangled Multi-Walled Carbon Nanotubes and Part B: Vertically-Aligned Carbon Nanotubes

S. Karthikeyan and P. Mahalingam

Additional information is available at the end of the chapter

http://dx.doi.org/10.5772/51073

1. Part A: Entangled Multi-walled carbon nanotubes

1.1. Introduction

Nanotechnology is a topic attracting scientists, industrialists, journalists, governments and even a common people alike. Carbon nanotubes (CNTs) and other carbon nanostructures are supposed to be a key component of this nanotechnology. Having realized its tremendous application potential in nanotechnology, a huge amount of efforts and energy is invested in CNT projects worldwide. Till date, the art of CNT synthesis lies in the optimization of parameters for selected group materials on a particular experimental set-up. As viewed from the perspective of green chemistry, sustaining the environment implies sustaining the human civilization. The long-term key of a sustainable society lies in stable economy that uses energy and resources efficiently. Therefore, it is high time to evaluate the existing CNT techniques on these parameters.

Let us examine three popular methods of CNT synthesis viz Arc discharge, Laser-vaporization and CVD method. Arc-discharge method, in which the first CNT was discovered [1], employs evaporation of graphite electrodes in electric arcs that involve very high temperatures around 4000º C. Although arc-grown CNTs are well crystallized, they are highly impure. Laser-vaporization technique employs evaporation of high-purity graphite target by high-power lasers in conjunction with high-temperature furnaces [2]. Although laser-grown CNTs are of high purity, their production yield is very low. Thus it is obvious that these two methods score too low on account of efficient use of energy and resources. Chemical vapor deposition (CVD), incorporating catalyst-assisted thermal decomposition of hydrocarbons, is the most popular method of producing CNTs, and it is truly a low-cost and scalable technique for mass production of CNTs [3]. Unfortunately, however, till date only purified petroleum products such as methane, ethylene, acetylene, benzene, xylene are in practice, as precursor, for synthesizing CNTs. Apart from petroleum based hydrocarbons, carbon nanotubes have been synthesized from polymers, metallocenes and domestic fuels such as kerosene and liquefied petroleum gas [4-7].

According to the principle of green chemistry, the feed stock of any industrial process must be renewable, rather than depleting a natural resource. Moreover, the process must be designed to achieve maximum incorporation of the constituent atoms (of the feed stock) into the final product. Hence, it is the time's prime demand to explore regenerative materials for CNT synthesis with high efficiency. This well-valued material of biotechnology research was successfully brought to nanotechnology research with the first report of CNTs from natural precursors in 2001. Since then, investigators involved with this environment-friendly source of CNTs and established the conditions for growing multiwalled nanotubes (MWNTs), single-wall nanotubes (SWNTs) and vertically aligned MWNTs on the suitable catalytic support by a simple and inexpensive CVD technique. Researchers prepared good quality of Multi-walled carbon nanotubes (MWNTs) and vertically aligned ones by thermal decomposition of turpentine oil and Camphor [8,9]. Andrews et al. synthesized pure Single-walled carbon nanotubes (SWNTs) by catalytic decomposition of camphor and its anologes [10]. Ghosh et al. prepared single-walled carbon nanotubes from turpentine oil and eucatlyptus oil [11]. We have succeeded in growing of CNTs from pine oil [12] and methyl ester of *Jatropha curcas* oil [13], a botanical hydrocarbon extracted from plant source, having boiling point around 200 - 280º C. Being a green-plant product, these oils are eco-friendly source and can be easily cultivated in as much quantity as required. Unlike any fossil/petroleum product, there is no fear of its ultimate shortage as it is a regenerative source. These sources are readily available, cheap and also user-friendly for chemical vapor deposition due to its volatile and non-toxic nature.

United states Environmental Protection Agency has developed 12 principles of green chemistry [14] that explain what the green chemistry means in practice. Using those definitions as a protocol, we can evaluate the CNT precursor materials such as Pine oil, Methyl ester of *Jatropha curcas* oil and Methyl ester of *Pongamia pinnata* oil, as follows.

Prevention : With the highest CNT-production efficiency, plant based precursor complies with waste-prevention rule significantly.

Atom Economy : Botanical hydrocarbons gets maximum incorporation of its constituent atoms into the final product, CNTs.

Less Hazardous Chemical Synthesis : All the substances involved in this technique (carbon source, catalyst and support material) possess little or no toxicity to human health and the environment.

Designing Safer Chemicals: Our final product is common-grade CNTs that are apparently safe.Safer Solvents and Auxiliaries : The only auxiliary used in our method is the metal catalyst that are apparently safe.

Design for Energy Efficiency : By virtue of a low-temperature atmospheric pressure CVD process, the energy requirements of our technique are significantly low.

Use of Renewable Feedstock : The raw material is purely a regenerative material; so there is no danger of depleting a natural resource.

Reduce Derivatives: No derivative is formed in this technique; solely catalyst-assisted in-situ decomposition of plant based precursor leads to CNTs.

Catalysis: It is a purely catalytic process, no stoichiometric reagents are involved.

Design for Degradation: CNTs as such are non-biodegradable; however, their intentional degradation can be achieved by introducing certain functional groups.

Real-time analysis for pollution Prevention: This technique as such should be fully compatible with real-time analysis for pollution prevention, if executed at an industrial laboratory.

Inherently Safer Chemistry for Accident Prevention : Substances and their form used in our process are chosen so as to minimize the potential for chemical accidents, including releases, explosions and fires.

In this chapter the authors discuss the morphology of the carbon nanostructures produced using Pine oil, Methyl ester of *Jatropha curcas* oil and Methyl ester of *Pongamia pinnata* oil as natural carbon precursors by spray pyrolysis method, a modified chemical vapor deposition method. It is similar to chemical vapor deposition method, the only difference is the vaporization and pyrolysis of carbon source occurs simultaneously in spray pyrolysis whereas in CVD it is a two step process. Since these precursors evaporate at relatively higher temperatures, spray pyrolysis method was adopted for synthesis of CNTs. The results show that the morphology closely correlates with precursor concentration and the authors propose that the effect is related to the competition between rate of decomposition of precursor and the diffusion of carbon species through the catalyst particle.

1.2. Experimental Methods

A catalytic supporting material Fe, Co and Mo with silica ($Fe:Co:Mo:SiO_2 = 1:0.5:0.1:4$) was prepared by wet impregnation method [15]. Appropriate quantities of metal salts (Merk) i.e. $Fe(NO_3)_3 \cdot 6H_2O$, $Co(NO_3)_3 \cdot 3H_2O$ and $(NH_4)_6Mo_7O_{24} \cdot 4H_2O$ were dissolved in methanol and mixed thoroughly with methanol suspension of silica (Merk). The solvent was then evaporated and the resultant cake heated to 90-100 °C for 3 h, removed from the furnace and ground to fine power. The fine powders were then calcined for 1 h at 450 °C and then re-ground before loading into the reactor. The prepared catalyst was directly placed in a quartz boat and kept at the centre of a quartz tube which was placed inside a tubular furnace. The carrier gas nitrogen was introduced at the rate of 100 mL per minute into the quartz tube to remove the presence of any oxygen inside the quartz tube. The temperature was raised from room temperature to the desired growing temperature. Subsequently, the carbon precursor was introduced into the quartz tube through spray nozzle and the flow was maintained at the rate of 0.1 - 0.5 ml/min. Spray pyrolysis was carried out for 45 minutes and thereafter furnace was cooled to room temperature. Nitrogen atmosphere was maintained throughout the experiment. The morphology and degree of graphitization of the as-grown nanostructures were characterized by high resolution transmission electron microscopy (JEOL-3010), Raman spectroscopy (JASCO NRS-1500W) green laser with excitation wavelength 532 nm) and thermo gravimetric analysis (TGA). The as-grown products were subjected to purification process as follows [16]. The sample material was added to 20 ml 1N HCl to form an acidic slurry. This slurry was heated to 60 °C and stirred at 600 rpm. To this heated acidic slurry 20 ml H_2O_2 was added to form oxidative slurry and stirred for 30 minutes. The sample was filtered and washed with 1N HCl and distilled water. The collected sample was calcined at 400 °C in nitrogen atmosphere for 2 hours. The experiment at optimum reaction conditions were repeated several times to ensure the reproducibility of formation of carbon nanostructures.

1.3. Result and Discussion

Pine oil, Methyl ester of *Jatropha curcas* oil and Methyl ester of *Pongamia pinnata* oil were used for synthesis of carbon nano structures by spray pyrolysis method. These oils were individually sprayed at different rate (10 mL, 20 mL and 30 mL per hour) over silica supported Fe, Co and Mo catalyst at different reaction temperatures (550 °C, 650 °C and 750 °C) to synthesis carbon nanostructures. The as-grown nanostructures were characterized by SEM, TEM, TGA and Raman spectroscopy techniques.

The result shows low yield of carbon nano structures with three types of precursors, Pine oil, Methyl ester of *Jatropha curcas* oil and Methyl ester of *Pongamia pinnata* oil, when sprayed individually at the precursor flow rate of 10mL per hour over silica supported Fe, Co and Mo catalyst at 550 °C. An improved carbon nanostructure formation at 650 °C for the precursor flow rate of 20 mL per hour was observed invariably for the three precursor materials. When the experimental condition was either 750 °C or precursor flow rate of 30 mL per hour, decrease in carbon nanotube yield observed.

Figure 1 illustrates the SEM image of the CNTs samples grown at different temperatures using pine oil as precursor at a flow rate of 20 mL per hour. At 550 °C few CNTs were found to grow because this temperature is not sufficient to pyrolyse the carbon source (Fig. 1a). On the other hand, at 650 °C (Fig. 1b) the formation of CNTs is high because at this temperature the carbon source decomposes effectively. At 750 °C, the quantity of CNTs has decreased and thick nanotubes have been formed (Fig. 1c). Experiments with methyl ester of *Jatropha Curcas* oil as a carbon precursor at different temperatures were conducted and the SEM images of the sample were recorded. The SEM images of sample synthesized at 650 °C revealed the formation of homogeneous and dense distribution of CNTs in a web-like network (Fig. 1e), while very few nanotubes formation at 550 °C (Fig. 1g).

At higher temperature (750 °C) an increase in diameter of the CNTs was observed (Fig. 1f). The morphology of carbon nanotube synthesized at different temperatures using methyl ester of *Pongamia pinnata* oil are recorded using SEM. Few carbon nanostructures were observed at a lower deposition temperature of 550 °C (Fig. 1g). When the temperature was increased to 650 °C chain like carbon nanostructure was dominant (Fig. 1h). At 750 °C, a more uniform distribution of MWNTs was observed with a diameter of around 150 nm (Fig. 1i).

http://dx.doi.org/10.5772/51073

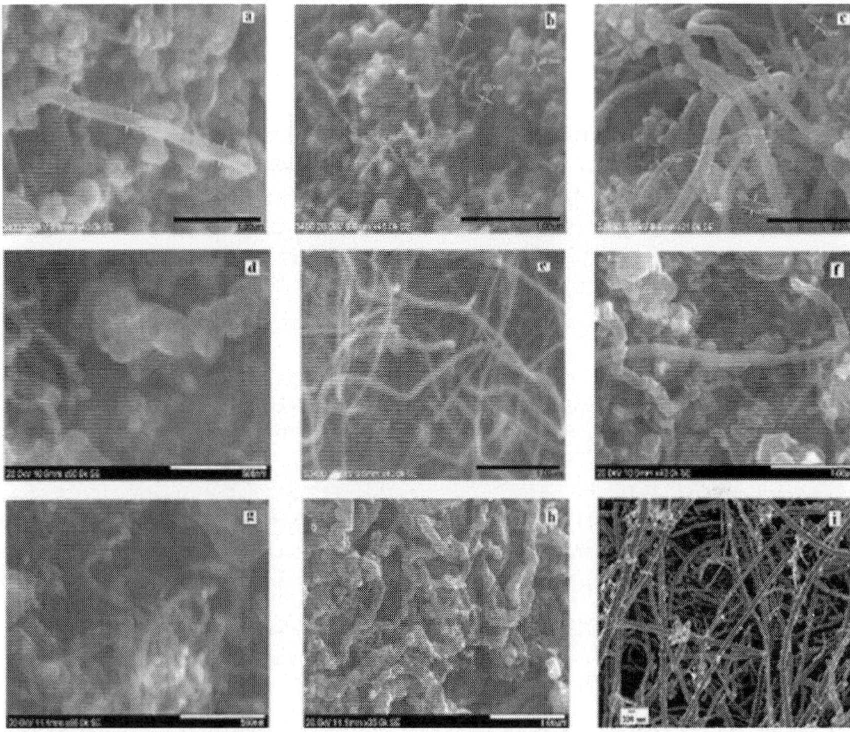

Figure 1. Representative SEM images of as-grown CNTs at different temperatures using Pine oil (a-c), Methyl ester of *Jatropha carcus* oil (d-f) and Methyl ester of *Pongamia pinnata* oil (g-i).

The TEM result shows varied morphologies of carbon nanostructures synthesized at different experimental conditions. All the three precursors used in this study produced mostly amorphous carbon at 550 °C for different flow rate of precursor materials. When the reaction temperature was 650 °C, Pine oil at a flow rate of 20 mL per hour, produced MWNTs of good quality (Fig.2a). The HRTEM result of sample obtained with pine oil as a precursor at a flow rate of 30 mL per hour and reaction temperature of 650 °C, shows (Fig.2b) that the inner and outer diameter of the synthesized MWNTs was about 6.6 and 14 nm and consist of 11 graphene layers with inter layer distance of 0.342 nm. The outer layer of MWNTs were covered with amorphous carbon. Methyl ester of *Jatropha curcas* oil produced well crystalline MWNTs of inner and outer diameter of 3 nm and 9 nm respectively, when the precursor sprayed at a flow rate of 20 mL per hour over silica supported Fe, Co and Mo catalyst at reaction temperature of 650°C (Fig. 2d). However, an increase of precursor flow rate to 30 mL per hour at the same temperature produced largely amorphous carbon and small quantity of metal encapsulated MWNTs of size around 40 nm (Fig. 2c). A metal filled MWNTs of inner and outer diameter of 4 nm and 24 nm respectively was observed when methyl ester of *Pongamia Pinnata* oil, at a flow rate of 20 mL per hour, was spray pyrolysed at 650 °C over silica supported Fe, Co and Mo catalyst. When flow rate was increased to 30 mL per hour, a short growth of metal encapsulated carbon nanostructures was observed (Fig. 2e). The TEM and TGA studies revel that the carbon nanostructures are not well graphitized.

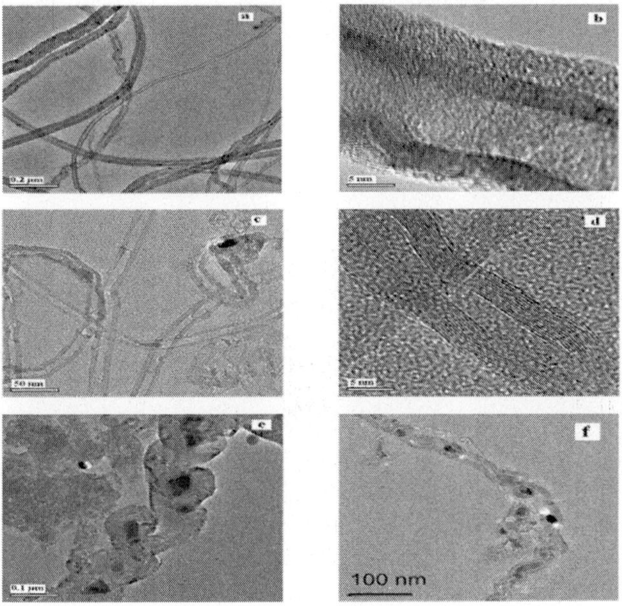

Figure 2. Representative TEM images of as-grown CNTs under constant reaction temperature of 650 °C for flow rate of precursor: Pine oil (a) 20 mL per hour (b) 30 mL per hour; Methyl ester of *Jatropha carcus* oil (c) 30 mL per hour (d) 20 mL per hour and Methyl ester of *Pongamia pinnata* oil (e) 30 mL per hour (f) 20 mL per hour.

The Fig. 2a shows TEM images of sample prepared using Pine oil as carbon source at a flow rate of 20 mL per hour over silica supported Fe, Co and Mo catalyst at a reaction temperature of 650°C. The TEM images indicate that the inner and outer diameter was uniform over entire length of the tube. It is also evident that carbon nanotubes do not contain encapsulated catalyst but amorphous carbon at the outer wall of the tube. The well crystallization of the graphene layer was confirmed from the I_G/I_D ratio of 2.5 (Fig.3a). The I_G/I_D ratio increases with increase of precursor flow rate from 10mL per hour to 20 mL per hour, but further increase of precursor flow rate to 30mL per hour results in decrease of I_G/I_D ratio to 0.5 (Fig.3b). The improved quality can be attributed to increase of precursor concentration in the tube. The higher flow rate of precursor (20 mL per hour) increases precursor concentration in the tube which leads to increase in decomposition of precursor over catalyst. Thus quality carbon nanotubes were formed at a reaction temperature of 650 °C. However, too high a concentration of precursor (30 mL per hour) or a reaction temperature of 750 °C leads to the formation of amorphous product as vapor phase decomposition of precursor is promoted than the decomposition over catalyst. The results are in good accordance with the reports by Afre et al. [17]. They have synthesized MWNTs using turpentine oil as a precursor by spray pyrolysis method in the temperature range of 600 to 800 °C. While Ghosh and co workers [18] reported synthesis of single walled carbon nanotube at a reaction temperature of 850 °C and abundant amount of MWCNTs at lower temperatures using turpentine oil as precursor by CVD method. No lower frequency RBM peaks in Raman spectra of our samples shows the absence of SWNTs.

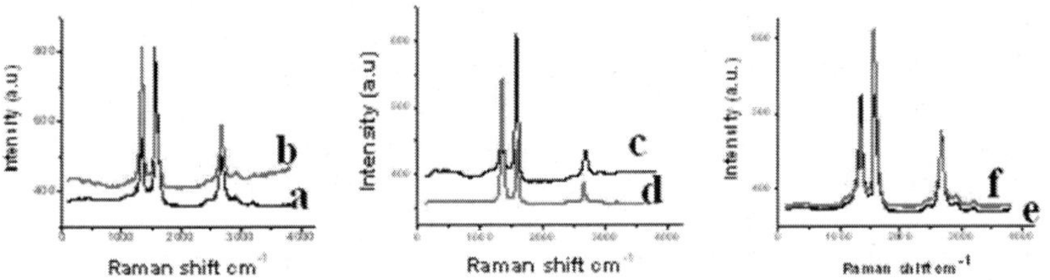

Figure 3. Typical Raman spectrum (green laser with excitation wavelength 532 nm) of as-grown MWNTs at different flow rate of precursors: Pine oil (a) 20 mL per hour (b) 30 mL per hour; Methyl ester of *Jatropha carcus* oil (c) 20 mL per hour (d) 30 mL per hour and Methyl ester of *Pongamia pinnata* oil (e) 20 mL per hour (f) 30 mL per hour.

Thermogravimetric analysis was performed to characterize the thermal behavior of the carbon nanotube synthesized using Pine oil as precursor. The Thermogravimetric graph shows the wt% vs. temperature. Upto a certain period of time there is no weight loss. The oxidation of carbon deposit starts after this point which is indicated by a dip in curve. With increase in temperature the weight loss increases till all the carbon get burnt. The residue is the catalyst and support. The TGA curves gives the temperature corresponds to the maximum mass decrease, which is considered to be a measure of the level of crystallinity of carbon nanotube. The TGA results (Fig.4a,b,c) shows the variation in decomposition temperature for the products synthesized at 650 °C for different flow rate of Pine oil precursor. It was found that the decomposition temperature varied between 568 and 591 °C. The relatively high decomposition temperature of 591 °C for the product synthesized at a reaction temperature of 650 °C for precursor flow rate of 20 mL per hour indicates well crystalline structure formation.

A HRTEM images (Fig. 2d) of sample prepared using Methyl ester of *Jatropha curcas* oil as carbon precursor at a flow rate of 20 mL per hour over silica supported Fe, Co and Mo catalyst at a reaction temperature of 650°C clearly shows well graphitized layers of a typical MWNTs with uniform inner and outer diameter. The TEM image also revels that encapsulated catalyst or amorphous carbon is rarely seen in the sample. The image indicates that the MWNTs are composed of around 26 walls and layers grow perpendicular to the growth axis of the tube. An increase of precursor flow rate to 30 mL per hour at the reaction temperature of 650 °C produced largely amorphous carbon and small quantity of metal incorporated MWNTs of size around 40 nm is evident from the TEM images (Fig. 2c). An additional confirmation for high degree graphitization and formation of metal filled MWNTs for sample prepared using the precursor flow rate of 20 mL per hour and 30 mL per hour respectively is shown by Raman spectra (Fig. 3c,d). The G band at 1571 cm^{-1} was attributed to well crystallized carbon structure, while the D band at 1359 cm^{-1} was attributed to defects in the structure [19]. The decrease in relative intensity of the G band and D band (I_G/I_D ratio) for sample prepared with precursor flow rate of 30 mL per hour indicates more defects in as-grown sample (Fig. 3d). The defects in MWNTs can be attributed to increase of precursor concentration in the tube and encapsulation of catalyst particles. An increase of precursor concentration in the tube leads to increase in decomposition of precursor over catalyst. Above the critical concentration of precursor, rate of decomposition of precursor exceeds rate of diffusion of carbon into the catalyst particle and thus encapsulation of metal particle occurs. The TGA results of MWNTs sample grown using methyl ester of *Jatropha curcas* oil are shown in (Fig.4d,e,f). Higher decomposition temperature and 37% residue observed in the TGA studies for the product synthesized at a reaction temperature of 650 °C for precursor flow rate of 20 mL per hour shows the sample contain around 60% MWNTs with well crystallized structure. A decrease in decomposition temperature and high residue shows defects in structure and metal encapsulation for the sample prepared at 650 °C for the precursor flow rate of 30 mL per hour.

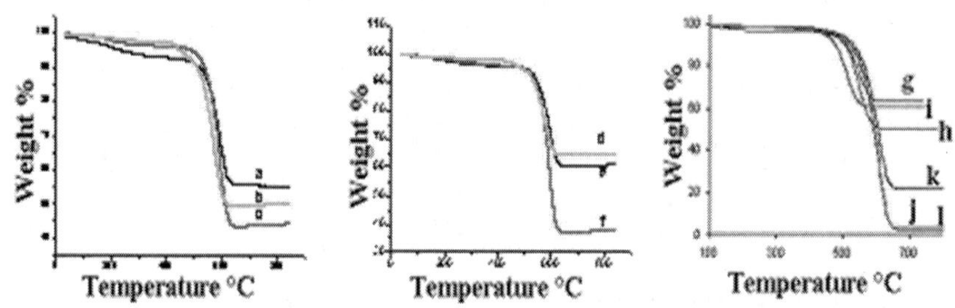

Figure 4. TGA curves of as-grown CNTs samples at different flow rate of precursors: Pine oil (a) 10 mL per hour (b) 20 mL per hour (c) 30 mL per hour, Methyl ester of *Jatropha carcus* oil (d) 10 mL per hour (e) 20 mL per hour (f) 30 mL per hour, Methyl ester of *Pongamia pinnata* oil, as-grown, (g) 10 mL per hour (h) 20 mL per hour (i) 30 mL per hour and purified (j) 10 mL per hour (k) 20 mL per hour (l) 30 mL per hour.

The TEM images of sample synthesized for 20 mL per hour flow rate of methyl ester of *pongamia pinnata* oil over silica supported Fe, Co and Mo catalyst at a reaction temperature of 650 °C was shown in Fig. 2f. The average outer diameter of the tube synthesized varied randomly as the reaction temperature was changed. At 550 °C, the diameter of the arborization-like nanostructures were around 65 nm, whilst at 650 °C, the formed MWNTs has inner and outer diameter in the range of 7nm & 33nm. The HRTEM image of the sample synthesized at 650 °C (Fig.2f) indicates the metal particles, seen as a dark spot on the image, were tightly covered by carbon layers with a thickness of a few nanometers. When precursor flow rate was reduced from 30 to 20 mL per hour, due to effective decomposition of precursor and the fluid nature of catalyst particle, the morphology of the product changed from arborization-like nanostructure to magnetic nanopartical encapsulated in multi-walled carbon nanotubes.

A similar observation was reported by Kang et al. [20]. Also, morphology change from magnetic nanopartical encapsulated in multi-walled carbon nanotubes to multi-walled carbon nanotubes structure was attributed to the high fluid nature of catalyst at 750 °C. The inner diameter of the carbon structure formed is same as that observed for products synthesized at 650 °C indicates the prevention of agglomeration of catalyst particles by Mo [21].

Amorphous carbon formation in large quantity, at 750 °C or at precursor flow rate of 30mL per hour, may be due to increased thermal decomposition of precursor material. The TGA results are shown in Fig.4g,h,i. It is evident that the weight loss continues to increase rapidly with temperature until reaches a constant value. Residue of the as-grown sample for precursor flow rate of 10 mL, 20 mL and 30 mL per hour at 650 °C was found to be 63.2, 60.9 and 50.5% respectively by weight (Fig.4g,h,i). The TGA results of same samples after purification shows (Fig. 4j,k,l) weight of residue as 3.5, 22.5 and 2.5% mass fraction. The more decline in mass fraction was caused by the acid leaching of catalyst particles that was not encapsulated by carbon. The low residue observed for products synthesized at 550 °C and 750 °C are attributed to low catalytic activity and high thermal decomposition of precursors respectively, which leads to formation of high amorphous carbon and low encapsulated products. The products synthesized at 650 °C shows minimum mass loss in TGA studies, even after purification, is due to better encapsulation of metal particles by carbon layers. The I_G/I_D value of 0.9 (Fig.3e) for samples prepared at 650 °C with flow rate of 20 mL per hour indicates that magnetic nanoparticals encapsulated in carbon nanotubes structure had defects and moderate crystallization of graphene planes [22,23]. This supports the HRTEM results. The increased I_G/I_D value of 1.87 (Fig. 3f) for the same sample, after purification indicates the removal of amorphous carbon and defective structures during purification. The removal of amorphous carbon and defective structures were further supported by higher ignition temperature 610 °C in TGA studies. According to TGA curves, weight of residue for the purified sample decreased to about 13 to 31% mass fraction comparing to unpurified sample due to leaching of metal particle and amorphous carbon removal during purification process. This shows that carbon layers covering the metal particles prevent their dissolution during purification process.

The mechanism of CNT nucleation and growth is one of the challenging and complex topics in current scientific research. Presently, various growth models based on experimental and quantitative studies have been proposed. It is well established, that during CNT nucleation and growth the following consecutive steps were taken place [24].

1. Carbon species formation by decomposition of precursor over the catalyst
2. Diffusion of carbon species through the catalyst particle
3. Precipitation of the carbon in the form of CNTs

The first step involves formation of carbon species by catalytic vapor decomposition of vapors of the precursor material over the catalyst. In the second step the diffusion of carbon species through the catalyst particle takes place. The catalyst surface may exert a diffusion barrier. It is still unclear whether carbon species diffuse on the particle surface [25], on the particle bulk [24] or whether surface and diffusion compete. The most accepted growth model suggests bulk diffusion of carbon species into the metal particles [26]. The third step is the precipitation of the carbon in the form of CNTs from the saturated catalyst particle.

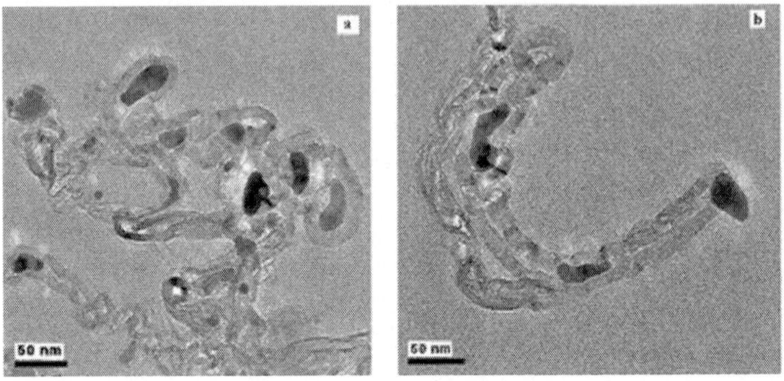

Figure 5. TEM images of as-grown CNTs sample synthesized under constant reaction temperature of 650 °C usingr Methyl ester of *Pongamia pinnata* oil at a flow rate of 20 mL per hour. (a) indicates the reshaping of catalyst particle (b) metal particle at the tip of tube.

Based on experimental results, a possible growth mechanism of MWNTs was proposed. It is known from the fact that Catalytic centers on catalyst particle act as nucleation site for the growth of MWNTs [22]. The precursor vapor decomposed on surface of the catalyst particle produces carbon. As the reactivity between the catalyst and the carbon exceeds the threshold value, carbon atoms loose their mobility in the solid solution, forming metal carbides [23]. These meta stable Fe and Co carbides decomposed and produce carbon which dissolve in these metal particles. The dissolved carbon diffuses through the metal particle and gets precipitated in the form of crystalline graphene layer. This carburized surface acts as a barrier for further carbon transfer from the gas phase to the bulk of the catalyst since carbon diffusion is slower through metal carbides [27]. The saturated metal carbide have lower melting point and they are fluid like during the growth process [28].

If the rate of precursor decomposition and the rate of diffusion of carbon are equal, then the metal raise through a capillary action and tube growth occurs. The fact that long carbon nanotubes observed have their catalyst particles partially exposed indicates that the direct contact of catalyst surface with carbon precursor is essential for continuous CNT growth (Fig. 5a). This is consistent with the growth mechanism proposed by Rodriguez [29]. In case the decomposition rate exceeds the diffusion rate, more of carbon produced forms a thick carbide layer over the surface of metal which acts as a barrier for further carbon transfer from the gas phase to the bulk of the catalyst. However, the thick carbide layer crystallizes out as graphene layer which encapsulate the metal particle. When a catalyst particle is fully encapsulated by layers of graphene sheets, the carbon supply route is cut and CNT growth stops resulting in short MWCNTs. The catalyst particle undergoes several mechanical reshaping during the tip growth of multi-walled nanotubes [30, 31]. This gives the impression that the catalyst is in liquid state during reaction. The catalyst particle seen inside and at the tip of tube could be the solidified form of the liquid phase metal particle. Thus the growth process is by the vapor–liquid–solid (VLS) mechanism [32].

The CNTs grow with either a tip growth mode or a base growth mode. Base growth mode is suggested when the catalyst particle remain attached to the support, while tip growth happens when the catalyst particle lifts off the support material. These growth modes depends on the contact forces or adhesion forces between the catalyst particle and support [33], while a weak contact favors tip-growth mechanism, a strong interaction promotes base growth [34]. Catalyst particle seen at tip of CNTs (Fig. 5b) indicate tip growth mode. These catalyst particles have lifted off the support and elongated due to the flow nature and stress induced by the carbon surrounding the catalyst.

2. Part B: Vertically-aligned carbon nanotubes

2.1. Introduction

Aligned carbon nanotubes were first reported by Thess et al.[2]. In the same year the Chinese academy of science reported that a 50 μm thick film of highly aligned nanotubes had successfully grown by chemical vapor deposition (CVD) [35]. Vertically aligned CNTs are quasi-dimensional carbon cylinders that align perpendicular to a substrate [36]. Vertically aligned with high aspect ratios [37] and uniform tube length made it easy spinning into macroscopic fibres [38] Aligned CNTs are widely used in nano electronics, composite materials as reinforcing agents and self-cleaning applications [39-41]. Aligned CNTs are ideal electrode material for biosensors over entangled CNTs, may be due to its high electrical conductive property [42]. Large CNT arrays have successfully been grown on different substrates, such as mesoporous silica [43] planar silicon substrate [44] and quartz glass plate [45]. Substrate provides a solid foundation for growing aligned CNTs. The substrate must able to inhibit the mobility of the catalyst particles in order to prevent agglomeration. The most commonly used active catalyst for growing CNTs are magnetic elements such as Fe, Co or Ni. Gunjishima et al. [46] used Fe-V bimetallic catalyst for synthesize of aligned DWCNTs. Recently, there have been appreciable attempt of using ferrocence as a catalyst for synthesis of aligned carbon nanotubes[47]. Here we report fabrication of aligned CNTs by spray pyrolysis on silicon wafer using mixture of Pine oil, Methyl ester of *Jatropha curcas* oil and Methyl ester of *Pongamia pinnata* oil with ferrocence.

2.2. Experimental Methods

The syntheses of aligned CNTs were carried out using the spray pyrolysis method. In this spray pyrolysis method, pyrolysis of the carbon precursor with a catalyst take place followed by deposition of aligned CNTs occur on silicon substrate. Pine oil, Methyl ester of *Jatropha curcas* oil and Methyl ester of *Pongamia pinnata* oil were used as carbon source and ferrocene [Fe $(C_5H_5)_2$] (Sigma Aldrich, high purity 98 %) was used as a source of Fe which acts as a catalyst for the growth of CNTs. n type silicon wafer (100) of size (1x1cm ^{2}) was used as a substrate and kept inside the quartz tube. In a typical experiment, the quartz tube was first flushed with argon (Ar) gas in order to eliminate air from the quartz tube and then heated to a reaction temperature. The precursor mixture was sprayed into the quartz tube, using Ar gas. The concentration of ferrocene in carbon precursor was ~25 mg/ml. The flow rate of Ar was 200 sccm/min. The experiments were conducted at 650 ºC with reaction time of 45 min was maintained for each deposition. After deposition, the furnace was switched off and allowed to cool down to room temperature under Ar gas flow. A uniform black deposition on the silicon substrate was observed. Finely, the substrate containing aligned CNTs was removed from the quartz tube for characterization. The experiments were repeated several times to ensure the reproducibility of the formation of vertically aligned carbon nanotubes.

2.3. Result and Discussion

The morphology of carbon sample grown on silicon substrate using a mixture of Pine oil and ferrocene at 650 °C can be observed in Figure 6a. The image revel the formation of high abundance of carbon nanotubes which are forest like and vertically-aligned to the substrate surface. The growth of carbon nanotubes seems to be uniform and reaches up to a length of 10μm. Figure 6b shows the SEM image of carbon sample grown on silicon substrate using a mixture of Methyl ester of *Jatropha curcas* oil and ferrocene at 650 °C. The dense, aligned but non-uniform growth of carbon nanotubes was observed. The length of carbon nanotubes grown was found to be varied from 12.5 to 7.5μm. Figure 6c illustrates the SEM image of the carbon naotubes grown at 650 °C using Methyl ester of *Pongamia pinnata* oil. A thick carbon nanotube with poor structure and alignment was observed.

Figure 6. Representative SEM images of as-grown vertically-aligned carbon nanotubes at 650 °C using Pine oil (a), Methyl ester of *Jatropha carcus* oil (b) and Methyl ester of *Pongamia pinnata* oil (c).

From the experimental results we suggest that the synthesis of aligned CNTs is very sensitive to the carbon precursors used. Ferrocene on thermal decomposition at high temperature forms Fe nano particles on the silicon substrate surface. During the chemical vapor deposition process, the carbon precursor is catalytically decomposed and the carbon fragments formed diffuse into the Fe catalyst. The Fe particles may thus easily become saturated or supersaturated with carbon atoms, and the precipitation of the carbon from the surface of the Fe particle leads to the formation of dense carbon nanotubes [48]. The high surface density of the growing nanotubes serves as an additional advantage for the constituent nanotubes to be "uncoiled". The Vander waals forces between the tube keep them aligned. Thus, the Fe catalysts can effectively catalyze the growth of highly dense vertically aligned carbon nanotubes on silicon substrate. Further in-

http://dx.doi.org/10.5772/51073

vestigation is going on in our laboratory for a better understanding of the actual growth mechanism of vertically aligned carbon nanotubes.

3. Conclusions, challenges and future prospects

In view of the perspective of green chemistry, we attempt to explore regenerative materials for CNT synthesis with high efficiency. In this research work a well graphitized MWNTs were synthesized from Pine oil and Methyl ester of *Jatropha curcas* oil using silica supported Fe, Co and Mo catalyst by spray pyrolysis method. The optimum reaction conditions for synthesis of MWNTs were 650 °C and precursor flow rate of 20 mL per hour. Spray pyrolysis of Methyl ester of *Pongamia pinnata* oil over silica supported Fe, Co and Mo catalyst results in formation of MWNTs filled with magnetic nanoparticles, which find potential applications in magnetic recording, biomedical and environmental protection. Vertically aligned carbon nanotubes were obtained by spray pyrolysis of Pine oil and Methyl ester of *Jatropha curcas* oil and ferrocene mixture, at 650 ºC on silicon substrate under Ar atmosphere. The use of natural precursors gives sensible yield and makes the process natural world friendly as well. A thick carbon nanotube with poor structure and alignment was observed with mixture of Methyl ester of *Pongamia pinnata* oil and ferrocene.

The studies in this work demonstrate that the carbon materials are potential precursor for CNTs production under suitable experimental conditions and comply with green chemistry principles. It is clear that specific carbon nanostructures can be synthesized by suitably altering the experimental parameters. However, it is a challenge to consistently reproduce CNT of same quality and quantity form the precursor of inconsistent composition. Designing of catalyst material and optimization of reaction parameters which is suitable for synthesis of specific morphological CNTs from a renewable natural precursor of inconsistent chemical composition is one of the future prospects in this area of research.

Acknowledgements

The authors acknowledge the UGC New Delhi for financial support, the Institute for Environmental and Nanotechnology for technical support and IITM for access to Electron microscopes.

Author details

S. Karthikeyan[1*] and P. Mahalingam[2]

*Address all correspondence to: skmush@rediffmail.com

1 Chikkanna Government Arts College, TN, India

2 Arignar Anna Government Arts College, TN, India

References

[1] Iijima, S. (1991). Helical microtubules of graphitic carbon. *Nature*, 354, 56-58.

[2] Thess, A., Lee, R., Nikolaev, P., Dai, H., Petit, P., Robert, J., Xu, C., Smalley, R., et al. (1996). Crystalline ropes of metallic carbon nanotubes. *Science*, 273, 483-487.

[3] Karthikeyan, S., Mahalingam, P., & Karthik, M. (2008). Large scale synthesis of carbon nanotubes. *E-Journal of Chemistry*, 6, 1-12.

[4] Parathasarathy, R. V., Phani, K. L. N., & Martin, C. R. (1995). Template synthesis of graphitic nanotubules. *Advanced Materials*, 7, 896-897.

[5] Sen, R., Govindaraj, A., & Rao, C. N. R. (1997). Carbon nanotubes by the Metallocene Route. *Chemical Physics Letters*, 267, 276-280.

[6] Pradhan, D., & Sharon, M. (2002). Carbon nanotubes, Nanofilaments and Nanobeads by thermal Chemical Vapor Deposition Process. *Material Science and Engineering: B*, 96, 24-28.

[7] Qian, W., Yu, H., Wei, F., Zhang, Q., & Wang, Z. (2002). Synthesis of carbon nanotubes from liquefied petroleum gas containing sulfur. *Carbon*, 40, 2968-2970.

[8] Aswasthi, K., Kumar, R., Tiwari, R. S., & Srivastava, O. N. (2010). Large scale synthesis of bundles of aligned carbon nanotubes using a natural precursor: Turpentine oil. *Journal of Experimental Nanoscience*, 5(6), 498-508.

[9] Ghosh, P., Soga, T., Tanemura, M., Zamri, M., Jimbo, T., Katoh, R., & Sumiyama, K. (2009). Vertically aligned carbon nanotubes from natural precursors by spray pyrolysis method and their field electron emission properties. *Applied Physics A: Materials Science & Processing*, 94, 151-156.

[10] Andrews, R. J., Smith, C. F., & Alexander, A. J. (2006). Mechanism of carbon nanotube growth from camphor and camphor analogs by chemical vapor deposition. *Carbon*, 44(2), 341-347.

[11] Ghosh, P., Afre, R. A., Soga, T., & Jimbo, T. (2007). A simple method of producing single-walled carbon nanotubes from a natural precursor: Eucalyptus oil. *Materials Letters*, 61(17), 3768-3770.

[12] Karthikeyan, S., & Mahalingam, P. (2010). Studies of yield and nature of multi-walled carbon nanotubes synthesized by spray pyrolysis of pine Oil at different temperatures. *International Journal of Nanotechnology and Applications*, 4, 189-197.

[13] Karthikeyan, S., & Mahalingam, P. (2010). Synthesis and characterization of multi-walled carbon nanotubes from biodiesel oil: green nanotechnology route. *International Journal of Green Nanotechnology: Physics and Chemistry*, 2(2), 39-46.

[14] Anastas, P., & Warner, J. C. (1998). *Green Chemistry: Theory and Practice*, Oxford University Press, Oxford, 30.

[15] Sabelo, D. M., Kartick, C. M., Robin, C., Michael, J. W., & Neil, J. C. (2009). The effect of synthesis parameters on the catalytic synthesis of multiwalled carbon nanotubes using Fe-Co / CaCO3 catalysts. *South African Journal of Chemistry*, 62, 67-76.

[16] Smalley, R. E., Marek, I. M., Wang, Y., & Hange, R. H. (2007). Purification of carbon nanotubes based on the chemistry of fenton's reagent. *USPTO Patent Application No. 20070065975*.

[17] Afre, R. A., Soga, T., Jimbp, T., Mukul, Kumar., Anto, Y., Sharon, M., Prakash, R., Somani, , & Umeno, M. (2006). Carbon nanotubes by spray pyrolysis of turpentine oil at different temperatures and their studies. *Microporous and Mesoporous Materials*, 96, 184-190.

[18] Ghosh, P., Soga, T., Afre, R. A., & Jimbo, T. (2008). Simplified synthesis of single-walled carbon nanotubes from a botanical hydrocarbon: Turpentine Oil. *Journal of alloys and Compounds*, 462-92008.

[19] Dresselhausa, M. S., Dresselhausc, G., Joriob, A., Souza Filhob, A. G., & Saito, R. (2002). Raman spectroscopy on isolated single wall carbon nanotubes. *Carbon*, 40, 2043-2061.

[20] Kang, J. L., Li, J. J., Du, X. W., Shi, C. S., Zhao, N. Q., Cui, L., & Nash, P. (2008). Synthesis and growth mechanism of metal filled carbon nanostructures by CVD using Ni/Y catalyst supported on copper. *Journal of Alloys and Compounds*, 456(1-2), 290-296.

[21] Abello, M. C., Gomez, M. F., & Ferretti, O. (2001). Mo/γ-Al2O3 catalysts for the oxidative dehydrogenation of propane: Effect of Mo loading. *Applied Catalysis A*, 207(1-2), 421-431.

[22] Lee, M. H., & Park, D. G. (2003). Preparation of MgO with High Surface Area, and Modification of Its Pore Characteristics. *Korean Chemical Society*, 24(10), 1437-1443.

[23] Sinclair, R., Itoh, T., & Chin, R. (2002). In situ TEM studies of metal-carbon reactions. *Microscopy and Microanalysis*, 8(4), 288-304.

[24] Brukh, R., & Mitra, S. (2006). Mechanism of carbon nanotube growth by CVD. *Chemical Physics Letters*, 424, 126-32.

[25] Hofmann, S., Csanji, G., Ferrari, A. C., Payne, M. C., & Robertson, J. (2005). Surface diffusion: The low activation energy path for nanotube growth. *Physical Review letters*, 95, 036101-036104.

[26] Ducati, C., Alexandrou, I., Chhowalla, M., Robertson, J., & Amaratunga, G. A. J. (2004). The role of the catalytic particle in the growth of carbon nanotubes by plasma enhanced chemical vapor deposition. *Journal of Applied Physics*, 95(11), 6387-6391.

[27] Ozturk, B., Fearing, Vl., Ruth, J. A., & Simkovich, G. (1982). Self-Diffusion Coefficients of Carbon in Fe3C at 723 K via the Kinetics of Formation of This Compound. *Metallurgical and Materials Transactions A*, 13(10), 1871-1873.

[28] Chakraborty, A. K., Jacobs, J., Anderson, C., Roberts, C. J., & Hunt, M. R. C. (2005). Chemical vapor deposition growth of carbon nanotubes on Si substrates using Fe catalyst: What happens at the nanotube/Fe/Si interface. *Journal of Applied Physics*, 100(8), 084321.

[29] Rodriguez, N. M. (1993). A review of catalytically grown carbon nanofibers. *Journal of Materials Research*, 8, 3233-3250.

[30] Hofmann, S., Sharma, K., Ducati, C., Du, G., Mattevi, Cepek C., Cantoro, M., Pisana, S., Parvez, A., Cervantes-sodi, F., Ferrari, A. C., & Dunin-Borkowski, R. E. (2007). In situ Observations of Catalyst Dynamics during Surface-Bound Carbon Nanotube Nucleation. *Nano Letters*, 7(3), 602-608.

[31] Helveg, S., Lopez-Cartes, C., Sehested, J., Hensen, P. L., Clausen-Nielsen, Rostrup., Abild-Pederson, J. R., , F., & Norskov, J. K. (2004). Atomic-scale imaging of carbon nanofibre growth. *Nature*, 427, 426-429.

[32] Kukovitsky, E. F., L'vov, S. G., & Sainov, N. A. (2000). VLS-growth of carbon nanotubes from the vapor. *Chemical Physics Letters*, 317, 65-70.

[33] Leonhardt, A., Hampel, S., Büchner, B., et al. (2006). Synthesis, properties and applications of ferromagnetic filled carbon. *Chemical Vapor Deposition*, 12(6), 380-387.

[34] Bower, C., Otto, Z., Wei, Z., Werder, D. J., & Jin, S. (2000). Nucleation and growth of carbon nanotubes by microwave plasma chemical vapor deposition. *Applied Physics Letters*, 77(17), 2767-2768.

[35] Li, W. Z., Xie, S. S., Qian, L. X., Chang, B. H., Zou, B. S., Zhou, W. Y., Zhao, R. A., & Wang, G. (1996). Large-scale synthesis of aligned carbon nanotubes. *Science*, 274(5293), 1701-1703.

[36] Feng, W., Bai, X. D., Lian, Y. Q., Liang, J., Wang, X. G., & Yoshino, K. (2003). Well aligned polyaniline/carbon-nanotube composite films grown by in-situ aniline polymerization. *Carbon*, 41(8), 1551-1557.

[37] Zhang, Q., Zhao, M. Q., Huang, J. Q., Liu, Y., Wang, Y., Qian, W. Z., & Wei, F. (2009). Vertically aligned carbon nanotube arrays grown on a lamellar catalyst by fluidized bed catalytic chemical vapor deposition. *Carbon*, 47(11), 2600-2610.

[38] Zhang, Q., Zhou, W. P., Qian, W. Z., Xiang, R., Huang, J. Q., Wang, D. Z., & Wei, F. (2007). Synchronous growth of vertically aligned carbon nanotubes with pristine stress in the heterogeneous catalysis process. *Journal of Physical Chemistry*, 111(40), 14638-14643.

[39] Afre, R. A., Soga, T., Jimbo, T., Kumar, M., Ando, A., & Sharon, M. (2005). Growth of vertically aligned carbon nanotubes on silicon and quartz substrate by spray pyrolysis of a natural precursor: Turpentine oil. *Chemical Physics Letters*, 414(1-3), 6-10.

[40] Teo, K. B. K., Singh, C., Chhowalla, M., & Milne, W. I. (2003). Catalytic synthesis of carbon nanotubes and nanofiber. *In: Nalwa H S, eds. Encyclopedia of nanoscience and nanotechnology*, American Scientific Publishers, 1-22.

[41] Bu, I. Y. Y., & Oei, S. P. (2010). Hydrophobic vertically aligned carbon nanotubes on Corning glass for self cleaning applications. *Applied surface science*, 256(22), 6699-6704.

[42] Yang, L., Xu, Y., Wang, X., Zhu, J., Zhang, R., He, P., & Fang, Y. (2011). The application of b-cyclodextrin derivative functionalized aligned carbon nanotubes for electrochemically DNA sensing via host-guest recognition. *Analytica Chimica Acta*, 689(1), 39-46.

[43] Pan, Z. W., Zhu, H. G., Zhang, Z. T., Im, H. J., Dai, S., Beach, D. B., & Lowndes, D. H. (2003). Patterned growth of vertically aligned carbon nanotubes on pre-patterned iron/silica substrates prepared by sol-gel and shadow masking. *Journal of physical chemistry B*, 107(6), 1338-1344.

[44] Jung, Y. J., Wei, B. Q., Vajtai, R., & Ajayan, P. M. (2003). Mechanism of selective growth of carbon nanotubes on SiO2/Si patterns. *Nano Letters*, 3(4), 561-564.

[45] Gong, Q. M., Li, Z., Li, D., Bai, X. D., & Liang, J. (2004). Fabrication and structure: a study of aligned carbon nanotube/carbon nanocomposites. *Solid State Communications*, 131(6), 399-404.

[46] Gunjishima, I., Inoue, T., Yamamuro, S., Sumiyama, K., & Okamoto, A. (2007). Synthesis of vertically aligned, double-walled carbon nanotubes from highly active Fe-V-O nanoparticles. Carbon ., 45(6), 1193-1199.

[47] Kumar, M., & Ando, Y. (2003). A simple method of producing aligned carbon nanotubes from an unconventional precursor- Camphor. *Chemical Physics Letters*, 374(5-6), 521-526.

[48] Liu, Y., Yang, X. C., Pu, Y., & Yi, B. (2010). Synthesis of aligned carbon nanotube with straight-chained alkanes by nebulization method. *Transaction of Nonferrous Metals Society of China*, 20(6), 1012-1016.

Chapter 12

Toroidal and Coiled Carbon Nanotubes

Lizhao Liu and Jijun Zhao

Additional information is available at the end of the chapter

http://dx.doi.org/10.5772/51125

1. Introduction

The perfect graphite and carbon nanotube (CNT) are composed of hexagonal rings of carbon atoms. However, non-hexagonal rings like pentagons and heptagons usual exist in the realistic CNT. Due to the change of topology, different arrangements of the pentagons and heptagons would lead to various structures, such as CNTs with Stone-Wales defects [1], CNT junctions [2], toroidal CNTs [3], and coiled CNTs [4, 5]. Each type of these CNT-based structures has its unique physical and chemical properties; as a consequence, the diversity in morphology extends the applications of CNTs. In this chapter, we will review the current progress on two important members of the CNT family, i.e., the toroidal CNTs at the first and coiled CNTs in the second.

The toroidal CNT (also known as carbon nanotorus or carbon nanoring) is a kind of zero-dimensional CNT-based nanostructure. In other words, a carbon nanotorus can be considered as a giant molecule and directly used as a nanoscale device. As for the synthesis of the toroidal CNTs, numerous methods have been proposed, including laser-growth method, ultrasonic treatments, organic reactions, and chemical vapour deposition (CVD), which will be illustrated in the following. In addition to experimental synthesis, various theoretical efforts have been devoted to construct the structural models of the toroidal CNTs. In general, there are two kinds of toroidal CNTs: one is formed by pristine nanotube with pure hexagon networks, and the other contains certain amount of pentagon and heptagon defects. Due to the circular geometry of the carbon nanotorus and incorporation of pentagon/heptagon defects, it may exhibit novel mechanical, electronic and magnetic properties different from the straight CNTs.

Another kind of curved CNT-based nanostructure is the coiled CNT, which is also known as carbon nanocoil or carbon nanospring. Different from the zero-dimensional toroidal CNT, the coiled CNT is a kind of quasi one-dimensional CNT-based nanostructures with a certain spiral angle. Intuitionally, a carbon nanocoil is like a spring in geometry. Therefore, mechanic properties of the coiled CNTs attract lots of attentions. Among various methods to produce the coiled CNTs, CVD approach is predominant due to the high quality and good controllability. Besides, several methods have been proposed to build the structural models of the coiled CNTs. An important feature of the carbon nanocoil models is the periodic incorporation of pentagons and heptagons in the hexagonal network. In addition, due to the excellent properties of the coiled CNTs, they have promising applications in many fields, such as sensors, electromagnetic nano-transformers or nano-switches, and energy storage devices.

2. Toroidal CNTs

In this section, we summarize experimental fabrication and theoretical modelling of the toroidal CNTs, as well as their physical and chemical properties. The toroidal CNTs are predicted to be both thermodynamic and kinetically stable. Due to the circular geometry, the toroidal CNTs possess excellent properties, especially the electronic and magnetic properties.

2.1. Fabrication

Synthesis and characterization of the toroidal CNTs are of key importance in the carbon nanotorus related fields. Early in 1997, Liu et al. reported synthesis of the toroidal CNTs with typical diameters between 300 and 500 nm by using the laser growth method [3]. From the measurement of scanning force microscopy (SEM) and transmission electron microscopy (TEM), it was shown that the toroidal CNTs were form by single-walled carbon nanotube (SWNT) ropes consisting of 10 to 100 individual nanotubes. Soon after, the toroidal CNTs were also found in the CNT samples prepared by catalytically thermal decomposition of hydrocarbon gas [6] and an ultrasound-aided acid treatment [7, 8]. Later, a variety of experimental approaches were developed to fabricate carbon nanotori, such as organic reactions [9, 10], chemical vapor deposition (CVD) [11, 12], and depositing hydrocarbon films in Tokamak T-10, the facility for magnetic confinement of high-temperature plasma [13]. In addition, incomplete toroidal CNTs [13, 14], large toroidal CNTs with diameters of~200–300 nm, sealed tubular diameters of 50–100 nm [15], and patterning of toroidal CNTs [16, 17] were also achieved in laboratory. In particular, the tubular diameter of a carbon nanotorus is now controllable. Toroidal CNTs from single-walled [7-10, 18, 19], double-walled [20], triple-walled [21], and multi-walled CNTs [6] have been achieved. Combining the experimental measurements and a simple continuum elastic model, formation of the toroidal CNTs was supposed to involve a balance between the tube-tube van der Waals adhesion, the strain energy resulting from the coiling-induced curvature and the strong interaction with the substrate [8, 14]. Various kinds of the toroidal CNTs are presented in Figure 1.

http://dx.doi.org/10.5772/51125

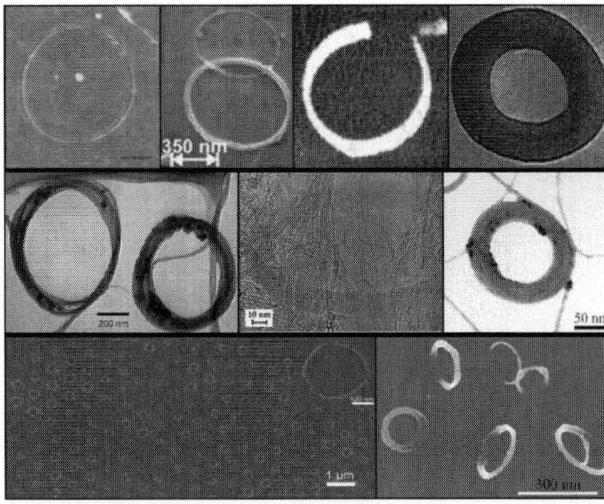

Figure 1. Experimental fabrications of various kinds of toroidal CNTs.

2.2. Structural models and thermodynamic stabilities

Prior to the experimental synthesis, Dunlap proposed to construct the structural model of a carbon nanotorus by connecting two CNTs with different diameters [22]. Almost at the same time, researchers in Japan built a C_{360} nanotorus from C_{60} fullerene [23] and then generated a series of toroidal CNTs with 120 to 1920 carbon atoms using the prescription of Goldberg [24, 25]. So far, there have been six major approaches to construct the structural models of toroidal CNTs: (1) bending a finite CNT and connecting its ends together [26-29]; (2) connecting CNTs with different diameters by introducing pentagons and heptagons [22, 30-32]; (3) constructing from fullerenes [23-25] by employing the prescription of Goldberg [33]; (4) built through the connection of one zigzag-edged chain of hexagons and another armchair-edged chain of hexagons [34]; (5) sewing the walls of a double-walled CNT at both ends [35], (6) constructing from only pentagons and heptagons [36]. To summarize, there are two kinds of toroidal CNTs: one is formed by pure hexagonal networks and the other is a hexagonal structure with pentagon-heptagon defects. In a more detailed way, Itoh et al. classified the toroidal CNTs into five types using the parameters of the inner (r_i) and outer (r_o) diameters, and the height (h) [37]. As depicted in Figure 2, type (A) indicates a nanotorus with $r_i \approx r_o$, $h \ll r_i$, and $h \approx (r_o - r_i)$, type (B) is the case of $r_i \sim r_o \sim h$ and $h \approx (r_o \circledcirc r_i)$, type (C) denotes $h \ll (r_o \circledcirc r_i)$, type (D) is the case of $r_i < r_o$, $r_o \sim h$, and $h \sim (r_o \circledcirc r_i)$, and type (E) means $(r_o \circledcirc r_i) \ll h$, respectively.

After establishing the structural models, one important issue is to examine the thermodynamic stabilities of the toroidal CNTs. Many groups demonstrated that toroidal CNTs are more stable than C_{60} fullerene through comparing their binding/cohesive energies calculated by means of empirical potential methods [22-25]. Besides, molecular dynamics (MD) simulations also demonstrated that toroidal CNTs can survive under high temperature [23, 29, 38,

39]. Generally, the thermodynamic stability of a carbon nanotorus depends on its geometric parameters, such as ring and tubular diameter, symmetry, curvature, and position of the pentagons and heptagons. Ihara et al. showed that the cohesive energy of a carbon nanotorus derived from C_{60} fullerene decreased with increasing number of carbon atoms in the carbon nanotorus [24]. The ring and tubular diameter can also affect the thermodynamic stability of a carbon nanotorus [40-42]. At a fixed tubular diameter, there was a preferable ring diameter where the nanotorus possesses the lowest formation energy [40]. Besides, dependence of the stability on the rotational symmetry was also reported for the toroidal CNTs [32, 37]. Among the toroidal CNTs constructed from (5, 5), (6, 6), and (7, 7) armchair CNTs, the one with D_{6h} symmetry is energetically favourable [32]. Despite the dependence on the geometric details, it was believed [43, 44] that for the toroidal CNTs with large ring diameters, the pure hexagonal structure is energetically more stable, but for the ones with small ring diameters, the mixture of hexagonal networks and pentagon-heptagon defects is energetically more favourable. In [44], this critical ring diameter is given by the equation $R_c = \pi r^2 Y/(4\sigma)$, where r is the tubular diameter of the initial CNT, Y is the Young's modulus of the initial CNT, and σ is the surface tension of graphite perpendicular to the basal planes. For example, taking the Y = 1.0 TPa, a R_c of 90 nm can be obtained for a carbon nanotorus made of a (10,10) nanotube (r = 0.68 nm).

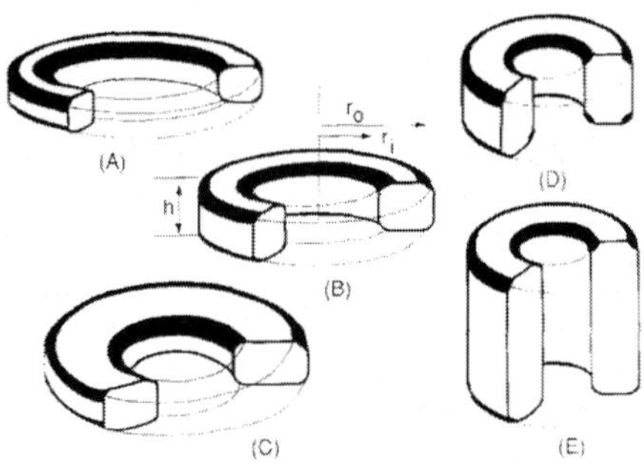

Figure 2. Schematic diagram for five types of toroidal CNTs classified by the parameters of the inner diameter r_i, the outer diameter r_o, and the height h, respectively. Reprinted with permission from [37]. Copyright (1995) Elsevier.

2.3. Mechanical properties

Mechanical property is of fundamental importance for the applications of a material. Employing MD simulation with a reactive force field, Chen et al. investigated the mechanical properties of zero-dimensional nanotorus, one-dimensional nanochain and two-dimensional nanomaile constructed from toroidal CNTs [45]. For a nanotorus constructed from bending a (5, 5) CNT, its Young's modulus increases monotonically with tensile strain from 19.43 to 121.94 GPa without any side constraints and from 124.98 GPa to 1560 GPa with side constraints, respectively, where the side constraint means fixing the position of small regions of carbon atoms at left and right sides. Besides, the tensile strength of the unconstrained and constrained nanotorus was estimated to be 5.72 and 8.52 GPa, respectively. In addition, the maximum elastic strain is approximate 39% for the nanochain and 25.2% for thenanomaile. For a nanotorus obtained from bending a (10, 10) CNT, its Young's modulus along the tube axis was 913 GPa by taking [46]. Later, buckling behavior of toroidal CNTs under tension was investigated using the molecular mechanics (MM) computations, including the toroidal CNTs formed from (5, 5), (8, 8) and (9, 0) CNTs [47, 48]. It was found that the buckling shapes of the toroidal CNTs constructed from both armchair and zigzag CNTs with an odd number of units are unsymmetrical, whereas those with an even number of units are symmetrical. Recently, reversible elastic transformation between the circular and compressed nanotorus in a colloid has been observed under TEM [17]. This geometric reversibility was also predicted theoretically by using a nonlinear continuum elastic model [49, 50], suggesting the potential application of toroidal CNTs as ultrasensitive force sensors and flexible and stretchable nanodevices.

2.4. Electronic properties

It is well-known that a CNT can be expressed by a chiral vector C_h (n, m) and a translation vector T (p, q) and can be either metallic or semiconducting, depending on its chirality [51]. Since a carbon nanotorus can be considered as a bended CNT or a CNT incorporated with pentagons and heptagons, it would be interesting to explore will the bending behavior or inclusion of pentagons and heptagons affect the electronic properties of the pristine CNT. For a carbon nanotorus formed by bending a (n, m) CNT, it can be divided into three types: (1) if $m - n = 3i$, and $p - q = 3i$ (i is an integral), the carbon nanotorus is metallic; (2) if $m - n = 3i$, and $p - q \neq 3i$, the carbon nanotorus is semiconducting; (3) if $m - n \neq 3i$, and $p - q = 3i$, the carbon nanotorus is insulating [52]. This classification was partly confirmed by the tight-binding (TB) calculation that a metallic carbon nanotorus can be constructed by bending a metallic CNT and also follows the rule of divisibility by three on the indices of chiral and twisting vectors [53]. Moreover, delocalized and localized deformations play different roles on the electronic properties of a carbon nanotorus built bending a CNT [27]. The delocalized deformations only slightly reduce the electrical conductance, while the localized deformations will dramatically lower the conductance even at relatively small bending angles. Here the delocalized deformation means the deformation induced by the mechanical bending of the CNT, and the localized deformation indicates the deformation induced by the pushing action of the tip of AFM. In addition, Liu et al. reported the oscillation behavior of the energy gap during increasing size of the nanotorus and the gap was eventually converged to that of the infinite CNT [54].

Meanwhile, in the case of incorporation of pentagons and heptagons, a HOMO-LUMO gap can be expected for the carbon nanotorus. For a carbon nanotorus C_{1960} constructed by connecting (6, 6) and (10, 0) CNTs, a gap of 0.05 eV was calculated by a TB approach [44]. Using both the TB and semiempirical quantum chemical approaches, a series of toroidal CNTs with total number of atoms ranging from 120 to 768 were investigated and most of them have HOMO-LUMO gaps [55]. Besides, employing the extended-Hückel method, energy gaps of 0.4-0.32μB eV were predicted for the toroidal CNTs of C_{170}, C_{250}, C_{360}, C_{520}, and C_{750} [56]. Further accurate DFT examination also showed that the nanotorus C_{444} has a gap of 0.079 eV and the nanotorus C_{672} has a gap of 0.063 eV, respectively [57].

2.5. Magnetic properties

The unique circular geometry endows its advantage to study the magnetic response when ring current flows in a carbon nanotorus. Early in 1997, Haddon predicted that the nanotorus C_{576} has an extremely large and anisotropic ring-current diamagnetic susceptibility, which can be 130 times larger than that of the benzene molecule [58]. Afterwards, colossal paramagnetic moment was also reported in the metallic toroidal CNTs, which was generated by the interplay between the toroidal geometry and the ballistic motion of the π-electrons [28], as shown in Figure 3. For example, the nanotorus C_{1500} built from a (5, 5) CNT possesses a large paramagnetic moment of 88.4 $_{\mu}$B at 0 K. Similarly, the nanotorus C_{1860} built from a (7, 4) CNT has a giant zore-temperature magnetic moment of 98.5 $_{\mu}$B. In addition to the paramagnetic moments, existence of ferromagnetic moments at low temperatures in the toroidal CNTs without heteroatoms was also predicted by using a π-orbital nearest-neighbor TB Hamiltonian with the London approximation, which is attributed to the presence of pentagons and heptagons [59]. Another important phenomenon, i.e., the Aharonov–Bohm effect can be also observed in the toroidal CNTs [60-64]. Indeed, the magnetic properties of the toroidal CNTs are affected by many factors. Liu et al. pointed out that the paramagnetic moments of the toroidal CNTs decrease distinctly as temperature increases [28]. Such temperature dependence was also confirmed by several successive studies [65-68]. Moreover, the magnetic properties of a toroidal CNT also rely on its geometric parameters, such as ring diameter, curvature, chirality, and the arrangement of pentagons and heptagons [65-67].

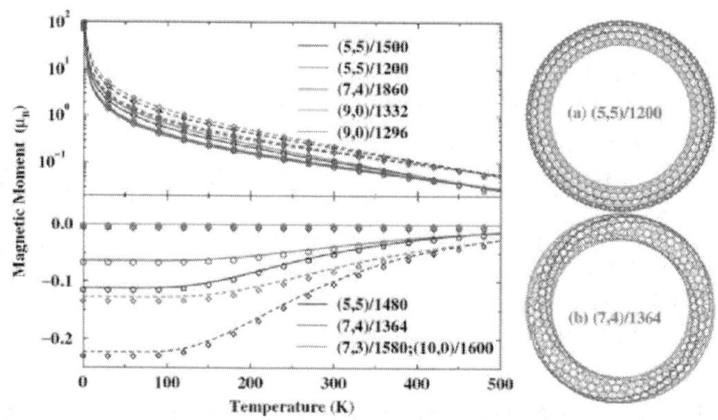

Figure 3. Induced magnetic moment as a function of temperature for various toroidal CNTs in a perpendicular magnetic field of 0.1 T (solid line) and 0.2 T (dashed line), respectively. Reprinted figures with permission from [Liu L, Guo GY, Jayanthi CS, Wu SY. Colossal Paramagnetic Moments in Metallic Carbon Nanotori. 88, 217206 (2002)]. Copyright (2002) by the American Physical Society. http://prl.aps.org/abstract/PRL/v88/i21/e217206.

2.6. Modification of the toroidal CNTs

Chemical modification is an important approach to tailor the properties of materials. A common approach of chemical modification is doping. It was found that doping electrons or holes into a carbon nanotorus could vary its magnetic properties through altering the band-filling configuration [69]. Our previous work also demonstrated that substitutional doping of boron or nitrogen atoms could modify the electronic properties of the toroidal CNTs due to change of the six π-electron orbitals [32]. Moreover, compared with the hexagonal rings, existence of pentagons favours the doping of nitrogen atoms and existence of heptagons prefers the doping of boron atoms. Besides, the toroidal CNTs coated with beryllium can be used as candidates for hydrogen storage. Each beryllium atom can adsorb three H_2 molecules with moderate adsorption energy of 0.2-03 eV/H_2 [70].

Since the toroidal CNTs also have the hollow tubular structures similar to the CNTs, atoms or molecules can be encapsulated into the toroidal CNTs. Early in 2007, Hilder et al. examined the motion of a single offset atom and a C_{60} fullerene inside a carbon nanotorus to explore its application as high frequency nanoscale oscillator [71]. They demonstrated that the C_{60} fullerene encapsulated carbon nanotorus can create high frequency up to 150 GHz, which may be controlled by changing the orbiting position. By inserting the chains of Fe, Au, and Cu atoms into a carbon nanotorus, Lusk et al. investigated the geometry, stability and electronic magnetic properties of this nano-composite structure [72]. Reduced HOMO-LUMO gap and ferromagnetism of the nanotorus were predicted by encapsulating chains of metal atoms. In addition, diffusion behavior of water molecules forming two oppositely polarized chains in a carbon nanotorus was studied by MD simulations. It was demonstrated that Fickian diffusion is in the case of a single chain and the diffusion for two or more chains is consistent with single-file diffusion [73].

3. Coiled carbon nanotube

Similar to the case of the toroidal CNTs, we first introduce the experimental synthesis and theoretical methods to construct the structural models, as well as their formation mechanism and stabilities. Then the mechanic properties and electronic properties of the coiled CNTs are summarized. Finally, the promising applications of coiled CNTs in various fields compared with their straight counterparts owing to their spiral geometry and excellent properties will be discussed in the end of this section.

3.1. Fabrication and formation mechanism

The coiled CNTs were first experimentally produced through catalytic decomposition of acetylene over silica-supported Co catalyst at 700 °C in 1994 [4, 5]. Afterwards, numerous methods have been proposed to synthesize the coiled CNTs, including the laser evaporation of the fullerene/Ni particle mixture in vacuum [74], opposed flow flame combustion of the fuel and the oxidizer streams [75], electrolysis of graphite in fused NaCl at 810 °C [76], self-assembly from π-conjugated building blocks [77, 78], and CVD method [79-83]. Among these various methods, the CVD approach is predominant due to its high quality, which has been reviewed by several literatures [84-86]. To fabricate the coiled CNTs, CVD process involves the pyrolysis of a hydrocarbon (e.g. methane, acetylene, benzene, propane) over transition-metal catalysts (e.g. Fe, Co, Ni) at high temperatures. Compared to the high growth temperature (> 2000 °C) of CNT through arc discharge and laser evaporation process, the relatively low growth temperature of CVD method (500–1000 °C) allows carbon atoms move slowly and form non-hexagonal carbon rings [84]. In 2006, Lau et al. reviewed the three major CVD-based methods to fabricate the coiled CNTs, including the catalyst supported CVD growth, on substrate CVD growth and template-based CVD growth [84]. Later, synthetic parameters of CVD growth of the coiled CNTs, such as catalyst, gas atmosphere and temperature, were introduced and catalogued by Fejes et al. [85] and Shaikjee et al. [86], respectively. Moreover, Shaikjee et al. [86] presented different types of the coiled CNTs with non-linear morphology, which are shown in Figure 4.

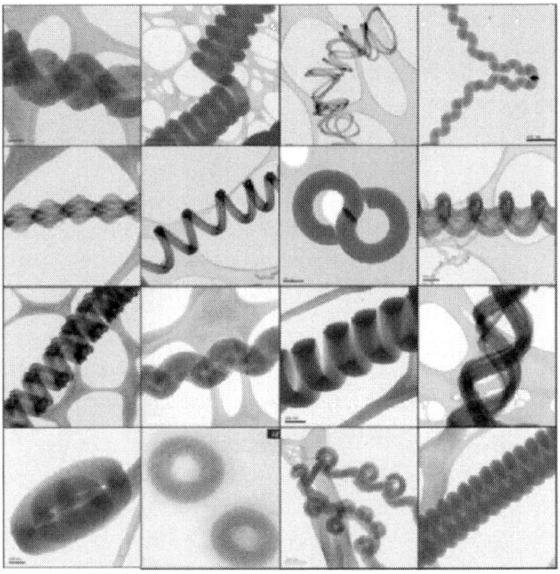

Figure 4. Experimental fabrications of various kinds of the coiled CNTs. Reprinted with permission from [86]. Copyright (2011) Elsevier.

As for the formation mechanism of the coiled CNTs, Fonseca et al. presented a formation of (chiral and achiral) knees on a catalyst particle to further form toroidal and coiled CNTs, which can be described by a simple formalism using the heptagon-pentagon construction [87]. In addition, formation of the coiled CNTs is closely related to the catalyst. Pan et al. suggested that the catalyst grain is crucial to the geometry of a carbon nanocoil and the non-uniformity of carbon extrusion speed in the different parts of the catalyst grain leads to the helical growth of the coiled CNTs [88]. Chen et al. pointed out that the driving force of coiling straight CNTs was the strong catalytic anisotropy of carbon deposition between different crystal faces [89]. For growth of carbon microcoils, the catalyst grain rotates around the coil axis which is on the symmetric face of the deposition faces; while for the twisted carbon nanocoils, the catalyst grain rotates around the axis which is perpendicular to the symmetric face of the deposition faces. Taking both the energy and entropy into account, Bandaru et al. proposed a mechanism that for a given volume of material, the helical form occupies the least amount of space and the entropy of the ambient conditions should increase to compensate for the close packing of the helices, which in turn is facilitated by nonwetting catalyst particles or induced by catalyst/ambient agitation in the CVD growth [90].

3.2. Structural models and thermodynamic stabilities

An important feature of a carbon nanocoil is incorporation of pentagons and heptagons in the hexagonal network. Dunlap [22, 91] showed that connecting two CNTs with pentagons and heptagons could result in a curved structure or knee structure. Based on the knee structure, Fonseca et al. was able to construct the toroidal and coiled CNTs using the knee segments as building blocks, where the former is an in-plane structure and the latter is out of plane [92]. In addition, researchers in Japan proposed two kinds of methods to construct structures of the coiled CNTs. One approach is to cut the toroidal CNTs into small pieces and recombine them to form the coiled CNTs with one pitch containing one nanotorus [37, 93]. For the coiled CNTs built from toroidal segments, Setton et al. suggested that the toroidal segments were only feasible for single-shell or at best two-shell nanocoils [94]. The other way is to insert pentagons and heptagons into a perfect graphene network and then roll up this structure to form a carbon nanocoil [95, 96]. Similarly, Biró et al. proposed to build the coiled CNTs from rolling up the Haeckelite structure, a graphite sheet composed of polygonal rings [97]. Recently, we were able to construct the carbon nanocoils from segment of CNTs in which the tube chirality is maintained [98]. Through introducing a pair of pentagons in the outer side and another pair of heptagons in the inner side into the segment of an armchair CNT, a curved structure can be obtained. Using this curved structure as a building block, a carbon nanocoil can be formed by connecting the building blocks with a rotate angle. This method was also employed to construct the structural models of the toroidal CNTs, as mentioned above [32]. A simple schematic diagram of this method is presented in Figure 5. Usually, a carbon nanocoil can be expressed by the parameters of inner coil diameter (D_i), outer coil diameter (D_o), tubular diameter (D_t) and coil pitch (λ) [84, 86], as illustrated in Figure 6.

In addition to the structural models of the coiled CNTs, several works have been devoted to investigating their thermodynamic stabilities. Employing MD simulation, Ihara et al. [93] obtained the cohesive energies of 7.41, 7.39 and 7.43 eV/atom for C_{360}, C_{540} and C_{1080} nanocoils, respectively, which are close to that of graphite sheet (7.44 eV/atom) and lower than that of the C_{60} fullerene (7.29 eV/atom). Therefore, these carbon nanocoils are more stable than C_{60} fullerene. Moreover, these carbon nanocoils can maintain the coiled geometry without collapse at a temperature up to 1500 K, which further confirms their thermodynamical stability. By taking into account the volume free energy, the surface energy, and the curvature elastic energy, it was found that there is a threshold condition for the formation of straight multiwall CNTs [99]. Below that the straight multiwall CNTs become unstable and would undergo a shape deformation to form the coiled CNTs.

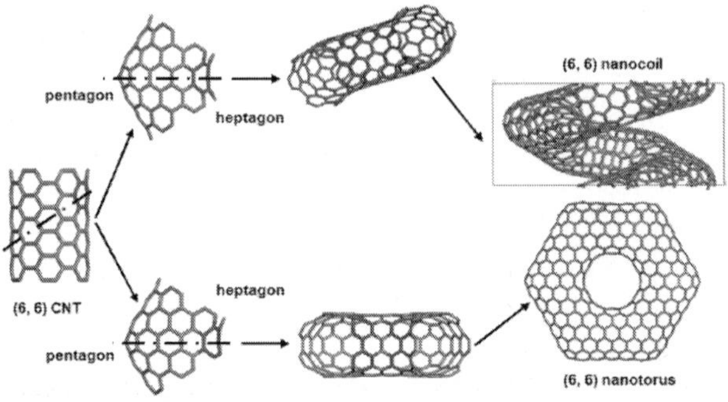

Figure 5. Schematic diagram for constructing the structural models of (6, 6) carbon nanotorus and nanocoil by introducing pairs of pentagons (highlighted in blue) and heptagons (highlighted in red).

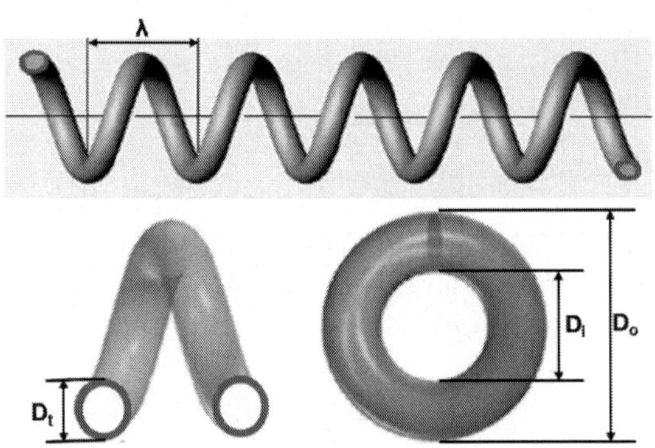

Figure 6. Parameters of inner coil diameter D_i, outer coil diameter D_o, tubular diameter D_t and coil pitch λ to describe a carbon nanocoil.

3.3. Mechanical properties

Intuitively, a carbon nanocoil is similar to a spring in geometry. It is well-known that spring exhibits excellent mechanic properties and is very useful in the mechanics-based devices. Therefore, mechanical properties of the coiled CNTs as "nanospring" have attracted lots of attentions. Early in 2000, Volodin et al. measured the elastic properties of the coiled CNTs with atomic force microscopy (AFM) and showed that the coiled CNTs with coil diameters

(> 170 nm) possess high Young's modulus of 0.4–0.9 TPa [100]. Using a manipulator-equipped SEM, Hayashida reported the Young's modulus of 0.04–0.13 TPa and the elastic spring constants of 0.01–0.6 N/m for the coiled CNTs with coil diameters ranging from 144 to 830 nm [101]. Remarkable spring-like behavior of an individual carbon nanocoil has been demonstrated by Chen et al. [102], as presented in Figure 7. A spring constant of 0.12 N/m in the low-strain regime and a maximum elastic elongation of 33% were obtained from AFM measurement. In contrast to the high measured Young's modulus, the shear modulus of the coiled CNTs is extremely low. Chen et al. [102] considered the coiled CNTs with a D_o of ~126±4 nm but different D_i. For the case of $D_i = 3/4\ D_o$, a shear modulus of ~2.5±0.4 GPa was estimated; if $D_i = 1/2\ D_o$, the corresponding shear modulus was ~2.3±0.4 GPa; and if $D_i = 0$, a shear modulus of ~2.1±0.3 GPa can be obtained. Afterwards, Huang [103] studied the coiled CNTs under uniaxial tension in simple explicit expressions and obtained a maximum elastic elongation of ~30%, a shear modulus of 2.8–3.4 GPa and a spring constant of ~0.1–0.4 N/m for the double nanocoils formed by twisting two single nanocoils, which is comparable to the experimental result [102]. Later, Chang et al. reported a shear modulus of 3±0.2 GPa for the double coiled CNTs [104]. In addition, Poggi et al. demonstrated the compression behavior of the coiled CNTs and presented that repeated compression/buckling/decompression of the nanocoil was very reproducible with a limiting compression of 400 nm [105].

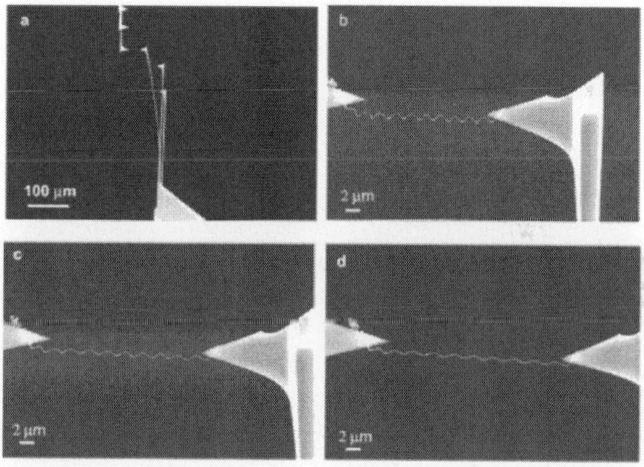

Figure 7. Measurement of the mechanical properties of a carbon nanocoil using the AFM cantilevers: (b) the initial state, (c) at a tensile strain of 20%, and (d) at a tensile strain of 33%. Reprinted with permission from [102]. Copyright (2003) American Chemical Society.

In addition to the experimental measurements, numerous theoretical simulations were carried out to investigate the mechanical properties of the coiled CNTs. Using the Kirchhoff rod model, Fonseca et al. derived a series of expressions to obtain the Young's modulus and Poisson's ratios for the coiled CNTs. Taking the parameters for the carbon nanocoil reported by Chen et al. [102], Fonseca et al. estimated the Young's modulus of 6.88 GPa for a nanocoil with a coil diameter of 120 nm, a Poisson's ratio of 0.27 and a shear modulus of 2.5 GPa [106,

107]. Besides, equations were derived to calculate the elastic constants of the forests of the coiled CNTs, which shows that the entanglement among neighboring nanocoils will contribute to the mechanical properties of the nanocoil forests [108]. Employing the DFT and TB calculations, we computed the Young's modulus and elastic constant of a series of single-walled carbon nanocoils built from the armchair CNTs [98]. The Young's modulus ranges from 3.43 to 5.40 GPa, in good agreement with the Fonseca's reports [106, 107] and the elastic constant lies between 15.37 to 44.36 N/m, higher than the experimental values [100, 102]. Furthermore, superelastic behavior of the coiled CNTs was also predicted from our computations where the coiled CNTs can undertake an elastically tensile strain up to ~60% and compressive strain up to ~20–35%. Such superelasticity is due to the invariance of bond length under strain associated with the strong covalent C-C bonding. In a recent computation on the mechanic properties of the single-walled carbon nanocoils using the finite element ANSYS code, spring constants ranging from 15–30 N/m were obtained for the armchair carbon nanocoils with different tubular diameters [109]. As the tubular diameter increases, the spring constant increases accordingly. Generally speaking, the calculated Young's modulus and elastic constants for the coiled CNTs are more or less different from that of the experimental measurements. This difference may be attributed to the structural details of the synthesized carbon nanocoils, especially the larger sizes of experimental nanocoil samples.

3.4. Electronic and transport properties

Similar to the toroidal CNTs, pentagons and heptagons exist in the coiled CNTs, which may lead to different electric properties with regard to that of the pristine CNTs. Using the two and four probes methods, Kaneto et al. measured the electric conductivity of the micro carbon nanocoils, which lies in 30–100 S/cm [110]. Later, it was found that for a carbon nanocoil with a coil diameter of 196 nm and a length of 1.5 mm, the conductivity is about 180 S/cm [101], which is much smaller than a straight CNT (~10^4 S/cm) [111]. Recently, Chiu et al. reported a very high conductivity of 2500 S/cm and an electron hopping length of ~5 nm for the single carbon nanocoils measured at low temperature [112]. An even higher electron hopping length of 5–50 nm was predicted by Tang et al. [113]. Moreover, the temperature dependence of the electric resistance was also observed where resistivity of the carbon nanocoil decreases as the annealing temperature increases [114]. Therefore, the measured electric properties of the coiled CNTs are closely related to the temperature and details of the samples.

Theoretically, employing a simple TB model, Akagi et al. [95, 96] calculated the band structures and electron density of states of the carbon nanocoils and suggested that the coiled CNTs could be metallic, semiconducting and semimetallic, depending on the arrangement of the pentagons and heptagons. Compared with the pristine CNTs, the semimetal property is unique for the carbon nanocoil [96]. Recently, we investigated the electric conductance of a series of armchair carbon nanocoils through using a π-orbital TB model combined with the Green's function approach [115]. Using the metallic armchair CNTs as the electrodes, we calculated the quantum conductance of the (5, 5), (6, 6) and (7, 7) carbon nanocoils, as presented in Figure 8. Clearly, there is a transport gap in the conductance spectrum. Further analysis of the electronic states indicates that only incorporation of pentagons and heptagons (such as Stone-Wales defects) can not lead to gap opening, and thus creation of the band gap should be attributed to the existence of sp^3 C-C bonds caused by coiling the CNTs. In addition, change of quantum conductance for the armchair carbon nanocoils under uniaxial elongation or compression is not distinct due to the nearly invariant bond length under strain, i.e. superelasticity [98].

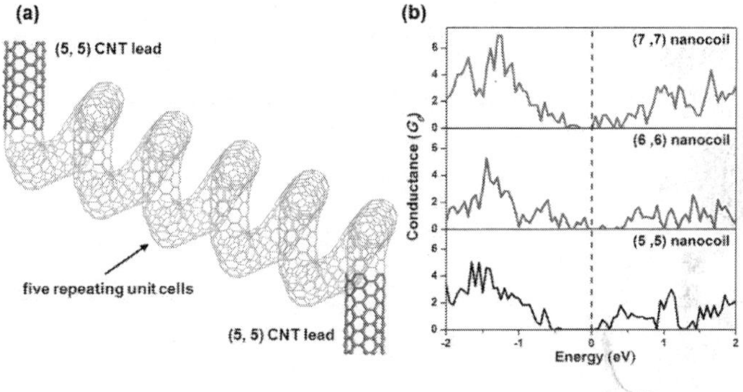

Figure 8. Structural model to calculate conductance of the (5, 5) carbon nanocoil (a) and conductance of the (5, 5), (6, 6) and (7, 7) carbon nanocoils (b). Reprinted with permission from [115]. Copyright (2011) Science China Press and Springer-Verlag Berlin Heidelberg.

3.5. Applications

Owing to the spiral geometry and unusual properties, the coiled CNTs have promising applications in various fields compared with their straight counterparts [84-86, 116]. One important application of the carbon nanocoils is to act as the sensors. In 2004, Volodin et al. [117] reported the use of coiled nanotubes as self-sensing mechanical resonators, which is able to detect fundamental resonances ranging from 100 to 400 MHz, as illustrated in Figure 9. The self-sensing carbon nanocoil sensors are sensitive to mass change and well suited for measuring small forces and masses in the femtogram range. After measuring the mechanical response of the coiled CNTs under compression using AFM, Poggi et al. pointed out that a nonlinear response of the carbon nanocoil can be observed, which is associated with compression and buckling of the nanocoil [105]. Bell et al. demonstrated that the coiled CNTs can be used as high-resolution force sensors in conjunction with visual displacement measurement as well as electromechanical sensors due to the piezoresistive behavior without an additional metal layer [118]. Besides, the applications of carbon nanocoils as magnetic sensors [119], tactile sensors [120], and gas sensors [121] were also exploited.

Another kind of major applications of the carbon nanocoils is to form composites with other materials. It was found that incorporation of carbon nanocoils in epoxy nanocomposites can enhance the mechanic properties of the epoxy nanocomposites [122-125]. Besides, the coiled CNT/silicone–rubber composites show high resistive sensitivity, relying on the density of the carbon nanocoil [126, 127]. In addition, metal-coated carbon nanocoils can also display some properties different from the pristine coiled CNTs. Tungsten-containing carbon nanocoils can expand and contract as flexibly as macro-scale springs and the elastic constants of the tungsten-containing carbon nanocoils rises along with increasing content of tungsten [128]. Bi et al. [129] found that the coiled CNTs coated with Ni have enhanced microwave absorption than the uncoated ones, which is results from stronger dielectric and magnetic losses.

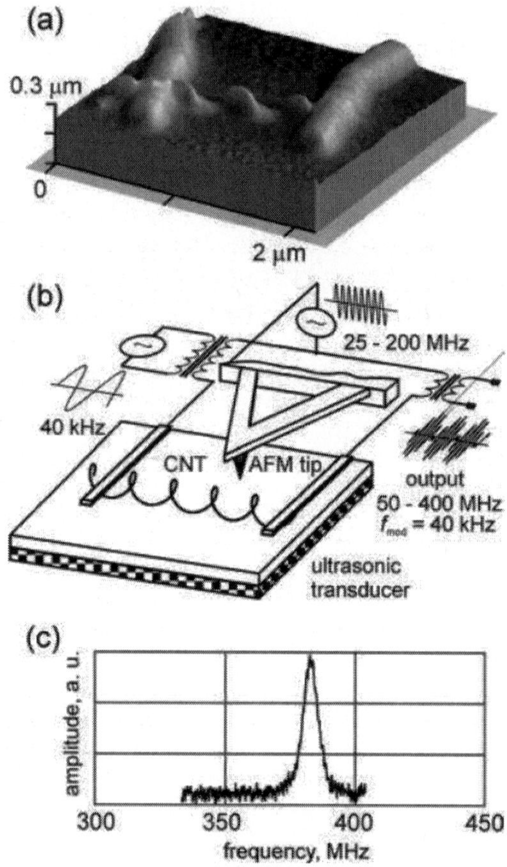

Figure 9. The carbon nanocoil to act as a mechanical resonator: (a) AFM image of the carbon nanocoil, (b) circuit contains two broad-band radio frequency transformers and the carbon nanocoil, and (c) resonant response of the carbon nanocoil device to electromechanical excitation. Reprinted with permission from [117]. Copyright (2004) American Chemical Society.

In addition, field emission [79, 130], energy storage [131, 132] and biological applications [133] of the coiled CNTs were also reported. Nowadays, the coiled CNT have been used as sensors [117-121], flat panel field emission display [79], microwave absorbers [134] and additives in the cosmetic industry [86].

4. Conclusion

Experimental fabrication and theoretical modelling of the toroidal and coiled CNTs were reviewed in this chapter. Compared with the pristine CNTs, the zero-dimensional toroidal CNTs exhibit excellent electromagnetic properties, such as persistent current and Aharonov–Bohm effect. Moreover, electronic properties of the toroidal CNTs can be tuned by chemical modification. In contrast to the toroidal CNTs, the coiled CNTs are quasi one-dimensional CNT-based nanostructures. Due to the spring-like geometry, the coiled CNTs possess fascinating mechanical properties, which are known as superelastic properties. This superelasticity allows the carbon nanocoils to act as electromechanical, electromagnetic, and chemical sensors. In addition, the coiled CNTs have been used commercially to fabricate flat panel field emission display, microwave absorbers and cosmetics.

As mentioned above, the toroidal CNTs synthesized experimentally are usually formed by the bundle of single-walled CNTs and have large ring diameters. Therefore, fabrications of the single-walled toroidal CNTs, as well as the toroidal CNTs of controllable ring diameters, are great challenges. Moreover, achievement of inserting atoms/molecules into the toroidal CNTs is another key issue under solution. On the other hand, since the formation mechanism of the coiled CNTs depends closely on the catalysts, searching for the optimal catalysts is significant for the quality and quantity of the nanocoil samples. Besides, finding appropriate geometry and concentration of the coiled CNTs is also necessary to improve performance of nanocomposites with the carbon nanocoils. Further experimental and theoretical works are expected to carry out to solve these problems.

Acknowledgements

This work was supported by the National Natural Science Foundation of China (No. 11174045, No. 11134005).

Author details

Lizhao Liu and Jijun Zhao*

*Address all correspondence to: zhaojj@dlut.edu.cn

Key Laboratory of Materials Modification by Laser, Ion and Electron Beams (Dalian University of Technology), Ministry of Education, Dalian 116024, China

References

[1] hang, P., Lammert, P. E., & Crespi, V. H. (1998). Plastic Deformations of Carbon Nanotubes. *Physical Review Letters*, 81(24), 5346-5349.

[2] Yao, Z., Postma, H. W. C., Balents, L., & Dekker, C. (1999). Carbon nanotube intramolecular junctions. *Nature*, 402(6759), 273-276.

[3] Liu, J., Dai, H. J., Hafner, J. H., Colbert, D. T., Smalley, R. E., Tans, S. J., & Dekker, C. (1997). Fullerene'crop circles. *Nature*, 385(6619), 780-781.

[4] Amelinckx, S., Zhang, X. B., Bernaerts, D., Zhang, X. F., Ivanov, V., & Nagy, J. B. (1994). A Formation Mechanism for Catalytically Grown Helix-Shaped Graphite Nanotubes. *Science*, 265(5172), 635-639.

[5] Zhang, X. B., Zhang, X. F., Bernaerts, D., Tendeloo, G. v., Amelinckx, S., Landuyt, J. v., Ivanov, V., Nagy, J. B., Ph, L., & Lucas, A. A. (1994). The Texture of Catalytically Grown Coil-Shaped Carbon Nanotubules. *Europhysics Letters*, 27(2), 141-146.

[6] Ahlskog, M., Seynaeve, E., Vullers, R. J. M., Van Haesendonck, C., Fonseca, A., Hernadi, K. B., & Nagy, J. (1999). Ring formations from catalytically synthesized carbon nanotubes. *Chemical Physics Letters*, 202.

[7] Martel, R., Shea, H. R., & Avouris, P. (1999). Rings of single-walled carbon nanotubes. *Nature*, 398(6725), 299.

[8] Martel, R., Shea, H. R., & Avouris, P. (1999). Ring Formation in Single-Wall Carbon Nanotubes. *The Journal of Physical Chemistry B*, 103(36), 7551-7556.

[9] Sano, M., Kamino, A., Okamura, J., & Shinkai, S. (2001). Ring Closure of Carbon Nanotubes. *Science*, 293(5533), 1299-1301.

[10] Geng, J., Ko, Y. K., Youn, S. C., Kim, Y. H., Kim, S. A., Jung, D. H., & Jung, H. T. (2008). Synthesis of SWNT Rings by Noncovalent Hybridization of Porphyrins and Single-Walled Carbon Nanotubes. *The Journal of Physical Chemistry C*, 112(32), 12264-12271.

[11] Song, L., Ci, L. J., Sun, L. F., Jin, C., Liu, L., Liu, W., Zhao, D., Luo, X., Zhang, S., Xiang, Z., Zhou, Y., Zhou, J., Ding, W., Wang, Y., , Z. L., & Xie, S. (2006). Large-Scale Synthesis of Rings of Bundled Single-Walled Carbon Nanotubes by Floating Chemical Vapor Deposition. *Advanced Materials*, 18(14), 1817-1821.

[12] Zhou, Z., Wan, D., Bai, Y., Dou, X., Song, L., Zhou, W., Mo, Y., & Xie, S. (2006). Ring formation from the direct floating catalytic chemical vapor deposition. *Physica E: Low-dimensional Systems and Nanostructures*, 33(1), 24-27.

[13] Kukushkin, A. B., Neverov, V. S., Marusov, N. L., Semenov, I. B., Kolbasov, B. N., Voloshinov, V. V., Afanasiev, A. P., Tarasov, A. S., Stankevich, V. G., Svechnikov, N. Y., Veligzhanin, A. A., Zubavichus, Y. V., & Chernozatonskii, L. A. (2011). Few-nano-

meter-wide carbon toroids in the hydrocarbon films deposited in tokamak T-10. *Chemical Physics Letters*, 265 -268 .

[14] Wang, X., Wang, Z., Liu, Yq., Wang, C., Bai, C., & Zhu, D. (2001). Ring formation and fracture of a carbon nanotube. *Chemical Physics Letters*, 339(1-2), 36 -40.

[15] Lyn, M. E., He, J., & Koplitz, B. (2005). Laser-induced production of large carbon-based toroids. *Applied Surface Science*, 246(1-3), 44-47.

[16] Motavas, S., Omrane, B., & Papadopoulos, C. (2009). Large-Area Patterning of Carbon Nanotube Ring Arrays. *Langmuir*, 25(8), 4655-4658.

[17] Chen, L., Wang, H., Xu, J., Shen, X., Yao, L., Zhu, L., Zeng, Z., Zhang, H., & Chen, H. (2011). Controlling Reversible Elastic Deformation of Carbon Nanotube Rings. *Journal of the American Chemical Society*, 133(25), 9654-9657.

[18] Komatsu, N., Shimawaki, T., Aonuma, S., & Kimura, T. (2006). Ultrasonic isolation of toroidal aggregates of single-walled carbon nanotubes. *Carbon*, 44(10), 2091-2093.

[19] Guo, A., Fu, Y., Guan, L., Zhang, Z., Wu, W., Chen, J., Shi, Z., Gu, Z., Huang, R., & Zhang, X. (2007). Spontaneously Formed Closed Rings of Single-Wall Carbon Nanotube Bundles and Their Physical Mechanism. *The Journal of Physical Chemistry C*, 111(9), 3555-3559.

[20] Colomer, J. F., Henrard, L., Flahaut, E., Van Tendeloo, G., Lucas, A. A., & Lambin, P. (2003). Rings of Double-Walled Carbon Nanotube Bundles. *Nano Letters*, 3(5), 685-689.

[21] Yu, H., Zhang, Q., Luo, G., & Wei, F. (2006). Rings of triple-walled carbon nanotube bundles. *Applied Physics Letters*, 89(22), 223106.

[22] Dunlap, B. I. (1992). Connecting carbon tubules. *Physical Review B*, 46(3), 1933-1936.

[23] Itoh, S., Ihara, S., & Kitakami, J. I. (1993). Toroidal form of carbon C_{360}. *Physical Review B*, 47(3), 1703-1704.

[24] Ihara, S., Itoh, S., & Kitakami, J. I. (1993). Toroidal forms of graphitic carbon. *Physical Review B*, 47(19), 12908-12911.

[25] Itoh, S., & Ihara, S. (1993). Toroidal forms of graphitic carbon II. Elongated tori. *Physical Review B*, 48(11), 8323-8328.

[26] Kirby, E. C., Mallion, R. B., & Pollak, P. (1993). Toroidal polyhexes. *Journal of the Chemical Society, Faraday Transactions*, 89(12), 1945-1953.

[27] Liu, L., Jayanthi, C. S., & Wu, S. Y. . (2001). Structural and electronic properties of a carbon nanotorus: Effects of delocalized and localized deformations. *Physical Review B*, 64(3), 033412 .

[28] Liu, L., Guo, G. Y., Jayanthi, C. S., & Wu, S. Y. (2002). Colossal Paramagnetic Moments in Metallic Carbon Nanotori. *Physical Review Letters*, 88(21), 217206 .

[29] Hod, O., Rabani, E., & Baer, R. (2003). Carbon nanotube closed-ring structures. *Physical Review B*, 67(19), 195408.

[30] Cox, B. J., & Hill, J. M. (2007). New Carbon Molecules in the Form of Elbow-Connected Nanotori. *The Journal of Physical Chemistry C*, 111(29), 10855-10860.

[31] Baowan, D., Cox, B. J., & Hill, J. M. (2008). Toroidal molecules formed from three distinct carbon nanotubes. *Journal of Mathematical Chemistry*, 44(2), 515-527.

[32] Liu, L., Zhang, L., Gao, H., & Zhao, J. (2011). Structure, energetics, and heteroatom doping of armchair carbon nanotori. *Carbon*, 49(13), 4518-4523.

[33] Klein, D. J., Seitz, W. A., & Schmalz, T. G. (1986). Icosahedral symmetry carbon cage molecules. *Nature*, 323(6090), 703-706.

[34] Itoh, S., & Ihara, S. (1994). Isomers of the toroidal forms of graphitic carbon. *Physical Review B*, 49(19), 13970-13974.

[35] Nagy, C., Nagy, K., & Diudea, M. (2009). Elongated tori from armchair DWNT. *Journal of Mathematical Chemistry*, 45(2), 452-459.

[36] László, I., & Rassat, A. (2001). Toroidal and spherical fullerene-like molecules with only pentagonal and heptagonal faces. *International Journal of Quantum Chemistry*, 84(1), 136-139.

[37] Ihara, S., & Itoh, S. (1995). Helically coiled and toroidal cage forms of graphitic carbon. *Carbon*, 33(7), 931-939.

[38] Taşcı, E., Yazgan, E., Malcıoğlu, O. B., & Erkoç, Ş. (2005). Stability of Carbon Nanotori under Heat Treatment: Molecular-Dynamics Simulations. *Fullerenes, Nanotubes and Carbon Nanostructures*, 13(2), 147-154.

[39] Chen, C., Chang, J. G., Ju, S. P., & Hwang, C. C. (2011). Thermal stability and morphological variation of carbon nanorings of different radii during the temperature elevating process: a molecular dynamics simulation study. *Journal of Nanoparticle Research*, 13(5), 1995-2006.

[40] Yang, L., Chen, J., & Dong, J. (2004). Stability of single-wall carbon nanotube tori. *Physica Status Solidi (b)*, 241(6), 1269-1273.

[41] Feng, C., & Liew, K. M. (2009). Energetics and structures of carbon nanorings. *Carbon.*, 47(7), 1664-1669.

[42] Liu, P., Zhang, Y. W., & Lu, C. . (2005). Structures and stability of defect-free multiwalled carbon toroidal rings. *Journal of Applied Physics*, 113522 .

[43] Han, J. (1998). Energetics and structures of fullerene crop circles. *Chemical Physics Letters*, 282(2), 187-191.

[44] Meunier, V., Lambin, P., & Lucas, A. A. (1998). Atomic and electronic structures of large and small carbon tori. *Physical Review B*, 57(23), 14886-14890.

[45] Chen, N., Lusk, M. T., van Duin, A. C. T., & Goddard, W. A. I. I. I. (2005). Mechanical properties of connected carbon nanorings via molecular dynamics simulation. *Physical Review B*, 72(8), 085416.

[46] Çağin, T., Gao, G., & Goddard, I. I. I. W. A. (2006). Computational studies on mechanical properties of carbon nanotori. *Turkish Journal of Physics*, 30(4), 221-229.

[47] Feng, C., & Liew, K. M. (2009). A molecular mechanics analysis of the buckling behavior of carbon nanorings under tension. *Carbon*, 47(15), 3508-3514.

[48] Feng, C., & Liew, K. M. (2010). Buckling Behavior of Armchair and Zigzag Carbon Nanorings. *Journal of Computational and Theoretical Nanoscience*, 7(10), 2049-2053.

[49] Zheng, M., & Ke, C. (2010). Elastic Deformation of Carbon-Nanotube Nanorings. *Small*, 6(15), 1647-1655.

[50] Zheng, M., & Ke, C. (2011). Mechanical deformation of carbon nanotube nano-rings on flat substrate. *Journal of Applied Physics*, 109(7), 074304-074310.

[51] Saito, R., Dresselhaus, G., & Dresselhaus, M. S. (1998). Physical properties of carbon nanotubes. London, Imperial College Press.

[52] Zhenhua, Z., Zhongqin, Y., Xun, W., Jianhui, Y., Hua, Z., Ming, Q., & Jingcui, P. (2005). The electronic structure of a deformed chiral carbon nanotorus. *Journal of Physics: Condensed Matter*, 17(26), 4111-4120.

[53] Ceulemans, A., Chibotaru, L. F., Bovin, S. A., & Fowler, P. W. (2000). The electronic structure of polyhex carbon tori. *The Journal of Chemical Physics*, 112(9), 4271-4278.

[54] Liu, C. P., & Ding, J. W. (2006). Electronic structure of carbon nanotori: the roles of curvature, hybridization, and disorder. *Journal of Physics Condensed Matter*, 18(16), 4077-4084.

[55] Oh, D. H., Mee, Park. J., & Kim, K. S. (2000). Structures and electronic properties of small carbon nanotube tori. *Physical Review B*, 62(3), 1600-1603.

[56] Yazgan, E., Taşci, E., Malcioğlu, O. B., & Erkoç, Ş. (2004). Electronic properties of carbon nanotoroidal structures. *Journal of Molecular Structure: THEOCHEM*, 681(1-3), 231-234.

[57] Wu, X., Zhou, R., Yang, J., & Zeng, X. C. (2011). Density-Functional Theory Studies of Step-Kinked Carbon Nanotubes. *The Journal of Physical Chemistry C*, 115(10), 4235-4239.

[58] Haddon, R. C. (1997). Electronic properties of carbon toroids. *Nature*, 388(6637), 31-32.

[59] Rodríguez-Manzo, J. A., López-Urías, F., Terrones, M., & Terrones, H. (2004). Magnetism in Corrugated Carbon Nanotori: The Importance of Symmetry, Defects, and Negative Curvature. *Nano Letters*, 4(11), 2179-2183.

[60] Lin, M. F., & Chuu, D. S. (1998). Persistent currents in toroidal carbon nanotubes. *Physical Review B*, 57(11), 6731-6737.

[61] Latil, S., Roche, S., & Rubio, A. (2003). Persistent currents in carbon nanotube based rings. *Physical Review B*, 67(16), 165420.

[62] Shyu, F. L., Tsai, C. C., Chang, C. P., Chen, R. B., & Lin, M. F. (2004). Magnetoelectronic states of carbon toroids. *Carbon*, 42(14), 2879-2885.

[63] Margańska, M., Szopa, M., & Zipper, E. (2005). Aharonov-Bohm effect in carbon nanotubes and tori. *Physica Status Solidi (b)*, 242(2), 285-290.

[64] Zhang, Z. H., Yuan, J. H., Qiu, M., Peng, J. C., & Xiao, F. L. (2006). Persistent currents in carbon nanotori: Effects of structure deformations and chirality. *Journal of Applied Physics*, 99(10), 104311 .

[65] Tsai, C. C., Shyu, F. L., Chiu, C. W., Chang, C. P., Chen, R. B., & Lin, M. F. (2004). Magnetization of armchair carbon tori. *Physical Review B*, 70(7), 075411.

[66] Liu, C. P., Chen, H. B., & Ding, J. W. (2008). Magnetic response of carbon nanotori: the importance of curvature and disorder. *Journal of Physics: Condensed Matter*, 20(1), 015206.

[67] Liu, C. P., & Xu, N. (2008). Magnetic response of chiral carbon nanotori: The dependence of torus radius. *Physica B: Condensed Matter*, 403(17), 2884-2887.

[68] Zhang, Z., & Li, Q. (2010). Combined Effects of the Structural Deformation and Temperature on Magnetic Characteristics of the Single-walled Chiral Toroidal Carbon Nanotubes. *Chinese Journal of Electronics*, 19(3), 423-426.

[69] Rodríguez-Manzo, J. A., López-Urías, F., Terrones, M., & Terrones, H. (2007). Anomalous Paramagnetism in Doped Carbon Nanostructures. *Small*, 3(1), 120-125.

[70] Castillo-Alvarado, F. d. L., Ortíz-López, J., Arellano, J. S., & Cruz-Torres, A. (2010). Hydrogen Storage on Beryllium-Coated Toroidal Carbon Nanostructure C_{120} Modeled with Density Functional Theory. *Advances in Science and Technology*, 72, 188-195.

[71] Hilder, T. A., & Hill, J. M. (2007). Orbiting atoms and C_{60} fullerenes inside carbon nanotori. *Journal of Applied Physics*, 101(6), 64319 .

[72] Lusk, M. T., & Hamm, N. (2007). Ab initio study of toroidal carbon nanotubes with encapsulated atomic metal loops. *Physical Review B*, 76(12), 125422.

[73] Mukherjee, B., Maiti, P. K., Dasgupta, C., & Sood, A. K. (2010). Single-File Diffusion of Water Inside Narrow Carbon Nanorings. *ACS Nano*, 4(2), 985-991.

[74] Koós, A. A., Ehlich, R., Horváth, Z. E., Osváth, Z., Gyulai, J., Nagy, J. B., & Biró, L. P. (2003). STM and AFM investigation of coiled carbon nanotubes produced by laser evaporation of fullerene. *Materials Science and Engineering: C*, 23(1-2), 275 -278.

[75] Saveliev, A. V., Merchan-Merchan, W., & Kennedy, L. A. (2003). Metal catalyzed synthesis of carbon nanostructures in an opposed flow methane oxygen flame. *Combustion and Flame*, 135(1-2), 27 -33.

[76] Bai, J. B., Hamon, A. L., Marraud, A., Jouffrey, B., & Zymla, V. (2002). Synthesis of SWNTs and MWNTs by a molten salt (NaCl) method. *Chemical Physics Letters*, 365(1-2), 184 -188.

[77] Ajayaghosh, A., Vijayakumar, C., Varghese, R., & George, S. J. (2006). Cholesterol-Aided Supramolecular Control over Chromophore Packing: Twisted and Coiled Helices with Distinct Optical, Chiroptical, and Morphological Features. *Angewandte Chemie*, 118(3), 470-474.

[78] Yamamoto, T., Fukushima, T., Aida, T., & Shimizu, T. (2008). Self-Assembled Nanotubes and Nanocoils from ss-Conjugated Building Blocks. *Advances in Polymer Science*, 220, 1-27.

[79] Lujun, P., Hayashida, T., Mei, Z., & Nakayama, Y. (2001). Field emission properties of carbon tubule nanocoils. *Japanese Journal of Applied Physics*, 40(3B), L235-L237.

[80] Jining, X., Mukhopadyay, K., Yadev, J., & Varadan, V. K. (2003). Catalytic chemical vapor deposition synthesis and electron microscopy observation of coiled carbon nanotubes. *Smart Materials and Structures*, 12(5), 744-748.

[81] Hou, H., Jun, Z., Weller, F., & Greiner, A. (2003). Large-Scale Synthesis and Characterization of Helically Coiled Carbon Nanotubes by Use of $Fe(CO)_5$ as Floating Catalyst Precursor. *Chemistry of Materials*, 15(16), 3170-3175.

[82] Zhong, D. Y., Liu, S., & Wang, E. G. (2003). Patterned growth of coiled carbon nanotubes by a template-assisted technique. *Applied Physics Letters*, 83(21), 4423-4425.

[83] Tang, N., Wen, J., Zhang, Y., Liu, F., Lin, K., & Du, Y. (2010). Helical Carbon Nanotubes: Catalytic Particle Size-Dependent Growth and Magnetic Properties. *ACS Nano*, 4(1), 241-250.

[84] Lau, K. T., Lu, M., & Hui, D. (2006). Coiled carbon nanotubes: Synthesis and their potential applications in advanced composite structures. *Composites Part B: Engineering*, 37(6), 437-448.

[85] Fejes, D., & Hernádi, K. (2010). A Review of the Properties and CVD Synthesis of Coiled Carbon Nanotubes. *Materials*, 3(4), 2618-2642.

[86] Shaikjee, A., & Coville, N.J. (2012). The synthesis, properties and uses of carbon materials with helical morphology. *Journal of Advanced Research*, 3(3), 195-223.

[87] Fonseca, A., Hernadi, K., Nagy, J. B., Lambin, P., & Lucas, A. A. (1996). Growth mechanism of coiled carbon nanotubes. *Synthetic Metals*, 77(1-3), 235 -242.

[88] Pan, L. J., Zhang, M., & Nakayama, Y. (2002). Growth mechanism of carbon nanocoils. *Journal of Applied Physics*, 91(12), 10058-10061.

[89] Chen, X., Yang, S., Takeuchi, K., Hashishin, T., Iwanaga, H., & Motojiima, S. (2003). Conformation and growth mechanism of the carbon nanocoils with twisting form in comparison with that of carbon microcoils. *Diamond and Related Materials*, 12(10-11), 1836-1840.

[90] Bandaru, P. R., Daraio, C., Yang, K., & Rao, A. M. (2007). A plausible mechanism for the evolution of helical forms in nanostructure growth. *Journal of Applied Physics*, 101(9), 094307.

[91] Dunlap, B. I. (1994). Relating carbon tubules. Physical Review B , 49(8), 5643-5651.

[92] Fonseca, A., Hernadi, K., Nagy, J. b., Lambin, P., & Lucas, A. A. (1995). Model structure of perfectly graphitizable coiled carbon nanotubes. *Carbon*, 33(12), 1759-1775.

[93] Ihara, S., Itoh, S., & Kitakami, J. i. (1993). Helically coiled cage forms of graphitic carbon. *Physical Review B*, 48(8), 5643-5647.

[94] Setton, R., & Setton, N. (1997). Carbon nanotubes: III. Toroidal structures and limits of a model for the construction of helical and S-shaped nanotubes. *Carbon*, 35(4), 497-505.

[95] Akagi, K., Tamura, R., Tsukada, M., Itoh, S., & Ihara, S. (1995). Electronic Structure of Helically Coiled Cage of Graphitic Carbon. *Physical Review Letters*, 74(12), 2307-2310.

[96] Akagi, K., Tamura, R., Tsukada, M., Itoh, S., & Ihara, S. (1996). Electronic structure of helically coiled carbon nanotubes: Relation between the phason lines and energy band features. *Physical Review B*, 53(4), 2114-2120.

[97] Biro, L. P., Mark, G. I., & Lambin, P. (2003). Regularly coiled carbon nanotubes. *Nanotechnology, IEEE Transactions on*, 2(4), 362-367.

[98] Liu, L., Gao, H., Zhao, J., & Lu, J. (2010). Superelasticity of Carbon Nanocoils from Atomistic Quantum Simulations. *Nanoscale Research Letters*, 5(3), 478-483.

[99] Zhong-can, O. Y., Su, Z. B., & Wang, C. L. (1997). Coil Formation in Multishell Carbon Nanotubes: Competition between Curvature Elasticity and Interlayer Adhesion. *Physical Review Letters*, 78(21), 4055-4058.

[100] Volodin, A., Ahlskog, M., Seynaeve, E., Van Haesendonck, C., Fonseca, A., & Nagy, J. B. (2000). Imaging the Elastic Properties of Coiled Carbon Nanotubes with Atomic Force Microscopy. *Physical Review Letters*, 84(15), 3342-3345.

[101] Hayashida, T., Pan, L., & Nakayama, Y. (2002). Mechanical and electrical properties of carbon tubule nanocoils. *Physica B: Condensed Matter*, 323(1-4), 352-353.

[102] Chen, X., Zhang, S., Dikin, D. A., Ding, W., Ruoff, R. S., Pan, L., & Nakayama, Y. (2003). Mechanics of a Carbon Nanocoil. *Nano Letters*, 3(9), 1299-1304.

[103] Huang, W. M. (2005). Mechanics of coiled nanotubes in uniaxial tension. *Materials Science and Engineering: A*, 408(1-2), 136 -140.

[104] Neng-Kai, C., & Shuo-Hung, C. (2008). Determining Mechanical Properties of Carbon Microcoils Using Lateral Force Microscopy. *IEEE Transactions on Nanotechnology*, 7(2), 197-201.

[105] Poggi, M. A., Boyles, J. S., Bottomley, L. A., Mc Farland, A. W., Colton, J. S., Nguyen, C. V., Stevens, R. M., & Lillehei, P. T. (2004). Measuring the Compression of a Carbon Nanospring. *Nano Letters*, 4(6), 1009-1016.

[106] Fonseca, A. F. d., & Galvão, D. S. (2004). Mechanical Properties of Nanosprings. *Physical Review Letters*, 92(17), 175502 .

[107] Fonseca, A. F. d., Malta, C. P., & Galvão, D. S. (2006). Mechanical properties of amorphous nanosprings. *Nanotechnology*, 17(22), 5620-5626.

[108] Coluci, V. R., Fonseca, A. F., Galvão, D. S., & Daraio, C. (2008). Entanglement and the Nonlinear Elastic Behavior of Forests of Coiled Carbon Nanotubes. *Physical Review Letters*, 100(8), 086807 .

[109] Ghaderi, S. H., & Hajiesmaili, E. (2010). Molecular structural mechanics applied to coiled carbon nanotubes. *Computational Materials Science*, 55(0), 344-349.

[110] Kaneto, K., Tsuruta, M., & Motojima, S. (1999). Electrical properties of carbon micro coils. *Synthetic Metals*, 103(1-3), 2578 -2579.

[111] Ebbesen, T. W., Lezec, H. J., Hiura, H., Bennett, J. W., Ghaemi, H. F., & Thio, T. (1996). Electrical conductivity of individual carbon nanotubes. *Nature*, 382(6586), 54-56.

[112] Chiu, H. S., Lin, P. I., Wu, H. C., Hsieh, W. H., Chen, C. D., & Chen, Y. T. (2009). Electron hopping conduction in highly disordered carbon coils. *Carbon*, 47(7), 1761-1769.

[113] Tang, N., Kuo, W., Jeng, C., Wang, L., Lin, K., & Du, Y. (2010). Coil-in-Coil Carbon Nanocoils: 11 Gram-Scale Synthesis, Single Nanocoil Electrical Properties, and Electrical Contact Improvement. *ACS Nano*, 4(2), 781-788.

[114] Fujii, M., Matsui, M., Motojima, S., & Hishikawa, Y. (2002). Magnetoresistance in carbon micro-coils obtained by chemical vapor deposition. *Thin Solid Films*, 409(1), 78-81.

[115] Liu, L., Gao, H., Zhao, J., & Lu, J. (2011). Quantum conductance of armchair carbon nanocoils: roles of geometry effects. *SCIENCE CHINA Physics, Mechanics & Astronomy*, 54(5), 841-845.

[116] Motojima, S., Chen, X., Yang, S., & Hasegawa, M. (2004). Properties and potential applications of carbon microcoils/nanocoils. *Diamond and Related Materials*, 13(11-12), 19895-1992.

[117] Volodin, A., Buntinx, D., Ahlskog, M., Fonseca, A., Nagy, J. B., & Van Haesendonck, C. (2004). Coiled Carbon Nanotubes as Self-Sensing Mechanical Resonators. *Nano Letters*, 4(9), 1775-1779.

[118] Bell, D. J., Sun, Y., Zhang, L., Dong, L. X., Nelson, B. J., & Grützmacher, D. (2006). Three-dimensional nanosprings for electromechanical sensors. *Sensors and Actuators A Physical*, 130-131(0), 54 -61.

[119] Kato, Y., Adachi, N., Okuda, T., Yoshida, T., Motojima, S., & Tsuda, T. (2003). Evaluation of induced electromotive force of a carbon micro coil. *Japanese Journal of Applied Physics*, 42(8), 5035-5037.

[120] Shaoming, Y., Xiuqin, C., Aoki, H., & Motojima, S. (2006). Tactile microsensor elements prepared from aligned superelastic carbon microcoils and polysilicone matrix. *Smart Materials and Structures*, 15(3), 687-694.

[121] Greenshields, M. W., Hummelgen, I. A., Mamo, M. A., Shaikjee, A., Mhlanga, S. D., van Otterlo, W. A., & Coville, N. J. (2011). Composites of Polyvinyl Alcohol and Carbon (Coils, Undoped and Nitrogen Doped Multiwalled Carbon Nanotubes) as Ethanol, Methanol and Toluene Vapor Sensors. *Journal of Nanoscience and Nanotechnology*, 11(11), 10211-10218.

[122] Lau, K. t., Lu, M., & Liao, K. (2006). Improved mechanical properties of coiled carbon nanotubes reinforced epoxy nanocomposites. *Composites Part A: Applied Science and Manufacturing*, 37(10), 18375-18405.

[123] Yoshimura, K., Nakano, K., Miyake, T., Hishikawa, Y., & Motojima, S. (2006). Effectiveness of carbon microcoils as a reinforcing material for a polymer matrix. *Carbon*, 44(13), 2833-2838.

[124] Li, X. F., Lau, K. T., & Yin, Y. S. (2008). Mechanical properties of epoxy-based composites using coiled carbon nanotubes. *Composites Science and Technology*, 68(14), 2876-2881.

[125] Sanada, K., Takada, Y., Yamamoto, S., & Shindo, Y. (2008). Analytical and Experimental Characterization of Stiffness and Damping in Carbon Nanocoil Reinforced Polymer Composites. *Journal of Solid Mechanics and Materials Engineering*, 2(12), 1517-1527.

[126] Katsuno, T., Chen, X., Yang, S., Motojima, S., Homma, M., Maeno, T., & Konyo, M. (2006). Observation and analysis of percolation behavior in carbon microcoils/silicone-rubber composite sheets. *Applied Physics Letters*, 88(23), 232115-232113.

[127] Yoshimura, K., Nakano, K., Miyake, T., Hishikawa, Y., Kuzuya, C., Katsuno, T., & Motojima, S. (2007). Effect of compressive and tensile strains on the electrical resistivity of carbon microcoil/silicone-rubber composites. *Carbon*, 45(10), 1997-2003.

[128] Nakamatsu, K., Igaki, J., Nagase, M., Ichihashi, T., & Matsui, S. (2006). Mechanical characteristics of tungsten-containing carbon nanosprings grown by FIB-CVD. *Microelectronic Engineering*, 83(4-9), 808 -810 .

[129] Bi, H., Kou, K. C., Ostrikov, K., Yan, L. K., & Wang, Z. C. (2009). Microstructure and electromagnetic characteristics of Ni nanoparticle film coated carbon microcoils. *Journal of Alloys and Compounds.*, 478(1-2), 796 -800 .

[130] Zhang, G. Y., Jiang, X., & Wang, E. G. (2004). Self-assembly of carbon nanohelices: Characteristics and field electron emission properties. *Applied Physics Letters*, 84(14), 2646-2648.

[131] Wu, X. L., Liu, Q., Guo, Y. G., & Song, W. G. (2009). Superior storage performance of carbon nanosprings as anode materials for lithium-ion batteries. *Electrochemistry Communications*, 11(7), 1468-1471.

[132] Raghubanshi, H., Hudson, M. S. L., & Srivastava, O. N. (2011). Synthesis of helical carbon nanofibres and its application in hydrogen desorption. *International Journal of Hydrogen Energy*, 36(7), 4482-4490.

[133] Motojima, S. (2008). Development of ceramic microcoils with 3D-herical/spiral structures. *Journal of the Ceramic Society of Japan*, 116(1357), 921-927.

[134] Motojima, S., Hoshiya, S., & Hishikawa, Y. (2003). Electromagnetic wave absorption properties of carbon microcoils/PMMA composite beads in W bands. *Carbon*, 41(13), 2658-2660.

Section 2

Electrical and Biomedical Applications of Carbon Nanotubes

Chapter 13

Carbon Nanotubes for Use in Medicine: Potentials and Limitations

Wei Shao, Paul Arghya, Mai Yiyong,
Laetitia Rodes and Satya Prakash

Additional information is available at the end of the chapter

http://dx.doi.org/10.5772/51785

1. Introduction

Structurally, Carbon nanotubes (CNTs) can be viewed as wrapped from graphene sheets. Single-walled carbon nanotubes (SWNTs) have one layer of graphene sheet, whereas, the multiwalled carbon nanotubes (MWNTs) contain multi layers of graphene sheets. The well-ordered molecular structure brings CNTs many remarkable physical properties, such as, excellent mechanic strength, ultrahigh surface area, high aspect ratio, distinct optical properties [1], and excellent electrical conductivity [2]. In last decade, CNTs are intensively explored for in-vitro and in-vivo delivery of therapeutics, which was inspired by an important finding that CNTs can penetrate cells by themselves without apparent cytotoxic effect to the cells [3]. The high aspect ratio makes CNTs outstanding from other types of round nanoparticles in that the needle-like CNTs allow loading large quantities of payloads along the longitude of tubes without affecting their cell penetration capability. With the adequate loading capacity, the CNTs can carry multifunctional therapeutics, including drugs, genes and targeting molecules, into one cell to exert multi-valence effects. In the other side, owing to the ultrahigh surface area along with the strong mechanical properties and electrically conductive nature, CNTs are excellent material for nanoscaffolds and three dimensional nanocomposites. In recent year, CNT-based devices have been successfully utilized in tissue engineering and stem cell based therapeutic applications, including myocardial therapy, bone formation, muscle and neuronal regeneration. Furthermore, owing to the distinct optical properties of CNTs, such as, high absorption in the near-infrared (NIR) range, photoluminescence, and strong Raman shift [4], CNTs are excellent agents for biology detection and imaging. Combined with high surface area of CNTs for attaching molecular recognition molecules, CNT-based, targeted nanodevices have been developed for selective imaging and sensing. There are many areas where CNTs are extremely useful. Given the scope in this chapter, We describe strategies for preparation of CNTs for their use in medicine. Specifically, we focus and highlight the important biomedical applications of CNTs in the field of drug delivery, gene delivery, stem cell therapy, thermal therapy, biological detection and imaging (figure 1). The methods for formulating CNT-based therapeutics to suit different routes of drug administration are also described. The limitations with emphasis on toxicity and over all future directions are discussed.

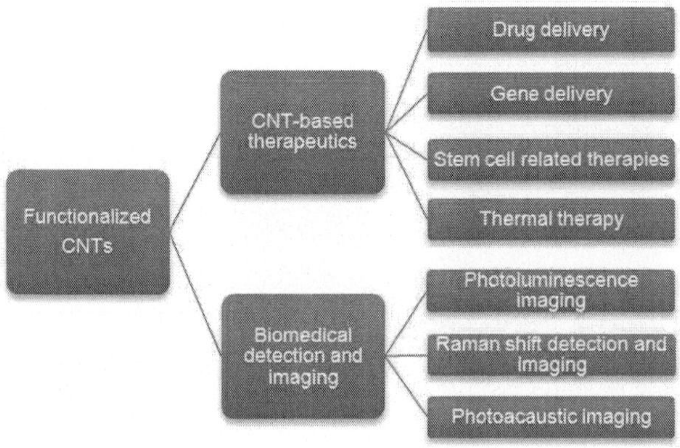

Figure 1. Functionalized CNTs in major biomedical applications.

2. Preparation of CNTs for use in medicine

Raw CNTs, persisting metallic nature, are highly hydrophobic. Therefore, surface modification of CNTs, or CNT functionalization, so as to disperse them into aqueous solutions becomes a key step for their biomedical applications. The CNT modification methods are involved in non-covalent and covalent strategies. The non-covalent modification utilizes the hydrophobic nature of CNTs, especially, π-π interactions for coating of amphiphilic molecules. The covalent modification generates chemical bonds on carbon atoms on CNT surface via chemical reactions followed by further conjugation of hydrophilic organic molecules or polymers rendering CNTs better solubility. These modifications not only offer CNTs water solubility, but also produce functional moieties that enable linking of therapeutic agents, such as genes , drugs, and recognition molecules for biomedical applications.

2.1. Non-covalent modification of carbon nanotube surface

The non-covalent modification approaches typically use amphiphilic molecules ranged from small molecules to polymers. The amphiphilic molecules associate with CNTs by either adsorbing onto or wrapping the CNTs [5]. The non-covalent modifications of CNTs are easy to perform. The process is only involved in sonication of CNTs with amphiphilic molecules in solvent at room temperature. Since it is a mild condition, CNTs molecular structure is not affected, and therefore their optical and electrical conductive properties are conserved.

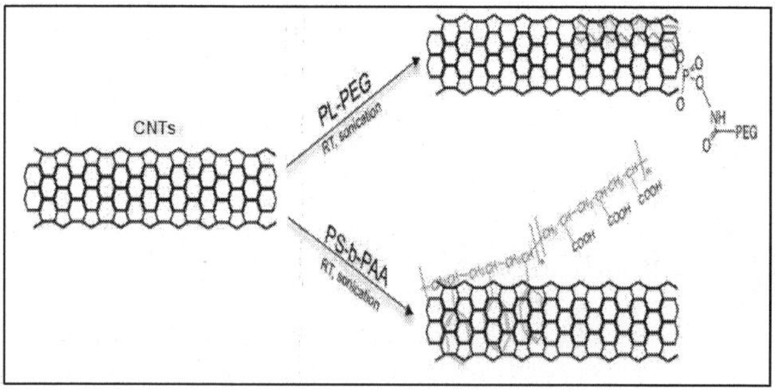

Figure 2. Schematic representation of adsorption of amphiphilic molecules onto carbon nanotube surface by π-π stacking and other hydrophobic interactions. Abbreviations: PL-PEG, phospholipid-polyethylene glycol; PS-b-PAA, polystyrene-block-polyacrylic acid.

Adsorption of amphiphilic molecules, such as surfactants, amphiphilic copolymers or others, onto CNT surfaces is one of the simplest and most effective way to disperse CNTs without destruction of their *sp* ^{2} hybridization [5]. The hydrophilic portions of surfactants interact with the polar solvent molecules, whereas, the hydrophobic portions adsorb onto the nanotube surface [5, 6]. The dispersity depends strongly on the length of the hydrophobic regions and the types of hydrophilic groups in the amphiphilic molecule. For example, surfactants with ionic hydrophilic head groups, such as sodiumdodecylsulfate (SDS) [7] or cetyltrimethyl ammonium bromide (CTAB) [8, 9], can stabilize a nanotube by electrostatic repulsion between micellar domains [7]. Nonionic surfactants, such as Triton X-100 [8], disperse CNTs mainly by forming a large solvation shell around a nanotube [8]. Figure 2 illustrates the manor of adsorbing amphiphilic molecules onto CNT surfaces, in which, hydrophobic alkyl chains or aromatic rings lay flat on graphitic tube surfaces. For example, an synthetic biocompatible lipid-polymer conjugate, phospholipid-polyethylene glycol (PL-PEG) has been applied for surface modification of CNTs, which gives rise to a variety of biomedical applications ranged from drug delivery, biomedical imaging, detection and biosensors [10]. The ionic surfactants, particularly those based on alkyl-substituted imidazolium cationic surfactants [11], can effectively disperse CNTs in organic or aqueous media by the counter anion [12, 13]. Polyaromatic derivatives carrying hydrophilic moieties can also effectively disperse CNTs in aqueous media by forming specific directional π–π stacking with the graphitic surfaces of nanotubes [6, 14]. In this context, pyrene, a polyaromatic molecule, demonstrated a high affinity toward CNT surfaces [6]. Interactions of the polyaromatic-moitie of pyrene with CNTs are strong enough to be irreversible, and therefore, the pyrene derivatives are used to anchor proteins or biomolecules on nanotube surface [6, 14, 15]. Other classes of polyaromatic molecules, such as substituted anthracenes, heterocyclic polyaromaticporphyrins [16] and phthalocyanines [17], disperse CNTs via the same mechanism.

The polymers containing hydrophobic backbone and hydrophilic side groups, eg. poly[p-{2,5-bis(3-propoxysulfonic acid sodium salt)}phenylene] ethynylene (PPES), can effectively disperse CNTs in water, in which, the strong π– π interactions between CNTs and aromatic backbone of the polymers drive the wrapping of CNTs, and the water-soluble side groups impart solubility of CNTs in water [18]. DNA and siRNA can disperse CNTs by wrapping. DNA or siRNA are made of hydrophobic bases and alternative hydrophilic phosphates and riboses. Such structure facilitates CNT dispersion by the bases wrapping to CNTs and the hydrophilic sugar-phosphate groups extending to water phase [19].

2.2. Covalent modification of CNT surface

The covalent modification, namely the chemical modification of CNTs is an emerging area in materials science. Among the various strategies, the most common ones are:

i. esterification and amidation of oxidized CNTs,

ii. generation of functional groups on CNT sidewalls by cycloaddition reactions.

Oxidation of CNTs is a purification method for raw CNTs. Oxidation of CNTs is carried out by reflexing raw CNTs in strong acidic media, e.g. HNO_3/H_2SO_4. Under this condition, the end caps of the CNTs are opened, and carboxylic groups are formed at these ends caps and at some defect sites on nanotube sidewalls (Figure 3a) [20]. The carboxylic groups provide opportunities for further derivatization of the CNTs through esterification or amidation reactions. For example, some organic molecules with amine groups can be directly condensed with the carboxylic groups present on the surface of the CNTs [6, 14]. Alternatively, the carboxyl moieties can be activated with thionyl chloride and subsequent react with amine groups (Figure 3b) [6, 14]. These reactions are widely applied for conjugation of water-soluble organic molecules, hydrophilic polymers, nucleic acid (DNA or RNA), or peptides to the oxidized CNTs, which result in multifunctional CNTs [6, 14]. In most cases, the length of nanotubes is often shortened [20] ,but the electronic properties of such functionalized CNTs remain intact. Oxidation reaction only generates carboxyl groups on cap ends and defect sites on CNTs. To generate chemical bonds on sidewall and cap ends of CNTs, cycloaddition reactions are used [21](Figure 4). Cycloaddition reaction is a very powerful methodology, in which the 1,3-dipolar cycloaddition of azomethineylides can easily attach a large amount of pyrrolidine rings on sidewalls of nanotubes. Thus, the resulting functionalized CNTs are highly soluble in water [22]. In addition, pyrrolidine ring can be substituted with many functional groups for different applications.

http://dx.doi.org/10.5772/51785

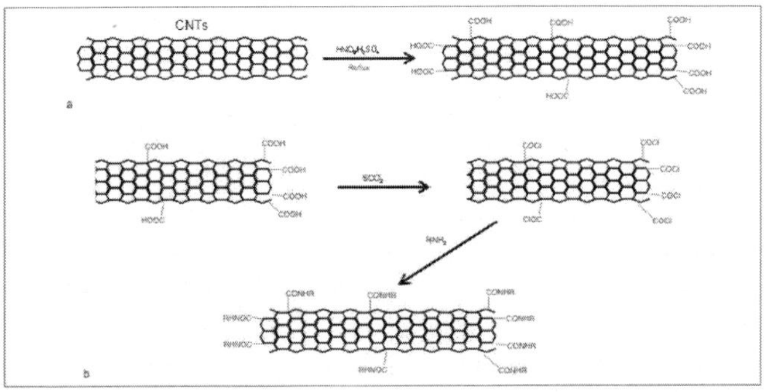

Figure 3. Covalent modification of carbon nanotubesby a) Oxidation reaction of cap end of CNTs and b) further attaching hydrophilic molecules by amidation reactions.

In contrast to non-covalent surface modifications, which do not locally disrupt *sp* [2] hybridization, or create defects, the covalent surface modifications disrupt CNT *sp* [2]-conjugated structures and therefore, could affect the electronic and optical performances [5].

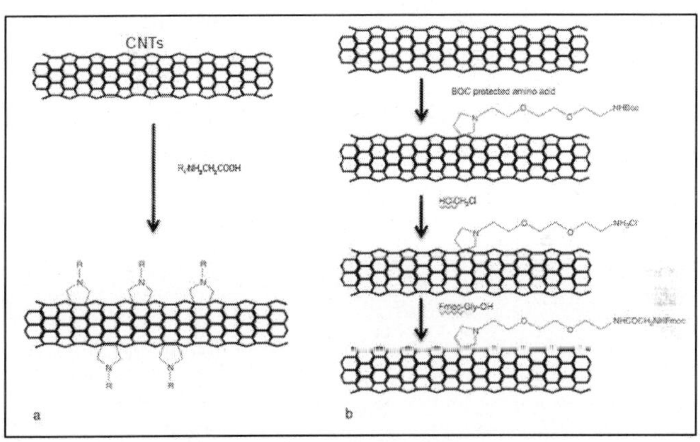

Figure 4. Sidewall covalent modification of carbon nanotubes a) a general scheme of 1,3-dipolar cycloaddition reaction b) preparation of amino-functionalized CNTs by 1,3-dipolar cycloaddition.

3. Carbon nanotube based therapeutics

3.1. Carbon nanotubes for chemotherapy drug delivery

Cancer is one of the most common causes of death worldwide. Chemotherapy in addition to the surgical removal of tumors is a conventional treatment for cancers. However, the effectiveness of chemotherapy drugs is often limited by the toxicity to other tissues in the body. This is because most chemotherapy drugs do not specifically kill cancer cells, they act to kill all cells undergoing fast division. Nanoparticles have been applied to drug delivery and showed improved drug efficiency and reduced off-target tissue toxicity due to accumulation in tumor tissues. Nanoparticles target tumor tissues by two mechanisms: passive targeting and active targeting. As fast growing tissues, tumors display enhanced vascular permeability due to high demand for nutrients and possible oxygen. The features of the leaky vasculature are employed for delivery of nanoparticle drugs since the size of nanoparticle allows them to accumulate in tumor tissues [23]. The phenomenon is termed as tumor-selective *enhanced permeability and retention* (EPR) effect. More efficient tumor targeting can be achieved through active targeting approaches, in which, targeting molecules can recognize tumor biomarkers on cancer cell surface. The properties of CNTs are beneficial for cancer drug delivery, firstly, like other nanoparticles, the size of functionalized CNTs is preferable for accumulation in tumor tissues; secondly, CNTs contain ultrahigh surface area of CNTs facilitate loading of drugs and targeting molecules; thirdly, the hydrophobic benzene ring structure of CNTs can be used for loading drugs that contain benzene ring structure, eg. doxorubicin (DOX), epirubicin (EPI), and daunorubicin (DAU).

Preparation of tumor-targeted devices using carbon nanotubes

A range of tumor targeting molecules has been discovered, including tumor specific antibodies, peptides, and others. Antibodies have been developed to specifically binding to biomarkers on cancer cell surface, eg, Trastuzumab recognizes Human Epidermal Growth Factor Receptor 2 (HER-2) positive cancer cells[24] and anti-CD20 for CD20 biomarker on B cell lymphoma [25]. These antibodies have therapeutic effects on their own, and can also serve as tumor targeting probes. Alpha V beta 3 ($\alpha v\beta 3$) integrin is a heterodimerictransmembrane glycoprotein found on a variety of tumor cells, including osteosarcomas, neuroblastomas, glioblastomas, melanomas, lung, breast, prostate cancers. $\alpha v\beta 3$) integrin is a wellrecognized target for cancers. The amino acid sequence of Arg-Gly-Asp (RGD) is identified to be responsible for tight binding to $\alpha v\beta 3$ integrin, which leads to the development of short tumor targeting peptide RGD [26]. Similar to RGD, another type of peptide contains Asn-Gly-Asp (NGR) triad that binds to the endothelium cells on neoangiogenic vessels. NGR-tagged delivery systems have been developed to deliver cytokines, nanoparticles, and imaging agents to tumor blood vessels [27]. Folic acid, a small molecule vitamin, binds to folate receptor overexpressed in a variety of cancer cells, including breast, colon, renal and lung tumors[28]. As described in section 2 of this chapter, a variety of chemical and physical methods have been developed for functionalization of CNTs. The above listed tumor targeting molecules are mostly proteins or peptides, which contains sulfhydryl groups that can be easily conjugated to amino-functionalized CNTs [14] using heterbifunctional linker molecules that contain NHS ester on one end and Maleimide on the other end [29]. These conjugation reactions are usually carried out under mild conditions [10]. Thus, the molecular structure of CNTs is not disturbed, and therefore, the optical properties are preserved. The CNT-based targeted devices developed by this methods are good for potential tumor detection and imaging applications. To date, all above-mentioned tumor-targeting strategies have been applied for construction of CNT-based, which will be discussed separate in the following sub-sections.

Targeted delivery of chemotherapy drugs by physically absorbed on carbon nanotubes

Supermolecular benzene ring structure of CNTs affords surprisingly high degree of aromatic molecules by π-π stacking. DOX, an important chemotherapeutic agent, has been efficiently loaded onto SWNTs-PL-PEG for tumor-targeted delivery [30, 31]. Binding to and release of drug molecules from nanotubes could be controlled by adjusting pH.The appropriate diameter of nanotube for drug loading was used because the strength of π-π stacking of aromatic molecules was dependent on nanotube diameter. In-vivo study with SWNTs-PL-PEG/DOX showed significantly enhanced therapeutic efficacy in a murine breast cancer model [30]. With further attaching tumor targeting molecules, eg. folic acid (FA) or RGD, the targeted SWNTs-DOX could more effectively inhibit the growth of cancer cells in-vitro and in-vivo [31-34]. Similar physical absorption method was applied for drug DAU using SWNTs, in which, sgc8c Aptamer was used to target leukemia biomarker protein tyrosine kinase-7 [33]. It has been shown that Aptamer-SWNTs-DAU was able to selectively target leukemia cells. The release of DAU was pH-dependent.Other hydrophobic drug molecules, such as paclitaxel (PTX), docetaxel (DTX), can also be absorbed on CNTs surface for delivery [32, 35], however, their loading efficiency and stability were much lower due to their comparatively bulky structure.

Targeted delivery of chemotherapy drugs by covalently linked to functionalized carbon nanotubes

Non-aromatic small molecule drugs can be chemically conjugated to CNTs for delivery. However, the drugmolecules have to be released from the CNTs to take effect, so the linkages between the drugs and CNTs have to be cleavable. Preferably, the active drugs are released inside of the target cells to reduce toxic effect to the neighbouring healthy cells. The common linkers that are used for drug delivery include ester, peptide, and disulfide bonds. These linkers can be cleaved by the 7enzymes present in the routes of delivery. Specifically designed linkers allow controlled release of drug into desired sites. For example, drug cisplatin has been conjugated directly to oxidized SWNTs via a peptide linker [36]. This specific peptide linkage has been shown tobe selectively cleaved by proteases overexpressed in tumor cells. Further conjugation of epidermal growth factor (EGF), a growth factor that selective binding to EGF receptor overexpressed on cancer cells, to SWNTs-cisplatin led to more efficient tumor inhibition compared to both free cisplatin and non-targeted SWNTs-cisplatin [36]. Alternative to conjugation of drugs to CNTs directly, drugs can also be conjugated to the molecules, eg. polymers, that are used to disperse CNTs. The end functional groups in the polymers are used for drug linkage. This method is very useful for delivery of bulky, hydrophobic drug molecules. In one example, SWNTs was dispersed using a biocompatible polymer PL-PEG-NH_2[10] and drug PTX was conjugated to SWNT-PL-PEG-NH_2 via via ester bonding for delivery [37]. PTX is one of the most important drugs for metastatic breast cancer. However, currently available formulations for PTX have to be infused intravenously over long periods of time due to the side effects. In addition, due to poor water solubility of the drug, necessity of use organic solvent, such as Cremophor in clinical formulation Taxol® causes sever side effects and hypersensitivity reactions [38, 39]. Conjugation of PTX to SWNTs-PL-PEG-NH_2 enable removing of solvent in delivery. Indeed, the SWNTs-PL-PEG-PTX displayed increased tumor inhibition effect and reduced side effects in a murine breast cancer model compared with Taxol® formulation [37].

Drugs	Targeting Moieties	Cancer Biomarkers	Type of CNTs	References
Cisplatin	EGF	EGF Receptor	SWNTs	[36]]
Daunorubicin	Sgc8c Aptamer	Tyrosine Kinase-7	SWNTs	[33]
Docetaxel	NGR	Endothelial Cells	SWNTs	[32]
Doxorubicin	Folate /Magnetic	Folate Receptor	MWNTs	[34]
Doxorubicin	RGD	Integrin αvβ3	SWNTs	[30]
Doxorubicin	Folate	Folate Receptor	SWNTs	[31, 40]
Gemcitabine	Magnetic	/	MWNTs	[41]
Platinum (IV)	Folate	Folate Receptor	SWNTs	[42]

Table 1. . Abbreviations: SWNTs, single walled carbon nanotubes; MWNTs, multiwalled carbon nanotubes, EGF, epidermal growth factor; RGD, peptide with arginine-glycine-aspartate sequence; NGR, peptide with asparagineglycine - arginine sequence; Sgc8c, oligonucleotide sequence.

3.2. Carbon nanotubes for gene delivery

Gene therapy is an important treatment for cancer and other genetic diseases. However, the effects of gene therapy are limited by the efficiencies of transfection and system delivery. Since DNA and siRNA are macromolecules, they cannot pass through cell membrane by themselves, carriers are needed to take them inside of cells to take effects. Structurally, both DNA and siRNA contain anionic phosphodiester backbone that and be complexed with cationic reagents, such as cationic lipids and polymers, etc. For system delivery, the DNA or siRNA can be loaded into cationic nanoparticles made from cationic lipids or polymers [43, 44]. The nanoparticles could protect them from nucleases degradation. Since CNTs are able to penetrate cells [3], they are investigated for gene delivery. Typically, two methods are used for loading nucleic acids to CNTs:

i. electrostatic association with cationic molecule functionalized CNTs [45, 46];

ii. chemical conjugation of nucleic acids to functionalized CNTs via cleavable chemical bonds [47];

iii. DNA or siRNA are directly wrap to raw or oxidized CNTs.

Gene delivery using cationic molecule functionalized carbon nanotubes via electrostatic interactions

As discussed early, cationic molecules, such as, ammonium-containing molecules and polyethylene imine (PEI), can be covalently linked to chemically modified CNTs by oxidation or 1,3-cycloadditions reactions[47-50]. In one application, DNA was loaded into CNTs conjugated with ammonium-terminated oligoethylene glycol(CNTs-OEG-$NH_3{}^+$) for delivery [47,

48]. Using this deliver vehicle, expression of test plasmid pCMV-βgal was examined in-vitro.Result showed that the transfection efficiency of CNTs carrier was 5-10 times higher than naked DNA; but, much lower than that of liposome [47]. It has been shown that charge ratio (ammonium groups on CNTs *vs* phosphate groups of the DNA backbone) is a determination factor for gene expression [48]. In contrast to DNA delivery, the same CNTs carrier for delivery of cyclin A2 siRNA demonstrated pronounced silencing effect in-vitro[51]. Surprisingly, In-vivo delivery of SOCS1 significantly inhibited SOCS1 expression and retarded the tumor growth in murine B16 tumor model [52]. The studies with PEI functionalized CNTs also showed very positive results. PEI is an efficient gene delivery reagent by its own, however, high amount of PEI is toxic to cells. The siRNA delivery by PEI-grafted MWNTs showed improved gene expression to the equivalent amounts of PEI polymer alone but with reduced cytotoxicity [46, 53].

Gene delivery by covalently conjugation to carbon nanotubes via cleavable chemical bonds

Alternatively, genes can be conjugated to amphiphilic polymers that are used for non-covalent CNT functionalization [10, 54,55]. Incorporation of cleavable chemical bonds facilitates releasing of DNA or siRNA cargos from CNTs in a controlled manner [54]. Thiol-modified DNA or siRNA were covalently conjugated to amino group of SWNT-PL-PEG- NH_2via cleavable disulfide bond [55]. The genes were released by the cleavage of disulfide bonds by thiol digesting enzymes upon cellular internalization of CNT-PL-PEG-siRNA. The CNT-mediated siRNA delivery showed better gene transfection efficiency than liposome-based delivery system in hard-to-transfect human T cells and primary cells lines [54].

Gene delivery by wrapping directly on carbon nanotubes

Nucleic acids, DNA or siRNA, contains alternative amphiphilic motifs, which can be used to disperse CNTs in water. The nucleic acids form helical wrapping around the CNTs with the bases binding to the hydrophobic CNTs and the hydrophilic sugar-phosphate groups extending to the water phase [19]. In this way, DNA or siRNA serves both CNT dispersing agent and the cargo. It has been shown that the siRNA functionalized SWNTs readily enter cells and exerts its biological activity in cell culture [19]. Studies with intratumoralinjection of siRNA functionalized SWNTs showed significantly inhibition effect in-vivo [56].

3.3. Carbon nanotubes for stem cell related therapies

There has been an increasing trend in attempts to design and develop different CNT based tools and devices for tissue engineering and stem cell therapy applications. In particular, CNT impregnated nanoscaffolds have shown multiple advantages over currently available scaffolds. This includes its strong mechanical properties, resemblance of structure with collagen fibrils and extracellular matrix and electrically conductive nature. These attributes of the CNT based scaffolds and three dimensional nanocomposites have led to their diverse therapeutic applications in the field of myocardial therapy, bone formation, muscle and neuronal regeneration. These applications are mainly based on one principle and that is to modulate the stem cell growth and differentiation in a more controlled and desirable manner.

Carbon nanotubes for stem cell based heart therapy

Over the past two decades there has been significant advancement in stem cell therapy to repair and replace damaged tissues, such as heart muscle[57]. This is because of their ability to divide and differentiate into diverse specialized cell types. Recently, there has been increasing body of evidence indicating that the extracellular matrix plays a critical role in stem cell viability, proliferation and differentiation [58, 59]. Hence, designing a microenvironment prepared from polymeric scaffolds which imitate the physical characteristics of natural biomatrix has been the central strategy in tissue engineering. The emergence of nanomaterials such as nanotubes provide opportunities to design such biocompatible scaffolds for hosting and directing stem cell differentiation [60].

Preliminary studies demonstrate that neonatal rat ventricular myocytes cultured on substrates of multiwall carbon nanotubes can interact with the nanofibres by forming tight contacts and show significantly improved mitotic and chemotactic effects [61]. Moreover, such mode of culture also altered the electrophysiological properties of cardiomyocytes, indicating that CNts are able to promote cardiomyocyte maturation. Further investigations with a nanocomposite of PLGA:CNF show that cardiomyocyte density increases with greater amounts of CNF in PLGA [62]. The study also showed similar trends with neurons. The immense potential of this technology for myocardial therapy roots from the fact that this cardiac patch can not only promote myocardial cells, but also induce the nerve cell growth that help the cardiac cells to contract. In addition, it also supports endothelial cells that make the inner lining of the blood vessels supplying oxygen to the heart.

Carbon nanotubes for stem cell based bone regeneration

In order to direct stem cell differentiation towards bone regeneration, there has been increasing interest by the researchers to explore topographical features of the cell culture substrate. Physical factors, such as rigidity of the extracellular environment, can influence stem cell growth and differentiation. Such differentiation of human stem cells can be detected by altering the size of the nanotubes on which the cells are grown [63]. It has been reported that 70- to 100-nm diameter nanotubes can initiate rapid stem cell elongations, which induce cytoskeletal stress and selective differentiation into osteoblast-like cells, offering a promising route for quicker and better recovery, for example, for patients who undergo orthopedic surgery. The group also showed that the differentiated stem cells express osteopontin and osteocalcin, the two important osteogenetic protein markers.

Moreover CNTs are promising materials for nanaoscaffold and implantation purposes due to the fact that CNTs are conductive, have excellent mechanical properties and their nanostructured dimensions mimic the 3D structure of proteins found in extracellular matrices. Their dimensions resembles closely with that of the triple helix of collagen fibrils which can promote for nucleation and growth of hydroxyapatite, the major inorganic component of bone. A newly developed nanocomposite scaffold of CNFs/CNTs has been shown to influence the cell behaviour [64]. In-vitro study demonstrated that, smaller dimension CNFs dispersed in polycarbonate urethane promoted osteoblast adhesion but did not promote the adhesion of fibroblasts, chondrocytes, and smooth muscle cells. But the mechanisms that guide such cell functions are yet to be understood.

Surface functionalizing the nanotube surface with bone morphogenetic protein-2 (BMP-2) further accelerates chondrogenic and osteogenic differentiation of MSCs [65, 66]. This stimulation is a combined effect of the surface nanoscale geometry of the substrate nanostructures and their BMP-2 coating efficiency. In a similar kind of study, the system also exhibited higher cell proliferation rate, apart from enhanced differentiation [66]. Nanotubes can also be used for extended drug release as has been demonstrated by Hu et al, where drug loaded nanotubes, in combination with multilayers of gelatin and chitosan, have been shown as a new way to use nanotubes as reservoir for storing drugs [67]. The system effectively promoted osteoblastic differentiation of MSCs. Further studies in this direction can be beneficial in order to develop potential bone implants for improved bone osteointegration.

Carbon nanotubes for stem cell based neuronal regeneration

The unique abilities of human embryonic stem cells (hESCs), such as their self-renewal and potency, hold great promise in the field of regenerative medicine and stem cell based therapy. The derivation of neuronal lineages from hESCs holds promise to treat neurological pathologies of the central and peripheral nervous system such as Parkinson's disease, spinal cord injury, multiple sclerosis and glaucoma [68, 69]. CNT based substrates have been shown to promote neuronal differentiation [70]. It has also been proposed that neurons grown on a CNT meshwork displayed better signal transmission, due to tight contacts between the CNTs and neural membranes conducible to electrical shortcuts [71]. It was demonstrated that the MSCs and the neurosphere of cortex-derived neural stem cells (NSCs) can grow on the CNT array and both MSCs and NSCs interacted with the aligned CNTs. The results suggest that CNTs assist in the proliferation of MSCs and aid differentiation of cortex-derived NSCs [72]. However, due to the harsh external environment in the host body and lack of supportive substrates during transplantation, much of the transplanted cells lose its viability resulting in reduced therapeutic efficacy [73]. It has been reported that two dimensional thin film scaffolds, composed of biocompatible poly(acrylic acid) polymer grafted carbon nanotubes (CNTs), can selectively differentiate human embryonic stem cells into neuron cells while maintaining the viability of transplanted cells [74]. Even multiwalled carbon nanotube (MWNT) sheets showed to significantly enhance neural differentiation of hMSCs grown on the CNT sheets. Axon outgrowth was also controlled using nanoscale patterning of CNTs [75]. Recently, silk-CNT-based nanocomposite scaffolds are shown to protect and promote neuronal differentiation of hESCs [76]. Silks are natural polymers (protein) that have been widely used as biomaterials for many years. Fibroin, comprising the major portion of the silk protein fibre, consists of 90% of amino acids including glycine, alanine, and serine. Due to its strong mechanical and flexible nature in thin film form, biocompatibility, and *in-vivo*bioresorbable properties, fibroin protein has been used as the building block for scaffolds. As confirmed by scanning electron microscope (Figure5 A-C), similar results were obtained with the developed silk-CNT scaffold where cells grown on the silk substrate exhibited denser complex three-dimensional axonal bundle networks as well as better spatial density distribution of the networks compared to other scaffolds. Overall, the silk-CNT nanocomposite provided an efficient three-dimensional supporting matrix for stem cell-derived neuronal transplants, offering a promising opportunity for nerve repair treatments for patients with neurological disorder. In-vitro analysis showed that β-

III tubulin, representing the mature differentiated neurons and nestin, representing the neuron precursors, were highly expressed in hESCs grown on the silk-CNT substrate compared to the expression level of cells grown on the control poly-L-ornithine substrate (figure 5D). In addition, hESCs cultured on the silk-CNT scaffold exhibited higher maturity along with dense axonal projections.

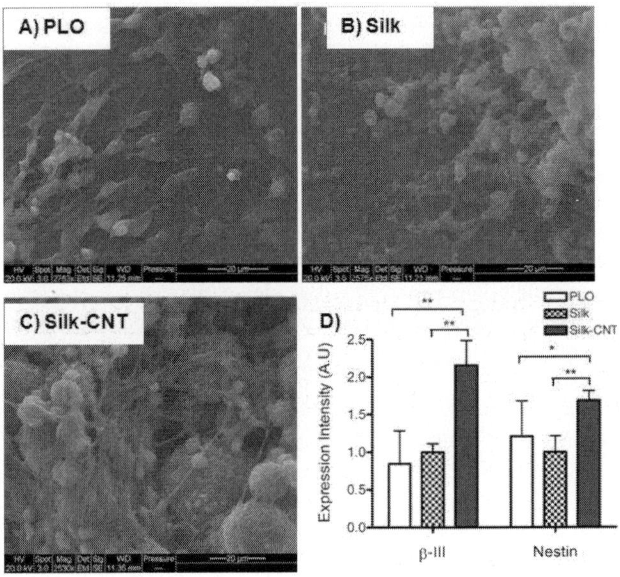

Figure 5. Scanning Electron Microscope images of hESCs on various substrates. SEM images of (A) cells cultured on PLO exhibiting a flat morphology and two-dimensional axonal connections, (B) cells cultured on silk scaffolds demonstrating three-dimensional structures and cell migration, and (C) cells cultured on silk-CNT scaffolds demonstrating three-dimensional axonal connections and silk-CNT matrix degradation. (D) Two neuronal markers (β-III tubulin and nestin) were used to further determine the hESC differentiation efficiency.Expression intensity of β-III tubulin and nestin was observed with fluorescence microscopy. Silk-CNT scaffolds exhibited maximum β-III tubulin expression, while nestin expression exhibited a similar trend. *=P< 0.01, and **= P< 0.001[76]. Abbreviations: PLO, Poly-L-ornithine; hESCs, human embryonic stem cells; CNT, carbon nanotube.

3.4. Carbon nanotubes for thermal destruction of tumors

Tissues are known to be highly transparent to 700- to 1,100-nm near-infrared (NIR) light, whereas, SWNTs display strong optical absorbance in this special spectral window. When constantly absorb energy in NIR region, SWNTs emit heat [77]. Continuous heating leads to killing of the cells. SWNTs have been engineered with tumor recognition molecules for selective enteringcancer cells. Upon NIR radiation, the cancer cells were killed by thermal ablation [78-83]. Previous studies have shown that folic acid decorated SWNTs more effectively killed folate receptor positive cancer cells [83]; monoclonal antibody (mAb) against

http://dx.doi.org/10.5772/51785

human CD22 conjugated SWNTsonly targeted CD22(+)CD25(-) Daudi cells; whereas, anti-CD25 mAb coupled SWNTs only target CD22(-)CD25(+) activated peripheral blood mononuclear cells [81]. The thermal ablation effects can be combined with other therapies, eg. chemotherapy, by loading drugs on CNTs for synergic effect [32].

Tumors, in general, contain a small population of tumor initiating stem-like cells, termed as cancer stem cells. These cells are unmanageable by standard treatment modalities such as chemotherapy and radiotherapy and tend to persist after treatment [84]. Heat-based cancer treatments are increasingly becoming a potential alternative to approach this problem. Combining CNTs with such hyperthermia based therapies can further enhance its efficacy by simultaneously eliminating both the stem cells and bulk cancer cells that constitute a tumor. In fact, CNTs offer several properties that make them promising candidates for such thermal therapy. This includes their ability for thermal conductance and strong absorbance of electromagnetic radiation. It generates significant amounts of heat upon excitation with near-infrared light which is transparent to biological systems including skins. Such a photothermal effect can be employed to induce thermal cell death in a noninvasive manner. Thus, if CNTs can be localized to tumors, they can be stimulated with near-infrared radiation or radiofrequency energy to generate site-specific heat [85]. Preliminary in-vivo results show that a combination of multiwalled carbon nanotubes (MWNTs) and NIR can be useful for tumor regression and long-term survival in a mouse model [86]. Such CNT-mediated thermal therapy addresses the limitations of presently available medical strategies. This includes the minimally invasive site-specific heating which will greatly diminish the off-target toxicities, generation of uniform temperature distribution throughout the tumor mass by the activated CNTs, its compatibility with concurrent MRI temperature mapping techniques. It has also been recently reported that breast cancer stem cells, highly resistant to conventional thermal treatments, can be successfully treated with CNT-based photothermal therapies by promoting necrotic cell death [84]. Further studies in this direction shows that DNA-encased MWNTs are more efficient at converting NIR irradiation into heat compared to non-encased MWNTs and that this method can be effectively used in-vivo for the selective thermal ablation of cancer cells [87].

Glioblastomamultiforme is the most common and aggressive malignant primary brain tumor involving glial cells and accounting for a large percentage of brain and intracranial tumor [88, 89]. It is also known for its recurrence and overall resistance to therapy. CD133+ stem cells occurring among GBM cells are responsible for such huge recurrence risk [90]. Research has been focused on developing strategies to efficiently deliver CNTs to these target sites, harboring CD133+ cancer stem cells [80]. In-vitro studies show that such targeted elimination of CD133 (+) cancer stem cells are possible by adding SWNTs functionalized with CD133 monoclonal antibody, followed by irradiation with NIR laser light. In a separate study, embryonic stem cells, once administered with MWNTs, have shown to induce an enhanced immune boost and provide subsequent anticancer protection in mice with colon cancer by suppressing the proliferation and development of malignant colon tumors [91].

4. Carbon nanotubes for biomedical imaging and detection

The well-ordered molecular structure attributes CNTs with multiple distinct optical properties, include strong NIR absorption, photoluminescence and Raman shift [92]. Structurally, SWNTs can be viewed as a cylinder rolled up by one layer of graphene sheet. TThere are infinite numbers of ways to roll a graphene sheet into a cylinder. Depending on different ways of wrapping, the particular nanotube could be metallic or semi-conductive. Individual semi-conductive SWNTs with appropriate chirality can generate a small band gap fluorescence of 1 eV, which corresponds to NIR range (900-1600 nm), where biological tissues have very low absorption, scattering, and autofluorescence, and therefore, are very useful for biological imaging. In the other side, the inherent graphene structure provides SWNTs with specific Raman scattering signature [93], which is strong enough for use in-vivo imaging. All these optical properties offer opportunities for SWNTs as contrast agents for near-infrared (NIR) photoluminescence imaging [94, 95], Raman imaging and optical absorption agent for photoacoustic imaging [96-98].

4.1. Photoluminescence imaging

NIR photoluminescence of micelle encapsulated SWNTs was firstly discovered by O.Connel et al [7]. The single-particle dispersion of individual nanotubes was prepared by ultrasonically agitating of raw SWNTs in SDS. The tube bundles, ropes, and residual catalyst were removed by ultracentrifugation, since the aggregation of nanotubes would quench fluorescence. One advantage of the photoluminescence of SWNTs over organic fluorescence dyes is that SWNTs have no apparent photobleaching, and therefore, the SWNTs could be a powerful tool for tracking changes in living system. Researchers have applied NIR photoluminescence of SWNTs for tracking endocytosis and exocytosis of SWNTs in NIH-3T3 cells in real time [94, 95]. Moreover, conjugation of antibodies to SWNTs surface allowed specific cell targeting. They have shown that, with conjugation of anti-CD20, SWNTs selectively recognized CD20 cell surface receptor on B-cells with little binding to receptor negative T-cells. Similarly, with conjugation of Herceptin, SWNTs only recognize HER2/neu positive breast cancer cells. The selective binding of SWNTs was detected by intrinsic NIR photoluminescence of nanotubes. This technique allows deep tissue penetration and high-resolution intravital microscopy imaging of tumor vessels beneath thick skin [99, 100].

4.2. Raman shift imaging

Raman spectroscopy is a sensitive analytical tool for biological samples. It also has advantages of resistance to autofluorescence and photobleaching, high spatial resolution, and small sample size [101]. CNTs exhibit strong resonance Raman scattering with several distinctive scattering features including the radial breathing mode (RBM) and tangential mode (G-band) [93, 102]. Both RBM and G-band of CNTs are sharp and strong peaks, which can be easily distinguished from autofluorescence of tissue samples, Recently, Raman microspectroscopy of SWNTs has been applied for imaging of tissue samples, live cells, and small animal models [96-98]. Tumor targeted delivery by RGD peptide functionalized SWNTs has been investigated in murine tumor model [85]. Raman spectroscopy image of excised tissues confirmed efficient targeting of $\alpha v\beta 3$ integrin positive U87MG tumor by RGD [85]. This study also disclosed that CNTs have relatively long circulation time, and rapid renal clearance, which makes SWNTs an attractive diagnostic and therapeutic delivery vehicle. Zevaleta et al further developed a Raman microscope capable of noninvasive in-vivo evaluation of tumors in mice with RGD-labeled SWNTs. Using the dynamic Raman microscope, pharmacokinetics of SWNTs in the tumor was evaluated immediately following an intravenous injection of SWNTs. Raman spectral analysis revealed effectiveness of the RGD nanotubes to the integrin expressing U87MG tumor. The noninvasive Raman imaging results were compared with excised tissues and shows consistency [97].

4.3. Photoacaustic imaging

Photoacoustic imaging is an optical imaging technique that combines high optical absorption contrast with diffraction-limited resolution of ultrasonic imaging, which allows deeper tissues to be viewed in living subjects. In photoacoustic imaging, short pulses of stimulating radiation are absorbed by tissues, resulting a subsequent thermal expansion and ultrasonic emission that can be detected by highly sensitive piezoelectric devices. However, many diseases, for example cancer, in their early stages, do not exhibit a natural photoacoustic contrast, therefore administering an external photoacoustic contrast agent is necessary. Owing to the strong light absorption characteristic [77], the CNTs can be utilized as photoacoustic contrast agents. De la Zerdaet. al, in the first time, applied SWNTs for in-vivo imaging of tumors in mice [103]. In this study, SWNTs was surface modified by PL-PEG and further conjugated with cyclic RGD peptides for targeting $\alpha v\beta 3$ integrin on cancer cells. Intravenous injection of these cyclic RGD functionalized CNTs to mice bearing tumors showed eight times stronger photoacoustic signal in the tumors than mice injected with non-targeted CNTs. This study suggested that photoacoustic imaging using targeted SWNTs could contribute to non-invasive in-vivo cancer imaging [103]. Similarly, in another study, SWNTs functionalized with antibody against $\alpha v\beta 3$ integrin for photoacoustic imaging of human glioblastoma tumors in nude mice [104].

5. Selected examples for preparation of carbon nanotube based therapeutics

In the above CNTs for drug delivery section, we described the functionalization of CNTs with drugs or targeting molecules. These preparations are usually applied directly via intravenous delivery route, which is the most widely used route of drug administration. Alternative to the intravenous drug administration, some other routes of drug administrations are also important for certain specific applications. Different formulations of CNT-based therapeutics have also been developed to suit the specific routes of administration. Here, we report several novel cases of CNTs applications for oral and transdermal delivery routes.

5.1. Carbon nanotube based therapeutics delivery using artificial cells: oral delivery

For many therapeutics, oral and targeted delivery are challenge. One way to deliver them at the targeted site is by novel methods of encapsulating them in polymeric artificial cells. Artificial cells are vesicles made by polymeric membranes. They can mimic certain functions of biological cells. The size of artificial cells ranged from nanometer to hundreds of micrometer [105]. The membranes of artificial cells are usually semi-permeable that allows for exchange of small molecules and prevention of passage of large substances across it. Up to date, artificial cells have been applied for encapsulation biologically active agents, including enzymes, hormones, drugs, even live bacteria cells for in-vivo delivery. Currently a couple of artificial cells have been applied clinically [106]. The advantages of artificial cells include protection of the cargos from immune elimination in the body, targeted delivery of cargos to desired sites and increasing cargo solubility [106]. In a first feasibility study, functionalized CNTs have been encapsulated in artificial cell made from biocompatible polymeric membrane for target specific delivery. The polymeric membrane was assembled with three layers of polymers, alginate-poly-L-lysine-alginate (APA), via electrostatic association (figure 6). Artificial cells protected CNT therapeutics from degradation by the harsh environments [107]. PH degradation profile of the polymeric membrane of artificial cells can be adjusted by composition of polymers, which allows the breakdown of the artificial cells and release of CNT therapeutics to desired sites. This system is ideal for oral delivery, and can be used for other delivery routes as well.

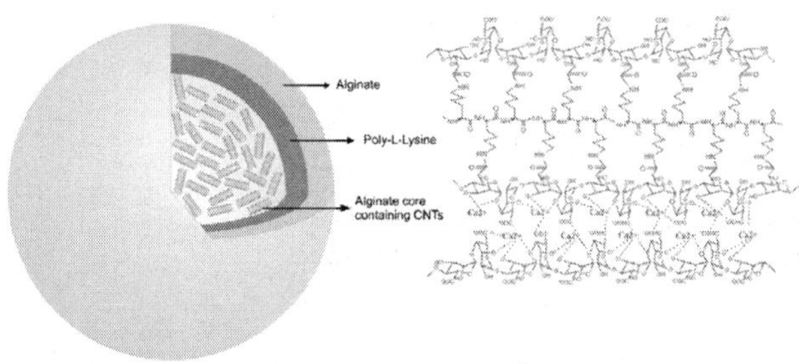

Figure 6. Alginate-poly-L-lysine-alginate (APA) microcapsules encapsulatingcarbon nanotubes. The calcium ions are responsible for cross-linking of the alginate monomeric units trapping the carbon nanotubes into the core of microcapsule [108].

5.2. Carbon nanotube based membrane: transdermal drug delivery

One of the most important areas of transdermal drug delivery (TDD) is in addiction treatment. Nicotine TDD has been widely used for smoking cessation programs. However, these traditional transdermal patches could not provide variable drug delivery rates. Some TDD has the capability to provide variable and programmable delivery rates, however, it needs a strong electric current across the human skin, which can cause serious skin irritation. Recently, the membranes prepared by functionalized CNTs have been employed in transdermal drug delivery. It has been shown that this CNT-membrane was very effective for treatment of drug abuse and addiction [109]. To prepare the CNT-membrane, MWNTs were firstly functionalized with negatively charged molecule containing sulphonate ($-SO^{3-}$) to have a high charge density on the surface of CNTs, which is necessary to get efficient electro-osmosis pumping effect. By this functionalization strategy, CNT-membrane achieves dramatically fast flow through CNT cores, high charge density, and highly efficient electrophoretic pumping effect. These membranes were the integrated with a nicotine formulation to obtain switchable transdermal nicotine delivery rates on human skin (in vitro). The transdermal nicotine formulated CNT-membrane was able to successfully switch between high level and low level fluxes that coincide with therapeutic demand levels for nicotine cessation treatment. These programmable devices cause minimal skin irritation [109].

6. Potentials and limitations of carbon nanotubes in medicine

CNTs are being highly explored in the fields of targeted drug delivery, nanoscaffold for tissue engineering, biomedical imagining and detecting for disease treatment and health monitoring. The use of CNTs in drug delivery, detection and tissue engineering has shown the potentials to revolutionize medicine. CNTs affords for a large amounts of payloads for specific-targeting and drug delivery. With their intrinsic properties, the CNTs have potential for building-up multifunctional nanodevices for simultaneous therapeutic delivery and detection. Current cancer therapies (eg. radiation therapy, and chemotherapy) are usually painful and less efficient since they kill normal cells in addition to cancer cells, and therefore, producing adverse side effects and resistance. The CNT-based drug delivery systems have shown efficient tumor-targeting, and they can effectively kill cancer cells with a dosage lower than conventional drugs used, however significantly reduces side effects. Current CNT-based nanoscaffolds are very advantages for stem cells therapy in that they can modulate the stem cell growth and differentiation in a more controlled and desirable manner. Thus, CNTs have recently gained much interest in the field of medicine.

Although very useful, CNTs exist some limitations. Firstly, pristine CNTs, being metallic in their nature, are insoluble and they form large bundles or ropes in many solvents, including water and most solvents, so they cannot be used directly in biomedical applications. Much work has been done to prepare them for use in medicine. Secondly, the CNTs are not homogenous in their sizes (both diameters and lengths), which could be a problem for generation of reproducible results that allows evaluation of the biological activity relating to specific structures. Up to date, tremendous efforts have been put in surface functionalization of CNTs for use in medicine. This includes numerous effective methods for covalent or non-covalent modification of CNTs as to disperse them into aqueous solutions and to attach functional molecules for therapeutic applications. However, in terms of homegenecity of CNTs. Not much work has been done so far. We propose that attentions are needed to develop the methods for generation of CNTs with homogeneous size, which is very important for future clinical applications of CNT-based therapeutics.

As a novel nanomaterial of great potentials in medicine, the toxicology of CNTs has received much attention in recent years. Pristine CNTs are very light powders and they can enter the body through inhalation via the respiratory tract, ingestion via the gastrointestinal tract or, dermal absorption via the skin. Following entering, CNTs distribute rapidly in the central and peripheral nervous system, lymphatic and blood circulation and potentially cause toxic effects in a variety of tissues and organs that they reach, such as heart, spleen, kidney, bone marrow and liver, etc. Toxicity of CNTs has been evaluated in a variety of cell or animal models for assessing pulmonary, dermal and immune effects. However, the published results have not led to any consensus on the toxicity profile of pristine or functionalized SWNTs and MWNTs. Some investigators reported that pristine SWNTs that were purified by acid treatment demonstrated no acute toxicity, as opposed to non-purified CNTs, however, they induced reactive oxygen species (ROS) in human lung carcinoma epithelial A549 cells and NR8383 cells [110]. Others demonstrated that pristine SWNTs, either acid-treated or non-treated, were capable of increasing chromosome and DNA damage, and oxidative stress in macrophage cell lines [111, 112].

In contrast to raw or acid treated CNTs, the well-dispersed CNTs with high levels of surface functionalization can reduce the toxicity of MWNTs. One study demonstrated that taurine-MCWNTs in low and medium doses induced slight and recoverable pulmonary inflammation in mice, and are less toxic than raw MCWNTs [113]. This is supported by other studies indicating that the damage caused by non-PEGylated MWNTs is slightly more severe than that of PEGylated MWNTs [114]. Furthermore, administrations of high doses of PL-PEG functionalized SWNTs following intravenous injection did not lead to acute or chronic toxicity in nude mice, albeit SWNTs persisted within liver and spleen macrophages for 4 months in mice without apparent toxicity [115] and the SWNTs-PL-PEG were excreted from mice via the biliary and renal pathways [116]. It is hypothesized that the van der Waals forces on the surfaces of pristine CNTs cause hydrophobic interactions between CNTs, resulting in aggregation and network formation, which further induce prolonged toxicity. Thus, functionalization of CNTs overcomes the aggregate-forming surface properties of CNTs, and therefore, reduces toxicity.

7. Conclusions:

CNTs have exhibited diverse physical, chemical and mechanical properties suitable for a variety of applications. In last decade, biomedical applications of CNTs have undergone rapid progress. Their unique properties, such as, ultrahigh surface area, high aspect ratio, distinct optical properties have been applied to develop innovative, multi-functional CNT-based nanodevices for broad applications. This chapter have described the chemical and physical methods to prepare CNTs for used in medicine. With these methods, targeting molecules are attached on CNTs for targeted drug delivery, selective imaging, and other therapies. As a new type of nanomaterial, the toxicity of CNTs has been extensively investigated. To date, tremendous toxicity studies on CNTs have been published. However, the published data are inconsistent. The reason is that CNTs used in these studies vary in dispersion status, size and length of tubes, metal impurities and functionalization methods etc. Moreover, different analysis methods used in the evaluation CNTs toxicity studies also cause disparities. Despite these disparities, there is a broad agreement that well-dispersed CNTs have little or no toxicity both in-vitro and in-vivo, and therefore are useful for biomedical applications. Finally, an urgent need has been proposed for long-term studies on the absorption, deposition, metabolism and excretion (ADME) of CNTs. Only after the uncertainty on CNT toxicity is resolved, the CNT-based therapeutics can be possible applied clinically.

Acknowledgements

This work is partially supported by research grant to Satya Prakash from Canadian Institutes of Health Research (CIHR) (MOP 93641). W. Shao and L. Rodes acknowledges Doctor Training Award from Fonds de Research Sante (FRSQ). A. Paul acknowledges Post-Doctoral Award from FRSQ.

Author details

Wei Shao[1], Paul Arghya[1], Mai Yiyong[2], Laetitia Rodes[1] and Satya Prakash[1*]

*Address all correspondence to: satya.prakash@mcgill.ca

1 Biomedical Technology and Cell Therapy Research Laboratory, Department of Biomedical Engineering and Artificial cells and Organs Research Centre, Faculty of Medicine, McGill University, Canada

2 Department of Chemistry, McGill University, Canada

References

[1] Chen, Z., et al. (2011). Single-walled carbon nanotubes as optical materials for biosensing. *Nanoscale*, 3(5), 1949-1956.

[2] Bekyarova, E., et al. (2005). Electronic properties of single-walled carbon nanotube networks. *Journal of the American Chemical Society*, 5990-5995.

[3] Kostarelos, K., et al. (2007). Cellular uptake of functionalized carbon nanotubes is independent of functional group and cell type. *Nat Nanotechnol*, 2(2), 108-113.

[4] Ando, Y. (2010). Carbon nanotube: the inside story. *J Nanosci Nanotechnol*, 10(6), 3726-3738.

[5] Sang Won Kim, T.K, Yern Seung, Kim, Hong Soo, Choi, Hyeong, Jun Lim, Seung Jae, Yang, & Chong Rae, Park. (2012). Surface modifications for the effective dispersion of carbon nanotubes in solvents and polymers. *Carbon*, 50(1), 30.

[6] Tasis, D., et al. (2003). Soluble carbon nanotubes. *Chemistry*, 9(17), 4000-4008.

[7] O'Connell, M. J., et al. (2002). Band gap fluorescence from individual single-walled carbon nanotubes. *Science*, 297(5581), 593-596.

[8] Valerie, C., Moore, M. S. S., Erik, H., Haroz, Robert. H., & Hauge, Richard E. (2003). Smalley, Individually Suspended Single-Walled Carbon Nanotubes in Various Surfactants. *Nano Lett*, 3(10), 3.

[9] Strano, M. S., et al. (2003). The role of surfactant adsorption during ultrasonication in the dispersion of single-walled carbon nanotubes. *J Nanosci Nanotechnol*, 3(1-2), 81-86.

[10] Liu, Z., et al. (2009). Preparation of carbon nanotube bioconjugates for biomedical applications. *Nat Protoc*, 4(9), 1372-1382.

[11] Fukushima, T, et al. (2003). Molecular ordering of organic molten salts triggered by single-walled carbon nanotubes. *Science*, 300(5628), 2072-2074.

[12] Dong, B., et al. (2011). Dispersion of carbon nanotubes by carbazole-tailed amphiphilic imidazolium ionic liquids in aqueous solutions. *Journal of colloid and interface science*, 190-195.

[13] Fu, Q., & Liu, J. (2005). Effects of ionic surfactant adsorption on single-walled carbon nanotube thin film devices in aqueous solutions. *Langmuir : the ACS journal of surfaces and colloids*, 1162-1165.

[14] Tasis, D., et al. (2006). Chemistry of carbon nanotubes. *Chem Rev*, 106(3), 1105-1136.

[15] Petrov, P., et al. (2003). Noncovalent functionalization of multi-walled carbon nanotubes by pyrene containing polymers. *Chem Commun (Camb)* [23], 2904-2905.

[16] Li, H., et al. (2004). Selective interactions of porphyrins with semiconducting single-walled carbon nanotubes. *Journal of the American Chemical Society*, 1014-1015.

[17] Xianbao Wang, Y.L, Wenfeng, Qiu, & Daoben, Zhu. (2002). Immobilization of tetra-tert-butylphthalocyanines on carbon nanotubes: a first step towards the development of new nanomaterials. *J. Mater. Chem*, 12(6), 3.

[18] Kang, Y. K., et al. (2009). Helical wrapping of single-walled carbon nanotubes by water soluble poly(p-phenyleneethynylene). *Nano letters*, 9(4), 1414-1418.

[19] Zheng, M., et al. (2003). DNA-assisted dispersion and separation of carbon nanotubes. *Nature materials*, 338-342.

[20] Hu, H., et al. (2003). Nitric Acid Purification of Single-Walled Carbon Nanotubes. *The Journal of Physical Chemistry B*, 107(50), 13838-13842.

[21] Karousis, N., Tagmatarchis, N., & Tasis, D. (2010). Current progress on the chemical modification of carbon nanotubes. *Chem Rev*, 110(9), 5366-5397.

[22] Georgakilas, V., et al. (2002). Amino acid functionalisation of water soluble carbon nanotubes. *Chem Commun (Camb)* [24], 3050-3051.

[23] Ojima, I. (2008). Guided molecular missiles for tumor-targeting chemotherapy--case studies using the second-generation taxoids as warheads. *Accounts of chemical research*, 108-119.

[24] Krauss, W. C., et al. (2000). Emerging antibody-based HER2 (ErbB-2/neu) therapeutics. *Breast Dis*, 11, 113-124.

[25] Abramson, J. S., & Shipp, M. A. (2005). Advances in the biology and therapy of diffuse large B-cell lymphoma: moving toward a molecularly targeted approach. *Blood*, 1164-1174.

[26] Garanger, E., Boturyn, D., & Dumy, P. (2007). Tumor targeting with RGD peptide ligands-design of new molecular conjugates for imaging and therapy of cancers. *Anticancer Agents Med Chem*, 7(5), 552-558.

[27] Corti, A., & Curnis, F. (2011). Tumor Vasculature Targeting Through NGR Peptide-Based Drug Delivery Systems. *Current Pharmaceutical Biotechnology*, 1128-1134.

[28] Lu, Y., & Low, P. S. (2002). Folate-mediated delivery of macromolecular anticancer therapeutic agents. *Advanced drug delivery reviews*, 675-693.

[29] Hermanson, G. T. (2008). Bioconjugate Techniques,2nd Edition. Academic Press, Inc., 1202, pages.

[30] Liu, Z., et al. (2007). Supramolecular chemistry on water-soluble carbon nanotubes for drug loading and delivery. *ACS nano*, 50-56.

[31] Ji, Z., et al. (2012). Targeted therapy of SMMC-7721 liver cancer in vitro and in vivo with carbon nanotubes based drug delivery system. *Journal of colloid and interface science*, 143-149.

[32] Wang, L., et al. (2011). Synergistic enhancement of cancer therapy using a combination of docetaxel and photothermal ablation induced by single-walled carbon nanotubes. *Int J Nanomedicine*, 6, 2641-2652.

[33] Taghdisi, S. M., et al. (2011). Reversible targeting and controlled release delivery of daunorubicin to cancer cells by aptamer-wrapped carbon nanotubes. *European journal of pharmaceutics and biopharmaceutics official journal of Arbeitsgemeinschaft fur Pharmazeutische Verfahrenstechnik e.*, 77, 200-206.

[34] Lu, Y. J., et al. (2012). Dual targeted delivery of doxorubicin to cancer cells using folate-conjugated magnetic multi-walled carbon nanotubes. *Colloids Surf B Biointerfaces*, 89, 1-9.

[35] Lay, C. L., et al. (2010). Delivery of paclitaxel by physically loading onto poly(ethylene glycol) (PEG)-graft-carbon nanotubes for potent cancer therapeutics. *Nanotechnology*, 21(6), 065101.

[36] Bhirde, A. A., et al. (2009). Targeted killing of cancer cells in vivo and in vitro with EGF-directed carbon nanotube-based drug delivery. *ACS nano*, 307-316.

[37] Liu, Z., et al. (2008). Drug delivery with carbon nanotubes for in vivo cancer treatment. *Cancer research*, 68(16), 6652-6660.

[38] Tsavaris, N. B., & Kosmas, C. (1998). Risk of severe acute hypersensitivity reactions after rapid paclitaxel infusion of less than 1-h duration. *Cancer chemotherapy and pharmacology*, 509-511.

[39] Sendo, T., et al. (2005). Incidence and risk factors for paclitaxel hypersensitivity during ovarian cancer chemotherapy. *Cancer chemotherapy and pharmacology*, 91-96.

[40] Zhang, X., et al. (2009). Targeted delivery and controlled release of doxorubicin to cancer cells using modified single wall carbon nanotubes. *Biomaterials*, 30(30), 6041-6047.

[41] Yang, F, et al. (2011). Magnetic functionalised carbon nanotubes as drug vehicles for cancer lymph node metastasis treatment. *Eur J Cancer*, 47(12), 1873-1882.

[42] Dhar, S., et al. (2008). Targeted single-wall carbon nanotube-mediated Pt(IV) prodrug delivery using folate as a homing device. *Journal of the American Chemical Society*, 130(34), 11467-11476.

[43] Sun, X., & Zhang, N. (2010). Cationic polymer optimization for efficient gene delivery. *Mini reviews in medicinal chemistry*, 108-125.

[44] Shao, W., et al. (2012). A novel polyethyleneimine-coated adeno-associated virus-like particle formulation for efficient siRNA delivery in breast cancer therapy: preparation and in vitro analysis. *Int J Nanomedicine*, 7, 1575-1586.

[45] Yang, F., et al. (2009). Pilot study of targeting magnetic carbon nanotubes to lymph nodes. *Nanomedicine*, 317-330.

[46] Nunes, A, et al. (2010). Hybrid polymer-grafted multiwalled carbon nanotubes for in vitro gene delivery. *Small*, 6(20), 2281-2291.

[47] Pantarotto, D., et al. (2004). Functionalized carbon nanotubes for plasmid DNA gene delivery. *Angew Chem Int Ed Engl*, 43(39), 5242-5246.

[48] Singh, R., et al. (2005). Binding and condensation of plasmid DNA onto functionalized carbon nanotubes: toward the construction of nanotube-based gene delivery vectors. *Journal of the American Chemical Society*, 4388-4396.

[49] Wang, X., et al. (2009). Knocking-down cyclin A(2) by siRNA suppresses apoptosis and switches differentiation pathways in K562 cells upon administration with doxorubicin. *PloS one*, 4(8), e6665.

[50] Zhang, Z., et al. (2006). Delivery of telomerase reverse transcriptase small interfering RNA in complex with positively charged single-walled carbon nanotubes suppresses tumor growth. *Clinical cancer research : an official journal of the American Association for Cancer Research*, 4933-4939.

[51] Wang, X., Ren, J., & Qu, X. (2008). Targeted RNA interference of cyclin A2 mediated by functionalized single-walled carbon nanotubes induces proliferation arrest and apoptosis in chronic myelogenous leukemia K562 cells. *ChemMedChem*, 3(6), 940-945.

[52] Yang, R., et al. (2006). Single-walled carbon nanotubes-mediated in vivo and in vitro delivery of siRNA into antigen-presenting cells. *Gene therapy*, 13(24), 1714-1723.

[53] Liu, Y., et al. (2005). Polyethylenimine-grafted multiwalled carbon nanotubes for secure noncovalent immobilization and efficient delivery of DNA. *Angew Chem Int Ed Engl*, 44(30), 4782-4785.

[54] Liu, Z, et al. (2007). siRNA delivery into human T cells and primary cells with carbon-nanotube transporters. *Angew Chem Int Ed Engl*, 46(12), 2023-2027.

[55] Kam, N. W., Liu, Z., & Dai, H. (2005). Functionalization of carbon nanotubes via cleavable disulfide bonds for efficient intracellular delivery of siRNA and potent gene silencing. *Journal of the American Chemical Society*, 127(36), 12492-12493.

[56] Bartholomeusz, G., et al. (2009). In Vivo Therapeutic Silencing of Hypoxia-Inducible Factor 1 Alpha (HIF-1alpha) Using Single-Walled Carbon Nanotubes Noncovalently Coated with siRNA. *Nano Res*, 2(4), 279-291.

[57] Paul, A., et al. (2009). Microencapsulated stem cells for tissue repairing: implications in cell-based myocardial therapy. *Regen Med*, 4(5), 733-745.

[58] Harrison, B. S., & Atala, A. (2007). Carbon nanotube applications for tissue engineering. *Biomaterials*, 28(2), 344-353.

[59] Chao, T. I., et al. (2010). Poly(methacrylic acid)-grafted carbon nanotube scaffolds enhance differentiation of hESCs into neuronal cells. *Adv Mater*, 22(32), 3542-3547.

[60] Mooney, E, et al. (2008). Carbon nanotubes and mesenchymal stem cells: biocompatibility, proliferation and differentiation. *Nano Lett*, 8(8), 2137-2143.

[61] Martinelli, V, et al. (2012). Carbon nanotubes promote growth and spontaneous electrical activity in cultured cardiac myocytes. *Nano Lett*, 12(4), 1831-1838.

[62] Stout, D. A., Basu, B., & Webster, T. J. (2011). Poly(lactic-co-glycolic acid): carbon nanofiber composites for myocardial tissue engineering applications. *Acta Biomater*, 7(8), 3101-3112.

[63] Oh, S, et al. 2009. Stem cell fate dictated solely by altered nanotube dimension. *Proc Natl Acad Sci U S A*, 106(7), 2130-2135.

[64] Tran, P. A., Zhang, L., & Webster, T. J. (2009). Carbon nanofibers and carbon nanotubes in regenerative medicine. *Adv Drug Deliv Rev*, 61(12), 1097-1114.

[65] Park, J., et al. (2012). Synergistic control of mesenchymal stem cell differentiation by nanoscale surface geometry and immobilized growth factors on TiO2 nanotubes. *Small*, 8(1), 98-107.

[66] Lai, M., et al. (2011). Surface functionalization of TiO2 nanotubes with bone morphogenetic protein 2 and its synergistic effect on the differentiation of mesenchymal stem cells. *Biomacromolecules*, 12(4), 1097-1105.

[67] Hu, Y., et al. (2012). TiO2 nanotubes as drug nanoreservoirs for the regulation of mobility and differentiation of mesenchymal stem cells. *Acta Biomater*, 8(1), 439-448.

[68] Connick, P., Patani, R., & Chandran, S. (2011). Stem cells as a resource for regenerative neurology. *Pract Neurol*, 11(1), 29-36.

[69] Levenberg, S., et al. (2005). Neurotrophin-induced differentiation of human embryonic stem cells on three-dimensional polymeric scaffolds. *Tissue Eng*, 11(3-4), 506-512.

[70] Jan, E., & Kotov, N. A. (2007). Successful differentiation of mouse neural stem cells on layer-by-layer assembled single-walled carbon nanotube composite. *Nano Lett*, 7(5), 1123-1128.

[71] Mazzatenta, A., et al. (2007). Interfacing neurons with carbon nanotubes: electrical signal transfer and synaptic stimulation in cultured brain circuits. *J Neurosci*, 27(26), 6931-6936.

[72] Nho, Y., et al. (2010). Adsorption of mesenchymal stem cells and cortical neural stem cells on carbon nanotube/polycarbonate urethane. *Nanomedicine (Lond*, 5(3), 409-417.

[73] Zhang, S. C., et al. (2001). In vitro differentiation of transplantable neural precursors from human embryonic stem cells. *Nat Biotechnol*, 19(12), 1129-1133.

[74] Chao, T. I., et al. (2009). Carbon nanotubes promote neuron differentiation from human embryonic stem cells. *Biochem Biophys Res Commun*, 384(4), 426-430.

[75] Kim, J. A., et al. (2012). Regulation of morphogenesis and neural differentiation of human mesenchymal stem cells using carbon nanotube sheets. *Integr Biol (Camb*, 4(6), 587-594.

[76] Chen, C. S., et al. (2012). Human stem cell neuronal differentiation on silk-carbon nanotube composite. *Nanoscale Res Lett*, 7(1), 126.

[77] Berber, S., Kwon, Y. K., & Tomanek, D. (2000). Unusually high thermal conductivity of carbon nanotubes. *Phys Rev Lett*, 84(20), 4613-4616.

[78] Marches, R., et al. (2009). Specific thermal ablation of tumor cells using single-walled carbon nanotubes targeted by covalently-coupled monoclonal antibodies. *Int J Cancer*, 125(12), 2970-2977.

[79] Gannon, C. J., et al. (2007). Carbon nanotube-enhanced thermal destruction of cancer cells in a noninvasive radiofrequency field. *Cancer*, 110(12), 2654-2665.

[80] Wang, C. H., et al. (2011). Photothermolysis of glioblastoma stem-like cells targeted by carbon nanotubes conjugated with CD133 monoclonal antibody. *Nanomedicine*, 7(1), 69-79.

[81] Chakravarty, P., et al. (2008). Thermal ablation of tumor cells with antibody-functionalized single-walled carbon nanotubes. *Proceedings of the National Academy of Sciences of the United States of America*, 105(25), 8697-8702.

[82] Xiao, Y., et al. (2009). Anti-HER2 IgY antibody-functionalized single-walled carbon nanotubes for detection and selective destruction of breast cancer cells. *BMC cancer*, 9, 351.

[83] Kam, N. W., et al. (2005). Carbon nanotubes as multifunctional biological transporters and near-infrared agents for selective cancer cell destruction. *Proceedings of the National Academy of Sciences of the United States of America*, 102(33), 11600-11605.

[84] Burke, A. R., et al. (2012). The resistance of breast cancer stem cells to conventional hyperthermia and their sensitivity to nanoparticle-mediated photothermal therapy. *Biomaterials*, 33(10), 2961-2970.

[85] Liu, Z., et al. (2007). In vivo biodistribution and highly efficient tumour targeting of carbon nanotubes in mice. *Nat Nanotechnol*, 2(1), 47-52.

[86] Burke, A., et al. (2009). Long-term survival following a single treatment of kidney tumors with multiwalled carbon nanotubes and near-infrared radiation. *Proc Natl Acad Sci U S A*, 106(31), 12897-12902.

[87] Ghosh, S., et al. (2009). Increased heating efficiency and selective thermal ablation of malignant tissue with DNA-encased multiwalled carbon nanotubes. *ACS Nano*, 3(9), 2667-2673.

[88] Singh, S. K., et al. (2003). Identification of a cancer stem cell in human brain tumors. *Cancer Res*, 63(18), 5821-5828.

[89] Lanzetta, G., et al. (2003). Temozolomide in radio-chemotherapy combined treatment for newly-diagnosed glioblastoma multiforme: phase II clinical trial. *Anticancer Res*, 23(6D), 5159-5164.

[90] Zeppernick, F., et al. (2008). Stem cell marker CD133 affects clinical outcome in glioma patients. *Clin Cancer Res*, 14(1), 123-129.

[91] Mocan, T, & Iancu, C. (2011). Effective colon cancer prophylaxis in mice using embryonic stem cells and carbon nanotubes. *Int J Nanomedicine*, 6, 1945-1954.

[92] Xie, S. S., et al. (2000). Mechanical and physical properties on carbon nanotube. *Journal of Physics and Chemistry of Solids*, 61(7), 1153-1158.

[93] Tasis, D., et al. (2008). Diameter-selective solubilization of carbon nanotubes by lipid micelles. *J Nanosci Nanotechnol*, 8(1), 420-423.

[94] Strano, M. S., & Jin, H. (2008). Where is it heading? Single-particle tracking of single-walled carbon nanotubes. *ACS nano*, 2(9), 1749-1752.

[95] Jin, H., Heller, D. A., & Strano, M. S. (2008). Single-particle tracking of endocytosis and exocytosis of single-walled carbon nanotubes in NIH-3T3 cells. *Nano letters*, 8(6), 1577-1585.

[96] Keren, S., et al. (2008). Noninvasive molecular imaging of small living subjects using Raman spectroscopy. *Proceedings of the National Academy of Sciences of the United States of America*, 105(15), 5844-5849.

[97] Zavaleta, C., et al. (2008). Noninvasive Raman spectroscopy in living mice for evaluation of tumor targeting with carbon nanotubes. *Nano letters*, 8(9), 2800-2805.

[98] Liu, Z., et al. (2010). Multiplexed Five-Color Molecular Imaging of Cancer Cells and Tumor Tissues with Carbon Nanotube Raman Tags in the Near-Infrared. *Nano Res*, 3(3), 222-233.

[99] Welsher, K., et al. (2008). Selective probing and imaging of cells with single walled carbon nanotubes as near-infrared fluorescent molecules. *Nano letters*, 8(2), 586-590.

[100] Welsher, K., et al. (2009). A route to brightly fluorescent carbon nanotubes for near-infrared imaging in mice. *Nat Nanotechnol*, 4(11), 773-780.

[101] Hanlon, E. B., et al. (2000). Prospects for in vivo Raman spectroscopy. *Physics in medicine and biology*, 45(2), R1-59.

[102] Karmakar, A., et al. (2011). Raman spectroscopy as a detection and analysis tool for in vitro specific targeting of pancreatic cancer cells by EGF-conjugated, single-walled carbon nanotubes. *J Appl Toxicol.*

[103] De la Zerda, A., et al. (2008). Carbon nanotubes as photoacoustic molecular imaging agents in living mice. *Nat Nanotechnol*, 3(9), 557-562.

[104] Xiang, L., et al. (2009). Photoacoustic molecular imaging with antibody-functionalized single-walled carbon nanotubes for early diagnosis of tumor. *J Biomed Opt*, 14(2), 021008.

[105] Chang, T. M. (2004). Artificial cell bioencapsulation in macro, micro, nano, and molecular dimensions: keynote lecture. *Artificial cells, blood substitutes, and immobilization biotechnology*, 32(1), 1-23.

[106] Prakash, S. (2007). Artificial cells, cell engineering and therapy:. *Woodhead.*

[107] Kulamarva, A., et al. (2009). Microcapsule carbon nanotube devices for therapeutic applications. *Nanotechnology*, 20(2), 025612.

[108] Prakash, S., et al. (2011). Polymeric nanohybrids and functionalized carbon nanotubes as drug delivery carriers for cancer therapy. *Advanced drug delivery reviews*, 63(14-15), 1340-1351.

[109] Wu, J., et al. (2010). Programmable transdermal drug delivery of nicotine using carbon nanotube membranes. *Proceedings of the National Academy of Sciences of the United States of America*, 107(26), 11698-11702.

[110] Pulskamp, K., Fau-Diabate, S., Diabate, H. F., Fau-Krug, S., & Krug, H. F. Carbon nanotubes show no sign of acute toxicity but induce intracellular reactive oxygen species in dependence on contaminants. 0378-4274, Print.

[111] Migliore, L., Fau-Saracino, D., et al. Carbon nanotubes induce oxidative DNA damage in RAW 264.7 cells. 1098-2280, Electronic.

[112] Kagan Ve Fau- Tyurina, Y.Y, et al. Direct and indirect effects of single walled carbon nanotubes on RAW 264.7 macrophages: role of iron. 0378-4274, Print.

[113] Wang, X., Fau-Zang, J. J., et al. Pulmonary toxicity in mice exposed to low and medium doses of water-soluble multi-walled carbon nanotubes. 1533-4880, Print.

[114] Zhang, D., Fau-Deng, X., et al. Long-term hepatotoxicity of polyethylene-glycol functionalized multi-walled carbon nanotubes in mice. 1361-6528, Electronic.

[115] Schipper, Ml., Fau-Nakayama-Ratchford, N., et al. A pilot toxicology study of single-walled carbon nanotubes in a small sample of mice. 1748-3395, Electronic.

[116] Liu, Z., Fau-Davis, C., et al. Circulation and long-term fate of functionalized, biocompatible single-walled carbon nanotubes in mice probed by Raman spectroscopy. 1091-6490, Electronic.

Chapter 14

Carbon Nanotube Transparent Electrode

Jing Sun and Ranran Wang

Additional information is available at the end of the chapter

http://dx.doi.org/10.5772/51783

1. Introduction

In the modern world, transparent conductive films (TCF) are extremely common and critically important in electrical devices. In our homes or offices, they are found in flat panel displays such as in TVs, laptops and in touch panels, of phones, tablet computers, E-readers and digital cameras [1]. Besides, they are also used as the electrodes for photovotaic devices such as solar cells [2] and organic light-emitting diodes (OLEDs) [3]. Liquid crystal display (LCD) is by far the largest user of transparent conductive films but many devices are showing rapid growth in popularity such as touch panels (362 million units in 2010 with annual growth of 20% through 2013), E-paper (30 fold growth expected from 2008 to 2014), and thin film solar cells (expected sales of over $13 billion by 2017) [4].

The dominant transparent conductive material used today is tin doped indium oxide (ITO) with a demand growing at 20% per annum [5]. ITO has been studied and refined for over 70 years, and as a result, the material offers many beneficial properties. However, ITO has certain drawbacks, mainly reflected on the depleted supply of raw material and their brittleness. The supply of indium is constrained by both mining and geo-political issues, which leads to dramatic price fluctuations over the last decades, from $ 100- $ 900, as shown in Figure 1. The high price of indium determined the high cost of ITO, since they compose nearly 75wt % of a typical ITO film [6]. In addition to the raw materials, the expense of setting up and maintaining a sputtering deposition line, as well as the low deposition yield (3-30%) [7] also increases the cost of ITO. Though current devices are typically based on rigid substrates, there is a continued trend toward flexible devices. As ITO tend to fracture at strains of 2%, they are completely unsuitable for using in flexible electronics. Therefore, new transparent electrode materials have rapidly emerged in recent years, including carbon nanotubes (CNTs), graphene and metal nanowires. The intrinsically high conductivity coupled with high aspect ratio yields films with high transmittance, adequately low sheet resistance, and superior mechanical flexibility. These material properties, combined with inexpensive material and deposition costs make these emerging nanomaterials very attractive for as transparent electrodes. Of the three dominant nanoscale materials, CNTs are perhaps the most promising and mature intensively investigated.

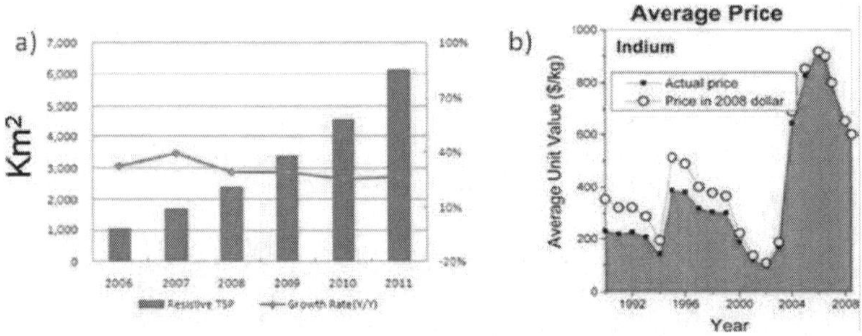

Figure 1. (a) Global demand for resistive style touch panels by area; (b) Average price of Indium over the last several decades; Reprinted with permission from reference [4] copyright 2011 Wiley.

This review will focus on transparent electrode made of CNTs, and six main parts will be covered.

1. At first, some basic theories and parameters for characterizing transparent conductive materials will be presented so that the following parts of the review can be profoundly understood.
2. CNTs prepared from different methods or modified under various conditions have diverse physical and chemical properties, which will yield films with distinct performance. Therefore, in the second part, CNTs of different types will be investigated, and the performance of the as prepared thin films will be compared.
3. One of the major advantages in using CNTs is their ability to be applied to substrates from solution, which opens up many alternative deposition techniques. Therefore, one of the primary research areas for making transparent conductive films is to process the CNT material into printable inks.The third part will outline major approaches to disperse CNTs and focus on the most important details with regards to making transparent conductive films.
4. In the fourth part, a variety of techniques for making transparent conductive CNT films will be presented and evaluated.
5. During the solubilization step, non-conducting dispersants are induced, which sacrifice the conductance of the films a lot. Therefore, post-treatment needs to be done to remove them for enhancing the performance of the films. In the fifth part, various methods used to improve the performance of the transparent conductive films after their preparation will be discussed.

http://dx.doi.org/10.5772/51783

Finally, the latest progress on CNT transparent conductive films and their applications on electrical devices will be summarized.

2. Optoelectronic properties

The two most important features for a transparent conductingmaterial are its sheet resistance (Rs) and optical transparency. The sheet resistance is defined as Rs = R(W/L), where R is DC resistance, W and L are width and length of the film. Grüner et al. [8] developed a suitable merit, the DC conductivity/optical conductivity (σ_{dc}/σ_{op}), to compare the performance of various transparent conductors based on the standard percolation theory, in which each bundle of nanotubes was counted as one conducting stick. They assumed the conductivity ratio σ_{dc}/σ_{op} remains constant for nanotube networks with different densities in the measured optical frequency range. By plotting RsvsT and fitting the data to equation 1, one can estimate the value ofσ_{dc}/σ_{op}. This value is often used as a Figure of Merit for transparent conductors since high values of σ_{dc}/σ_{op} leads to films with high T and low Rs.

$$T = \frac{1}{1 + \frac{2\pi\sigma_{OP}}{cR_s\sigma_{dc}}} \tag{1}$$

Geng et al. [9] found that this equation can be fitted well to the curve of single-walled carbon nanotube TCFs, nevertheless can not be fitted well with carbon nanotubes of other types. They modified the equation as follows:

$$T = t \bullet \left(1 + \frac{188\ (\Omega)}{R_s}\frac{\sigma_{op}}{\sigma_{dc}}\right)^{-2} \tag{2}$$

The parameter t may represent the optical property of CNT films. A high t value gives a high transmittance for the CNT films. The t value of SWCNT films is 0.999, while that of MWCNT is much lower, around 0.884.

Recently, Coleman et al. [5] modified this model to evaluate thinner (more transparent) films. They found that the data tend to deviate severely from the fits for thinner films, as seen from Figure 2. This deviation has been observed before [10-12] and tends to occur for films with T between 50% and 92%.Thus,σ_{dc}/σ_{ac} fails to describe the relationship between T and Rsin the relevant regime. The deviation from bulk-like behavior as described in Equation1, can be explained by percolation effects [13]. Such effects become important for very sparse networks of nano conductors. When the number of nanoconductors per unit isvery low, a continuous conducting path from one side of the sample to the other will generally not exist.

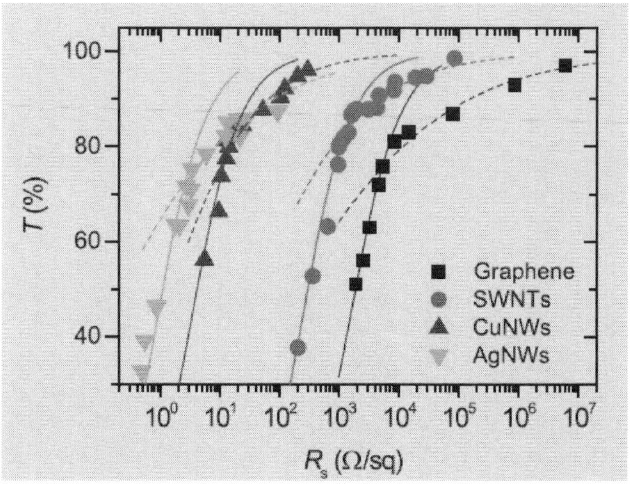

Figure 2. Typical graph of transmittance (generally measured at 550 nm) plotted versus sheet resistance for thin films of nanostructured materials. Reprinted with permission from reference [5] copyright MRS.

As more nanoconductors are added, at some point (the percolation threshold) the first conducting path will be formed. As more material is added, more conductive paths are formed, and the conductivity of the network increases rapidly. Eventually it reached a "bulk-like" value above which it remains constant. Percolation theory describes how the dc conductivity of sparse networks depends on network thickness and predicts a non-linear, power law dependence:

$$\sigma_{dc} \propto (t - t_c)^n \tag{3}$$

where tis the estimated thickness of the network, t_c is the thickness associated with the percolation thres hold, and nis the percolation exponent. This leads to a new relationship between T and Rs, which applies to thin, transparent networks:

$$T = \left[1 + \frac{1}{\Pi}\left(\frac{Z_0}{R_s}\right)^{1/(n+1)}\right]^{-2} \tag{4}$$

where Π is the percolative FoM:

$$\Pi = 2\left[\frac{\sigma_{dc} / \sigma_{op}}{(Z_0 t_{min} \sigma_{op})^n}\right]^{1/(n+1)} \tag{5}$$

Here, t_{min} is the thickness below which the dc conductivity becomes thickness dependent. It scales closely with the nanostructures' smallest demision, tmin ≈ 2.33 D. The high Tportion

http://dx.doi.org/10.5772/51783

of data in Figure 1 was fitted using Equation 4, and good fits allow the calculation of both nand Π. Analysis of these equations shows that large values of Π but low values of nare desirable to achieve low Rs coupled with high T, which are used to evaluate the performance of CNT films with high performance.

In addition to their sheet resistance and optical transparency, the stability and mechanical durability are also critical criteria to evaluate the performance of transparent conductors. Undoped CNT films exhibit excellent stability upon exposure to atmospheric conditions, as seen in Figure 3 [14]. Doping with nitric acid or $SOCl_2$ could decrease the sheet resistance significantly, however at the expense of sacrificing their stability [15-17]. The sheet resistance of undoped SWCNT films decreases slightly with increasing temperature, which is consistent with the electrical behavior of semiconductors. Thermal stability of doped CNTs is dependent on dopants since elevated temperatures may increase chemical reactions or enhance the desorption of dopants out of the films. CNT-PET thin films are significantly more flexible than commercial ITO/PET films. They can be bent all the way to 180° without a significant change in resistance, [18] and the conductivity of the films can be retained after 500 bending cycles [19].

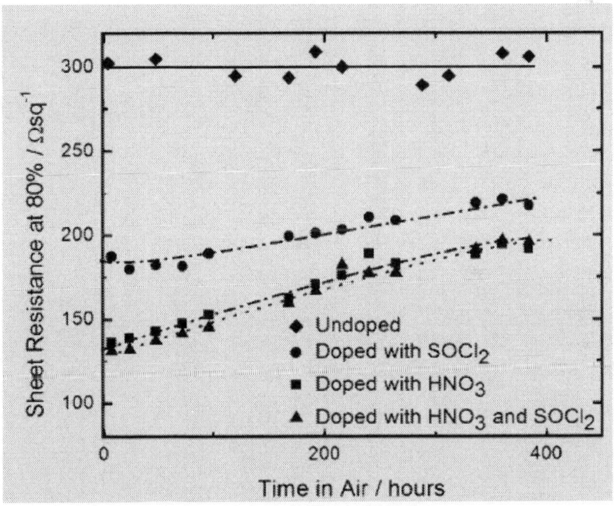

Figure 3. Absolute sheet resistance versus time in air of four SWNT films. Reprinted with permission from reference [14] copyright Wiley.

3. The choice of Carbon Nanotubes

Carbon nanotubes synthesized from different methods or processes have diverse material qualities, such as the degree of purity, the defects, their length and diameters, and the chiralities, which are presumably important factors in determining the film performance.Therefore, the choice of CNTs as well as their further treatment is markedly important. Young Hee Lee group [9,20] did systematical analysis to investigate the CNT quality dependence. In their work, single-walled carbon nanotubes (SWCNT), double-walled carbon nanotubes (DWCNT), thin multiwalled carbon nanotubes (t-MWCNT) and multiwalled carbon nanotubes (MWCNT) powders were separately dispersed in deionized water with sodium dodecyl sulfate (SDS) and dichloroethane (DCE) by sonication and sprayed onto poly (ethylene terephthalate) (PET) substrates to fabricate thin films. The sheet resistance and transmittance of each film was measured and compared. As seen in Figure 4, the film's performance changes dramatically for different types of CNTs dispersed in deionized water with SDS, as well as in DCE. The TCFs fabricated with SWCNTs show the best film performance among all the selected CNTs. The trends of film performances are similar for the TCFs fabricated by using the CNT solution dispersed in deionized water and in DCE, which is SWCNTTCF>DWCNTTCF> t-MWCNTTCF>MWCNTTCF. Furthermore, they analyzed the defects and metallicity by Raman spectra, and found that CNTs with fewer defects and high content of metallic tubes leads to TCFs with higher conductivity. Nevertheless, in Li's report, [21]. MWCNTTCFs exhibit better performance than SWCNTTCFs. They indicated that MWCNT have more conductive π channels than SWCNTs does, therefore MWCNTs have better electronic transportability. In the case of a MWCNT where conduction occurs through the outer most shell, the large diameter of the outernanotube causes the gap to approach 0 eV and the nanotubeto become basically metallic. On the contrary, 2/3 of SWCNTs are semiconducting. The other reason they mentioned is that the MWCNTs they used are longer than SWCNTs, which could decrease the contacts numbers. Another point needs to be addressed is that dimethylformamide (DMF) which was chosen as the solvent in their work is actually not efficient to exfoliate SWCNTs. Therefore, SWCNTs bundled together which would open up an energy gap or pseudo gap owing to intertube interactions. We believe this is a critical reason for the worse performance of SWCNTTCFs in their work.

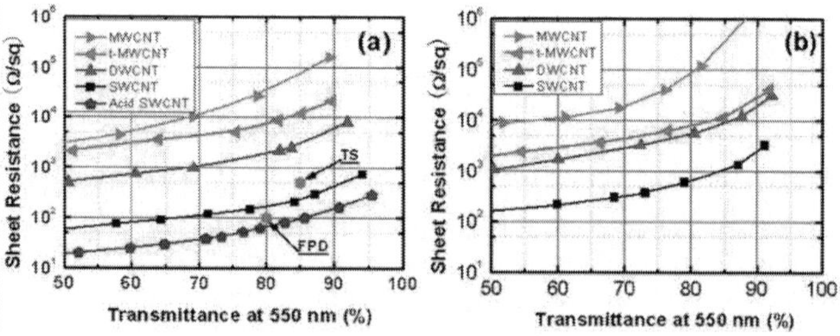

Figure 4. Characteristic sheet resistance-transmittance curves for various CNT-films. Each curve contains several data points from films with different numbers of sprays by a CNT solution dispersed in (a) deionized water with SDS and (b) DCE without dispersant. Reprinted with permission from reference [9].

http://dx.doi.org/10.5772/51783

SWCNTs synthesized by different methods such as arc discharge (Arc), catalytic chemical vapor deposition (CVD), high pressure carbon monoxide (Hipco), and laser ablation (Laser) were also analyzed systematically [20]. After the SWCNT powder was characterized, each of them was dispersed in deionized water with sodiumdodecyl sulfate (SDS) by sonication followed by aspray process to fabricate the SWCNT film onto PET substrates.By analyzing the SWCNT film performance varying with the SWCNT parameters, they found that the metallicity of the SWCNTsextracted from G′-band intensity of Raman spectros copy and the degree of dispersion in the solutionare the most decisive factors in determining the film performance. Figure 5 shows that the film performance changes dramatically with different types of SWCNTs. The TCFs fabricated with Arc SWCNTs result in the best film performance, consistent with previous report [22]. The sheet resistance of the Arc TCF is ~160Ω/sq at a transmittance of 80%, which can be used in a wide range of applications from touch panels to electrodes for future flexible displays.

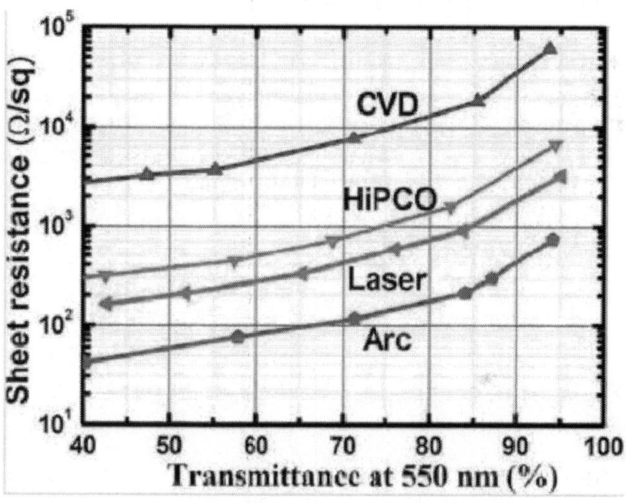

Figure 5. Characteristic curves of sheet resistance-transmittance of TCFs fabricated by various SWCNTs. Reprinted with permission from Ref. [20].

In order to investigate the underlying reason, CNTs were characterized with SEM, TEM, TGA and Rama spectra. TEM analysis showed that the diameter of individual nanotube synthesized with CVD and Hipco process were about 1nm, smaller than those (~1.4 nm) of Laser and Arc SWCNTs. The CVDSWCNTs had the smallest average bundle size, as estimated from the SEM images, where as the Laser sample exhibited the largest average bundle size among samples. Carbonaceous particleson the SWCNT bundles are present in the CVDSWCNTs. The Arc SWCNTs have relatively well-defined crystallinity without amorphous carbonson the tube walls, although the bundle size of the Arc sample is smaller than that of the Laser sample. Figure 6 disclosed that the influence of the purity of the SWCNT is less deterministic, particularlyin CVD and HiPCOSWCNTs, where as the diameter has a strong correlation to the sheet conductance of SWCNT film. The sheet conductance of the film increases consistently with increasing diameters of nanotubes, as shown in Figure 6. This can be attributed to the decreasing band gap with increasing diameters of semi-conducting SWCNTs. Although individual metallic tubes are independent of the diameters, there are usually a pseudogap induced by tube-tube interactions, which is also inversely proportional to thetube diameter. Thus, the conductivityof the metallic nanotubes reveals the similar diameter dependence to semiconducting ones.

The radical breathing modes (RBM) of Raman spectra were used to characterize the metallicity of SWCNTs [20]. At 514 nm, the Laserand Arc SWCNTs reveal the semiconducting behavior exclusively, on the other hand, CVD and HiPCOSWCNTs containboth metallic and semiconducting nanotubes. At 633 nm, the Laser and Arc SWCNTs pick up mostly metallic SWCNTs, where as the CVDSWCNTs retain mostly semiconducting properties (less prominent Fano line) and the HiPCOSWCNTscontain both the metallic and the semiconducting behaviors. Other than RBM mode, the G′-band intensity is strongly correlated with the metallicity of SWCNTs. Despite the abundance of metallicity, the presence of defects on then anotube walls that may act as scattering centersdegrades the conductivity of the SWCNT network [23]. The intensity of the D-band indicates the amount of defects on the nanotube walls. Therefore, anappropriate parameter to express conductivity of nanotubes for SWCNTs is the intensity ratio, G′-band/D-band. High content of metallicity and few defects on the nanotube walls will be desired for high conductivity of the SWCNT films.

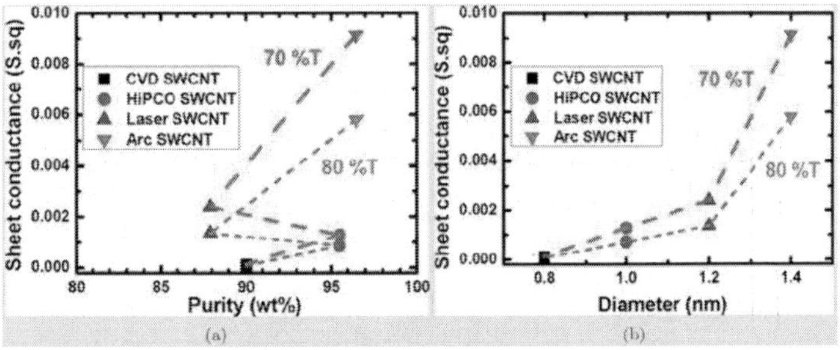

Figure 6. The sheet conductance of TCFs at transmittance of 70% and 80% versus (a) purity and (b) diameter of SWCNT powders. Reprinted with permission from Ref. [20].

The purity affects the conductivity.The diameter contributes to the conductivity via bandgap described in the previous paragraph. More defects reduce the *mean free path* of carriers and decrease the mobility of carriers in nanotubes.The conductivity is proportional to the metallicity of nanotubes and inversely proportional to the number of scattering centers or defects [24-26]. Considering all these factors, a material quality factor Q_m was defined to govern the conductivity of SWCNTs:

http://dx.doi.org/10.5772/51783

$$Q_m = P \times \left(e^{E_{pq}/2k_BT} \times \overline{\sum I_M} \right) + e^{-E_q/2k_BT} \times e^{\left(E_i - E_f\right)/k_BT} \times \overline{\sum I_S} \tag{6}$$

where Eg = 0.82/D (eV), Epg = 0.105/D (eV), D is the average diameter of individual SWCNTs, P is the purity of the sample, $E_i$is the intrinsic Fermi Level, E_f is the Fermi Level for the extrinsic semiconductors, k_B is the Boltzmann constant and T is the temperature of the system. Here I_S (I_M) is defined as

$$I_S\left(I_M\right) = I_{G'/D} \times \frac{A_S\left(A_M\right)}{A_M + A_S} \tag{7}$$

where $A_S(A_M)$ is the areal intensity of semiconducting (metallic) peaks of RBMs from Raman shift. After calculation, it was observed that the sheet conductance reveals a linear relationship with the material quality factor. Although this empirical formula is not rigorous, it can provide atleast a means to estimate material quality that governs the conductivity of the SWCNTTCFs. Forinstance, large diameter, higher purity, less defects (lower intensity of D-band), and more metallic nanotubes (higher intensity of G′-band) will give better conductivity of the SWCNTTC. From this point of view, the Arc TCF is the best sample providing the highest conductivity in comparison toTCFs made by other types of SWCNTs.

In addition to the material parameters discussed above, the length of SWCNTs is also crucial to the TCF performance. According to the percolation theory, a conducting path could be formed at a lower density for longer nanotubes, which means at the same sheet resistance, TCFs prepared with longer nanotubes should exhibit higher optical transparency. This conjecture has been confirmed by experiments [27,28]. In order to optimize the CNTs quality, such as their purity, their dispersibility and the content of metallic tubes, some pretreatments need to be done. Several attempts have been tried to purify the CNT powders.Generally, Gas phase reaction or thermal annealing in air or oxygen atmosphere is used to remove amorphous carbon [29,30]. The key idea with these approaches is a selective oxidative etching processes, based on the fact that the etching rate of amorphous carbons is faster than that of CNTs. Since the edge of the CNTs can be etched away as well as carbonaceous particles during the annealing, itis crucial to have a keen control of annealing temperatures and annealing times to obtain high yield. Liquid-phasereactions in various acids are always conducted to remove the transitionmetal catalysts [31-33]. Hydrochloric acid, nitric acid and sulfuric acid are the most commonly used acid, and the purification effect is dependent on the concentration, the reaction temperature and the reaction time. In addition to their reaction with metal catalysts, nitric acid and sulfuric acid could induce some carboxyl or hydroxyl groups onto the walls of nanotubes, which will improve their dispersibility in water [34,35]. However, some damages were introduced during this process. Therefore, subsequent annealing or ammonium treatment was sometimes carried out to repair the wall structures of the nanotubes to fulfill some special requests.[36]. In order to enhance the content of metallic tubes, discriminated adsorption and separation or ion change chromatography was generally used.

4. CNT Ink Preparation

One of the major advantages in using CNTs overmore conventional metal oxides is their ability to be applied to substrates from solution, which opens up many alternative deposition techniques. Therefore, one of the primary areas of research for making transparent conductive films is finding ways to process the CNT materials into printableinks.The first part of the ink making process is in finding suitable ways to disperse the CNT materialinto solution. Commercial SWCNTs always aggregated into thick bundles due to their high surface energy and strong van der Waals force between tubes. However, the conductivity of the SWCNTTCFs is inversely proportional to the bundle size considering tube-tube junction resistance [37]. Therefore, it is crucial to exfoliate SWCNT thick bundles into thinner or even individual ones.

There are three major approaches to dispersing CNTs:

a. dispersing CNTs in neat organic solvents [38,39];

b. dispersing CNTs in aqueous media with the assistant of dispersing agents such as surfactants and biomolecules [40];

c. introducing functional groups which will help draw the CNTs into solution [41].

Each of these methods have advantages and disadvantages in terms of making processable CNT based inks.

Direct solubilization of CNTs in a suitable solvent is perhaps the simplest and the most favorable method from a manufacturing point of view, since there are no solubilization agents involved which could create processing issues during manufacturing,and also lead to decreased conductivity in the as deposited film. A range of solvents have been tried to exfoliate SWCNTs, and exhibit tremendous differences on the efficiency. The major issue with using these organic solvents has beenthe inability to disperse CNTs at a concentration high enough to be useful for industrial applications (>0.1 g/L). Recently, workby Prof. Coleman's group [42] has shown that the solvent cyclohexylpyrrolidone (CHP) can disperse CNTs up to 3.5 g/L with high levels of individual tubes or small bundles and can keep stable for at least one month. However, the high boiling point of this solvent may be an issuein high speed roll-to-roll manufacturing on plastic. Continuing to search for optimal solvents which can disperse CNTs at high concentrations and have a reasonably low boiling point (150 °Cor below) could lead to a facile manufacturing process for high performance transparent conductive films.

Over the years, significant efforts have been devoted to finding a suitable parameter to guide the selection of good solvents. Three major theories have been proposed, which are non-hydrogen Lewis base theory, [43] polar π system and optimal geometry theory [44] and Hansen parameter [42]. According to non-hydrogen Lewis base theory, all of the solvents can be divided into three groups on the basis of their properties. Class 1consists of the best solvents, *N*-methylpyrrolidone (NMP),*N,N*-dimethylformamide (DMF), hexamethylphosphoramide(HMPA), cyclopentanone, tetramethylenesulfoxide andε-caprolactone (listed in decreasing order of optical densityof the dispersions), which readily disperse SWNTs, forminglight-grey, slightly scattering liquid phases. All ofthese solvents are characterized by high values for electron-pair donicityβ[45], negligible values for H-bond donation parameter α,[46] and high values for solvochromic parameter$\pi*$. Thus, *Lewis basicity* (availability of a free electronpair) without H-donors is key to good solvation of SWNTs.Class 2 contains the good solvents, toluene, 1,2-dimethylbenzene (DMB),CS_2, 1-methylnaphthalene, iodobenzene,CHCl3, bromobenzene and 1,2-DCB. They show $\alpha \approx \beta \approx 0$ and high valueof $\pi*$. Class 3 entails the badsolvents, *n*-hexane, ethylisothiocyanate, acrylonitrile, dimethyl sulfoxide (DMSO),water and 4-chloroanisole. Badsolvents would have $\alpha = \beta = \pi* \approx 0$. However, the high electron-pair donicity alone has proven tobe insufficient, as dimethyl sulfoxide (DMSO) is not an effectivesolvent for SWNTs even though it contains three lone pairs [47]. A systematic study of the efficiency of a series of amide solvents to disperse as-produced and purified laser-generated SWNTssuggested that the favorable interaction between SWNTs andalkyl amide solvents is attributable to the highly polar π systemand optimal geometries (appropriate bond lengths and bondangles) of the solvent structures [48]. However, this conclusion is some what undermined by the poor solubility of SWNTs intoluene [47]. Recently, Coleman et al found that the dispersibility of SWCNTs was intimately related with the Hansen parameters of the solvents and it is more sensitive to the dispersive Hansen parameter than thepolar or H-bonding Hansen parameter. The dispersion, polar, and hydrogen bonding Hansenparameter for the nanotubes is estimated to be$\langle\delta_D\rangle = 17.8$ MPa$^{1/2}$,$\langle\delta_P\rangle = 7.5$ MPa$^{1/2}$, and$\langle\delta_H\rangle = 7.6$ MPa$^{1/2}$. Success ful solvents exist in only a small volume of Hansen space, which is $17 < \delta_D < 19$ MPa$^{1/2}$, $5 < \delta_P < 14$ MPa$^{1/2}$, $3 < \delta_H < 11$ MPa$^{1/2}$. Hansen parameters have been used successfully to aid solvent discovery. Unfortunately they are not perfect. A number of non-solvents exist in the region of Hansen parameter space close to the solubility parameters of nanotubes.

Compared with organic solvent, it is more efficient to exfoliate SWCNTs into thin bundles or even individual tubes with the assistant of dispersants. The most common dispersants used in TCFs are anionic surfactants including sodium dodecyl sulphate (SDS) and sodium dodecylbenzenesulphonate (SDBS). They are preferable dispersants because nanotubes can be highly exfoliated by them at rather high concentrations [49]. Besides, they nearly have no absorption over the visible spectrum region. However, they are not without disadvantage. Large amount of them is needed to exfoliate nanotubes into thin bundles; usually the CMC (critical micelle concentration) value should be reached [50]. Their residue will increase the sheet resistance of nanotube films significantly since they are nonconductive. In recent years, a lot of research has been done on the dispersion of CNTs with biomolecules such as DNA and RNA [51-54]. There are a number of advantages using them as dispersants.First, they can coat, separate, and solubilize CNTs more effectively with their phosphate backbones interacting with water and many bases binding to CNTs [55]. DNA wrapped around

CNTs helically and there were strong π-π interactions between them [56]. Charges were transferred from the bases of DNA to CNTs leading to the change of their electron structures and electrical property [57]. 1 mgDNA could disperse an equal amount of as-producedHiP-COCNT in 1 ml water, yielding 0.2 to 0.4 mg/ml CNT solution after removal of non-soluble material by centrifugation. Such a CNTsolution could be further concentrated by ten-fold to give a concentration as high as 4 mg/ml [52]. Jeynes's research disclosed that total cellular RNA showed better dispersion ability than dT(30) which was the most effective oligonucleotide dispersants in previous reports [54]. Second, the amount of DNA needed to exfoliate CNTs into thin bundles was much less than common surfactants such as SDS. In Zheng's work, the weight ratio between SWCNTs and DNA was 1:1 [52] while the dosage of RNA in Jeynes's work was lower, only half amount of the nanotubes [54]. By contrast, ten fold of SDS was needed to exfoliate SWCNTs efficiently [11,58]. High dosage of dispersant is not preferred since they are nonconductive and their residue will decrease the conductivity of the films significantly. Third, they have little absorption over the visible range and will not decrease the transmittance of CNT films. Last but not least, as biomolecules, they are easily degraded and removed by acid, base or appropriate enzyme. Jeynes et al [54] have used RNA to disperse CNTs and digested them by RNase effectively.

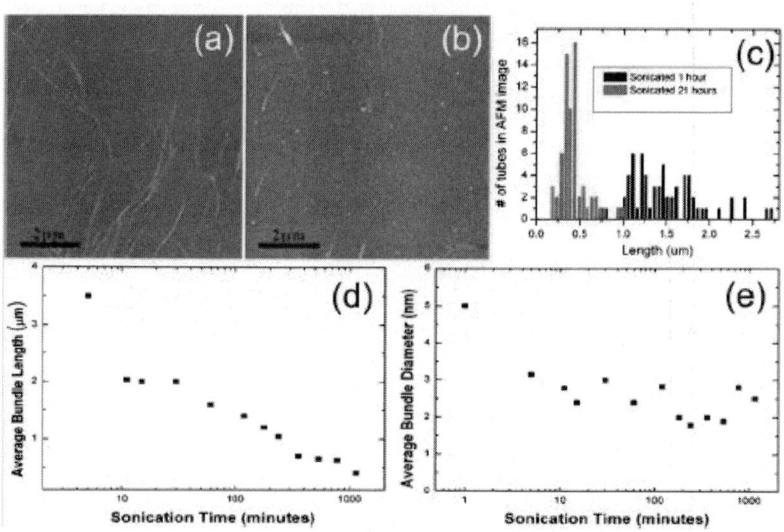

Figure 7. Effects of sonication on SWNT bundle length anddiameter. (a) and (b) AFMimage of SWNTs absorbed on a silicon waferafter (a) 1 h and (b) 21 h of sonication time. (c) Histogram of bundle length distribution taken from several AFM images for 1 h (black) and 21 h (red) of sonication. Plot of the (d) average bundle diameter and (e) average bundlelength for various sonication times measured from AFM images. Reprinted from Ref. [37] copyright AIP.

The final solubilization approach involves functionalizing CNT walls with covalently bonded molecules. The most commonly used process is introducing carboxyl groups by reacting with concentrated acid, such as nitric acid and sulfuric acid [59]. Although thismethod has been proven to lead to CNT solutions with high concentrations of thin bundles, the films made from these tubes tend to have extremely low conductivity values, as the functionalization procedure inducesdefects into the pristine CNTsp^2 bond structure.

For all solubilization approaches, energy must be imparted to the system to break the strong van der Wall force between tubes. This is commonly done by mixing techniques such as high-shear mixing, rotor-stator, three-roll milling, ball milling, homogenizers, and ultrasonication. Among these, ultrasonication is the most commonly used and the most efficient technique to prepare SWCNT water solution. The vibration of the sonicationtip in the solution causes pressure waves which expand and collapse dissolved gas in the liquid; the collapse of these bubbles causes temperature of local zones exceeding 10 000 °C, [60] which can impart enough energy to separate CNTs from each other, long enough for surfactants to surround the tubes and prevent them from aggregating. However, such high energy of sonication would introduce defects onto the walls of CNTs or even shorten them [37]. As seen from Figure 7, the diameter of the bundles decreases sharply from 5 to 3 nm in the first 5 min of sonication, and then remains 2-3 nm after that. However, the length of the tubes decreases exponentially with sonication time from 4μm initially, to 0.4μm after about 21 h of sonication. Therefore, suitable sonication powder and time needs to be chosen to make SWCNT inks with thin bundles and long length.

Figure 8. Freestanding SACNT film drawn out from a230-mm-high SACNT array on an 8-inch silicon wafer. The film in the visualfield is about 18cm wide and 30cm long. b) SEM image of the SACNT array on the silicon wafer in side view. c) SEM image of an SACNT film intop view. Reprinted with permission from Ref. [63] copyright Wiley

5. Film Fabrication

Many techniques have been developed to prepare CNT thin films, including both dry and solution-based methods. Although solution-based techniques are the mostly commonly used and industry preferred, dry method is negligible for preparing high performance TCFs. Direct growth of CNT films is one of the typical dry method. CVD can grow CNT films either randomly distributed or aligned by controlling the gas flow, catalyst patterns, or by using a substrate with a defined lattice structure [61]. Compared with a solution-based process, the direct growth method leads to films with individually separated tubes with fewer defects and better CNT-CNT contact, which leads to highly conductive films [62]. However, films directly grown on a substrate may have significant amounts of residual catalyst, imprecise density control, and substrate incompatibility for device integration. Furthermore,CVD is a high vacuum, high temperature process and is not compatible with substrates used in the emerging plastic electronics field.

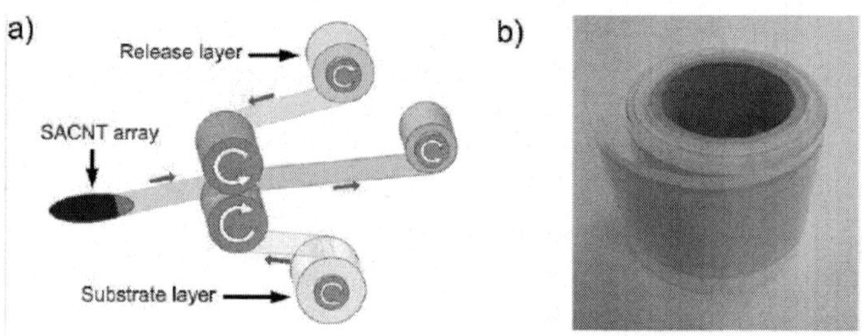

Figure 9. Production and performance of SACNTTCFs. a) Illustration of the roll-to-roll setup for producing composite TCFs. b) A reel of SACNT/PE composite TCF produced by the roll-to-rollsetup. The grey central region of the reel is the SACNT/PE composite TCF. Reprinted with permission from Ref. [63] copyright Wiley.

In 2002, a method was pioneered by Dr Fan's group [63] and involves drawing out MWCNT films directly fromas-grown super aligned CNT (SACNT) arrays. An example of such process and films are shown in Figure 8. An SACNT array is a special kind of vertically aligned MWCNT array having a higher surface density and better alignment of MWCNTs than an ordinary one.Typically, an SACNT array with an area of 0.01 m^2 can be totally converted to a SACNT film of ~6–10 m^2, depending on the height of the SACNT array. Unlike the solution-based process, an entire SACNT array can be converted to films without any significant loss by the drawing process, which will lower the cost. Another crucial advantage of this solution-free process is that it can be straight forwardly incorporated into a roll-to-roll process to make SACNT/polymer-sheet composite films. In a roll-to-roll process as shown in Figure 9a, aSACNT film is drawn out, then sandwiched by a release layer and a substrate layer, and pressed by two close rollers tightly, forming an SACNT/substrate composite film. The release layer, suchas a slick paper, protects the SACNT film from sticking to the roller, and can be peeled off when using the film.Figure 9b shows a reel of SACNT/polyethylene (PE) compositefilmthat is produced from anentire wafer of SACNT array. The width of the film in this reel is about 8 cm, and the length can be over 60 m. In principle, by periodically inserting a new SACNT source wafer, the composite film can be produced continuously by the roll-to-roll process. Unfortunately, the performance of such as-drawn films is far below our expectation. In order to improve their performance, the SACNTarrays were trimmed by the oxygen plasma to reduce their height, since lower arrays give rise to films without large bundles. Besides, the SACNT films were trimmed by lasers to burn the outmost CNTs of the bundles and to make the bundles thinner. After treatment, films with excellent performance (24 Ω/sq @ 83.4%, 208 Ω/sq @ 90%) were obtained, and successfully used as touch panels.

Compared with dry method, solution-based method is much easier to prepare CNT films with high reproducibility. Perhaps the simplest way to make CNT films is by filtering the solution of dispersed tubes over a porous filter membrane. Filtration leads to highly uniform and reproducible films, and has precisely control over density [64]. Therefore, this method is often used to evaluate CNT materials and dispersion quality. Deposition method does not have the issues on the wetting on various substrates and it works well with extremely dilute CNT solutions. Another merit deserve to be addressed is that some excess dispersants could be washed away during the filtering process, which could enhance the conductance of the films. To our experience, films prepared with filtration method always show higher conductance than films prepared with spray coating or rod-coating method, since all of the dispersants resided in the films in the later methods. Since the films are deposited onto filters, a transfer from filters to other substrates is generally needed. Accordingly, transfer methods such as PDMS method [65]. Laser transfer method and microwave assisted method were developed [66]. The limitation of this method is that the size of the films is constrained by the filter, and is difficult to scale up. It is likely that this method will continue to be restricted to academic research.

In addition to vacuum assisted filtration, there are other deposition techniques that are useful for small scalelab testing. These include spray coating, [11] spin coating, [67] dipcoating, [68] and draw-downs using a Mayer rod or Slot Die [69]. Spray coating is a simple and quick method to deposit CNT films. Typically, CNT ink is sprayed onto a heated substrate. The substrate is heated to facilitate the drying of the liquid. The set temperature for the substrate is adjusted by the choice ofsolvent. By using diluted solution and multiple spray coating steps, homogeneous films can be obtained. Bundling mayhappen during the drying process after the sprayed mist of CNT has hit the PET substrate. Thus, it is difficult to get good film uniformity. The most widespread deposition method involves depositing solution on a substrate by Mayer Rod or Slot Die, followed by controlled drying. Aheating bar is used to control the drying process.This technique can be used to coat directly onto polyethylene terephthalate (PET), glass, and other substrates at room temperature and in a scalable way. Inkjet printing is an old and popular technology due to its ability to print fine and easily controllable patterns, noncontact injection, solution saving, and high repeatability [62]. It is very prevalent inprinted electronics. In a typical ink jet printing process, the droplet size is around~10 pL and, on the substrate, has a diameter of around 20-50 μm. Printing on paper is much easier than printing on a plastic or glass substrate, due to the high liquid absorption of the paper, which avoids the dewetting of the liquid on substrates. The liquid droplet and substrate interactionis crucial for uniform drying of the liquid. The most useful deposition technique is roll to roll coating of CNT inks onto continuous rolls of plastics. This technique can coat film up to 2 m wide at speeds up to 500 m/min.One such roll-to-roll coating line running continuously would have the equivalent output of 30 traditional sputter coaters, and could produce enough film to satisfy half of the available touch panel market. Examples of various film fabrication methods were shown in Figure 10.

Figure 10. a) Transparent CNT film pulled from vertically grown CNT forest; b) CNT film transferred to PET using PDMS stamp. c) CNT film spray coated onto large areaplastic; d) Mayer rodcoating schematic. e) Image of CNT film being coated by slot die f) Roll of printed CNTfilm. g) Inkjet printed CNT lines. Reprinted with permissions from Ref. [4] copyright Wiley

6. Post-Treatment of CNT Films

During the preparation of CNT water solutions, dispersants are always introduced to assistant the exfoliation of CNT bundles. Since these dispersants are insulating, their residue decrease the conductance of CNT films significantly. Hence, post-treatments to remove these dispersants are necessary for preparing TCFs with high performance. In addition to remove the dispersants, doping is the other goal of post-treatment. In addition to rinsing with water, acid treatment is the most commonly used method to post-treat CNT films. As reported by Geng, [11] the sheet resistance of CNT films reduced by a factor of 2.5 times after treatment in concentrated nitric acid owing to the removal of surfactants SDS. Except their function on removing dispersants, concentrated nitric acid is often used to p-dope CNTs and enhances their conductivity [70]. Although nitric acid was effective to remove dispersants, they induced p-doping of CNTs, which will lead to instability of the films [71]. Besides, PET substrates will turn brittle after long time acid treatment. To solve this problem, Dr Sun's group developed a novel technique combing base treatment and short time acid treatment [72]. In their work, biomolecule RNA was chosen was the dispersant since they are easily degraded by base, acid and RNase. After depositing CNT films onto a PET substrate, they were immersed in the 5 wt % NaOH solution for one hour, and then treated with nitric acid for 10 min. The sheet resistance decreased significantly after treatment with NaOH solution owing to the removal of RNA molecules. After treatment with nitric acid, the RNA molecules were removed further and SWCNTs were slightly doped, therefore, the sheet resistance was reduced further. Base treatment combining short time acid treatment could remove RNA molecules efficiently as well as retaining the flexibility of PET substrates and the stability of the films.

7. Application of CNTTCFs

CNTTCFs have found a range of applications, among which we focus on the touch screens, plat panel displays, solar cells and OLEDs.

Touch screen is almost omnipresent in our daily life, such as in cell phones, tablet computers and many other electronics. Transparent electrodes are an essential component in most types of touch screens. High optical transmittance ($>$ 85%) and low sheet resistance Rs ($<$ 500 Ω/sq) are normally needed for touch screens. Meanwhile, extremely excellent durability, flexibility, and mechanical robustness are required given that the touch screen may be under indentation for millions of times. The mechanical robustness demonstrated by CNT touch panels give promises for increasing the lifetime and durability of current touch screens. There are a variety of touchscreen technologies that sense touch in different ways.Figure 11a shows the basic device structure and the transparent conductor arrangement for a 4-wire analog resistive touchpanel. These panels use two continuous electrodes separated by hemispheres of polymeric "spacer dots" that are10–100 μm in radius and 1–2.5 mm apart. Only at the edges (where electrode attachment occurs) is the transparent electrode patterned. Surface capacitive devices share the same type of continuous conductor whereas the projected capacitive deviceuses transparent conductors with specific patterning into predefined geometries. Resistive touch panels function by current driven measurements andcapacitive devices depend on capacitive coupling with the input device. Both panel types utilize signal processing controllers todetermine X-Y and sometimes Z position of inputs.

The mechanical durability of the transparent conductors is very important for resistive touch panels, since it involves compressive, sheer, and tensile stress every time it works. Their working process can be summarized as [4]:

1. Deformation of the touchside electrode–compressive, tensile
2. Contact of the touch sideand device side–compressive, shear
3. Contact of touch sideelectrode with spacer dots–compressive, shear
4. Extreme deformationof touch side electrode near edge seal–high tensile.

Compressive stress is not required to activatethe projected capacitive (ProCap) touch panels (of which theiPhone is a prime example). The ProCap touch panels are activated by a capacitive coupling with a suitable input device. Thus, there willnot be the mechanical flexing issues in ProCap devices. Still, the mechanical properties of the conducting layer are important since the conductors may be patterned to a size assmall as 10 μm in width. Metal oxides patterned to such small dimension become susceptible to cracking, fractures,and thermal cycling stress.

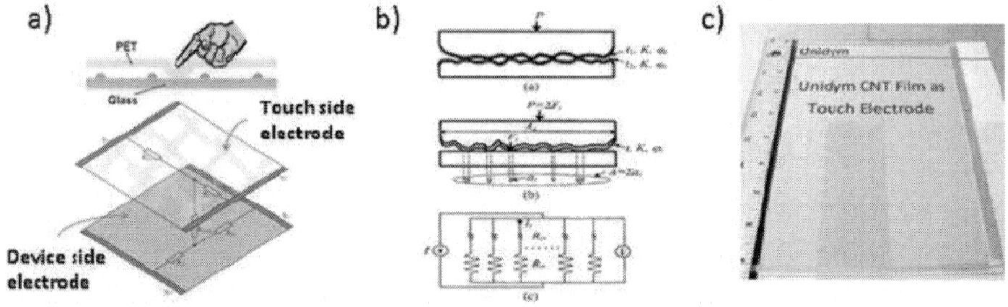

Figure 11. a) Schematic of four-wire resistive touch panel operation and functional layers; b) Schematic of the contact resistance experienced at the interface between two rough conductive layers separated by a very thin dielectric; c) Photograph of touch panel utilizing CNT film as touch electrode. Reprinted with permission form Ref. [4] copyright Wiley

Display panels are produced at nearly 1.7 billion unitsannually (1.2 billion mobile phones, 200 million televisions,150 million laptops, and 200 million desktop, machine interfaces, monitors etc. There are four common types of displays, which are electrowetting displays (EWD), electrochromic displays (ECD), electrophoretic displays (EPD) and liquid crystal displays (LCD). Currently, LCD devicesare manufactured in the greatest number and will be the mainsubject of this section. A transparent conductor'smajor role in LCD/EPD devices is to serve as pixel and common electrodes. An interesting advantage of using CNT films for LCD is the ability to use them possibly as both the transparent electrode and the alignment layer [73]. Recently, Lee et al demonstrated high performance TN-LC cells with ultra-thin and solution-processible SWNT/PS-*b*-PPP nanocomposite alignmentlayers. At an optimized SWNT density, a nanocompositegave rise to low power operation with a super-fast LC responsetime of 3.8 ms, which is more than four times faster than thaton a commercial polyimide layer due to the locally enhancedelectric field around individually networked SWNTs. Furthermore,TN-LC cells with their SWNT nanocomposite layers exhibited high thermal stability up to 200 °C without capacitance hysteresis.

Transparent electrodes are the essential components forphotovoltaic devices. The traditional electrodes for photovoltaic devices is ITO, which has high transmittance and low sheet resistance (~10-20 Ω/aq with the transmittance of 90%). However, their application was constrained by the high price of indium. Besides, the brittleness of ITO limited their usage in flexible devices, which will be a developing trend in the future. Therefore, replacing materials need to be developed. Carbon nanotubes are promising candidates since they have extremely high conductivity, high work function of 4.7-5.2 eV, relatively low cost and excellent flexibility. Besides, they are easy to be deposited into film via solution based process. Glatkowskiet al. [74] reported on the application of transparent CNT electrodesand found a PE-

DOT:PSS coating dramatically improves the device efficiency from 0.47% to 1.5%. The thin layer of PEDOT:PSS can smooth the CNT surface and enhance the charge transfer according to their investigation. In Hu's work, [75] flexible transparent electrodes were fabricated by printing SWCNT solutions on plastic substrates. The SWCNT films have a sheet resistance of 200 Ω/sq with a transmittance of 85%. The achieved efficiency of 2.5% (AM1.5G) approaches that of the controldevice made with ITO/glass (3%). Furthermore, the flexibility is far superior to devices using ITO coated on the same flexible substrate material. However, there are several aspects that need to be solved for CNT based electrodes.

1. Long termelectrical stability;

2. Occasional shorting betweenthe cathode and anode due to protruding CNTs;

3. Relatively high sheet resistance.

Light emitting diodes have an opposite light electricity coupling process as solar cells. Applications of nanoscale materials based transparent electrodes are mainly focused on organic light emitting diodes which hold great promise for the future electronics. In Aguirre's work, carbon nanotube anodes were implemented in small molecule OLED devices and achieved performance comparable to ITO-based anodes [76]. Recently, Feng et al [77] proposed a single walled carbon nanotubes-based anodes for organic light-emitting diodes (OLEDs) by spray-coating process without any use of surfactant or acid treatment. A layer of DMSO doped PEDOT:PSS was spray-coated on the SWCNT sheets to not only lessen the surface roughness to an acceptable level, but also improve the conductivity by more than three orders of magnitude. For the produced SWCNT-based OLEDs, a maximum luminance 4224 cd/m^2 and current efficiency 3.12 cd/A were achieved, which is close to the efficiency of ITO-based OLEDs.

Author details

Jing Sun* and Ranran Wang

*Address all correspondence to: Jingsun@mail.sic.ac.cn

State Key Lab of High Performance Ceramics and Superfine Microstructure, Shanghai Institute of Ceramics, Chinese Academy of Sciences, China

References

[1] Niu, C. M. (2011). *MRS Bull.*, 36, 766.

[2] Tyler, T. P., Brock, R. E., Karmel, H. J., Marks, T. J., & Hersam, M. C. (2011). *Adv. Energy Mater.*, 1, 785.

[3] Kauffman, D. R., Sorescu, D. C., Schofield, D. P., Allen, B. L., Jordan, K. D., & Star, A. (2010). *Nano Lett.*, 10, 958.

[4] Hecht, D. S., Hu, L. B., & Irvin, G. (2011). *Adv. Mater*, 23, 1482.

[5] De , S., & Coleman, J. N. (2011). *MRS Bull*, 36, 774.

[6] Tolcin, A. (2009). *Minerals Yearbook*, 35.

[7] Green, M. A. (2009). *Progress in Photovoltaics : Research and Apllications*, 17, 347.

[8] Hu, L., Hecht, D. S., & Gruner, G. (2004). *Nano Lett.*, 4, 2513.

[9] Geng, H. Z., Lee, D. S., Kim, K. K., Kim, S. J., Bae, J. J., & Lee, Y. H. (2008). *Journal of the Korean Physical Society*, 53, 979.

[10] De , S., Lyons, P. E., Sorel, S., Doherty, E. M., King, P. J., Blau, W. J., Nirmalraj, P. N., Boland, J. J., Scardaci, V., Joimel, J., & Coleman, J. N. (2009). *Acs Nano*, 3, 714.

[11] Geng, H. Z., Kim, K. K., So, K. P., Lee, Y. S., Chang, Y., & Lee, Y. H. (2007). *J. Am. Chem. Soc.*, 129, 7758.

[12] Scardaci, V., Coull, R., & Coleman, J. N. (2010). *Appl. Phys. Lett.*, 97.

[13] De , S., King, P. J., Lyons, P. E., Khan, U., & Coleman, J. N. (2010). *Acs Nano*, 4, 7064.

[14] Jackson, R., Domercq, B., Jain, R., Kippelen, B., & Graham, S. (2008). *Adv. Funct. Mater.*, 18, 2548.

[15] Zheng, Q. B., Gudarzi, M. M., Wang, S. J., Geng, Y., Li, Z. G., & Kim, J. K. (2011). *Carbon*, 49, 2905.

[16] Kim, K. K., Reina, A., Shi, Y. M., Park, H., Li, L. J., Lee, Y. H., & Kong, J. (2011). *Nanotechnology*, 21.

[17] Tantang, H, Ong, J. Y., Loh, C. L, Dong, X. C, Chen, P, Chen, Y, Hu, X, Tan, L P, & Li, L. J. (2009). *Carbon*, 47, 1867.

[18] Saran, N., Parikh, K., Suh, D. S., Munoz, E., Kolla, H., & Manohar, S. K. (2004). *J. Am. Chem. Soc.*, 126, 4462.

[19] Ko, W. Y, Su, J. W, Guo, C. H, & Lin, K. J. (2012). *Carbon*, 50, 2244.

[20] Geng, H. Z., Kim, K. K., Lee, K., Kim, G. Y., Choi, H. K., Lee, D. S., An, K. H., Lee, Y. H., Chang, Y., Lee, Y. S., Kim, B., & Lee, Y. J. (2007). *Nano*, 2, 157.

[21] Li, Z. R., Kandel, H. R., Dervishi, E., Saini, V., Biris, A. S., Biris, A. R., & Lupu, D. (2007). *Appl. Phys. Lett.*, 91.

[22] Zhang, D. H., Ryu, K, Liu, X. L, Polikarpov, E, Ly, J, Tompson, M. E, & Zhou, C. W. (2006). Nano Lett. , 6, 1880.

[23] Lyons, P. E., De , S., Blighe, F., Nicolosi, V., Pereira, L. F. C., Ferreira, M. S., & Coleman, J. N. (2008). *J Phys.*, 104.

[24] Maeda, Y., Hashimoto, M., Kaneko, S., Kanda, M., Hasegawa, T., Tsuchiya, T., Akasaka, T., Naitoh, Y., Shimizu, T., Tokumoto, H., & Nagase, S. (2008). *J. Mater. Chem.*, 18, 4189.

[25] Wang, W., Fernando, K. A. S., Lin, Y., Meziani, M. J., Veca, L. M., Cao, L., Zhang, P., Kimani, M. M., & Sun, Y. P. J. (2008). *Am. Chem. Soc.*, 130, 1415.

[26] Jackson, R. K., Munro, A., Nebesny, K., Armstrong, N., & Graham, S. (2010). *Acs Nano*, 4, 1377.

[27] Sorel, S., Lyons, P. E., De , S., Dickerson, J. C., & Coleman, J. N. (2012). *Nanotechnology*, 23.

[28] Park, J. G, Cheng, Q. F, Lu, J, Bao, J. W, Li, S, Tian, Y, Liang, Z. Y, Zhang, C, & Wang, B. (2012). *Carbon*, 50, 2083.

[29] Ebbesen, T. W., Ajayan, P. M., Hiura, H., & Tanigaki, K. (1994). *Nature*, 367, 519.

[30] Zimmerman, J. L., Bradley, R. K., Huffman, C. B., Hauge, R. H., & Margrave, J. L. (2000). *Chem. Mater.*, 12, 1361.

[31] Zhang, J., Gao, L., Sun, J., Liu, Y. Q., Wang, Y., Wang, J. P., Kajiura, H., Li, Y. M., & Noda, K. (2008). *J. Phys. Chem. C*, 112, 16370.

[32] Liu, Y. Q, Gao, L, Sun, J, Zheng, S, Jiang, L. Q, Wang, Y, Kajiura, H, Li, Y. M, & Noda, K. (2007). *Carbon*, 45, 1972.

[33] Wang, Y., Gao, L., Sun, J., Liu, Y. Q., Zheng, S., Kajiura, H., Li, Y. M., & Noda, K. (2006). *Chem. Phys. Lett*, 432, 205.

[34] Wang, R. R., Sun, J., Gao, L. A., & Zhang, J. (2010). *J. Mater. Chem.*, 20, 6903.

[35] Moon, J. M., An, K. H., Lee, Y. H., Park, Y. S., Bae, D. J., & Park, G. S. (2001). *J. Phys. Chem. B*, 105, 5677.

[36] Datsyuk, V., Kalyva, M., Papagelis, K., Parthenios, J., Tasis, D., Siokou, A., Kallitsis, I., & Galiotis, C. (2008). *Carbon*, 46, 833.

[37] Hecht, D., Hu, L. B., & Gruner, G. (2006). *Appl. Phys. Lett.*, 89.

[38] Bahr, J. L., Mickelson, E. T., Bronikowski, M. J., Smalley, R. E., & Tour, J. M. (2001). *Chem. Commun*, 193.

[39] Cheng, Q. H., Debnath, S., Gregan, E., & Byrne, H. J. (2010). *J. Phys. Chem. C*, 114, 8821.

[40] Moore, V. C., Strano, M. S., Haroz, E. H., Hauge, R. H., Smalley, R. E., Schmidt, J., & Talmon, Y. (2003). *Nano Lett.*, 3, 1379.

[41] Chen, J., Rao, A. M., Lyuksyutov, S., Itkis, M. E., Hamon, M. A., Hu, H., Cohn, R. W., Eklund, P. C., Colbert, D. T., Smalley, R. E., & Haddon, R. C. (2001). *J. Phys. Chem. B*, 105, 2525.

[42] Bergin, S. D, Sun, Z. Y, Rickard, D, Streich, P. V, Hamilton, J. P, & Coleman, J. N. (2009). *Acs Nano*, 3, 2340.

[43] Torrens, F. (2005). *Nanotechnology*, 16, S181.

[44] Landi, B. J., Ruf, H. J., Evans, C. M., Cress, C. D., & Raffaelle, R. P. (2005). *J. Phys. Chem. B*, 109, 9952.

[45] Kamlet, M. J., & Taft, R. W. (1976). *J. Am. Chem. Soc.*, 98, 337.

[46] Taft, R. W., & Kamlet, M. J. (1976). *J. Am. Chem. Soc.*, 98, 2886.

[47] Giordani, S., Bergin, S. D., Nicolosi, V., Lebedkin, S., Kappes, M. M., Blau, W. J., & Coleman, J. N. (2006). *J. Phys. Chem. B*, 110, 15708.

[48] Landi, B. J., Ruf, H. J., Worman, J. J., & Raffaelle, R. P. (2004). *J. Phys. Chem. B*, 108, 17089.

[49] Islam, M. F., Rojas, E., Bergey, D. M., Johnson, A. T., & Yodh, A. G. (2003). *Nano Lett.*, 3, 269.

[50] Ishibashi, A., & Nakashima, N. (2006). *Chem.-Eur. J.*, 12, 7595.

[51] Zheng, M., Jagota, A., Strano, M. S., Santos, A. P., Barone, P., Chou, S. G., Diner, B. A., Dresselhaus, M. S., Mc Lean, R. S., Onoa, G. B., Samsonidze, G. G., Semke, E. D., Usrey, M., & Walls, D. J. (2003). *Science*, 302, 1545.

[52] Zheng, M., Jagota, A., Semke, E. D., Diner, B. A., Mc Lean, R. S., Lustig, S. R., Richardson, R. E., & Tassi, N. G. (2003). *Nature Materials*, 2, 338.

[53] Cathcart, H., Quinn, S., Nicolosi, V., Kelly, J. M., Blau, W. J., & Coleman, J. N. (2007). *Journal of Physical Chemistry C*, 111, 66.

[54] Jeynes, J. C. G., Mendoza, E., Chow, D. C. S., Watts, P. C. R., Mc Fadden, J., & Silva, S. R. P. (2006). *Adv. Mater.*, 18, 1598.

[55] Wang, H., Lewis, J. P., & Sankey, O. F. (2004). *Phys. Rev. Lett.*, 93.

[56] Wang, H. M., & Ceulemans, A. (2009). Phys. Rev. B, , 79.

[57] Gowtham, S., Scheicher, R. H., Pandey, R., Karna, S. P., & Ahuja, R. (2008). Nanotechnology, , 19.

[58] Paul, S, & Kim, D. W. (2009). *Carbon*, 47, 2436.

[59] Marques, R. R. N., Machado, B. F., Faria, J. L., & Silva, A. M. T. (2010). *Carbon*, 48, 1515.

[60] Hopkins, S. D., Putterman, S. J., Kappus, B. A., Suslick, K. S., & Camara, C. G. (2005). *Phys. Rev. Lett.*, 95.

[61] Liu, P., Sun, Q., Zhu, F., Liu, K., Jiang, K., Liu, L., Li, Q., & Fan, S. (2008). *Nano Lett.*, 8, 647.

[62] Hu, L. B., Hecht, D. S., & Gruner, G. (2010). *Chem. Rev.*, 110, 5790.

[63] Feng, C., Liu, K., Wu, J. S., Liu, L., Cheng, J. S., Zhang, Y. Y., Sun, Y. H., Li, Q. Q., Fan, S. S., & Jiang, K. L. (2010). *Adv. Funct. Mater.*, 20, 885.

[64] Wu, Z. C., Chen, Z. H., Du, X., Logan, J. M., Sippel, J., Nikolou, M., Kamaras, K., Reynolds, J. R., Tanner, D. B., Hebard, A. F., & Rinzler, A. G. (2004). *Science*, 305, 1273.

[65] Cao, Q., Hur, S. H., Zhu, Z. T., Sun, Y. G., Wang, C. J., Meitl, M. A., Shim, M., & Rogers, J. A. (2006). *Adv. Mater.*, 18, 304.

[66] Allen, A. C., Sunden, E., Cannon, A., Graham, S., & King, W. (2006). *Appl. Phys. Lett*, 88.

[67] Manivannan, S., Ryu, J. H., Lim, H. E., Nakamoto, M., Jang, J., & Park, K. C. (2010). *Journal of Materials Science-Materials in Electronics*, 21, 72.

[68] Song, Y. I., Yang, C. M., Kim, D. Y., Kanoh, H., & Kaneko, K. (2008). *J. Colloid Interface Sci.*, 318, 365.

[69] Tenent, R. C., Barnes, T. M., Bergeson, J. D., Ferguson, A. J., To, B., Gedvilas, L. M., Heben, M. J., & Blackburn, J. L. (2009). Adv. Mater., , 21, 3210.

[70] Zhao, Y. L., & Li, W. Z. (2010). *Microelectron. Eng.*, 87, 576.

[71] Graupner, R., Abraham, J., Vencelova, A., Seyller, T., Hennrich, F., Kappes, M. M., Hirsch, A., & Ley, L. (2003). *PCCP*, 5, 5472.

[72] Wang, R. R., Sun, J., Gao, L. A., & Zhang, J. (2010). *Acs Nano*, 4, 4890.

[73] Fu, W. Q, Liu, L, Jiang, K. L, Li, Q. Q, & Fan, S. S. (2010). *Carbon*, 48, 1876.

[74] Van de Lagemaat, J., Barnes, T. M., Rumbles, G., Shaheen, S. E., Coutts, T. J., Weeks, C., Levitsky, I., Peltola, J., & Glatkowski, P. (2006). *Appl. Phys. Lett.*, 88.

[75] Rowell, M. W., Topinka, M. A., Mc Gehee, M. D., Prall, H. J., Dennler, G., Sariciftci, N. S., Hu, L. B., & Gruner, G. (2006). *Appl. Phys. Lett*, 88.

[76] Aguirre, C. M., Auvray, S., Pigeon, S., Izquierdo, R., Desjardins, P., & Martel, R. (2006). *Appl. Phys. Lett*, 88.

[77] Xue, F., Zhuo, W. Q., Yanc, L., Xue, H., Xiong, L. H., & Lia, J. H. (2012). *Org. Electron.*, 13, 302.

Chapter 15

Latest Advances in Modified/Functionalized Carbon Nanotube-Based Gas Sensors

Enid Contés-de Jesús, Jing Li and Carlos R. Cabrera

Additional information is available at the end of the chapter

http://dx.doi.org/10.5772/52173

1. Introduction

A gas sensor is a device that when exposed to gaseous species, is able to alter one or more of its physical properties, so that can be measured and quantified, directly or indirectly. These devices are used for applications in homeland security, medical diagnosis, environmental pollution, food processing, industrial emission, public security, agriculture, aerospace and aeronautics, among others. Desirable characteristics of a gas sensor are selectivity for different gases, sensitivity at low concentrations, fast response, room temperature operation (some applications may require high temperature), low power consumption, low-cost, low maintenance and portability. Traditional techniques like gas chromatography (GC), GC coupled to mass spectrometry (GC-MS), Fourier transmission infrared spectroscopy (FTIR) and atomic emission detection (AED) provide high sensitivity, reliability and precision, but they are also bulky, time consuming, power consuming, operate at high temperature, and the high maintenance and requirement of trained technicians translate in high costs. In an effort to overcome those disadvantages, research in the area has been focused on the search for functional sensing materials.

Carbon nanotubes (CNTs) have been have been focus of intense research as alternative sensing material because of their attractive characteristics like chemical, thermal and mechanical stability, high surface area, metallic and semi-conductive properties and functionalization capability [1]. CNTs are graphene sheets rolled in a tubular fashion. Different types of CNTs can be synthesized: single walled carbon nanotubes (SWCNT), double wall carbon nanotubes (DWCNTs) and multi walled carbon nanotubes (MWCNT).

The publication of the first CNT-based sensor for NH_3 and NO_2 detection using an individual semiconducting SWCNT [2] triggered the research activity in this area. Pristine CNTs have shown to be chemically inactive to gas molecules in general. However, their modification/functionalization capability has been exploited throughout the last years, especially for the development of devices with enhanced selectivity and sensitivity for the room temperature detection of a wide variety of gases. Numerous articles and reviews focused on different aspect of CNTs-based sensors and summarizing their progress and potential have been published throughout the years. Some of them are listed in references [3-34]. A most recent review [35] addressing the technological and commercial aspects of CNTs sensors presents evidence of the continuous active research in the area and that they have real potential to complement or substitute current technologies.

This chapter presents a summary of selected original research articles that have been published between 2010 and present in which the main subject are modified/functionalized SWCNTs, DWCNTs and MWCNTs and their use as gas sensing material. The majority of the references included in this chapter content are based on experimental results. However, theoretical studies based on computational science are also included because of their importance in the study of CNT-based sensors. The use of different methods of calculations and simulation has been useful to design new structures and materials and to study, evaluate and predict the interactions and adsorption energies between those materials and gaseous molecules. First, we present current research activities on pristine CNTs-based sensors and the different approaches used to improve their sensitivity and selectivity without modifying the CNTs structure, followed by the review of CNTs modified with conducting polymers, metallic nanoparticles (NPs), nanostructured oxides and sidewall modification, doping and others. Different modification/functionalization techniques like chemical deposition, plasma, sputtering and electrodeposition are discussed. Gas sensors based on changes of electrical conductivity caused by adsorption of gas molecules (resistors) are the most common sensor type discussed in this review. Other sensing platforms like surface acoustic wave (SAW) and quartz microbalance (QMB)are also included.

2. Unmodified carbon nanotubes

Pristine CNTs are known for their high stability because of their strong sp^2 carbon-carbon bonds and thus insensitive when used as sensing material for certain gases. However, the detection of NO, NO_2 and NH_3 has been previously reported. In order to improve their sensitivity and recovery time, different approaches like dispersion techniques to debundling the CNTs ropes, humidity assisted detection, application of an electric field, continuous use of ultraviolet (UV) light and even separation of semi-conductive types from conductive have been reported. The detection of NO, NO_2 and NH_3, as well as other gases like formaldehyde and dimethyl methylphosphonate (DMMP) with pristine SWCNTs are discussed in this section.

A MWCNTs based sensor was used for the detection of 50 ppm of nitrogen monoxide (NO) [36]. With the purpose to increase the sensitivity, an electric field was applied between two copper plates as electrodes, one of them containing MWCNTs-silicon wafer. It was found that when a positive potential was applied to the copper plate and the negative potential applied to the copper plate containing the MWCNTs-sensor, NO, being an electron acceptor, moves to the electron enriched zone, which is the one containing the MWCNTs-sensor and thus enhancement in the sensitivity is observed. The stronger the applied electric field, the better the sensitivity. However, applying a negative electric field was applied to the copper plate and a positive potential was applied to the copper plate containing the MWCNTs-sensor, the NO molecules moved away from the MWCNTs sensor and thus a decrease in the sensitivity is observed. The more negative the applied electric field, the lower the sensitivity of the sensor. Recovery of the sensors was achieved by applying reverse potential from the one used to perform the gas sensing experiments.

Cava, et al. proposed the use of a homogeneous film of MWCNTs prepared by the self-assembly technique and use it as an active layer for an oxygen gas sensor with increased sensitivity [37]. When the sensors were exposed to 10% O_2 in Nitrogen at 160 °C, the electrical resistance decreased and showed a better oxygen sensitivity when compared to sensors prepared under the same condition but using the drop-cast method. The reason for this is that the self-assembly technique provides a better distribution of the nanotubes and thus promoting a better gas adsorption between nanotubes (inter-tube contact).

The high van der walls attraction between CNTs causes them to remain in bundles or agglomerated. This can represent a problem for their application as gas sensors because it results in less adsorption/interaction (binding) sites, which translates in less sensitivity. Considering this, different dispersion techniques were used by Ndiaye and coworkers for the preparation of CNTs based sensors for NO_2 detection [38]. SWCNTs were dispersed in a surfactant, sodium dodecylbenzene sulfonate (NaDDBs) and an organic solvent, chloroform ($CHCl_3$), drop-casted in IDEs and tested for the detection of 50, 100, 120, 200 ppb of NO_2 at 80 °C. Sensors prepared with SWCNTs dispersed in NaDDBs showed better sensitivity than those with SWCNTs dispersed in chloroform. The explanation to this is that the surfactant was more effective in debundle the SWCNTs than the organic solvent. It was stated that even though the surfactant was not completely removed after several rinsing steps and heating treatment at 150°C, it does not have significant effect on the electronic behavior of the sensor. Both surfactant-dispersed and organic solvent-dispersed samples showed a decrease in resistance with increasing temperature, which demonstrate the semi-conductive behavior of the SWCNTs and thus no effect of the solvent.

A SWCNT-based gas sensor selective for NO_2 and SO_2 at room temperature and ambient pressure was developed by Yao *et al.* [39]. High sensitivity and selectivity for 2 ppm of each gas was achieved by controlling the humidity levels. For instance, at low humidity levels, the sensors showed to be selective for $NO_2$and insensitive to SO_2. At high humidity levels (92%), both gases were detected. However, NO_2 showed a decrease in resistance and SO_2 showed an increase in resistance.

Continuous *in situ* UV illumination on SWCNTs during gas sensing experiments was used to enhance the sensor's overall performance in the detection of NO, NO_2 and NH_3 (Figure 1) [40]. Changes in conductance ($\Delta G/G_0$) as function of time were recorded and used to prepare calibration curves in order to determine sensors sensitivity. It was found that the continuous exposure of the sensors to UV light under inert atmosphere (N_2, Ar) regenerating the surface, therefore, enhances their sensitivity for the detection of NO and NO_2. Linear responses were achieved at low concentrations and up to 50 ppm. Detection limits (DL), derived from the noise of the baseline and the slope obtained from the calibration curve, were found to be as low as 590 parts per quadrillion (ppq) and 1.51ppt for NO and NO_2, respectively. For NH_3 it was found not only that the *in situ* UV illumination reverses the direction of the changes in conductance, but it was also confirmed that it helps to improve the DL from 5.67 ppm to 27.8 ppt when tested under identical conditions. The achieved DL outperformed by several orders of magnitude the sensitivity of other CNTs-based NO, $NO_2$and NH_3 sensors that have been previously reported. This is attributed to the UV light inducing surface regeneration and actively removal of all gases adsorbed on SWCNTs surface. It is worth noticing that this *in situ* cleaning with continuous UV-light exposure without device degradation was just achieved under inert atmospheres.

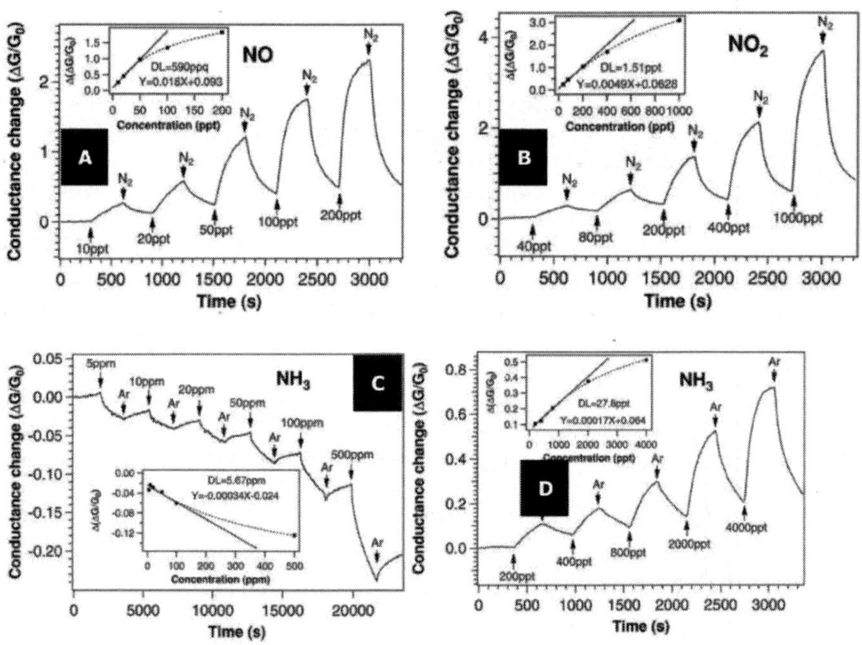

Figure 1. SWCNTs-sensor responses to (A) NO (10 – 200ppt) under *in situ* UV illumination, (B) NO_2 (40 – 1000 ppt) under *in situ* UV illumination, (C) NH_3 (5 – 500 ppm) without *in situ* UV illumination, and (D) NH_3 (200 – 4000 ppt) under *in situ* UV illumination. (From Chen *et al.* [40]. Copyright © 2012, with permission from Nature Publishing Group.)

Battie and coworkers used sorted semi-conducting SWCNTs as sensing film for the detection of NO_2 and NH_3 [41]. The density gradient ultracentrifugation (DGU) technique was used to separate semi-conducting from as produced SWCNTs. Films of as produced and sorted semi-conducting SWCNTs were exposed to NO_2 and NH_3 in air. Both films showed decrease in resistance and increase in resistance as exposed to NO_2 and NH_3, respectively. Full recovery was achieved by applying heat after NH_3 exposure and vacuum after NO_2 exposure. However, semi-conducting SWCNTs films were more sensitive for NH_3 than to NO_2 at different concentrations at ppm level.

Horrillo *et al.* used SWCNTs films for the room temperature detection of Chemical Warfare Agents (CWA) [42]. Changes in resistance were measured as samples were exposed to simulants of CWA at different ppm levels, DL of 0.01 ppm, 0.1 ppm and 50 ppm were achieved for DMMP, dipropylene glycol methyl ether (DPGME) and dimethylacetamide (DMA), respectively. The most remarkable advantage is that the sensors perform better and at more sensitive at room temperature, when tested at different temperatures.

3. Surface modified carbon nanotubes

CNTS modified with different functional groups have been used for the development of sensors for detection of volatile organic compounds (VOCs) in the environment as well as in exhaled breath. For the detection of VOCs in air, Wang *et al.* worked in the preparation of a sensor array based on MWCNTs covalently modified with different functional groups like propargyl, allyl, alkyltriazole, thiochain, thioacid, hexafluoroisopropanol (HFIP) [43] and Shrisat, *et al.* reported another one based on SWCNTs modified with different porphyrins (organic macrocyclic compounds) likeoctaethyl porphyrin (OEP), ruthenium OEP (RuOEP), iron OEP (FeOEP), tetraphenylporphyrin (TPP), among others [44]. Penza *et al.* also worked in the modification of MWCNTs with TPP for the room temperature detection of VOCs [45]. In this case, the TTP contained two different metals, Zn (CNT:ZnTPP) and Mn (CNT:MnTPP). Sensors were exposed to ethanol, acetone, ethylacetate, toluene and Triethylamine at ppm levels and all showed increase in resistance when exposed to the different gases. CNT: MnTPP showed the highest sensitivity towards all gases with respect to unmodified CNTs but for triethylamine and CNT: ZnTPP was more sensitive to ethylacetate.

Two different CNT-based sensor arrays have been reported for the detection and pattern recognition of VOCs present in exhaled breath samples for medical diagnosis, Tisch *et al.* presented a sensor array containing different nanomaterials including organically functionalized random networks of SWCNTs for the detection of VOCs related to Parkinson disease and that were present in exhaled breath collected from rats [46]. Ionescu and coworkers reported a sensor array based on bilayers of SWCNTS and polycyclic aromatic hydrocarbons (PAH) for the detection of multiple sclerosis in exhaled human breath [47]. In general, the incorporation of organic functional groups provided not only enhanced sensitivity but also provided better selectivity for each gas when compared to pristine CNTs. The use of statistical techniques like principal component analysis (PCA), discriminant factor analysis (DFA) and linear discriminant analysis (LDA) was possible to determine the discrimination capability of the sensors toward each VOC.

SWCNTs were functionalized with tetrafluorohydroquinone (TFQ) at the room temperature for detection of dimethyl methylphosponate (DMMP) at parts per trillion (ppt) levels (Figure 2) [48]. The conductance of the TFQ-SWCNTs samples increased as function of concentration when exposed to DMMP in N_2 in a concentration range from 20ppt to 5.4ppb. Sensors showed fast response and ultra sensitivity down to 20ppt when compared to unmodified SWCNTs sensor, which had a DL of nearly 1 ppm. The presence of TFQ clearly showed to improve the sensitivity and this is because it provides additional binding sites thru hydrogen bonds between hydroxyl groups in TQF and DMMP. In addition, TQF tailors the electronic properties of SWCNTs via hole doping.

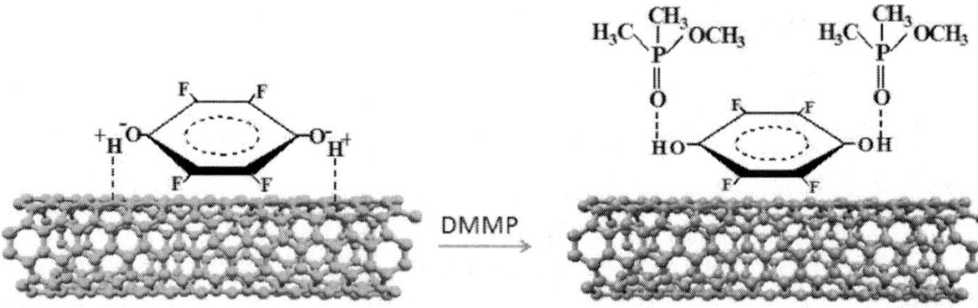

Figure 2. Representation of the possible sensing mechanism of DMMP with TFQ functionalized SWCNTs. (From Wei *et al.* [48]. Copyright © 2011, with permission from Institute of Physics Publishing.)

Wang *et al.* deposited a uniform network of SWCNT with carboxylic groups (-COOH) on a flexible poly (dimethyldiallylammonium chloride) (PDDA) modified polyimide (PI) substrate for DMMP detection at room temperature [49]. Changes in resistance as function of time were measured as the sensors were exposed to DMMP in N_2 at a concentration range of 1-40 ppm. Calibration curve showed a [17] linear increase of resistance as function of concentration and responses were found to be fast, stable and reproducible. Sensors showed to be selective to DMMP in presence of other volatile organic vapors like methanol, xylene and hexane, among others. Changing the carrier gas from N_2 to air caused the responses to DMMP to be lower which might be due to influence of oxygen and humidity contained in air. The apparent enhanced performance of this SWCNTs (-COOH) flexible sensor when compared to sensors prepared on Si/SiO_2 rigid substrates is attributed to the presence of PDDA, which is a polymer that absorbs DMMP. There is no information on the adsorption of DMMP by PDDA and its possible effect in the recovery of the sensors.

MWCNTs were oxidized with KMnO4 to add oxygenated functional groups, mainly (-COOH) for the detection of organic vapors [50]. Oxidized MWCNTs in form of a bucky paper were exposed to different concentrations of acetone. Variations in electrical resistance were recorded for both unmodified and oxidized MWCNTs. Oxidized MWCNTs showed higher sensitivity to acetone (2.3 vol. %) than unmodified ones, and good selectivity when sensing other oxygen containing vapors such as diethyl ether and methanol. The sensors also showed complete reversibility and high reproducibility for all tested vapors.

MWCNTs were chemically treated with acid to obtain hydroxyl groups (OH) and used as sensing material for humidity sensors [51]. Changes in resistance were measured as the RH was varied from 11% to 98%. It was observed that the resistance increased as sensors were exposed to the different humidity levels. It was found that acid treated SWCNTS were more sensitive to humidity than pristine MWCNTs. The higher sensitivity of Acid treated SWCNTs is attributed to their higher surface and thus more adsorption sites that result from the acid treatment. Sensors showed fast response and to be stable. As with most humidity sensors, the recovery time was longer than response time due to slow desorption of water molecules.

Purified MWCNTs were treated with oxygen plasma or fluorine plasma and used for the detection of ethanol (Figure 3) [52]. Changes in resistance as function of time were recorded for the sensors when exposed of 50-500 ppm of ethanol vapor in air. Samples treated with oxygen plasma for 30 sec and with fluorine plasma for 60 sec showed the highest sensitivity to 100 ppm of ethanol, compared to pristine MWCNTS and other oxygen and fluorine plasma treated for different duration time. However, fluorine plasma treated samples showed the better sensitivity and reduced response and recovery time. The improvement of the fluorine plasma treated samples is explained by the difference in electronegativity between oxygen and fluorine.

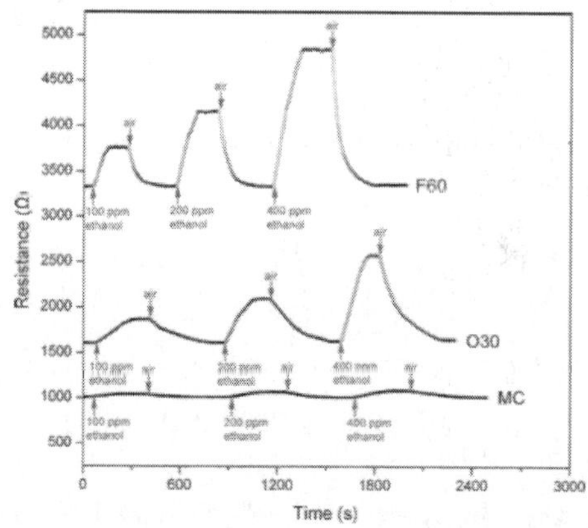

Figure 3. Responses of pristine (MC), oxygen plasma treated (O30) and fluorine plasma treated (F60) MWCNTs to 100-400 ppm of ethanol. (From Liu *et al.* [52]Copyright © 2012, with permission from Elsevier.)

Thermally fluorinated MWCNTs (TFC) were used for NO gas detection at room temperature [53]. The effect of thermal fluorination process was performed at various temperatures (100 -1000 °C) and 200 °C was found to be the optimum fluorination temperature. TFC samples prepared at temperatures higher than 200 °C showed a decrease of the fluorine functional groups, and even fluorine-assisted pyrolysis and fluorine-induced reorientation of the MWCNTs structure occurred at 1000 °C. TFC prepared at 200 °C showed high sensitivity, stability, reproducibility and full recovery, when their gas sensing properties were evaluated towards the detection of 50 ppm NO in dry air. Interestingly, the presence of fluorine reverses the electron transfer process, when compared to pristine MWCNTs, allowing them to go from NO to MWCNTs and thus causing an increase in resistance. The fluorination not only helped to enhance the sensitivity but also made the sensors insensitive to humidity changes.

MWCNTs were modified with amino groups for the detection of formaldehyde at ppb level [54]. Changes in resistance as function of time were measured as the sensors were exposed to formaldehyde in a concentration range between 20 and 200 ppb. Sensors containing MWCNTs with higher amino group content (18%) were 2.4 and 13 times more sensitive to formaldehyde than samples containing 5% amino groups and pristine CNTs, respectively. Short response times are due to a chemical reaction between the aldehyde and amino group. For the same reason, the recovery times are longer, since chemical desorption is a slow and irreversible process. SWCNTs with 18% amino groups showed to be selective to formaldehyde when tested against interferences like acetone, CO2, ammonia, methanol and ethanol.

Silicon (Si) nitrogen (N), and phosphorous-nitrogen (P-N) were used to modify MWCNTs and study their gas sensing properties for hydrogen peroxide, sodium hypochlorite and1, 2-dichloromethane, nitrogen, and ammonia [55]. Samples of Si-MWCNTs, N-SWCNTs, and P-N-SWCNTs were prepared by aerosol chemical vapor deposition. It is known that the incorporation of heteroatoms in the CNT structure changes its morphology and thus the reactivity. To evaluate the gas sensing properties of the prepared materials, changes in resistance as function of time were recorded when they were exposed to the different gases. Exposure to N_2 caused the removal of physisorbed water molecules and thus a decrease in the resistance values. Sodium hypochlorite and dichloroethane caused decrease in resistance of pristine MWCNTs and Si-MWCNTs due to charge transfer (electrons) from CNTs to Chlorine atoms and increase in resistance of N-P MWCNTs. Ammonia showed the opposite effect in resistance. These results demonstrate the p-type semiconductor behavior for pristine MWCNTs and Si-MWCNTs and n-type of N-MWCNTs and N-P-MWCNTs. All sensors recovered in 10 min for all gases with the exception of ammonia that exceeded 1 hour.

In an effort to enhance the selectivity of SWCNTs-based vapor sensors, Battie *et al.* worked in the preparation of SWCNTs covered with a mesoporous silica film [56]. Sensors were fabricated by covering a SWCNTs film with a mesoporous silica film via by sol-gel deposition technique. Characterization of the sensors was done by measuring changes in resistance when exposed to 200 ppm of NO_2, NH_3, and H_2O in dry air. A sensor of SWCNTs without the mesoporous silica film prepared and tested under the same conditions. While the SWCNTs sensor showed to be sensitive to the three gases, the sensor based on SWCNTs film covered with mesoporous silica film showed to insensitivity to H_2O, and its sensitivity for NH_3 was considerably reduced. These observations can be explained considering the polarization capabilities and dipole moments of the silanol groups contained in the mesoporous silica layer and the gas molecules. The silanol groups allow the mesoporous silica film to act as a diffusion barrier and allow the physical interaction and entrapment of highly polarized molecules like H_2O and NH_3, avoiding them to get in contact with the SWCNTs layer. On the other hand, the sensitivity to NO_2 was greatly enhanced, compared to the SWCNTs sensor. Compared to H_2O and NH_3, NO_2 has a weaker dipole moment and its diffusion thru the mesoporous silica gel and to the SWCNTs film results easier and thus its enhanced and selective detection.

Computational studies based on SWCNTs doped with heteroatoms have been also reported. *Ab initio* (ABINIT) simulations of CNTs doped with heteroatoms like boron, oxygen and nitrogen were performed to predict the behavior of the doped CNTs and to study their application as gas sensors for Cl_2, CO, NO and H_2 [57, 58]. Density functional theory (DFT) applied in the ABINIT code and the Generalized Gradient Approximation (GGA) were used to perform the calculations. The calculations demonstrated that doping the CNTs with B, O, and N causes a shift in the conduction band of the CNTs. For B and O, the conduction band shifts downward and creates a p-type semiconducting material. On the other hand, N dopant causes the conduction band to shift upward and create an n-type semiconducting material. Calculations also demonstrated that Cl_2, NO, $H_2$and CO considerably affects the NTs density of states (DOS) and Fermi level as the gases become close to their surface. B-doped CNTs can detect CO, NO and H_2 gas molecules, O-doped can detect H_2, Cl_2 and CO and N-doped can detect CO, NO and Cl_2.

Similarly, Hamadanian *et al.* presented a computational study of Al-substituted SWCNTs (10, 0) (2.5% and 25%) and their use as CO gas sensor [59]. DFT calculations (local density approximation with ultrasoft pseudopotential) were used to study the electronic properties of Al-substituted SWCNTs and how those properties are affected by the adsorption of CO molecules. Substitution of one carbon atom with an Al atom causes deformation of the 6-membered ring and increasing the bond length. Doping with Al also alters the DOS and band structure of the CNTs. Since Al has one electron less than C in the valence shell, introduces one electron holes in the band structure, therefore the tube is changed to p-type semiconductor. Calculations showed low adsorption energy for CO on pristine CNTs and that CO does not cause significant changes in the electronic band structure and DOS when adsorbed on pristine CNTs. These results confirm that pristine CNTs are insensitive to CO as result of their weak physical interaction. When CO is adsorbed on both 2.5% and 25% Al-substituted SWCNTs, it causes severe changes in the Band structure near Fermi level. Those changes strongly depended on the site of CNTs and the direction in which the CO molecule interacts. For instance, the most stable adsorption structure is when the C of the CO interacts with the middle point of a C-C bond of CNTs. Even when the adsorption energies of CO in 25% Al-substituted SWCNTs were higher than in 5% Al-substituted SWCNTs, the fact that the conductivity of the proposed material changes, makes them suitable for their use as CO gas sensing material.

Metal oxide	Nanotube Type	Op T. °C	Target (gas)	DL	Response Time	Ref.
ZnO SnO_2 TiO_2	MWNTs	RT	Ethanol	100ppm*	NS	[60]
Co_3O_4	SWNTs	RT, 250	NO_x	20ppm	NS	[61]
			H_2	4%		
ZnO	Pd-COOH SWCNTs	RT	NH_3	50ppm*	NS	[62]
	F-SWCNTs					
	N-SWCNTs					
SnO_2	SWNTs	200	NO_2	2ppm	NS	[63]
SnO_2	MWNTs	RT, 150	NO_2	1ppm	3min (150C) 4min (RT)	[64]
			CO	2ppm	5min	
WO_3	MWNTs	350	H_2	100ppm*	Ns	[65]
SnO_2	MWNTs	320	Ethanol			[66]
			LPG		21s	
SnO_2	O-doped	RT	NO_2	100ppb	7min	[67]
	N-doped				1min	
	B-doped				1min	

*Lowest tested concentration

Table 1. Summary of metal oxide NPs used to modify CNTs for gas sensor applications.

4. Conducting polymers and CNT composites

Conducting polymers have been widely used to enhance the sensing properties of CNTs-based sensors. CNTs unique characteristics combined with polymer's delocalized bonds, high permeability and low density have demonstrated that it is possible to detect many different gases with high sensitivity, fast response and good reproducibility. Previous reports on polymer/CNTs-based sensor have been summarized and reviewed [7-9, 12]. However, some challenges to overcome are aggregation or agglomeration of CNTs, thermal stability and selectivity, among others. Polymer/CNTs composites used in resistors, SAW and QMB type of sensors are discussed.

CNTs were used to improve Polyaniline (PANi) poor thermal stability (Figure 4) [68]. The proposed solution to this was the uniform incorporation of CNTs in the polymer network.

Considering that one of the problems of MWCNTs is their aggregation and agglomeration, they were oxyfluorinated under different conditions in order to obtain a better dispersion in aqueous solution and it was found that the better dispersion was obtained with the MWCNTs with the highest oxygen content. The oxyfluorinated MWCNTs were then mixed with aniline and ammonium persulfate (APS) and other chemicals, for in-situ polymerization. Changes in resistance as function of time were used to characterize and evaluate the resulting PANi/MWCNTs composite was for the detection of ammonia (NH_3) in a concentration range of 1-50 ppm. PANi/MWCNTs composite with the highest oxygen content had a uniform composition, improved thermal stability and highest and faster response for 50 ppm of NH_3. The composite was able to detect 1 ppm and showed excellent repeatability for cycling exposures to 50 ppm and it needed heat treatment to accelerate the NH_3 desorption and thus the recovery of the sensor. A possible drawback is the selectivity to NH_3 among gases that can extract protons (H^+) from PANi.

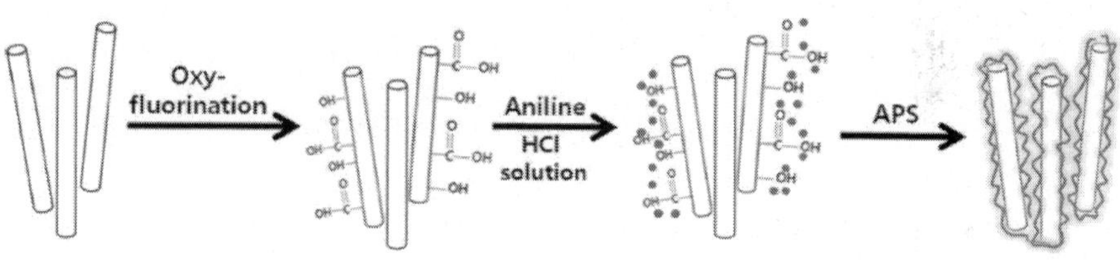

Figure 4. Illustration of the steps to obtain oxyfluorinated CNTs modified with PANI. (From Yun et al. [68]. Copyright © 2012, with permission from Elsevier.)

Mangu *et al.* also worked in the preparation of PANi-MCNTs as well as Poly (3, 4-ethylenedioxythiophene) poly (styrenesulfonate) (PEDOT:PSS)-MWCNTs composites for the detection of 100 ppm of NO_2 and NH_3 [69]. This group studied the effect of dissolving the polymers in different solvents for the gas sensing properties. PANi was dissolved in dimethyl sulfoxide (DMSO), *N, N*-dimethyl formamide (DMF), ethylene glycol (EG) and 2-propanol. PEDOT: PSS was dissolved in DMSO, DMF and 0.1M sodium hydroxide (NaOH). Each polymer solution was spin-coated in plasma treated MWCNTs and evaluated as sensing material. All PANi-MWCNTs composited showed an increase in resistance for NH_3 and a decrease in resistance for NO_2, which is typical of p-type semiconducting composite films. All PANi-MWCNTs composites were selective to NO_2. However, better sensitivities were achieved when PANi was dissolved in 2-propanol and DMSO for NH_3 and for NO_2, respectively. On the other hand, all PEDOT:PSS-MWCNTs composites were also excellent for the detection if both NO_2 and NH_3. PEDOT: PSS-MWCNTs (prepared without any solvent) showed to be more sensitive to NH_3 and PEDOT:PSS dissolved in NaOH to NO_2.

Sayago, *et al.* have worked on the preparation of different composites using polymers with small percentages of CNTs as sensitive layers for surface acoustic wave (SAW) gas sensors [70-72]. Composites of polyisobutilene (PIB), polyepichlorohydrin (PECH) and polyetherurethane (PEUT) with 2% and 5% of MWCNTs were prepared and used to detect volatile organic compounds (VOCs) at room temperature using SAW-sensor arrays. All the samples showed responses (frequency shifts) for octane and toluene (25-200 ppm), even though some samples were more sensitive than the others. For instance, samples with PIB/5%MWCNTs showed higher sensitivity for octane while PECH/2%MWCNTs and PEUT/2%MWCNTs were more sensitive to toluene. The difference in sensitivity is attributed to the difference in affinity between polymers and VOCs due to their respective polarities. The detection and recovery times were fast and fully reversible, which means that the main interaction is physisorption. The role of the MWCNTs is unknown. In general, their presence in the composite showed to improve sensitivity and reduce the limits of detection (LOD) but did not affect selectivity, response and recovery times.

Another SAW gas sensor was reported by Viespe *et al.* for the detection of methanol, ethanol, toluene using different polyethyleneimine (PEI)-based nanocomposite as sensitive layer, including MWCNTs-PEI [73]. In general, frequency shifts were proportional to the gas concentration and MWCNTs-PEI sensors showed better response time and higher sensitivity than the PEI-sensor. However, it did not show the best LOD when compared to the other PEI-based nanocomposite. The MWCNTs-PEI sensors showed higher sensitivity towards toluene and lower sensitivity to methanol when compared to ethanol.

Biopolymer/CNTs composites for chemical vapor sensors were produced by using two different biopolymers, cellulose, the most naturally abundant one and poly (lactic acid) (PLA). Considering that previous studies have shown that a homogeneous distribution of MCNTs in the cellulose matrix can improve the polymer's mechanical and electrical characteristics, MWCNTs were functionalized with imidazolide groups and covalently attached it to cellulose chains [74]. The resulting material, a paper-like film, was then used as sensing element for the detection of methanol, ethanol, 1-propanol and 1-butanol at ppm levels. Responses were measured by changes in resistance and were found to be reversible and consistent for all the tested vapors. However, the sensor only showed linear responses as function of concentration for 1-propanol in the range of 400-3600 ppm. The other composite, PLA/MWCNTs was prepared by doping the biopolymer with2 and 5% of MWCNTs and annealing, in order to understand the effect of MWCNTs in the crystallinity of the polymer and its performance in the detection of toluene, water, methanol and chloroform [75]. PLA/2%-MWCNTs showed highest responses for all gases, when compared to PLA/3%-MWCNTs. However, it was found that all samples were selective to chloroform. Moreover, annealing the samples showed a decrease in the responses that were significantly lower than the untreated ones. Annealing did not affect the selectivity to chloroform but it considerably affected its sensitivity.

Plasma-treated MWCNTs (p-MWCNTs) polyimide (PI) composite films (p-MWCNTs-PI) were developed by Yoo *et al.* in an effort to overcome some of the problems presented by PI-based resistive-type sensors, e. g. nonlinear sensitivity to relative humidity [33]. When tested as sensing material for 10-95% relative humidity (RH), p-MWCNTs-PI showed better sensitivity and linear response (resistance) as function of humidity, when compared to pristine-MWCNT-PI and pristine PI. The increase of p-MWCNTs content inp-MWCNTs-PI films not only showed improved the linearity and sensitivity but also lower resistance values. Lower resistance values might improve the performance at low RH range, a problem that PI resistive-type sensors present because of their high resistance values.

Yuana and coworkers prepared a sensor array based on polymer/MWCNTs composites for the selective detection of chloroform ($CHCl_3$), tetrahydrofuran (THF) and methanol (MeOH) [76]. Ethyl cellulose (EC), poly [methyl vinyl ether-alt-maleic acid] (PMVEMA), hydroxypropyl methyl cellulose (HPMC), poly (alpha-methylstyrene) (PMS), poly (vinyl benzyl chloride) (PVBC) and poly (ethyleneadipate) (PEA) were the polymers used to prepare the polymer/MWCNTs composites and to provide uniqueness to each sensor in terms of their physical and chemical characteristics like molecular structure, polymer length, polarity and intermolecular forces. Changes in resistance as function of time were recorded for the sensor array when exposed to the different gases at different temperatures (30 40, 50, 60 °C) and 50-60% R. H. The sensors showed to be selective to the three gases at the different temperatures when in presence of vapor molecules of chloride, cyclic oxide and hydroxide groups. The decreasing order of sensitivity concurred with the order of decreasing conductivity: PEA/MWCNTs >EC/MWCNTs >PMVEMA/MWCNTs >PVBC/MWCNTs >HPMC/MWCNTs >PMS/MWCNTs.

SWCNTs modified with Poly- (D) glucosamine (Chitosan) (SWCNTs-CHIT) were as high performance hydrogen sensor [77]. Three types of sensors were prepared: SWCNTs deposited on glass substrate (type 1), SWCNTs deposited over a glass substrate modified with CHIT-film (type 2), Chit-film deposited over SWCNTs deposited over a glass substrate. Each type of sensor showed a different changes in resistance when exposed to 4% H_2 in air. Increase in resistance was observed for three types of sensors and good recovery but for type (3) sensor. Sensors modified with CHIT showed better sensitivity. Moreover, the authors explained that the improved sensitivity was far higher than that reported for Pd-SWCNTs based sensors, which are commonly used for H_2 detection. The enhanced performance of SWCNTs-CHIT can be explained by the strong interaction/binding of the H_2 molecules with/to the –OH and –NH_3 groups contained in CHIT, which also explains the poor recovery of type 3 sensor.

SWCNTs were modified with Polypyrrole (PPy) and 5, 10, 15, 20-tetraphenylporphyrine (TPP) to prepare composites for the detection of 1-butanol in nitrogen using a quartz microbalance (QMB) [78]. The QMB gold electrodes were coated with PPy/SWCNTs-COOH, PPy/SWCNTs-COOH/SWCNTs-TPP via electropolymerization. Frequency shifts of the quartz crystal resonator as function of time were measured for the sensors when exposed to 1-butanol in ppb concentration range. Even though both composites showed good and higher response magnitudes than QMB prepared with other composites that did not contain CNTs, PPy/SWCNTs-COOH/SWCNTs-TPP showed better performance than PPy/SWCNTs-COOH. The results demonstrate that the incorporation of CNTs enhanced the sensitivity towards the detection of 1-butanol.

Lu *et al.* reported a sensor array containing pristine SWCNTs, Rh-loaded SWCNTs, PEI/SWCNTs and other CNTs with different coatings and loadings for the detection of hydrogen peroxide (H_2O_2) [79]. The measurements of changes in resistance as function of time were used to analyze the sensor array performance when exposed to H_2O_2. Pristine SWCNTs showed strong increases in resistance and fast responses for H_2O_2 and an estimated DL (by IUPAC definition) of 25 ppm. But when the sensor array was exposed to H_2O and CH_3OH in order to test its selectivity, pristine SWCNTs showed also good responses for both chemicals, which means that the discrimination capabilities towards H_2O_2 are limited. On the other hand, the PEI/SWCNTs sensors were sensitive to H_2O_2, showing decreases in resistance for each exposure. The PEI/SWCNTs did not show significant changes in resistance when exposed to H_2O and CH_3OH, which makes it selective to H_2O_2 under the tested conditions.

Polymer	CNT Type	Sensor Configuration	Target	DL	Ref.
PECH, PEUT, PIB	MWCNTs	SAW	Toluene	1.7-12.2ppm	[70 72]
			Octane	9.2-12.7ppm	
				NS	
Cellulose	MWCNTs	Resistor	Methanol	650ppm*	[74]
			Ethanol	672ppm*	
			1-propanol	635ppm*	
			1-butanol	687ppm*	
PANi	MWCNTs	Resistor	NH_3	1ppm	[68]
PEDOT:PSS	MWCNTs	Resistor	NH_3	100ppm*	[69]
			NO_2		
PI	MWCNTs	Resistor	Humidity	10%*	[33]
PEI	MWCNTs	SAW	Ethanol	176.5	[73]
			Methanol	184.2	
			Toluene	170.6	
PLA	MWCNTs	Resistor	VOC	NS	[75]
CHI	SWCNTs	Resistor	H_2	4%*	[77]
EC	MWCNTs	Resistor	THF, CH_3Cl_2, MeOH	NS	[76]
PMVEMA					
HPMC					
PMS					

PVBC					
PEA					
PPy	SWCNTs	QMB	1-butanol	46ppb*	[78]
PPT					
PEI	SWCNTs	resistor	H_2O_2	NS	[79 80]

* Lowest detected concentration

Table 2. Summary of polymers used for the preparation of polymer/CNTs-based sensors.

5. Metal nanoparticlesdecorated CNTs

Electronic, physical and chemical properties of metallic nanoclusters are usually sensitive to the changes in environment [81]. CNTs decorated with metallic nanoparticles (NPs) have been widely used to achieve selectivity and improve the sensitivity, response time and DLs for a variety of gas detections. Layer by layer, electrodeposition, chemical deposition, electrochemical deposition and sputtering are the methods used to prepare the metallic NP-CNTs composites discussed in this section.

SWCNTs films were modified with Pd NPs using sputtering method [82]. After apply different deposition times (40 – 160 s), it was found that 120 s was the optimum deposition time to obtain enhanced sensor response for 1% H_2 in dry air at 50 °C. A typical response curve (i_{gas}/i_{air} vs. time) showed differences in response and recovery between the first and following H_2 sensing cycles. FTIR studies were used to support and explain those differences and mechanisms of detection. The first cycle showed an overall larger electrical current in the presence of $H_2$and then it reached a new steady state. When the atmosphere was change to dry air, the current did not go to its original value but remained in the steady state, which is considered as an irreversible response. The explanation to this is that is atomized by the Pd NPs and spilled to the surface of the MWCNTs, occurring the chemical and irreversible reaction of hydrogenation of the carbonyl groups of the MWCNTs at the first cycle. The second cycle and following ones started at the steady state where the first cycle finished and the electrical current showed a decrease in the presence of H_2 and when the atmosphere was changed to dry air, the electrical current recovered back to where the cycle started. This reversible behavior is explained as physisorption of H_2 molecules onto Pd/SWCNTs.

Pd/MWCNTs and Pt-Pd/MWCNTs composites were tested for the detection of H_2 in a concentration range of 20 ppm– 2% in N_2 and 200 ppm – 2% in air [83]. Composites were prepared by growing CNTs yarns and then covered them with a layer of Pd NPs or sequentially deposited layers of Pd and Pt NPs, using a recently developed technique called self-fuelled electrodeposition (SFED). Exposure to 1% H_2using N_2 with 1% air as carrier gas. As with other Pd/CNTs-based sensors [82], an initial irreversible drop in resistance was observed, and after that, the sensor reached a steady state. A stable baseline was established just after a couple of exposure/recovery cycles. Pd-MWCNTs was not able to detect H_2 concentrations below 20 ppm but with the Pt-Pd/MWCNTs composites it was possible to detect concentrations as small as 5 ppm (0.0005%). The sensor saturated at 100 ppm and higher concentrations. When the same experiments were performed using air as carrier gas, it was found that the detection limit for both composites decreased.

A Pd/MWCNTs flexible substrate H_2 sensor was fabricated using the layer-by-layer technique [84]. For this, an Au IDE was sputtered over a polyester (PET) film, followed by the fabrication of a poly (4-styrenesulfonic acid-*co*-maleic acid)/poly (allylamine hydrochloride) (PSSMA/PAH) bilayer film. Then, a MWCNTs layer was deposited over the PSSMA/PAH bilayer film and decorated with Pd NPs using chemical deposition. The sensors were exposed to H_2 in a range of concentrations between 200 and 40000 ppm at room temperature and 53% RH. The rapid response and higher sensitivity of the Pd/MWCNTs when compared to plain MWCNTs is attributed to the well known catalytic effect of Pd NPs. Because the detection mechanism is the dissociation of the H_2 molecules on the surface of the Pd NPs, the linear relation was found to be between the sensor's response and the square root of the concentration. It also showed to be selective to H_2 at concentrations higher than 1000 ppm (vs. NH_3, CO, CH_4 and others), to be highly reproducible, to have long-term stability and comparable to similar sensors fabricated in rigid substrates.

Zilli *et al.* demonstrated that the H_2sensing capacity of the Pd/MWCNTs nanocomposite is affected by the different stages of purification of the CNTs [85]. Pd NPs were chemically deposited on pristine MWCNTs (Pd-CNT-P), gas-phase oxidized MWCNTs (Pd-CNT-O) and gas-phase oxidized/acid treated MWCNTs (Pd-CNT-A) and used as H_2 sensing material. All three nanocomposites showed increase in resistance as function of time in the presence of 500μL $_{(STP)}$ of H_2 and an immediate decrease when the H_2 disappeared from the testing chamber. However, Pd-CNT-O showed to be more sensitive than Pd-CNT-P and Pd-CNT-A, which both showed similar responses. The reason for these results is that for CNT-P, catalytic Fe NPs, used for the CNT growth, are encapsulated inside the CNTs and in CNT-A, the Fe NPs were removed the during the acid treatment. After the gas-phase oxidation process (CNT-O), the Fe NPs become exposed and at the same time, oxygen-containing groups are formed in the surface of CNTs, which act as additional anchoring site for Pd NPs and thus making the Pd-CNT-O sample more sensitive to H_2.

Pd/SWCNT composites were prepared by using a poly (amido amine) (PAMAM) dendrimer assisted synthesis, followed by a pyrolysis step to remove the dendrimers [86]. For H_2 sensing experiments, the samples were deposited over Ti/Au electrodes and changes in resistance as function of time were measured. The Pd/SWCNTs samples showed to be more sensitive to 10, 000 ppm of H_2 at room temperature, when compared to samples of chemically reduced Pd on SWCNTs (without the presence of dendrimer) and Pd/PAMAM-SWCNTs. Moreover, these samples were able to detect all of the concentrations in a 10-100 ppm range. It was concluded that not only the dendrimers provided more nucleation sites for the Pd NPs and thus higher NPs density, but also that the removal of the dendrimers thru the pyrolysis step reduces the distance between the NPs and the SWCNTs and consequently reduces the delay in electron transfer, allowing faster response times.

Double wall carbon nanotubes (DWCNTs) have shown to have longer length, compared to SWNTs, provide a better percolation behavior, the possibility of modifying the outer layer without modifying the inner one and flexibility are some of their attractive characteristics. Considering those characteristics, Rumiche *et al.* evaluated Pd NPs/ (DWCNTs) composites as room temperature H_2 sensor (Figure 5) [87]. Different amounts of DWNTs (15 and 20 μL) were deposited over silicon oxide substrates and decorated with 1, 3 and 6nm layers of Pd NPs. To evaluate their sensing properties, changes in resistance as function of time were recorded when exposed to 3%, 2%, 1%, 0.5%, 0.3%, 0.2%, 0.1%, and 0.05% of H_2 in dry air. Samples containing 1nm Pd layer were unsuccessful detecting any of the tested concentrations. On the other hand, samples containing 3 and 6nm Pd layer showed overall similar performance in terms of increases of resistance and recovery. When analyzing the results for the lowest tested H_2 concentration (0.05%)it was found that the increase in resistance was comparable for samples with same Pd layer thickness and different DWCNTs content (e. g. 3nm thick Pd layer deposited over 15μL DWCNTs was comparable to of 3nm thick Pd layer over 20μL DWCNTs). However, the increase in Pd coating thickness produced a reduction in the response. The obtained results confirmed that the combination of the amount of DWCNTs and the Pd-layer thickness directly affects the sensitivity of the sensors.

Another H_2 sensor based on N-doped MWCNTs electrochemically decorated with Au NPs (Au-NMWCNTs) was presented by Sadek and collaborators [88]. N-doped MWCNTs were chosen because they have enhanced surface reactivity and chemically active sites for the nucleation of Au NPs during the electrodeposition process. NPs of different sizes were obtained with variation of electrodeposition potential. Changes in resistance as function of time were used to evaluate the performance of the Au-NMWCNTs sensors when exposed to different H_2 concentration between 0.06% and 1%. Sensitivity, response time and recovery time highly depended on the size of the AuNPs: the smallest the size the better the sensitivity and the shorter the response and recovery times.

Penza *et al.* worked in the modification of SWCNTs with Pt, Ru and NiNPs to monitor toxic, landfill, and greenhouse gases (CO_2, CH_4, NH_3, and NO_2) [89, 90]. Pt NP layers with thickness of 8, 15 and 30 nm were sputtered over SWCNTs films and exposed to the different gases at an operation temperature of 120°C. Changes in resistance as function of time showed that Pt-SWCNTs had better sensitivity than unmodified SWCNTs. It was also found that the sensitivity depended on the layer thickness. For instance, 8 nm Pt layer-SWCNTs showed highest sensitivity for NO_2 and CH_4 and 15 nm Pt layer-SWCNTs showed better sensitivity for NH_3 and CO_2. In a similar study, a sensor array containing Pt, Ru, and Ag NPs sputtered over SWCNTs with a thickness of 5nmwas able to detect and selectively discriminate between landfill gases. Concentrations as low as 100 ppb of NO_2 were selectively detected at a temperature of 120 °C.

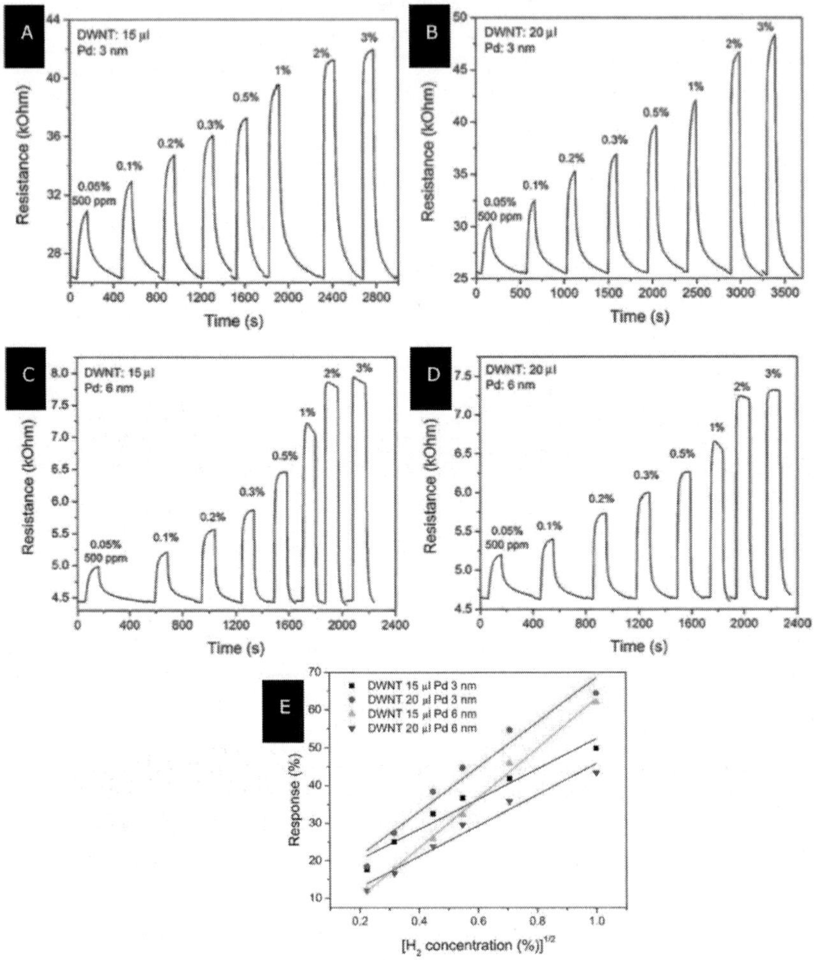

Figure 5. A-D) Responses of the different Pd/DWCNTs composites when exposed to H_2. (E) Calibration curves for the samples presented in figures (A-D). (From Rumiche *et al.* [87] Copyright © 2012, with permission from Elsevier.)

Lu *et al.* used pristine SWCNTs, fluorinated SWCNTs (F-SWCNTs) and rhodium doped SWCNTs (Rh-SWCNTs) and other various coatings and dopings on the SWCNTs for the room temperature detection of formaldehyde (HCOH) [80]. The measurements of changes in resistance as function of time when the array was exposed to the different concentrations of formaldehyde were used to analyze the sensor array performance. When exposed to 0.71 ppm formaldehyde, pristine SWCNTs showed the higher re-

http://dx.doi.org/10.5772/52173

sponse, followed by Rh/SWCNTs and F-SWCNTs, which were less sensitive. The other CNTs with different coating and loadings were insensitive to formaldehyde. However, when the array was exposed to formaldehyde at concentrations as low as 0.01 ppm (10ppb), Rh/SWCNTs sensor showed to be more sensitive and the presented an estimated DL (by IUPAC definition) of 10ppb. The DL for pristine SWCNTs and F-SWCNTs sensors were 15ppb and 20ppb, respectively. The three sensors presented very fast response (~18sec) and recovery time of approximately 1 minute.

Theoretical studies have been used to study the interaction between Pd and Pt NPs decoratedCNTs and a wide variety of gases for different applications. Zhou *et al.* used the density functional theory (DFT) to study the adsorption and interaction of SO_2, CH_3OH, and CH_4 with Pd-SWCNTs [91]. The replacement of a central C atom of the CNTs with a Pd atom causes structural deformations. As SO_2 is adsorbed, there is a charge transfer from Pd-SWCNT to SO_2. As for CH_3OH, the appropriate adsorption conformation is thru the lone par of the oxygen of CH_3OH and thus occurring an overall charge transfer from CH_3OH to Pd-SWCNT. The interaction between Pd-SWCNT and CH_4 is similar in that the charge transfer occurs from Pd-SWCNT to the gas. However, the interaction between Pd-SWCNTs and CH_4 is not as strong as the interaction between Pd-SWCNT and SO_2 or even CH_3OH.

NP	NT Type	Target	Sensor Configuration	DL	Ref.
Pd	MWCNT	H_2	Resistor	1%	[82]
Pd	MWCNTs	H_2	Resistor	10000ppm*	[84]
Pd	MWCNTs	H_2	Resistor	NS	[85]
Pd-Pt	MWCNTs	H_2	Resistor	5ppm	[83]
Pd	SWCNTs	H_2	Resistor	10ppm	[86]
Pd	DWCNTs	H_2	Resistor	0.05	[87]
Pt, Ru, Ag	SWCNTs	CO_2, CH_4, NH_3, NO_2	Resistor	100ppb NO_2	[89 , 90]
Au	N-doped MWCNTs	H_2	Resistor	0.06%*	[88]
Rh	SWCNTs	HCOH	Resistor	10ppb	[80]

*Tested concentration

^For concentrations >1%

Table 3. Metallic nanoparticles used to decorate SWCNTs for gas sensing applications

Li and co workers investigated the adsorption of CO and NO on SWCNTs-decorated with Pd and Pt using first-principle calculations [92]. It was found that the electronic properties of SWCNTs change upon modification with Pd or Pd atoms. The semi-conductive band gap is decreased compared to pristine SWCNTs. The reason for the observed decrease in the band gap is due to charge transfer from the Pd and Pt atoms to the surface of the SWCNTs. Different from pristine SWCNT that show poor adsorption, Pd-SWCNTs and Pt-SWCNTs showed to chemisorb CO molecules as well as NO. However, Pt-SWCNTs showed bigger binding energy and charge transfer than Pd-SWCNTs. The formation of C-Pd, N-Pd, C-Pt, and N-Pt bonds demonstrate that the metal atoms provide additional adsorptions sites for gases and open the possibility to use both materials as sensors for the detection of CO and NO.

6. Nanostructured oxides mixed with CNTs

Sensors made of metal oxides films have been used for a long time because of they provide high sensitivity for the detection of a wide variety of gases. However, their major drawback is their elevated operating temperatures. The development of metal-oxide NPs based films and nanocomposites has shown advantages like higher surface area and porosity, high catalytic activity, efficient charge transfer and adsorption capacity. However, it has been demonstrated that the improvements in gas detection at low temperature for CNTs/MO-based sensors is due to the introduction of CNTs in the nanocomposite.

Tin oxide (SnO_2)/MWCNTS were synthesized using different ratios of tin dioxide precursor and plasma treated MWCNTs (Figure 6) [64]. The composites was tested for 2, 10, 20 ppm for CO and 50 100, 500, 1000 ppb of NO_2 in dry air at both room temperature and 150 °C. Pure tin oxide films and pure plasma treated MWCNT were also tested for comparison purposes. Pure tin oxide films were unresponsive to all of the tested concentrations at the two different temperatures because both room temperature and 150 °C are too low when compared to the operation temperature for pure tin oxide-based sensors. On the other hand, pure plasma treated MWCNT responded to both gases at room temperature but not at 150 °C. As for the SnO_2/MWCNTS composite, higher sensor response for both gases was achieved from samples prepared with an intermediate ratio of tin dioxide precursor and plasma treated MWCNTs (i. e. 20mL and 12mg, respectively), especially when operated at room temperature. Response time for 1 ppm of NO_2 was 3minutesat 150 °C and 4 minutes at room temperature. Response time for 2 ppm of CO is stated to be 5 minutes but the temperature was not specified. SnO_2/MWCNTS showed higher sensor response to NO_2 than to CO, and was also sensitive to humidity changes.

Different composite synthesis temperature can affect the sensor performance. SnO_2/SWCNTs composites were synthesized at different oxidizing temperatures (300-600 °C)for testing the effect of temperature in their morphology, structure and gas sensing properties in the detection of NO_X [63]. The synthesized composites were exposed to 60 ppm of NO_X at 200 °C and it was found that the ones synthesized at 400 °C showed higher response. From here, composites synthesized at 400 °C were exposed to 30 ppm NO_X at different operating temperatures and it was determined that the optimum operation temperature was 200 °C. Concluding that the optimum oxidizing and operating temperatures were 200 °C and 400 °C, respectively, the samples were then exposed to different concentrations of NO_X. Under the aforementioned conditions, the SnO_2/SWCNTs composite showed improved performance for the detection of NO_X when compared to thin films of SWCNTs or SnO_2.

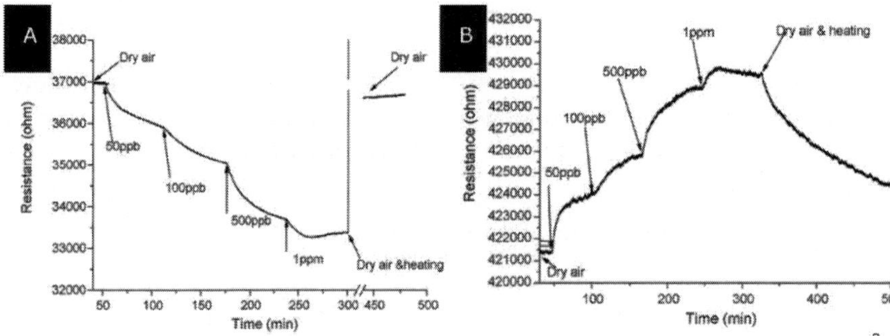

Figure 6. Response of SnO_2/MWCNTs to NO_2 at (A) room temperature and (B) 150 °C. (From Leghrib *et al.* [64]. Copyright © 2010, with permission from Elsevier.)

Wongchoosuk, *et al.* were the first to report the preparation of MWCNT-doped tungsten oxide (WO_3) thin films for H_2 sensing application [65]. The thin films of MWCNT-doped WO_3 and undoped WO_3 (for comparison purposes) were prepared using the electron beam (e-beam) evaporation technique and exposed to 1000 ppm of H_2 at different temperatures (200-400 °C). It was determined that 350 °C was the optimum operation temperature. Overall, MWCNT-doped WO_3 thin films showed higher responses for H_2 at any operating temperature when compared to the undoped WO_3 thin film. To demonstrate selectivity, the response of the MWCNT-doped WO_3 thin films was measured in presence of H_2, ethanol (C_2H_5OH), methane (CH_4), acetylene (C_2H_2) and ethylene (C_2H_4) at ppm level concentrations and operating temperature of 350 °C. It was concluded that the MWCNT-doped WO_3 thin filmswere selective to H_2 because they showed stronger response for H_2, much weaker responses for C_2H_5OH, CH_4, C_2H_2 and insensitivity ethylene (C_2H_4).

MWCNTs treated with nitric acid were used to fabricate MWCNTs-doped SnO_2sensors for the detection of ethanol and liquid petroleum gases (LPG) [66]. Sensors were tested to 100-1000 ppm of ethanol and 1000-10, 000 ppm of LPG at different operation temperatures in the range of 10-360 °C. The detection of both chemicals was improved when the operating temperature was 350 °C or lower and it was determined that the optimum operating temperature is 320 °C. The MCNTs-doped SnO_2 composite showed better selectivity for LGP than for ethanol and the calibration curve showed the sensors saturated at concentrations higher than 5000 ppm. The90% response and recovery time were 21s and 36s, respectively. When undoped SnO_2 sensors were exposed to 250 ppm of ethanol and 2500 ppm of LPG, they showed higher sensitivity to ethanol than to LPG at operation temperature range of 190 - 360 °C. Considering the obtained results, the selectivity of the MCNTs-doped SnO_2 composite for LGP can be attributed to the presence of MWCNTs but further studies are required.

Nanocomposite structures of cobalt oxide (Co_3O_4) and SWCNTs were prepared using a polymer assisted deposition (PAD) method [61]. For this, polyethyleneimine (PEI) was the polymer used to bind the cobalt ions from and adjust the viscosity of the solution during the deposition process in order to get a homogeneous distribution of the particles on the

SWCNTs thin film. The Co_3O_4/SWCNTs composite sensor was tested for the detection of NO_x in a concentration range of 20-100 ppm at room temperature. It showed proportional increases in response as function of concentration, poor recovery at room temperature and good recovery at 250 °C. Higher responses of the Co_3O_4/SWCNTs composite when compared to pristine CNTs are attributed to the high adsorption power of Co_3O_4 particles. The composite was also exposed to 4% of H_2 in air, and showed enhanced responses than pure SWCNTs at room temperature and than Co_3O_4 films at both room temperature and 250 °C.

MO NPs (ZnO, SnO_2, TiO_2) and MWCNTs composites were simultaneously grown on silicon and silica on silicon substrates by catalytic pyrolysis method and used for gas sensing [60]. Current differential-voltage (ΔI-V) curves were recorded for all the prepared composites, while to 100 ppm ethanol. TiO_2/MWCNTs showed better sensitivity (defined as ΔI/I-V) when compared to pure MWNT film, ZnO/MWCNTs and SnO_2/MWCNTs.

N-doped, B-doped and O-doped CNTs were used to prepare doped-CNTs/SnO_2hybrids [67]. All doped-CNTs and doped-CNTs/SnO_2hybrids were used to study the effect of functional groups on their gas sensing properties for 100, 200, 500, 1000ppb of NO_2 at room temperature. The responses were as follows: B-doped hybrid> N-doped hybrid>O-doped hybrid. All doped-CNTs/SnO_2hybrids responded better than N-doped and B-doped and O-doped CNTs. B-doped-CNTs/SnO_2 hybrids showed an improvement in the response time when compared to bare CNTs and recovered its baseline, which was not achieved with B-doped CNTs. The high sensitivity and improved performance achieved with the B-doped and N-doped-CNTs/ SnO_2 hybrids for low concentrations of NO_2 at room temperature are attributed to two main factors: the interaction of the N2 gas with the n-SnO_2/p-CNTs heterostructure that affects the conduction of the CNTs and the addition of new functionalities (i. e. B and N atoms) to the CNTs surface that affects the electronic density of states and Fermi level and consequently, its conductivity.

A combination of ZnO layer with functionalized MWCNTs for the room temperature detection of NH_3 has been reported by Tulliani and coworkers [62]. Samples of Pd-doped/COOH-MWCNTs, N-MWCNTs, and F-MWCNTs were deposited over a screen-printed ZnO layer. The materials were evaluated by measuring changes in resistance as the sensors were exposed to NH_3 at room temperature, in a concentration range 0-75 ppm and different relative humidity levels. The sensor based on ZnO with Pd-doped/COOH-MWCNTs was the only one that showed sensitivity to humidity. When exposed to NH_3, all sensors showed a decrease in electrical resistance but did not show better DL than other graphite-based sensors prepared under the same conditions.

7. Conclusion

Modification and functionalization of CNTs have shown to greatly improve the sensitivity and selectivity of CNTs-based sensors. For instance, great improvements for room temperature detection of different gases have been reported, especially when using metal oxide/ SWCNTs composites. Another subject of high interest is the development of deposition

http://dx.doi.org/10.5772/52173

methods and synthesis of Pd/CNTs for H_2 detection at room temperature. Interestingly, there is an increase in the tendency of combining other materials with modified CNTs. For example, CNTs decorated with metal NPs embedded in a polymer matrix or CNTs doped or CNTs doped with heteroatoms and decorated with NPs or metal oxides are some composites that have been successfully used as gas sensing materials. But not only the characteristics of the CNTs have contributed to these improvements. In fact, the reported improvements are attributed to the combination of materials and the intrinsic characteristics of the composites. This trend of combining materials demonstrates that the there is broad range of possibilities for the design of new materials to meet the requirements of an ideal sensor by showing selectivity for different gases, sensitivity at low concentrations, fast response, and room temperature operation among others.

Acknowledgements

The authors would like to acknowledge the NASA-URC Center for Advanced Nanoscale Materials (CANM) Grant # NNX08BA48A and NNX10AQ17A, NASA GSRP fellowship under Grant # NNX09AM23H and NASA Ames Research Center-Nanosensor Group

Author details

Enid Contés-de Jesús[1], Jing Li[2] and Carlos R. Cabrera[1*]

1 Department of Chemistry and NASA-URC Center for Advanced Nanoscale Materials, University of Puerto Rico, Puerto Rico

2 NASA-Ames Research Center, Moffett Field, California, USA

References

[1] Niyogi S, Hamon MA, Hu H, Zhao B, Bhowmik P, Sen R, et al. Chemistry of single-walled carbon nanotubes. Acc Chem Res. 2002 Dec;35(12):1105-1113.

[2] Kong J, Franklin NR, Zhou CW, Chapline MG, Peng S, Cho KJ, et al. Nanotube molecular wires as chemical sensors. Science. 2000 Jan;287(5453):622-625.

[3] Ciraci S, Dag S, Yildirim T, Gülseren O, Senger RT. Functionalized Carbon Nanotubes And Device Applications. J Phys: Condens Matter. 2004;16(29):R901-R960.

[4] Chu H, Wei L, Cui R, Wang J, Li Y. Carbon nanotubes combined with inorganic nanomaterials: Preparations and applications. Coord Chem Rev. 2010;254(9–10): 1117-1134.

[5] Liu S, Shen Q, Cao Y, Gan L, Wang Z, Steigerwald ML, et al. Chemical functionalization of single-walled carbon nanotube field-effect transistors as switches and sensors. Coord Chem Rev. 2010;254(9):1101-1116.

[6] Varghese OK, Kichambre PD, Gong D, Ong KG, Dickey EC, Grimes CA. Gas sensing characteristics of multi-wall carbon nanotubes. Sens Actuators, B. 2001;81(1):32-41.

[7] Wang Y, Yeow JTW. A Review of Carbon Nanotubes-Based Gas Sensors. Journal of Sensors. [electronic]. 2009;doi: 10.1155/2009/493904 (accessed 13 April 2012).

[8] Zhang T, Mubeen S, Myung NV, Deshusses MA. Recent progress in carbon nanotube-based gas sensors. Nanotechnology. 2008;doi: 332001 10.1088/0957-4484/19/33/332001 (accessed 13 April 2012).

[9] Zhang W-D, Zhang W-H. Carbon Nanotubes as Active Components for Gas Sensors. Journal of Sensors. 2009;doi: 10.1155/2009/160698 (accessed 13 April 2012).

[10] Di Francia G, Alfano B, La Ferrara V. Conductometric Gas Nanosensors. Journal of Sensors. 2009;doi:10.1155/2009/659275.

[11] Andzelm J, Govind N, Maiti A. Nanotube-based gas sensors - Role of structural defects. Chem Phys Lett. 2006 Apr;421 (1-3):58-62.

[12] Dai LM, Soundarrajan P, Kim T. Sensors and sensor arrays based on conjugated polymers and carbon nanotubes. Pure Appl Chem. 2002 Sep;74 (9):1753-1772.

[13] Goldoni A, Larciprete R, Petaccia L, Lizzit S. Single-wall carbon nanotube interaction with gases: Sample contaminants and environmental monitoring. JACS. [Article]. 2003 Sep;125 (37):11329-11333.

[14] Li J, Lu YJ, Ye Q, Cinke M, Han J, Meyyappan M. Carbon nanotube sensors for gas and organic vapor detection. Nano Lett. 2003 Jul;3 (7):929-933.

[15] Novak JP, Snow ES, Houser EJ, Park D, Stepnowski JL, McGill RA. Nerve agent detection using networks of single-walled carbon nanotubes. Appl Phys Lett. [Article]. 2003 Nov;83 (19):4026-4028.

[16] Valentini L, Armentano I, Kenny JM, Cantalini C, Lozzi L, Santucci S. Sensors for sub-ppm NO2 gas detection based on carbon nanotube thin films. Appl Phys Lett. 2003 Feb;82 (6):961-963.

[17] Lu YJ, Li J, Han J, Ng HT, Binder C, Partridge C, et al. Room temperature methane detection using palladium loaded single-walled carbon nanotube sensors. Chem Phys Lett. 2004 Jun;391 (4-6):344-348.

[18] Wang SG, Zhang Q, Yang DJ, Sellin PJ, Zhong GF. Multi-walled carbon nanotube-based gas sensors for NH_3 detection. Diamond Relat Mater. 2004 Apr-Aug;13 (4-8): 1327-1332.

[19] Lee CY, Strano MS. Understanding the Dynamics of Signal Transduction for Adsorption of Gases and Vapors on Carbon Nanotube Sensors. Langmuir. 2005 2005/05/01;21 (11):5192-5196.

[20] Penza M, Antolini F, Vittori-Antisari M. Carbon nanotubes-based surface acoustic waves oscillating sensor for vapour detection. Thin Solid Films. 2005 Jan;472 (1-2): 246-252.

[21] Lu YJ, Partridge C, Meyyappan M, Li J. A carbon nanotube sensor array for sensitive gas discrimination using principal component analysis. J Electroanal Chem. 2006 Aug;593 (1-2):105-110.

[22] Robinson JA, Snow ES, Badescu SC, Reinecke TL, Perkins FK. Role of Defects in Single-Walled Carbon Nanotube Chemical Sensors. Nano Lett. 2006 2006/08/01;6 (8): 1747-1751.

[23] Star A, Joshi V, Skarupo S, Thomas D, Gabriel J. Gas Sensor Array Based on Metal-Decorated Carbon Nanotubes. J Phys Chem B. 2006;110:21014 - 21020.

[24] Trojanowicz M. Analytical applications of carbon nanotubes: a review. TrAC, Trends Anal Chem. 2006 May;25 (5):480-489.

[25] Kuzmych O, Allen B, Star A. Carbon nanotube sensors for exhaled breath components. Nanotechnology. 2007; doi:10.1088/0957-4484/18/37/375502.

[26] Schlecht U, Balasubramanian K, Burghard M, Kern K. Electrochemically decorated carbon nanotubes for hydrogen sensing. Appl Surf Sci. 2007 Aug;253 (20):8394-8397.

[27] Valcarcel M, Cardenas S, Simonet BM. Role Of Carbon Nanotubes In Analytical Science. Anal Chem. 2007 Jul;79 (13):4788-4797.

[28] Endo M, Strano MS, Ajayan PM. Potential applications of carbon nanotubes. Carbon Nanotubes. 2008;111:13-61.

[29] Kauffman DR, Star A. Carbon nanotube gas and vapor sensors. Angewandte Chemie-International Edition. 2008;47 (35):6550-6570.

[30] Li C, Thostenson ET, Chou T-W. Sensors and actuators based on carbon nanotubes and their composites: A review. Compos Sci Technol. 2008;68 (6):1227-1249.

[31] Zanolli Z, Charlier JC. Defective carbon nanotubes for single-molecule sensing. Phys Rev B. 2009;doi:10.1103/PhysRevB.80.155447.

[32] Goldoni A, Petaccia L, Lizzit S, Larciprete R. Sensing gases with carbon nanotubes: a review of the actual situation. J Phys: Condens Matter. 2010; doi: 10.1088/0953-8984/22/1/013001.

[33] Yoo K-P, Lim L-T, Min N-K, Lee MJ, Lee CJ, Park C-W. Novel resistive-type humidity sensor based on multiwall carbon nanotube/polyimide composite films. Sens Actuators, B. 2010;145 (1):120-125.

[34] Lu Y, Meyyappan M, Li J. Fabrication of carbon-nanotube-based sensor array and interference study. Journal of Materials Research. 2011;26 (26):2017-2023.

[35] Fam DWH, Palaniappan A, Tok AIY, Liedberg B, Moochhala SM. A review on technological aspects influencing commercialization of carbon nanotube sensors. Sens Actuators, B. 2011;157 (1):1-7.

[36] Lee S-H, Im JS, Kang SC, Bae T-S, In SJ, Jeong E, et al. An increase in gas sensitivity and recovery of an MWCNT-based gas sensor system in response to an electric field. Chem Phys Lett. 2010;497 (4):191-195.

[37] Cava CE, Salvatierra RV, Alves DCB, Ferlauto AS, Zarbin AJG, Roman LS. Self-assembled films of multi-wall carbon nanotubes used in gas sensors to increase the sensitivity limit for oxygen detection. Carbon. 2012;50 (5):1953-1958.

[38] Ndiaye AL, Varenne C, Bonnet P, Petit â, Spinelle L, Brunet Jrm, et al. Elaboration of single wall carbon nanotubes-based gas sensors: Evaluating the bundling effect on the sensor performance. Thin Solid Films. 2012;520 (13):4465-4469.

[39] Yao F, Duong DL, Lim SC, Yang SB, Hwang HR, Yu WJ, et al. Humidity-assisted selective reactivity between NO2 and SO2 gas on carbon nanotubes. J Mater Chem. 2011;21 (12):4502 - 4508.

[40] Chen G, Paronyan TM, Pigos EM, Harutyunyan AR. Enhanced gas sensing in pristine carbon nanotubes under continuous ultraviolet light illumination. Sci Rep. 2012;2:1-7.

[41] Battie Y, Ducloux O, Thobois P, Dorval N, Lauret JS, Attal-Trétout B, et al. Gas sensors based on thick films of semi-conducting single walled carbon nanotubes. Carbon. 2011;49 (11):3544-3552.

[42] Horrillo MC, Martí J, Matatagui D, Santos JP, Sayago I, Gutiérrez J, et al. Single-walled carbon nanotube microsensors for nerve agent simulant detection. Sens Actuators, B. 2011;157 (1):253-259.

[43] Wang F, Swager TM. Diverse Chemiresistors Based upon Covalently Modified Multiwalled Carbon Nanotubes. JACS. 2011 2011/07/27;133 (29):11181-11193.

[44] Shirsat MD, Sarkar T, Kakoullis J, Myung NV, Konnanath B, Spanias A, et al. Porphyrin-Functionalized Single-Walled Carbon Nanotube Chemiresistive Sensor Arrays for VOCs. The Journal of Physical Chemistry C. 2012 2012/02/09;116 (5): 3845-3850.

[45] Penza M, Rossi R, Alvisi M, Signore MA, Serra E, Paolesse R, et al. Metalloporphyrins-modified carbon nanotubes networked films-based chemical sensors for enhanced gas sensitivity. Sens Actuators, B. 2010;144 (2):387-394.

[46] Tisch U, Aluf Y, Ionescu R, Nakhleh M, Bassal R, Axelrod N, et al. Detection of Asymptomatic Nigrostriatal Dopaminergic Lesion in Rats by Exhaled Air Analysis Using Carbon Nanotube Sensors. ACS Chemical Neuroscience. 2011;3 (3):161-166.

[47] Ionescu R, Broza Y, Shaltieli H, Sadeh D, Zilberman Y, Feng X, et al. Detection of Multiple Sclerosis from Exhaled Breath Using Bilayers of Polycyclic Aromatic Hydrocarbons and Single-Wall Carbon Nanotubes. ACS Chemical Neuroscience. 2011 2011/12/21;2 (12):687-693.

[48] Wei L, Shi D, Ye P, Dai Z, Chen H, Chen C, et al. Hole doping and surface functionalization of single-walled carbon nanotube chemiresistive sensors for ultrasensitive and highly selective organophosphor vapor detection. Nanotechnology. 2011;doi: 10.1088/0957-4484/22/42/425501

[49] Wang Y, Yang Z, Hou Z, Xu D, Wei L, Kong ES-W, et al. Flexible gas sensors with assembled carbon nanotube thin films for DMMP vapor detection. Sens Actuators, B. 2010;150 (2):708-714.

[50] Slobodian P, Riha P, Lengalova A, Svoboda P, Saha P. Multi-wall carbon nanotube networks as potential resistive gas sensors for organic vapor detection. Carbon. 2011;49 (7):2499-2507.

[51] Cao CL, Hu CG, Fang L, Wang SX, Tian YS, Pan CY. Humidity Sensor Based on Multi-Walled Carbon Nanotube Thin Films. Journal of Nanomaterials. 2011;doi: 10.1155/2011/707303.

[52] Liu C-K, Wu J-M, Shih HC. Application of plasma modified multi-wall carbon nanotubes to ethanol vapor detection. Sens Actuators, B. 2010;150 (2):641-648.

[53] Im JS, Kang SC, Bai BC, Bae T-S, In SJ, Jeong E, et al. Thermal fluorination effects on carbon nanotubes for preparation of a high-performance gas sensor. Carbon. 2011;49 (7):2235-2244.

[54] Xie H, Sheng C, Chen X, Wang X, Li Z, Zhou J. Multi-wall carbon nanotube gas sensors modified with amino-group to detect low concentration of formaldehyde. Sens Actuators, B. 2011;doi: 10.1016/j.snb.2011.12.112.

[55] Koós AA, Nicholls RJ, Dillon F, Kertész K, Biró LP, Crossley A, et al. Tailoring gas sensing properties of multi-walled carbon nanotubes by in situ modification with Si, P, and N. Carbon. 2012;50 (8):2816-2823.

[56] Battie Y, Ducloux O, Patout L, Thobois P, Loiseau A. Selectivity enhancement using mesoporous silica thin films for single walled carbon nanotube based vapour sensors. Sens Actuators, B. 2012;163 (1):121-127.

[57] Talla JA. First principles modeling of boron-doped carbon nanotube sensors. Physica B: Condensed Matter. 2012;407 (6):966-970.

[58] Talla JA. Ab initio simulations of doped single-walled carbon nanotube sensors. Chem Phys. 2012;392 (1):71-77.

[59] Hamadanian M, Khoshnevisan B, Fotooh FK, Tavangar Z. Computational study of super cell Al-substituted single-walled carbon nanotubes as CO sensor. Computational Materials Science. 2012;58:45-50.

[60] Liu H, Ma H, Zhou W, Liu W, Jie Z, Li X. Synthesis and gas sensing characteristic based on metal oxide modification multi wall carbon nanotube composites. Appl Surf Sci. 2012;258 (6):1991-1994.

[61] Li W, Jung H, Hoa ND, Kim D, Hong S-K, Kim H. Nanocomposite of cobalt oxide nanocrystals and single-walled carbon nanotubes for a gas sensor application. Sens Actuators, B. 2010;150 (1):160-166.

[62] Tulliani J-M, Cavalieri A, Musso S, Sardella E, Geobaldo F. Room temperature ammonia sensors based on zinc oxide and functionalized graphite and multi-walled carbon nanotubes. Sens Actuators, B. 2011;152 (2):144-154.

[63] Jang DM, Jung H, Hoa ND, Kim D, Hong S-K, Kim H. Tin Oxide-Carbon Nanotube Composite for NOX Sensing. Journal of Nanoscience and Nanotechnology. 2012;12 (2):1425-1428.

[64] Leghrib R, Pavelko R, Felten A, Vasiliev A, Cané C, Gracia I, et al. Gas sensors based on multiwall carbon nanotubes decorated with tin oxide nanoclusters. Sens Actuators, B. 2010;145 (1):411-416.

[65] Wongchoosuk C, Wisitsoraat A, Phokharatkul D, Tuantranont A, Kerdcharoen T. Multi-Walled Carbon Nanotube-Doped Tungsten Oxide Thin Films for Hydrogen Gas Sensing. Sensors. 2010;10 (8):7705-7715.

[66] Van Hieu N, Duc NAP, Trung T, Tuan MA, Chien ND. Gas-sensing properties of tin oxide doped with metal oxides and carbon nanotubes: A competitive sensor for ethanol and liquid petroleum gas. Sens Actuators, B. 2010;144 (2):450-456.

[67] Leghrib R, Felten A, Pireaux JJ, Llobet E. Gas sensors based on doped-CNT/SnO2 composites for NO2 detection at room temperature. Thin Solid Films. 2011;520 (3): 966-970.

[68] Yun J, Im JS, Kim H-I, Lee Y-S. Effect of oxyfluorination on gas sensing behavior of polyaniline-coated multi-walled carbon nanotubes. Appl Surf Sci. 2012;258 (8): 3462-3468.

[69] Mangu R, Rajaputra S, Singh VP. MWCNT–polymer composites as highly sensitive and selective room temperature gas sensors. Nanotechnology. 2011;doi: 10.1088/0957-4484/22/21/215502.

[70] Sayago I, Fernández MJ, Fontecha JL, Horrillo MC, Vera C, Obieta I, et al. New sensitive layers for surface acoustic wave gas sensors based on polymer and carbon nanotube composites. Procedia Engineering. 2011;25 (0):256-259.

[71] Sayago I, Fernández MJ, Fontecha JL, Horrillo MC, Vera C, Obieta I, et al. New sensitive layers for surface acoustic wave gas sensors based on polymer and carbon nanotube composites. Sens Actuators, B. 2012;doi:10.1016/j.snb.2011.12.031.

[72] Sayago I, Fernández MJ, Fontecha JL, Horrillo MC, Vera C, Obieta I, et al. Surface acoustic wave gas sensors based on polyisobutylene and carbon nanotube composites. Sens Actuators, B. 2011;156 (1):1-5.

[73] Viespe C, Grigoriu C. Surface acoustic wave sensors with carbon nanotubes and SiO2/Si nanoparticles based nanocomposites for VOC detection. Sens Actuators, B. 2010;147 (1):43-47.

[74] Yun S, Kim J. Multi-walled carbon nanotubes-cellulose paper for a chemical vapor sensor. Sens Actuators, B. 2010;150 (1):308-313.

[75] Kumar B, Castro M, Feller JF. Poly (lactic acid)-multi-wall carbon nanotube conductive biopolymer nanocomposite vapour sensors. Sens Actuators, B. 2012;161 (1): 621-628.

[76] Yuana CL, Chang CP, Song Y. Hazardous industrial gases identified using a novel polymer/MWNT composite resistance sensor array. Mater Sci Eng, B. 2011;176 (11): 821-829.

[77] Li W, Hoa ND, Kim D. High-performance carbon nanotube hydrogen sensor. Sens Actuators, B. 2010;149 (1):184-188.

[78] Lvova L, Mastroianni M, Pomarico G, Santonico M, Pennazza G, Di Natale C, et al. Carbon nanotubes modified with porphyrin units for gaseous phase chemical sensing. Sens Actuators, B. 2011:doi:10.1016/j.snb.2011.1005.1031.

[79] Lu Y, Meyyappan M, Li J. Trace Detection of Hydrogen Peroxide Vapor Using a Carbon-Nanotube-Based Chemical Sensor. Small. 2011;7 (12):1714-1718.

[80] Lu Y, Meyyappan M, Li J. A carbon-nanotube-based sensor array for formaldehyde detection. Nanotechnology. 2011;doi:10.1088/0957-4484/22/5/055502.

[81] Ruiz A, Arbiol J, Cirera A, Cornet A, Morante JR. Surface activation by Pt-nanoclusters on titania for gas sensing applications. Materials Science and Engineering: C. 2002;19 (1-2):105-109.

[82] Ghasempour R, Mortazavi SZ, Iraji zad A, Rahimi F. Hydrogen sensing properties of multi-walled carbon nanotube films sputtered by Pd. Int J Hydrogen Energy. 2010;35 (9):4445-4449.

[83] Randeniya LK, Martin PJ, Bendavid A. Detection of hydrogen using multi-walled carbon-nanotube yarns coated with nanocrystalline Pd and Pd/Pt layered structures. Carbon. 2012;50 (5):1786-1792.

[84] Su P-G, Chuang Y-S. Flexible H2 sensors fabricated by layer-by-layer self-assembly thin film of multi-walled carbon nanotubes and modified in situ with Pd nanoparticles. Sens Actuators, B. 2010;145(1):521-526.

[85] Zilli D, Bonelli PR, Cukierman AL. Room temperature hydrogen gas sensor nanocomposite based on Pd-decorated multi-walled carbon nanotubes thin films. Sens Actuators, B. 2011;157 (1):169-176.

[86] Ju S, Lee JM, Jung Y, Lee E, Lee W, Kim S-J. Highly sensitive hydrogen gas sensors using single-walled carbon nanotubes grafted with Pd nanoparticles. Sens Actuators, B. 2010;146 (1):122-128.

[87] Rumiche F, Wang HH, Indacochea JE. Development of a fast-response/high-sensitivity double wall carbon nanotube nanostructured hydrogen sensor. Sens Actuators, B. 2012;163 (1):97-106.

[88] Sadek AZ, Bansal V, McCulloch DG, Spizzirri PG, Latham K, Lau DWM, et al. Facile, size-controlled deposition of highly dispersed gold nanoparticles on nitrogen carbon nanotubes for hydrogen sensing. Sens Actuators, B. 2011;160 (1):1034-1042.

[89] Penza M, Rossi R, Alvisi M, Serra E. Metal-modified and vertically aligned carbon nanotube sensors array for landfill gas monitoring applications. Nanotechnology. 2010;doi:10.1088/0957-4484/21/10/105501.

[90] Penza M, Rossi R, Alvisi M, Suriano D, Serra E. Pt-modified carbon nanotube networked layers for enhanced gas microsensors. Thin Solid Films. 2011;520 (3):959-965.

[91] Zhou X, Tian WQ, Wang X-L. Adsorption sensitivity of Pd-doped SWCNTs to small gas molecules. Sens Actuators, B. 2010;151 (1):56-64.

[92] Li K, Wang W, Cao D. Metal (Pd, Pt)-decorated carbon nanotubes for CO and NO sensing. Sens Actuators, B. 2011;In Press, Corrected Proof:doi: 10.1016/j.snb.2011.1006.1068.

Chapter 16

Carbon Nanotube Composites for Electronic Interconnect Applications

Tamjid Chowdhury and James F. Rohan

Additional information is available at the end of the chapter

http://dx.doi.org/10.5772/52731

1. Introduction

1.1. Electronic interconnect

In the electronics industry interconnect is defined as a conductive connection between two or more circuit elements. The interconnect connects elements (transistor, resistors, etc.) on an integrated circuit or between components on a printed circuit board. The main function of the interconnect is to contact the junctions and gates between device cells and input/output (I/O) signal pads. These functions require specific material properties. For performance or speed, the metallization structure should have low resistance and capacitance. For reliability, it is important to have the capability of carrying high current density, stability against thermal annealing, resistance against corrosion and good mechanical properties.

Over the past 40 years the continuous improvements in microcircuit density and performance predicted by Moore's Law has led to reduced interconnect dimensions. According to Moore's law the number of transistors incorporated in a chip will approximately double every 18-24 months. The interconnect length increases with each generation, leading to higher resistances, while the distance between the adjacent interconnects decreases, leading to increase capacitance. Previously Al interconnect was used for VLSI processing [1]. Al and its alloys, suffer from the problems of high resistance-capacitance (RC) delay (a "time-delay" between the input and output, when a signal or voltage is applied to a circuit), poor electromigration resistance and poor mechanical properties for application in ultra-large-scale integrated (ULSI) circuits [2].

Table 1 shows the comparison of different metals resistivity at room temperature. It can be seen from the table that only three metals have lower resistivity than Al, namely Ag, Au and Cu. Ag has the lowest resistivity but it has poor electromigration reliability. Electromigration is the transport of material caused by the gradual movement of the ions in a conductor due to the momentum transfer between conducting electrons and diffusing metal atoms. The resistivity of Cu is 1.67 μΩ.cm, which is about 40% better than Al. The self-diffusivity (the spontaneous movement of an atom to a new site in a crystal of its own species) of Cu is also the smallest among the four metals, resulting in improved reliability [3, 4].

Metal	Bulk resistivity μΩ.cm
Ag	1.63
Cu	1.67
Au	2.35
Al	2.67
W	5.65

Table 1. Comparison of the bulk resistivity for different metals.

Table 2 shows a comparison of the activation energy (the minimum energy required for movement of an atom from a lattice position in a crystal) and melting temperature of Al vs. Cu. It can be seen from this table that Cu is a more reliable metal than Al with more energy required for diffusion of Cu atoms. The reason that Cu had not been used much earlier than its introduction in 1997 was because of device reliability concerns and processing difficulties. Cu diffuses rapidly through SiO_2 in the presence of an electric field [5]. This causes degradation of transistor reliability by increasing metallic impurity levels in the Si. Another problem with Cu is that it oxidises at low temperatures but without self-passivation [6]. Cu is also difficult to etch unlike Al. This means that the classical approach where metal is deposited over the entire surface, structures created in the metal and finally infilled with dielectric (oxide) cannot be followed with Cu.

Interconnect Metal	Melting Point °C	Ea for lattice diffusion eV	Ea for grain boundary diffusion eV
Al	660	1.4	0.4 – 0.8
Cu	1083	2.2	0.7 – 1.2

Table 2. Comparison of the active energy for diffusion of Al vs. Cu.

To overcome the problems of Cu integration the inter-level dielectric (ILD) is first deposited and patterned to define "trenches" into which the metal lines of the interconnect will be placed. A thin layer of barrier material (typically refractory metals or their alloys) is deposited generally using a physical vapour deposition (PVD) process. This layer covers the entire surface to act as a barrier to Cu diffusion. After the deposition of the barrier layer the Cu interconnect is deposited. This can be achieved by conventional methods such as physical vapour deposition (PVD) [7] and chemical vapour deposition (CVD) [8]. However, PVD presents poor step coverage in sub-micrometer dimension vias and trenches. This technique deposits blanket films which would require further patterning. CVD also requires the use of combustible and toxic precursors at elevated temperatures which has limited the development of Cu deposition by CVD. In 1997 IBM developed the electrodeposition technique (dual damascene) for Cu metallization [9]. Electrodeposition has become the standard method for Cu metallization with demonstrated uniformity, gap filling ability and low processing temperatures. In the dual damascene technique, lines and vias can be filled with electrodeposited Cu at the same time. Fig. 1 shows a schematic diagram of via filling with Cu and the requirement to achieve superfilling or bottom up deposition through the use of suitable additives in the plating bath rather than subconformal or conformal which result in voids or seams in the Cu. [10-12].

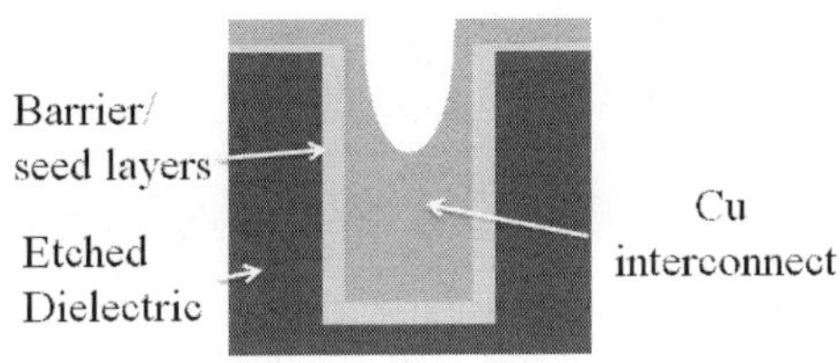

Figure 1. Cross section schematic of interconnect trench or via showing 'super-filling' or 'bottom-up filling' of features through the use of specific plating bath additives for optimum void-free profile evolution in damascene processing [9].

Semiconductor manufacturers have adopted the electroplating technique for Cu interconnect deposition in electronic devices and continue to work on miniaturization of device and feature sizes. Fig. 2 shows cross-sections from the International Technology Roadmap for Semiconductors (ITRS) of a typical microprocessor and application specific integrated circuit where the interconnect of different lines and vias between two adjacent layers are filled with Cu. As the feature sizes decrease and consequently the operating currents increase, electromigration becomes a serious issue once more [13].

The effect is important in applications where high direct current densities are used, such as in high performance processors. Grain boundaries are the fastest diffusion path for Al electromigration (activation energy 0.6 eV for grain boundary diffusion and 1 eV for interface diffusion) but an interface is the fastest diffusion path for Cu (activation energy 1.2 eV for grain boundary diffusion and 0.7 eV for interface diffusion) [14, 15]. The difference in electromigration mechanism drives different focus areas for Cu and Al reliability improvement. The damascene process requires the removal of overdeposited Cu by chemical mechanical polishing (CMP). The CMP produced top Cu surface is the fast Cu diffusion path which needs to be reliably capped. A nonconductive barrier layer is generally applied as the cap layer (silicon nitride, silicon carbide, nitride silicon carbide etc) is used to cover the top surface of the Cu line. However, there are some issues with using dielectric caps to passivate

Cu. As devices become smaller, the current density through the interconnect increases leading to the requirement for better electromigration resistance. The dielectric cap generally has a higher dielectric constant than the interlevel dielectric, resulting in an increase in line-to-line capacitance. Improved Cu electromigration resistance was reported when Cu lines were protected with thin conductive surface capping layers of self-aligned electrolessly deposited CoWP or CoSnP etc [16, 17]. Diffusion barrier layers such as Ta or TaN for Cu metallization act as redundant layers for current shunting as well as for uniform Cu seed deposition. It was reported that Cu vias are the weak link in the interconnect metallization [14]. The Cu via connects directly to the Cu metal below. If a void forms in the Cu underneath the via, there is no redundant layer available for current shunting. This is the primary cause of early failure distribution in Cu interconnects. For the 22 nm technology node or below, the interconnect metal should have current carrying capability of more than 10^7 A/cm^2 to overcome the electromigration issue but Cu is limited to 10^7 A/cm^2.

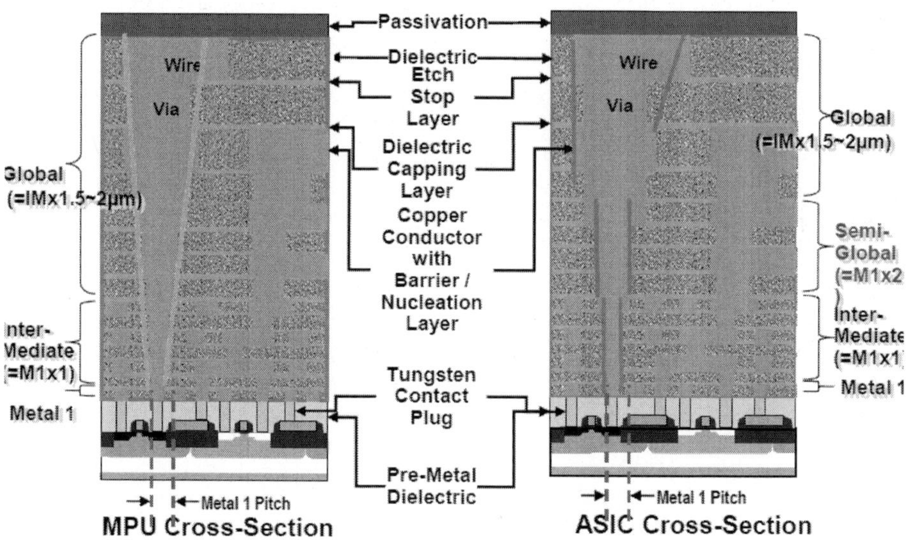

Figure 2. Typical cross section illustrating hierarchical scaling methodology [ITRS technology road map, 2011 update].

1.2. Carbon nanotubes

Carbon nanotubes (CNTs) have unique electrical, thermal and mechanical properties [18]. They can carry an electrical current density of ~ 4 × 10^9 A cm^{-2}, which is three orders of magnitude higher than Cu [19]. CNT's have high aspect ratio and the mean free path of the carriers (or the probability of an electron transmitted from at one end of the CNT to the other without phonon scattering or other thermal effects) in the CNT at 10 µm is much longer

http://dx.doi.org/10.5772/52731

than any metal (e.g. 40 nm for Cu). In addition the covalent C-C bonding between the neighbouring atoms in the CNT is one of the strongest bonds reported in the literature [20] and C atoms will not migrate even under very high current density (activation energy 7.7 eV for atom movement). Thus the electromigration resistance of CNT is much better than other interconnects material such as Al and Cu. Because of these advantages over Cu, CNTs require consideration as the next generation interconnect material for specific applications, such as through silicon vias (TSV) for stacked die.

However there are still significant scientific and engineering challenges to incorporate CNTs in devices. CNTs can deposit inside of vias on suitable catalyst like zeolite [21]. It is necessary to ensure the selective growth of metallic CNTs in vias and lines to achieve better electrical conductivity. Alternatively CNTs can be first synthesized in a powder form and metallic CNTs separated from bulk growth CNTs (mixture of metallic and semiconducting). After that metallic CNTs would need to be transferred onto specific wafer locations. The scale of this task is obvious when considering that there are billions of transistors in a microprocessor and the placement of CNTs inside of all vias and trenches on the wafer is unlikely.

	Single CNT	Cu at 22 nm node
Maximum Current density (A/cm^2)	1 x 10^9	1 x 10^7
Electrical conductivity (S/m)	10^6-10^7	6 x 10^7
Thermal coefficient of resistivity (/°C)	-1.5 x 10^{-3}	4 x 10^{-3}
Thermal conductivity (W/m K)	6,000	400
Coefficient of thermal expansion (ppm/°C)	-1.5	17
Activation energy (eV)	7	2

Table 3. Comparison of the properties of single walled CNTs vs. Cu.

The contact resistance between CNTs and metal is large (≥1 kΩ). The minimum resistance for a ballistic single-walled CNT is ~ 6.5 kΩ, Therefore, relatively dense arrays of nanotubes will be needed to replace Cu interconnects and these arrays will still only show reduced resistance by comparison with Cu interconnect for line lengths ≥ 1 μm. The ITRS [22] therefore predicts for the 22 nm node an estimated resistance for a 17 nm x 38.5 nm x 1.5 μm Cu interconnect is R_{Cu} 145 Ω. An ensemble of ~ 45, 1 nm diameter defect-free metallic CNTs with mean inter-tube separations ~ 4 nm in a trench of these dimensions would have the same total resistance as the Cu line.

Intrinsic voids between CNTs significantly reduce the electrical and thermal conductivity and bring reliability challenges for the use of CNTs as interconnects. Contact resistance between CNTs and interconnect metal like Cu becomes the dominant source of electrical and thermal resistance which significantly reduces the benefits of CNTs. The potential use of CNTs alone as interconnect in semiconductor manufacturing is still open to debate at this time.

1.3. Metal CNT composites

To overcome some of the interconnect issues described above metal-CNT composites can be an alternative candidate material for future interconnects. The composite material would increase the contact area between vias and interconnect lines. There is also less chance of intrinsic voids between CNTs as they would be metal filled. Among the different metals Cu is the best choice at this moment to use as a composite material with CNTs for interconnects applications because of superior electrical and thermal conductivity. Chai et al [23,24] and Yoo et al [25] reported that Cu fills the voids between neighbouring CNTs which results in a more densely packed structure. They reported that the addition of Cu increases the contact area between the nanotube (1 D) and the substrate (3 D contact) making it a mechanically strong material that can sustain high electrical or thermal stress cycling. To obtain superior properties of metal-CNT composites, it is necessary to achieve a homogeneous dispersion of CNT throughout the metal matrix. It is also necessary that the composites should be void free to obtain better electrical and thermal conductivity.

Cu/CNT composites can be prepared by powder metallurgy, electroless plating or electrodeposition techniques [25, 26]. Among these methods electrochemical routes are relatively straightfoward methods to produce defect free nanocomposites [24, 25]. Powder metallurgy requires sintering at elevated temperatures that may damage CNT's and the difficulty of composite placement remains an issue. Chen et al [26] observed a clear separation of CNTs and Cu matrix composites deposited by powder metallurgy. To achieve optimum performance, CNTs need to be well-dispersed and aligned parallel rather than randomly oriented in the Cu matrix. Hjortstam et al [27] estimated the increase of effective conductivity as a function of the volume fraction of CNT in a Cu matrix. Their calculation showed that 30-40% CNT is needed in the composite with a resistivity 50% lower than for Cu. Liu et al [28] found that electrical sheet resistance is lower in Cu/CNT composite films than Cu and also decreases due to annealing at 200 - 300ºC.

Improved electromigration resistance is expected to result from the location of the alloy element at grain boundaries to prevent movement of Cu at those vulnerable points, which may lead to wiring voids (opens) or hillocks (shorts) during operation [25]. Cu/CNT composites may also improve thermal conductivity of lines and vias which also increases electromigration resistance. Chai et al [24] reported that the Cu/CNT composite vias have lower electrical resistance than that of vias with CNT only. Their electromigration test results showed that the void growth rate for a Cu/CNT composite strip was four times lower than that of pure Cu strip. Their electromigration test of Cu and Cu/CNT composites which were carried out in the temperature range of 100 to 250ºC and current density from 5×10^5 to 2×10^6 A/cm^2 using a conventional Blech-Kinsborn test structure showed that longer strips had larger void length, while no void formation was detected in the strips below 40 μm. Below the critical length the electromigration flux is balanced by the opposing backflow generated by the stress gradient in the test strip.

Yoo et al [25] fabricated Cu/MWCNT composite films by a pulsed electrodeposition technique with additives and obtained a dense structure without any voids. Their microstructure analysis showed that most of the MWCNTs exist at the Cu grain boundaries and cross-linked each other. They reported that C content in the composite increased by increasing CNT concentration in the bath but it decreased with annealing. Chai [29] et al reported that the mechanical strength of Cu/CNT nanocomposite was three times higher than that of pure Cu. Chen et al [26] observed good interfacial bonding between CNT and Cu when the nanocomposites were codeposited by electrodeposition. They reported that for Cu/SWCNT nanocomposites, the radial breathing mode (RBM) in the Raman was absent and the tangential or G-band had shifted and widened. Recently several patents on the metal/CNT composites co-deposited by electrodeposition have been filed [30-32]. The comparison of electrical resistivity of Cu and Cu/SWCNT film which was reported by Chan [30] are shown in table 4.

	electrodeposited Cu (thickness = 10.5 μm)	**Cu/CNT composite (thickness = 22 μm)**
Resistivity (μΩ.cm)	1.72	1.22
Sheet resistance (mΩ/sq)	1.64	0.56

Table 4. Comparison of the electrical resistivity of electrodeposited Cu/SWCNT composite film vs. Cu alone [30].

1.4. Chlorosulphonic acid for CNT dissolution

Recently Davis et al [33] reported that CNTs can dissolve spontaneously in chlorosulphonic acid solution up to 0.5 wt % [5 g/l], which is much higher than previously reported in other acids (up to 80 mg/l). They reported that at higher concentrations, they form liquid-crystal phases that can be processed into fibres and sheets of controlled morphology. Their proposed phase diagram helps to identify the optimal starting fluid composition and determine micro and macrostructure of fibres and films such as plated fibres, straight fibres and smooth films. Plated fibres have potential application for hydrogen storage and sensors because of high surface area. Straight fibres are of interest for structural reinforcement and smooth, dense films for electrical applications such as electrically conductive thin films.

1.5. Purification and functionalization of carbon nanotubes

A significant problem in dealing with CNTs is the difficulty to separate them as the individual CNTs form bundles due to van der Waals attractive forces. Also in all of the synthesis techniques several impurities like catalyst particles, amorphous carbon etc. are also present in the bundles of CNTs. These impurities may deteriorate the properties of CNTs. To prepare stable and homogeneous dispersions of CNTs considerable efforts have been made [34-39] but the solubility of CNTs in water or organic solvent is relatively low. At room temperature the solubility of CNTs is in the range of 60 to 80 mg/l [34]. In order to achieve better stabilization, CNTs require additional hydrophilic groups directly on the CNT walls or provided by surfactant molecules to impart ionic charge on the CNTs [40-42]. The most common hydrophilic groups are –OH-, -COOH-, $-SO_3$-, $-NH_2$-. Functionalization of CNTs can be an important factor to manipulate the properties of CNTs. With functionalization CNTs mau be more easily separated. Several methods have been suggested for the purification and functionalization of CNTs mainly based on covalent and noncovalent functionalization. Functionalized CNTs are easily dispersed and highly ionized in contract with water [43].

1.5.1. Covalent functionalization of carbon nanotubes

Several methods have been suggested for covalent functionalization of CNTs. The most common technique is to functionalize CNTs in concentrated acid by refluxing. In this process raw materials are sonicated followed by refluxing at 120-130°C. This process requires long processing times. After cooling at room temperature, the mixture is then centrifuged, leaving a black precipitate and a clear brownish yellow supernatant acid. Ko et al [44] reported that the presence of metal impurities in the MWCNTs is reduced significantly using this method. The purification process usually requires two repeat processing steps. The first step is acid reflux which washes metal catalyst and carbon impurities and the second step is annealing which burns the defective tubes and carbon particles. Ko et al [44] also used a microwave oven technique to purify MWCNTs. Chen and Mitra [45] reported that MWCNTs were less reactive and had lower solubility than the SWNTs. Li and Grennberg [46] also found that microwave heating is highly useful for side wall functionalization of MWCNTs.

Lau et al [47] reported that the electrical conductivity of MWCNTs increased with different functionalization techniques such as oxidation, acid reflux, dry UV-ozonolysis. They explained that the new functionalized groups increase the number of bands near the Fermi level, promoting electron transfer between the carbon atoms. They have claimed that CNT functionalization by UV-ozonolyzed technique significantly increases the electrical conductivity of CNTs. Agarwal et al [48] reported that controlled defect creation could be an attractive strategy to induce an electrical conductivity increase in MWCNTs. They reported that the outermost shell of MWCNTs is semiconducting so it is difficult to make electrical contacts to the inner shells of MWCNTs. Functionalization of CNTs may promote cross-shell bridging via sp^3 bond formation. They proposed that intershell bridging facilitates charge carrier hopping to inner shells which can serve as additional charge carrier transport pathways. Tantang et al [49] also reported that acid treatment increases the conductivity of CNT electrodes.

1.5.2. Non-covalent functionalization of carbon nanotubes

Covalent functionalization may deteriorate the unique ionic properties of CNTs by the formation of new covalent bonds on the CNTs wall. To overcome this disadvantage non-covalent functionalization mainly based on polymer surfactant interaction was developed that can disperse nanotubes easily but not degrade the CNT's unique properties [50]. The proposed mechanism for this solubilisaiton is through an individual CNT being wrapped by the polymer which acts as a surfactant in the solution to achieve separation. A surfactant is a wetting agent which lowers surface tension of liquids. It is usually an organic compound that contains both hydrophobic and hydrophilic groups. As a result, they are soluble both in organic solvents and water. Surfactants are classified based on the presence of a charged group. The head of an ionic surfactant carries a net charge. If the charge is negative, the surfactant is more specifically called anionic; if the charge is positive, it is cationic. Ionic surfac-

http://dx.doi.org/10.5772/52731

tants not only separate individual CNTs but also carry charge to the surface of CNTs so that the CNTs can be codeposited with Cu by electrodeposition. Examples of surfactants which have been investigated are given below.

1. Nafion® as a surfactant

In our study we primarily used nafion, a polymer surfactant for the dispersion of CNT bundles. It is a sulfonated tetrafluorethylene co-polymer with ionic properties which bears a polar side chain (-SO_3H) and hydrophobic backbone (-CF_2-CF_2). It has unique ionic properties because of the incorporation of perfluorovinyl ether groups terminated with sulfonate groups onto a tetrafluoroethylene (Teflon) backbone. The hydrophobic backbone strongly anchors to the hydrophobic side-wall of CNTs. On the other hand the polar side-chain of the polymer imparts sufficient ionic charge to the CNT surface which enhances the solubility of CNTs in liquid solvents. Fig. 3 shows the chemical structure of nafion.

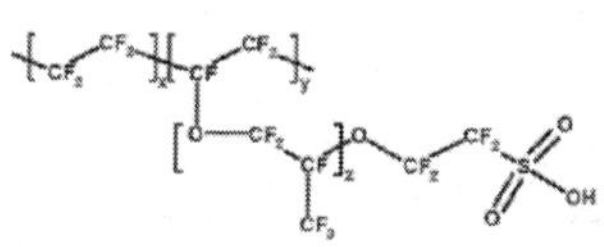

Figure 3. Chemical structure of Nafion.

2. CTAB as a surfactant

Cetyl trimethyl ammonium bromide (CTAB) is a cationic surfactant. Chen et al [51, 52] used CTAB for the dispersion of CNTs to prepare CNTs/Ni composites by an electroless deposition technique. The chemical structure of CTAB is shown in fig. 4 (a). As a cationic surfactant it can make the CNT surface positively charged to assist the codeposition of CNT on the cathodic surface [52]. The CNT with negative charge readily adsorbs the cationic surfactant. This adsorption develops a net positive charge on the CNT, which prevents them from agglomerating and leads to electrostatic attraction to the cathode surface with negative potential [51]. The net positive charge on the CNT increases the amount of CNT in the deposits. To calculate the volume fraction of CNTs, they dissolved the deposits in nitric acid. The CNTs in the deposits were filtered and the quantity of the CNTs in the deposits determined [52]. They reported that the content of CNTs in the deposit increases with an increase of CNT concentration in the bath, up to a maximum value at the CNT concentration of 1.1 g/l and then decreases. They explained this as a result of the CNT agglomeration in solution at higher concentration which reduces the content of CNT in the deposit. They also reported that the saturation concentration increases with decrease of length of CNTs because the longer CNTs tend to agglomerate more readily.

3. SDS as a surfactant

Sodium dodecyl sulfate (SDS) is an anionic surfactant used to improve the surface uniformity of the composite deposit [53]. The chemical structure of SDS is shown in fig. 4 (b).

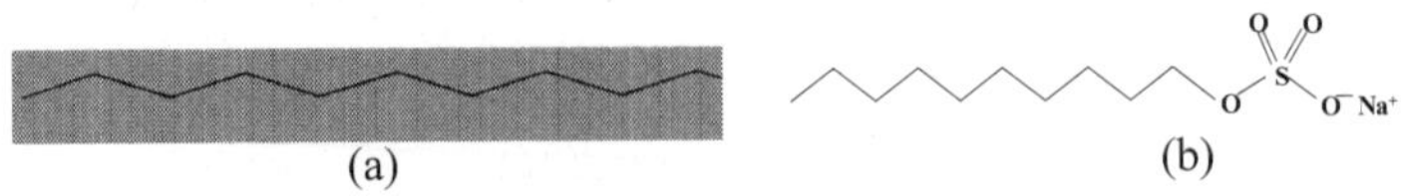

Figure 4. Chemical structure of (a) CTAB and (b) SDS.

A comparison of hardness and XRD patterns of Ni/CNT composites by using SDS and CTAB surfactant in the bath with Ni was performed [53]. The hardness changes for the composite films and depends on the concentration of CNTs in the bath as well as surfactant. The composite from the CTAB bath showed an increase of hardness unlike that of the composite from the SDS containing bath. It can be also seen from the XRD data that (111) is the preferred plane of Ni when the bath contains CTAB like pure Ni deposition. On the other hand, SDS in the bath reduces the preferred Ni (111) orientation significantly in the composite. A summary of surfactants which are commonly used for the dispersion of CNTs are reported in table 5.

CNT	CNT: g/l	Surfactant	Surfactant: g/l	Composites	References
MW	0.3	CTAB	0.6	Ni/CNT	[53]
MW	0.1	$Mg(NO_3)_2$		Cu/CNT	[26]
MW	6	PA	0.5	Cu/CNT	[23]
SW	3	CTAC	3	Cu/CNT	[54]
MW	1	CTAB	4	Cu/Zn	[55]
SW	0.002	Nafion	0.4	Nafion/CNT	[56]
MW	0.6	Gelatine	0.4	CNT/Cu	[57]
MW	2	CTAB		Ni/CNT	[58]
SW	2			Cu/CNT	[24]
MW	1	Nafion	0.01	Nafion/CNT	[59]
MW	6	PA	0.1	Cu/CNT	[60]
MW	1			Ni-Co/CNT	[61]
MW	1	Nafion	5	Nafion/CNT film	[62]
SW	0.2	Nafion		Nafion/CNT	[63]
SW	0.05	SDS	0.1	PVP/CNT	[50]

Table 5. Literature review of CNT dispersion surfactants.

1.6. Analysis of carbon nanotubes in deposit

There is very little in the literature on quantifying CNT content in the deposits [48, 51, 52]. Arai et al [60, 64, 65] measured the content of multi-walled CNTs in the electrodeposited composite film by dissolving the deposit in hot nitric acid. The CNTs in the nitric acid solution were filtered, dried and weighed. Osaka et al [66] reported the carbon content in the deposit was analyzed by the combustion infrared absorption method (CS 444, LECO) as this element analyzer is capable of analysing for carbon. The summary of CNT content in the deposit obtained from the literature is shown in table 2.5.

Bath	Surfactant	CNT g/l	Weight % CNT in deposit	Reference
Ni	PA	6	Up to 1	[64]
Cu	PA	2	0.4	[60]
Ni	SDS	0.3	Up to 7	[53]
Cu	PA	Up to 6	Up to 2.5	[25]

Table 6. Literature data of CNT content in the plating bath and deposit.

2. Cu and Cu/CNT pillars for flip chip interconnect assembly

Historically, IC chips have been electrically connected to the substrate by a wire bond method. In this method, the chip faces up and is attached to the package via wires. This connection has limited electrical performance and reliability problems in addition to requiring pad location at the edge of the die. Flip chip, also known as 'Controlled Collapse Chip Connection, C4', replaced the traditional wire bond method. In this method, solder bumps are deposited on the chip pads over the full area of the top side of the wafer during the final wafer processing step. In order to mount the chip to external circuitry (e.g., a circuit board or another chip or wafer), it is flipped over so that its top side faces down, aligned to the substrate and then the solder is reflowed to complete the interconnect. Generally, Sn-Pb based solder bumps have been used in flip chip packaging to connect chips to external circuitry. According to the International Technology Roadmap for Semiconductors the total number of I/Os will reach up to 10,000 cm^2 chip area by 2014 which require finer interconnect with a pitch size less than 20 μm. To fabricate such fine pitch interconnect, conventional solder bump requires fine solder deposition or paste particle which are not readily available [67]. It is also important to reduce lead-based solders for environmental concerns (RoHS compliance). As circuit density increases, devices are also more vulnerable to non-uniform thermal distribution.

Cu has higher thermal conductivity than most binary or ternary solders. Cu bumps in flip chip assembly offer increased reliability, extended temperature range capability, greater mechanical strength, higher connection density, improved manufacturability, better electrical and heat dissipating performance over Pb-Sn solder. It is also less expensive and decreases the amount of solder needed to create bumps. Cu pillars do not change shape during reflow so they do not encounter any volumetric redistribution which can lead to voids in the solder. Because of these advantages the semiconductor industry is adopting Cu pillar bump by electrodeposition for flip-chip attachment to replace the typical Pb solder [67, 68]. Power and thermal non-uniformity in devices are increasing steadily with each new device generation leading to serious concerns for the industry regarding thermal issues. Mechanical stress on Cu bumps generated by the difference in thermal expansion coefficients between the chip and the substrate materials can lead to device failures. This differential thermal expansion also creates shearing forces at the bump. As a result bumps are most vulnerable to damage. Repeated thermal expansion and contraction leads to fatigue cracking of the bump.

Cu/CNT composites could be a suitable candidate material to resolve these issues for next generation flip chip assembly. CNTs have high mechanical strength (10-60 GPa, c.f. Cu 70 MPa) and thermal conductivity (>3000 W/m.K, c.f. Cu 400 W/m.K) which may alleviate the issues related to die degradation and non-uniform temperature distribution in the pillars. CNTs have a negative temperature coefficient of resistivity (- 1.5 x10^{-3}/ºC, c.f. Cu + 4 x10^{-3}/ºC) and low coefficient of thermal expansion (- 1.5 ppm/ºC, c.f. Cu + 17 ppm/ºC) which can make the Cu/CNT composites material more reliable against thermal cycling and fatigue with less risk of stress induced failure. Typical photolithography techniques can be utilised to fabricate Cu/CNT pillar bumps on chip. Arai et al [64] recently demonstrated Cu/CNT pillar emitters deposited by electrodeposition on a patterned substrate.

3. Cu and Cu/CNT in through Si via (TSV) for 3D interconnect

Cu electrodeposition in TSV features is a key component of new 3D integration approaches that are of great interest in the semiconductor industry [69]. 3D integration increases performance and lowers power consumption due to reduced length of electrical connections. Cu has been selected as the TSV interconnect because of its low electrical resistance and compatibility with conventional multilayer interconnection in large-scale integration (LSI) and back-end processes. The key challenges for TSV plating processes are to fill the vias across the entire wafer and to complete the fill as fast as possible to minimize cost. TSV interconnect shortens the interconnect requirements and reduces signal delay. However, it is difficult to fill high aspect ratio vias without voids through conventional damascene electroplating. Perfect filling without voids is required to minimise interconnect failure and reliability issues. TSVs have been extensively studied because of their ability to achieve chip stacking for enhanced system performance. This is a very promising technology that may replace wire bonding in chips or single chip solder bumping. Metal filled TSVs allow devices to be connected using a 3D approach [69]. Cu is the best low cost conductor for TSV interconnect and an extension of the damascene plating in smaller features. Recently enormous attention has been given to bottom up filling of TSVs to fill high aspect ratio vias without voids like conventional damascene electroplating [70-72]. However, there are key issues that need to be resolved, such as process reliability, electrical continuity and thermal management. TSVs should have the ability to maintain operation over a wide range of temperatures and to withstand these temperatures in a cyclic manner. The TSV material proper-

http://dx.doi.org/10.5772/52731

ties must include mechanical strength, good thermal conductivity and stability with thermal cycling.

The key issues for 3D integration are process reliability, die degradation, electrical continuity, bump to pad electrical contact and thermal management. Temperature cycling and thermal shock accelerate fatigue failures. Also, non-uniform temperature distribution may influence the operation of circuits and sensing elements dramatically. Stress fields resulting from differential thermal expansion of Cu-based TSV may cause serious problems. The reliability problems of high aspect ratio TSV interconnect may be alleviated by the codeposition of carbon nanotubes (CNTs) with Cu as a suitable composite material. CNTs have high mechanical strength, thermal conductivity and low coefficient of thermal expansion which may alleviate the issues related to die degradation and non-uniform temperature distribution in the TSVs.

4. Experimental methods

In this work the Cu/CNT composites codeposition process was assessed and the deposited materials characterised. Electrochemical analysis of the deposition requires an analysis of the nucleation and growth characteristics for the candidate materials. MWCNTs have been added to the typical Cu sulphate plating bath to achieve homogeneous Cu/MWCNT composites. Here, we will report electrochemical analysis and kinetics of electrodeposited Cu when MWCNTs were present in the bath. Solubilisation or suspension of the CNTs in the Cu bath is also a key requirement. Composite plating bath chemistries for Cu/CNT deposition were investigated. The influence of typical additives in the Cu bath on the deposit characteristics was determined for optimised electrodeposition in vias and trenches. The influence of different surfactants on the deposition and electrical properties of composite films was also analyzed. Cu and Cu/CNT composites were electrodeposited on planar and structured substrates. Microstructure characterization of the deposit employed scanning electron microscopy (SEM), focussed ion beam microscopy (FIB) and x-ray diffraction (XRD). The sheet resistance of Cu/CNTs film and changes due to self-annealing and high temperature annealing were monitored by 4 point probe resistivity techniques. Cu/CNT composites were also deposited in test structures. After chemical mechanical polishing of the test structures, the line resistance was measured using a Cascade probe station.

The amount of CNTs in the deposit was determined by dissolving the deposit in a concentrate HNO_3 solution. The Cu/CNT films were deposited on 1 cm X 1 cm thin film sputtered Cu on Si. The deposition current was 1 A and deposition time was 1 h. The concentration of CNTs in the bath was 10 or 100 mg/l. After deposition, the sample was dipped in hot concentrate acidic solution (65% HNO_3, 65°C). The diluted acid solution was then vacuum filtered using PTFE filter paper. The filtration process was repeated at least 5 times to ensure all CNTs were left on filter residue. After filtration, the PTFE membrane was dried in an oven at 80°C for at least 30 minutes to ensure the membrane was completely dried. The weight difference of the PTFE membrane before and after filtration gives the amount of

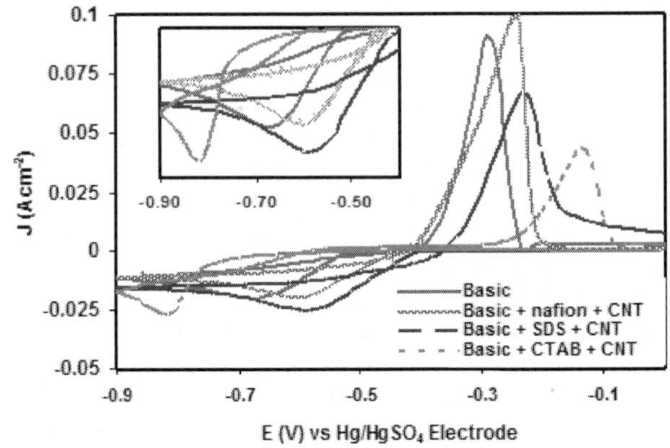

Figure 5. CVs of Cu and Cu/CNTs deposition on a glassy carbon electrode (scan rate: 0.1 V/s) from 0.24 mol dm^{-3} $CuSO_4$ + 1.8 mol dm^{-3} H_2SO_4 with/without CNTs and different surfactants in the bath.

CNTs in the deposit. The amount of CNTs in the deposit was approximately 2% by weight by using long CNTs (length 5-9 μm, diameter 110-170 nm) or short CNTs (length < 1 μm, diameter 9.5 nm) in the bath. The density of CNTs is close to 1.3 gm/cm^3 and pure Cu is 8.89 gm/cm^3 which indicates the CNTs in the deposit are up to 12% by volume.

5. Results

To utilise CNTs as a composite with Cu for interconnect applications it is necessary to verify the influence of the materials on the Cu plating chemistry. Fig. 5 shows the comparison of cyclic voltammetry of Cu and Cu/CNTs co-deposition from a simple $CuSO_4/H_2SO_4$ bath (hereafter referred to as the basic bath) with/without CNTs and surfactant at a scan rate of 0.01 V/s. It can be seen that the addition of CTAB and CNT results in a cathodic peak potential shift to a more negative value which represents a suppression influence on Cu deposition. On the other hand the Cu deposition occurs at lower overpotential when the bath contains either nafion or SDS which represents an acceleration influence. The diffusion coefficient for Cu^{2+} ions estimated from chronoamperometry data using the Cottrel equation in the basic bath (0.24M $CuSO_4$ + 1.8M H_2SO_4) is 4.5 x 10^{-6} cm^2/s. A similar value (4.6 x 10^{-6} cm^2/s) was found from the SDS containing Cu/CNTs bath. Upon addition of nafion or CTAB in the Cu/CNTs bath, the diffusion coefficient value of Cu^{2+} ions slightly increases to 5.1 x 10^{-6} cm^2/s and 5.3 x 10^{-6} cm^2/s, respectively. It is clear that CNTs and surfactants in the Cu bath do not have a significant influence on the diffusion coefficient value of Cu. These results indicate that the CNT + surfactant is compatible with the basic Cu sulphate/sulphuric acid bath chemistry. An assessment of the influence of the composite materials on baths that contain the basic constituents and the necessary additives to achieve bottom-up fill or superfilling of interconnect features in silicon technology is also required.

The kinetics of the metal nucleation and growth/dissolution can be analysed with a rotating disk electrode system. While acknowleding the limitations and complications in the kinetic analysis of Cu [12] the general trends indicated in the data are consistent with published data for the Cu sulphate system and those with typical damascene additives. It can be seen from the kinetic data analysis below that the exchange current density, i_0, for Cu nucleation and growth from the basic $CuSO_4$ bath without any additive is 7.24 mA/cm^2 and the E_0 value is - 406 mV. The exchange current value for Cu nucleation and growth in the literature varies from 1 to 15 mAcm^{-2} [73–75]. Addition of all typical additives in $CuSO_4$ bath decreases the exchange current density, i_0 for Cu nucleation and growth from 7.24 mA/cm^2 to 1.2 mA/cm^2 and increases the E_0 value from - 406 mV to - 417.5 mV. This result confirms that all additives together have a suppressor effect on Cu deposition. It can be observed that addition of 1% nafion also has a minor suppressor type behaviour on Cu nucleation and growth as it slightly decreases the exchange current density, i_0 from 7.24 mA/cm^2 to 7.07 mA/cm^2 and increases E_0 value from - 406 mV to - 410.5 mV. But addition of CNTs has an accelerator influence on Cu nucleation and growth increasing the exchange current density, i_0 from 7.24 mA/cm^2 to 10.23 mA/cm^2 and decreasing the E_0 value from - 406 mV to - 403.5 mV. It is also observed that all typical additives including nafion and nanotubes together in the solution have an overall suppressor effect on Cu deposition. The summary results are shown in table 7. It can be seen from the table that anodic slopes are in the range from 1/65 mV to 1/76 mV and cathodic slopes are in the range from 1/122 mV to 1/164 which are close to the theoretical values when the reactions are two separate single electron transfer steps. The above results show that baths containing nafion & CNTs are compatible with the existing typical $CuSO_4$ bath used in IC interconnect deposition.

Conditions	1/slope, mV Anode	Cathode	E_0/mV	I_0/ mAcm^{-2}
0.24 M $CuSO_4$ + 1.8 M H_2SO_4 (Basic bath)	154	67	- 406.0	**7.24**
Basic bath + 1% nafion	161	66	- 410.5	**7.07**
Basic bath + 1% nafion + 10 ppm CNTs	164	76	- 403.5	**10.23**
Basic bath + 50 ppm Cl^- + 300 ppm PEG + 1 ppm SPS	151	67	- 417.5	**0.54**
Basic bath + All additives + 1% nafion	128	72	- 420.0	**1.70**
Basic bath + All additives + 1% nafion + 10 ppm CNTs	122	76	- 421.0	1.66

Table 7. Summary of Tafel analysis obtained from rotating disk system.

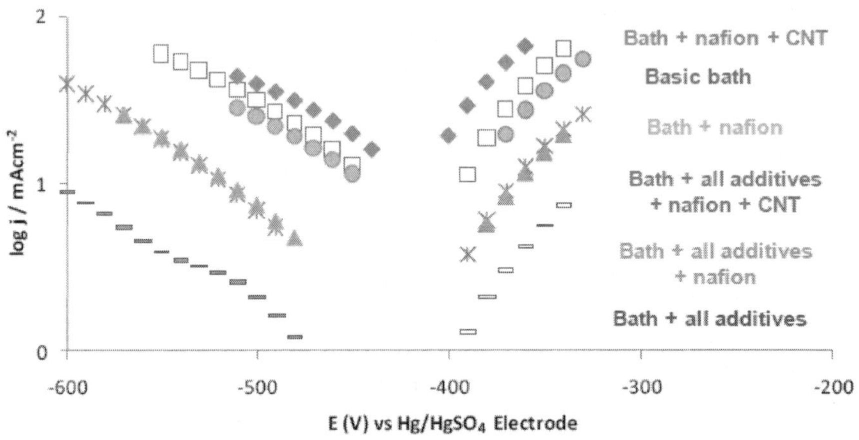

Figure 6. Comparison of Tafel plots for Cu deposition from with/without additives and CNTs in a standard Cu bath using a Cu rotating disk electrode which was rotated at 2000 rpm during experiments (Initial potential: 0 V, scan rate 0.1 V/s). Area of Cu RDE was 12.566 mm^2

On a 1 cm^2 Cu substrate (200 nm sputtered Cu on Si) the deposition current was 15 mA/cm^2 and deposition time was 1 h. The concentration of CNTs in the bath was 100 mg/l. The acid solution was then vacuum filtered using 5 μm PTFE filter paper. The length and diameter of the MWCNTs were 5-9 μm and 110-170 nm respectively. Fig. 7 shows the SEM images of PTFE membrane after filtration of Cu/CNT deposits. The CNTs are clearly observed.

Figure 7. SEM image of CNTs on PTFE membrane after filtration of dissolved Cu/CNT composites. Image magnification 5000X on left and 40000X on right.

The amounts of CNTs in the deposit are shown in table 8 which compares the percentage of CNTs in the deposit when long or short CNTs with different surfactants were added in the bath. We found the percentage of CNTs was slightly higher (CNT content 1.69% by weight) when SDS was added with long CNTs in the bath (CNT content 1.13% by weight when

http://dx.doi.org/10.5772/52731

CTAB was added). On the other hand, the percentage of CNTs was slightly higher (CNT content 1.64% by weight) when SDS or CTAB was used in the bath with short CNTs in the bath (CNT content 1.12% by weight when nafion was added). It can be seen from the table, the weight percentage of MWCNTs in the deposit was less than 2%. According to the literature [25, 60] the maximum CNT concentration in Cu/CNT composites achieved has been approximately 2.5 % by weight. So the value found in these experiments is quite reasonable. The density of MWCNTs is close to 1.3 g/cm^3 and pure Cu is 8.89 g/cm^3 which indicates that the CNT content in the deposit is up to 12% by volume.

MWCNTs Length µm	MWCNTs Diameter nm	**Surfactant**	**% CNTs**
5 – 9	110 - 170	Nafion	1.12
5 – 9	110 - 170	CTAB	1.64
5 – 9	110 - 170	SDS	1.64
<1	9.5	Nafion	1.56
<1	9.5	CTAB	1.13
<1	9.5	SDS	1.69

Table 8. The weight percentage comparison of CNTs in the composite films using different size of CNTs and surfactants in the bath. PTFE membrane was used for filtration purpose.

The electrical properties of Cu/CNT composites were assessed by determining the resistivity of submicron films. Room temperature self-anneal phenomenon is usually observed in electrodeposited Cu films [76-80]. Due to large grain growth at room temperature and annihilation of the defects (void, vacancy, stacking fault, impurities redistribution etc), the electrical resistivity of Cu may change with time. It is therefore necessary to monitor the resistivity changes of Cu and Cu/CNT composite films over time after electrodeposition. The resistivity of Cu and Cu/CNTs composite films at room temperature was monitored using a four point probe apparatus (Keithley 2400 four point probe). In each case we took 4 samples and recorded the average resistivity. The film was electrodeposited on a sputter Cu coated Si substrate. The deposition current density was 15 mAcm^{-2} and deposition time was 2 minutes. The thickness of film was measured by using surface profilometry to be approximately 660 nm.

The electrical resistivity results showed that at room temperature the resistivity of Cu/CNTs composite films (2.46 µΩ-cm) when nafion was used for the surfactant of CNTs is close to the resistivity of Cu film deposited (2.15 µΩ-cm). The resistivity of Cu/CNTs composite film was higher when SDS (3.03 µΩ-cm) or CTAB (4.19 µΩ-cm) was used as a surfactant. The results are summarised in table 9. There was a larger scatter in the distribution of resistivity data in the CTAB case. This is probably a result of a less uniform and void rich deposit from the CTAB containing bath which significantly increased the resistivity.

The resistivity of samples maintained at room temperature did not change significantly. The summary of the changes of the room temperature resistivity over time for Cu and Cu/CNT composite films deposited from different surfactant containing baths are shown in table 10. Osaka et al [66] reported that the resistivity of a deposit from an additive free bath and Cl^- + PEG containing bath was unchanged with time. But when SPS was present in the bath, the resistivity decreased over time due to self-annealing. Lee and Park [82] reported that self-annealing is caused by Cu grain boundary diffusion. They mentioned that locally high stress originated from the trapped large molecule PEG which can accelerate grain boundary diffusion of Cu. There is a lack of consensus about the cause of self-annealing [81-86]. Among the suggested possible causes for self-annealing of electrodeposited Cu film are bath compositions [83], additives [77, 81, 82], film thickness [79, 80, 84], barrier layers [86, 87], impurities [79, 85] and deposition current [80].

Bath	Resistivity / μΩ-cm
Basic (0.24 M $CuSO_4$ + 1.8 M H_2SO_4)	2.17
Basic + Nafion + CNT	2.43
Basic + SDS + CNT	3.03
Basic + CTAB + CNT	4.69

Table 9. Comparison of the resistivity of Cu and Cu/CNT composite film at room temperature 1 hour after deposition using different surfactants in the bath.

Bath	Time after deposition / hour	Resistivity / μΩ-cm
Basic	1	2.17
	312	2.15
Basic + Nafion + CNT	1	2.43
	311	2.47
Basic + CTAB + CNT	1	4.09
	313	4.19

Table 10. Comparison of the resistivity changes of Cu and Cu/CNT composite film at room temperature and 311 to 313 hours after deposition.

It is well known that through annealing at higher temperature a reduced defect Cu microstructure can be obtained [76-87]. Cu/CNT composite films (660 nm in thickness) were annealed in nitrogen at 215°C, 265°C and 315°C for 20 minutes. It can be seen that a clear decrease of sample resistivity was observed with increasing annealing temperature which is shown in table 11. The resistivity value of Cu film approaches that of bulk Cu value (1.67 μΩ-cm) after annealing at 315°C for 20 minutes. Also the resistivity of Cu/CNTs composite films decreased with increasing annealing temperature. The electrical resistivity of the Cu/CNTs composite films deposited from a nafion and CTAB containing bath became 1.88 μΩ-cm and 2.10 μΩ-cm respectively when the sample was annealed at 315 °C. The conductivity increase of the composite films was probably due to a decrease the interface resistance between CNTs and Cu matrix at the higher temperature, grain refinement and elimination of defects under high temperature annealing.

Annealing temperature °C	Basic bath	Resistivity / μΩ-cm Basic + Nafion + CNT	Basic + CTAB + CNT
No anneal	2.15	2.46	4.19
215	1.78	2.14	2.72
265	1.92	2.04	2.45
315	1.67	1.88	2.10

Table 11. Comparison of the resistivity of Cu and Cu/CNT composite films at room temperature and higher temperatures 312 hours after deposition using different surfactants in the bath.

The influence of CNT concentration in the Cu/CNT composite bath was investigated. The electrical resistivity results showed (fig. 8) that at room temperature the resistivity increased 10% when the concentration of CNT in the bath was increased from 10 mg/l to 100 mg/l. This data also shows no evidence of self annealing at room temperature for the composite material.

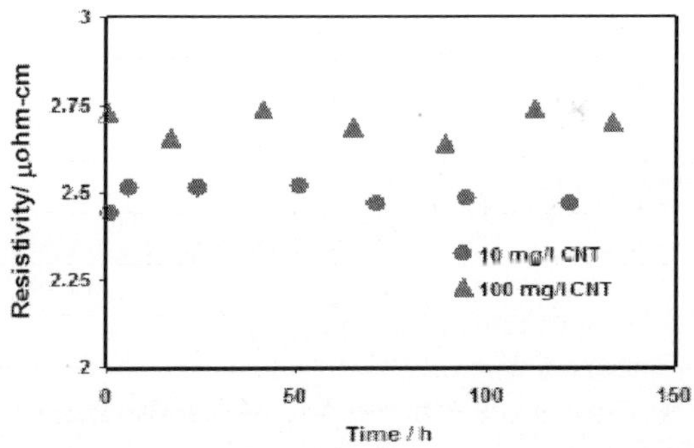

Figure 8. Comparison of the resistivity of Cu/CNT composite film at room temperature over time using 10 mg/l and 100 mg/l CNT in the bath.

It can be seen from fig. 9 that when the samples were annealed at higher temperature up to 315ºC for 20 minutes the resistivity decreased from 2.46 μΩ-cm to 1.89 μΩ-cm for 10 mg/l CNT and 2.7 μΩ-cm to 2.19 μΩ-cm for 100 mg/l CNT in the bath. It is expected that CNT content in the composite is higher when deposited from higher CNT concentration containing bath [25, 60]. The resistivity increase of the higher CNT content bath is probably due to increased CNTs content and higher contact resistance between CNTs and Cu in the composites.

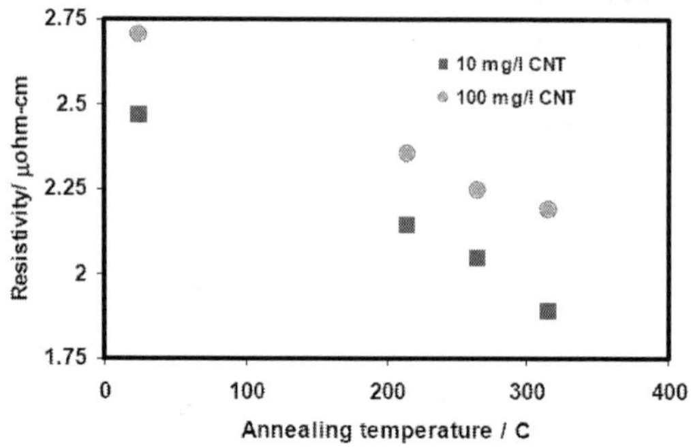

Figure 9. Comparison of the resistivity change of Cu/CNT composite film after annealing for 20 minutes at higher temperature using 10 mg/l and 100 mg/l CNT in the bath.

In summary, the Cu/CNT composites were codeposited with the aid of a surfactants at different current densities. As a comparative study, three surfactants (nafion, CTAB and SDS) were used separately to disperse CNTs in the bath and Cu/CNT composite films were electrodeposited. SDS in the bath results in a smoother deposition whereas CTAB leads to rougher deposition of the composite. The maximum CNT concentration in Cu/CNT composites achieved in our study was approximately 2 % by weight deposited from 100 mg/l CNT containing composite baths. The electrical resistivity results show that at room temperature the resistivity of Cu/MWCNT composite film (2.47 μΩ-cm) is close to the resistivity of Cu film (2.15 μΩ-cm) when nafion was used in the deposition bath for the surfactant of MWCNTs. With the use of CTAB or SDS in the bath, the resistivity of Cu/MWCNT film was higher [deposited from 10 mg/l CNT containing composite baths]. A clear decrease in sample resistivity was observed with increasing annealing temperature. The resistivity also increased when the concentration of MWCNTs was increased from 10 mg/l to 100 mg/l in the bath.

The line resistance of Cu filled and Cu/CNT filled test chip structures was measured using a Cascade probe station. The test chip structure consisted of 110 μm x 80 μm pads connected with metal lines of different widths. A Cu seed (12 nm) and a barrier Ta/TaN (25 nm) were PVD deposited. To achieve a uniform deposit the plated substrates were planarised with a CMP process. The test chip coupon was mounted on a 4 inch Si carrier-wafer. A Logitech CDP51 was used for the CMP. The polishing slurry used was a Cabot Microelectronics product, Eterpol 2362, which was mixed with H_2O_2 (30%), the ratio of H_2O_2 to slurry was 5% by volume. During CMP, the rotation of wafer holder and polishing pad was 50 rpm and the applied pressure was 2-3 psi. Fig. 10 shows the SEM image of the test structure after CMP for 2 minutes. It can be seen that excess deposits were completely removed by the developed CMP process. Before probing, the test sample was vacuum attached in a dedicated holder. Four micro-probes were placed on four pads in the structure. The pads were connected with Cu filled interconnect lines. The Cu was deposited from a damascene additive containing sulphate based bath. Fig. 11 shows the electrical measurement of 110 nm width Cu line in the test structure. It can be seen from the measurement that the I-V curves of Cu deposited line was linear and the resistance value was estimated to be 284 Ω which is to be expected for lines of that dimension.

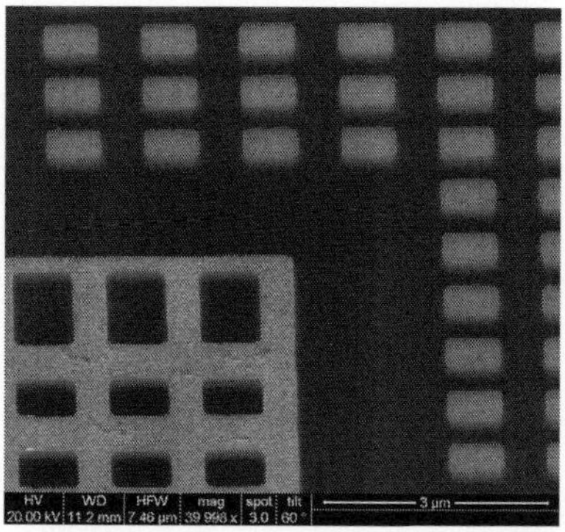

Figure 10. Plan view SEM image of test structure after complete CMP. The structure was filled by electrodeposited Cu (lighter colour in image). Image magnification 40000X.

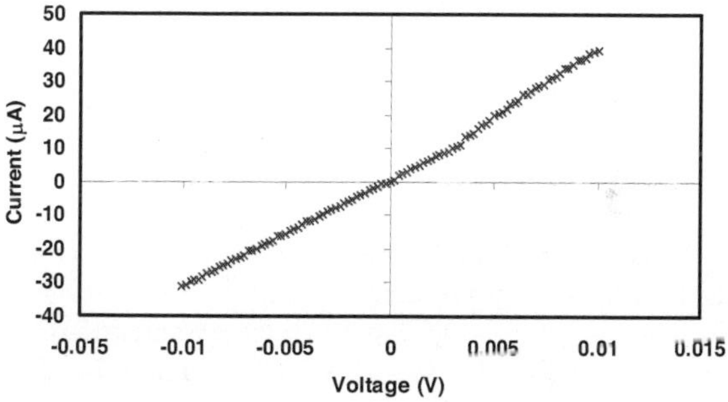

Figure 11. Current vs. voltage curve of the 110 nm line width connected with four 110 µm x 80 µm pads. The features were filled with electrodeposited Cu.

The next samples investigated contained CNTs in the Cu deposited from the nafion containing bath. The concentrations of CNTs and nafion were 50 mg/l and 0.5% respectively. Fig. 12 shows the electrical measurement of 110 nm line width filled with Cu/MWCNT in the test structure. It can be seen from the measurement that the I-V curve of the Cu/MWCNT deposited line was linear and the resistance value was 29.7 kΩ. The resistance of individual MWCNT is hundreds of kΩ (minimum resistance for a ballistic single-walled CNT is ~ 6.5 kΩ). The high resistance of individual CNTs is due to high contact and quantum resistance. Therefore, relatively dense arrays of CNTs will be needed to replace Cu interconnects. As Cu and MWCNTs were codeposited in the narrow line with 110 nm width, so the resistance in Cu/MWCNT composites is expected to be between the resistance value of Cu (284 Ω) and MWCNT (hundreds kΩ). The resistance of the line filled with Cu/CNTs could be improved by using SWCNT instead of the MWCNTs used in this composite.

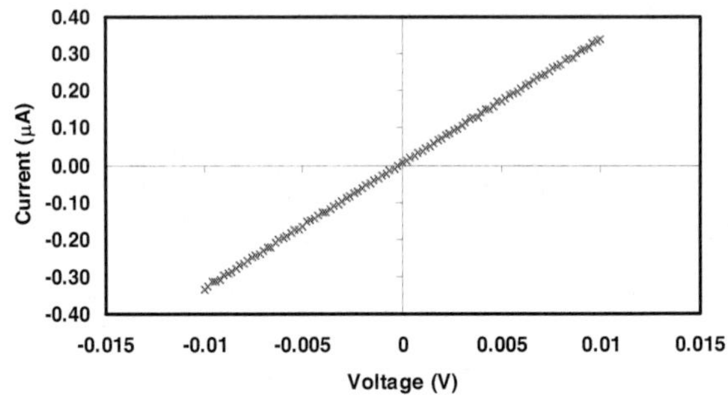

Figure 12. Current vs. voltage curve of the 110 nm line width which was filled with electrodeposited Cu/MWCNT composite. The concentrations of nafion and MWCNTs in the bath were 0.5% and 50 mg/l respectively.

6. Conclusion

In this chapter we have reported the influence of surfactants on the properties of Cu/CNT composites on Si substrates. Cu/CNTs composite films were co-deposited by electrodeposition. Before electrodeposition, CNTs were dispersed by a suitable surfactant. Electrochemical data shows that nafion or SDS accelerates the co-deposition where as CTAB suppresses the deposition. Nafion and SDS surfactants result in a relatively smooth deposit whereas CTAB surfactant leads to rougher deposition of the composite. The amount of CNTs in the deposit was up to 2 % by weight using different surfactants and different length/diameter of CNTs. Our electrical analysis showed that for Cu/CNT composite samples maintained at room temperature, the resistivity over time did not change significantly. The electrical resistivity results also showed that at room temperature the resistivity of the Cu/CNT composites film (2.43 μΩ cm) is close to the resistivity of Cu film (2.17 μΩ cm) when nafion was used in the bath to disperse the CNTs. The resistivity of Cu/CNTs film was higher when CTAB or SDS were used instead of nafion as a surfactant. The electrical resistivity results showed that at room temperature the resistivity increased 10% when the concentration of CNT in the bath was increased from 10 mg/l to 100 mg/l. A clear decrease of sample resistivity of composite films was observed with increasing annealing temperature. Cu/CNT composites deposited at a test structure with submicron lines and vias with Cu/CNT composites was only possible from the nafion surfactant containing damascene. The electrical measurement of 110 nm line width filled with Cu/MWCNT showed that the I-V curves of the Cu/MWCNT deposited line was linear and the resistance value was 29.7 kΩ which was significantly higher that the resistance value of Cu (284 Ω) deposited. Improvements on these val-

http://dx.doi.org/10.5772/52731

ues will require lower resistance SWCNTs or the improvement of the density of aligned nanotubes in the composite structure. This may be more feasible in larger dimension features such as those required for TSV interconnect at the chip scale rather the use of composites for IC interconnect at deep sub micron dimensions.

Acknowledgements

This research was supported by the Irish Research Council for Science, Engineering and Technology (IRCSET) postgraduate scholarship Enterprise Partnership scheme in collaboration with Intel Ireland Ltd., funded under the National Development Plan.

Author details

Tamjid Chowdhury and James F. Rohan

Tyndall National Institute, University College Cork, Lee Maltings, Cork, Ireland

References

[1] L. L. Vadasz, A. S. Grove, T. A. Rowe and G. E. Moore, Silicon gate technology. IEEE Spectrum 6, 28 (1969).

[2] R. Solanki and B. Pathangey, Electrochem. Solid St., 3, 479 (2000).

[3] S. Venkatesan, A. Gelatos, S. Hisra, B. Smith, R. Islam, J. Cope, B. Wilson, D. Tuttle, R. Cardwell, S. Anderson, M. Angyal, R. Bajaj, C. Capasso, P. Crabtree, S. Das, J. Farkas, S. Filipiak, B. Fiordalice, M. Freeman, P. Gilbert, M. Herrick, A. Jain, H. Kawasaki, C. King, J. Klein, T. Lii, K. Reid, T. Saaranen, C. Simpson, T. Sparks, P. Tsui, R. Venkatraman, D. Watts, E. Weitzman, R. Woodruff, I. Yang, N. Bhat, G. Hamilton and Y. Yu, Proc. IEEE-IEDM, 97, 769 (1997).

[4] D. Edelstein, J. Heidenreich, R. Goldblatt, W. Cote, C. Uzoh, N. Lustig, P. Roper, T. McDevitt, W. Motsiff, A. Simon, J. Dukovic, R. Wachnik, H. Rathore, R. Schulz, L .Su, S. Luce and J. Slattery, Proc. IEEE-IEDM, 97, 773 (1997).

[5] M. T. Wang, Y. C. Lin and M. C. Chen, J. Electrochem. Soc., 145 (7), 2538 (1998).

[6] C. K. Hu, B. Luther, F. B. Kaufman, J. Hummel, C. Uzah and D. J. Pearson, Thin Solid Films, 262, 84 (1995).

[7] K. Abe, Y. Harada and H. Onoda, Proc. 13th Intern VLSI Multilevel Interconnect Conf. 308 (1995).

[8] J. S. H. Cho, H. K. Kang, I. Asano and S. S. Wang. IEDM Tech Digest 297 (1992).

[9] P. C. Andricacos, C. Uzoh, J. O. Dukovic, J. Horkans and H. Deligianni, IBM J. Res. Dev., 42, 567 (1998).

[10] K. Kondo, K. Hayashi, Z. Tanaka and N. Yamakawa, ECS Proceedings on Electrochemical Processing in ULSI Fabrication, 8, 76 (2000).

[11] J. Kelly and A. West, Electrochem. Solid-St., 2, 561 (1999).

[12] J. P. Healy, D. Pletcher and M. Goodenough, J. Electroanal. Chem., 338, 179 (1992).

[13] C. K. Hu and J. M. E. Harper, Mater. Chem. Phys., 52, 5 (1998).

[14] B. Li, T. Sullivan, T. Lee and D. Badami, Microelectron. Reliab., 44. 365 (2004).

[15] C. Hu, L. Gignac and R. Rosenberg, Microelectron. Reliab., 46. 213 (2006).

[16] C. Hu, L. Gignac and R. Rosenberg, E. Liniger, J. Rubino and C. Sambucetti, Microelectron. Eng., 70, 406 (2003).

[17] J. Gambino, J. Wynne, J. Gill, S. Mongeon, D. Meatyard, B. Lee, H. Bamnolker, L. Hall, N. Li, M. Hernandez, P. Little, M. Hamed, I. Ivanov and C. Gan, Microelectron. Eng., 83, 2059 (2006).

[18] S. Iijima, Nature, 354, 56 (1991).

[19] H. Dai, A. Javery, E. Pop, D. Mann and Y. Lu, Nano: Brief Reports and Reviers, 1, 1 (2006).

[20] N. Srivastava and K. Banerjee, Proc 21st Int Multilevel Interconnect, 393 (2004).

[21] P. Rapposelli, B. Capraro, J. Dijon, G. Groeseneken, D. Cott, J. Pinson, X. Joyeux, J. Amadou, J. Noyen and B. Sels, ECS Trans., 25 (10), 63 (2009).

[22] International Technology Roadmap for Semiconductors, 2011 update, http://www.itrs.net/Links/2011Winter/11_Interconnect.pdf (accessed 13 July 2012)

[23] Y Chai, K. Zhang, M. Zhang, P. Chen and M. Yuen., Electronic Components and Technology Conference, IEEE, 1224 (2007).

[24] Y Chai, P. Chan and Y. Fu., Electronic Components and Technology Conference, IEEE, 412 (2008).

[25] J. Yoo, J. Song, J. Yu, H. Lyeo, S. Lee and J. Hahn., Electronic Components and Technology Conference, ECTC, 1282 (2008).

[26] Q. Chen, G. Chai and Bo Li., Proc. IMechE, 219, 67 (2006).

[27] O. Hjorstam, P. Isberg, S. Soderholm and H. Dai., Appl. Phys. A., 78, 1175 (2004).

[28] P. Liu, D. Xu, Z. Li, B. Zhao, E. Kong and Y. Zhang, Microelectron. Eng., 85, 1984 (2008)

[29] G. Chai, Y. Sun, J. Sun and Q. Chen., J. Micromech. Microeng., 18, 35013 (2008)

http://dx.doi.org/10.5772/52731

[30] Q. Chan, US patent application 11/437,180, filed May, 2006.

[31] Y. Son, J. Yoo and J. Yu, US patent application 11/589,305, filed Oct, 2006.

[32] P. Lo, J. Wei, B. Chen, J. Chiang and M. Kao, US patent application 11/289,523, filed Dec, 2005.

[33] V. Davis, A. Parra-Vasquez, M. Green, P. Rai, N. Behabtu, V. Prieto, R. Booker, J. Schmidt, E. Kesselman, W. Zhou, H. Fan, W. Adams, R. Hauge, J. Fischer, Y. Cohen, Y. Talmon, R. Smalley and M. Pasquali, Nat. Nanotechnol., 4, 830 (2009).

[34] F. Pompeo and D. Resasco, Nano Lett., 2, 369 (2002).

[35] L. Feng, H. Li, F. Li, Z. Shi and Z. Gu, Carbon, 41, 2385 (2003).

[36] Y. Lin, F. Allard and Y. Sun, J. Phys. Chem. B, 108, 3760 (2004).

[37] K. Matsuura, K. Hayashi and N. Kimizuka, Chem. Lett., 32, 212 (2003).

[38] W. Huang, S. Taylor, K. Fu, Y. Lin, D. Zhang, T. Hanks, A. Rao and Y. Sun, Nano Lett., 32, 212 (2003).

[39] H. Peng, L. Alemany, J. Margrave and V. Khabashesku, J. Am. Chem. Soc., 125. 15174 (2003).

[40] Y. Wang, Z. Iqbal, S. Mitra, J. Am. Chem. Soc., 128, 95 (2006).

[41] D. W. Schaefer, J. M. Brown, D. P. Anderson, J. Zhao, K. Chokalingam, D. Tomlin and J. Ilavsky, J. Appl. Crystallogr. 36, 553 (2003).

[42] J. Li, Q. Ye, A. Cassell, H. Tee Ng, R. Stevens, J. Han and M. Meyyappan, Appl. Phys. Lett., 82, 2491(2003).

[43] J. Lee, U. Paik, J. Choi, K. Kim, S. Yoon, J. Lee, B. Kim, J. Kim, M. Park, C. Yang, K. An, Y. Lee, J. Phys. Chem. C, 111, 2477 (2007).

[44] F. Ko, C. Lee, C. Ko and T. Chu, Carbon, 43, 727 (2005).

[45] Y. Chen and Y. Mitra, J. Nanosci. Nanotechnol.., 11, 5770 (2008).

[46] J. Li and H. Greenburg, Chem. Eur. J, 12, 3869 (2006).

[47] C. Lau, R. Cervini, S. Clarke, M. Markovic, J. Matisons, S. Hawkins, C. Huynh and G. Simon, J. Nanopart Res, 10, 77 (2008).

[48] S. Agarwal, M. Raghuveer, H. Li and G. Ramanath, Appl. Phys. Lett., 90, 193104 (2007).

[49] H. Tantang, J. Ong, C. Loh, X. Dong, P. Chen, Y. Chen, X. Hu, L. Tan and L. Li, Carbon, 47, 1467 (2009).

[50] M. O'Connell, P. Boul, L. Ericson, C. Huffman, Y. Wang, E. Haroz, C. Kuper, J. Tour, K. Ausman and R. Smalley, Chem. Phys. Lett., 342, 265 (2001)

[51] X. Chen, C. Cheng, H. Xiao, H. Liu, L. Zhou, S. Li and G. Zhang, Tribol. Int., 39, 22 (2006).

[52] X. Chen, F. Cheng, S. Li, L. Zhou and D. Li, Surf. Coat. Technol., 155, 274 (2002).

[53] C. Guo, Y Zuo, X. Zhao, J. Zhao and J. Xiong, Surf. Coat. Technol., 202, 3385 (2008)

[54] Y.Yang, Y. Wang, Y. Ren, C. He, J. Deng, J. Nan, J. Chen and L. Zuo. Mat. Lett., 62, 47 (2008)

[55] B. Praveen, T. Venkatesha, Y. Naik and K. Prashantha, Surf. Coat. Technol., 201, 5836 (2007)

[56] B. I. Yakobson and R.E. Smalley, Am. Sci, 85, 324 (1997).

[57] L. Xu, X. Chen, W. Pan, W. Li, Z. Yang and Y. Pu., Nanotechnol, 18, 435607 (2007)

[58] X. Chen, C. Cheng, H. Xiao, H. Liu, L. Zhou, S. Li and G. Zhang, Tribol. Int., 39, 22 (2006).

[59] N. Tuerui, B. Fugetsu and S. Tanaka, Anal. Sci., 22, 895 (2006).

[60] S. Arai, M. Endo, T. Sato and A. Koide, Electrochem. Solid-St., 9, C131 (2006)

[61] L. Shi, C. Sun, P. Gao, F. Zhou and W. Liu., Surf. Coat. Technol, 200, 4870 (2006)

[62] Y Tsai, S. Li and J. Chen, Proc IEEE Conf on Nanotechnol, (2005)

[63] C. Engtrakul, M. Davis, T. Gennett, A. Dillon, K Jones and M. Heben, J. Am. Chem. Soc., 127, 17548 (2005)

[64] S. Arai, A. Fujimori, M. Murai and M. Endo, Mat. Lett., 62, 3545 (2008).

[65] S. Arai, T. Saito and M. Endo, Electrochem. Solid-St., 11, D72 (2008).

[66] T. Osaka, N. Yamachika, M. Yoshino, M. Hasegawa, Y. Negishi and Y. Okinaka, Electrochem Solid-St., 12, D15 (2009).

[67] P. Dixit, C Tan, L. Xu, N. Lin, J. Miao, J. Pang, P. Backus and R. Preisser, J. Micromech. Microeng., 17, 1078 (2007).

[68] A. Yeoh, M. Chang, C. Pelto, T. Huang, S. Balakrishnan, G. Leatherman, S. Agraharam, W. Guota, Z. Wang, D. Chiang, P. Stover and P. Brandenburger, IEEE 56th Electronic Components and Technology Conference, 1611 (2006).

[69] A. Braun, Semiconductor International, 3D Integration in Design and Test Support. Article ID CA6615469.

[70] O. Luhn, A. Radisic, P. M. Vereecken, C. van Hoof, W. Ruythooren and J. P. Celis, Electrochem. Solid-St., 12 (5), D39 (2009).

[71] R. Beica, C. Sharbono and T. Ritzdorf, DTIP of MEMS & MOEMS, ISBN: 978-2-355500-006-5, 2008.

[72] A. Radisic, O. Lühn, J. Vaes, S. Armini, Z. Mekki, D. Radisic, W. Ruythooren, and P. Vereecken, ECS Trans., 25 (38), 119 (2010).

[73] E. Mattsson and J. O'M Bockris., Trans. Faraday. Soc., 55, 1586 (1958)

[74] C. H. Yang, Y. Y. Wang and C. C. Wan, J. Electrochem. Soc., 146 (12), 4473 (1999).

[75] S. Varvara, L. Muresan, I. C. Popescu and G. Maurin, J. Appl. Electrochem., 33, 685 (2003).

[76] K. Pantleoan and M. A. J. Somers, J. Appl Phys., 100, 114319 (2006).

[77] V. A. Vasko, I. Tabakovic, S. C. Riemer and M. T. Kief, Electrochem., Solid-St., 6, 100 (2003).

[78] P. M. Vereecken, R. A. Binstead, H. Deligianni and P. C. Andricacos, IBM J. Res & Dev., 49, 3 (2005).

[79] S. Lagrange, S. H. Brongersma, M. Judelewicz, A. Saerens, I. Vervoort, E. Richard, R. Palmans and K. Maex, Microelectron. Eng., 50, 449 (2000).

[80] J. M. E. Harper, C. Cabral, P. C. Andricacos, L. Gignac, I. C. Noyan, K. P. Rodbell and C. K. Hu, J. Appl. Phys., 86, 2516 (1999).

[81] T. Osaka, N. Yamchika, M. Yoshino, M. Hasegawa, Y. Neigishi and Y. Okinaka, Electrochem. Solid-St., 12, D15 (2009).

[82] C. Lee and C. Park, Jpn. J. Appl. Phys., 42, 4484 (2003).

[83] T. Hara, H. Toida and Y. Shimura, Electrochem. Solid-St., 6, G98 (2003) 204

[84] W. H. Teh, L. T. Kon, S. M. Chen, J. Xie, C. Y. Li and P. D. Foo, Microelectron J., 32, 579 (2001).

[85] S. H. Brongersma, E. Richard, I. Vervoot, H. Bender, W. Vandervost, S. Lagrange, G. Beyer and K. Maex, J. Appl. Phys., 86, 3642 (1999).

[86] S. Balakumar, R. Kumar, Y. Shimura, K. Namiki, M. Fujimoto, H. Toida, M. Uchida and T. Hara, Electrochem. Solid-St., 7, G68 (2004).

[87] H. Lee, S. S. Wong and S. D. Lopatin, J. Appl. Phys., 93, 3796 (2003).

Section 3

Carbon Nanotubes for Green Technologies

Chapter 17

Carbon Nanotubes Influence on Spectral, Photoconductive, Photorefractive and Dynamic Properties of the Optical Materials

Natalia V. Kamanina

Additional information is available at the end of the chapter

http://dx.doi.org/10.5772/50843

1. Introduction

It is well known that optoelectronics, telecommunication systems, aerospace, and correction of amplitude-phase aberration schemes, as well as laser, display, solar energy, gas storage and biomedicine techniques are searching for the new optical materials and for the new methods to optimize their properties. So many scientific and research groups are involving in this process and are opening the wide aspects of different applications of new materials, especially optical ones. It has been going on last century that simple manufacturing, design, ecology points of view, etc. indicate good advantage of the nanostructured materials with improved photorefractive parameters among other organic and inorganic systems.

Really, it should be tell that photorefractive properties change is correlated with the spectral, photoconductive and dynamics ones. The change in nonlinear refraction and cubic nonlinearity reveals the modification of barrier free electron pathway and dipole polarizability. From one side it is connected with the change of the dipole moment and the charge carrier mobility, from other side, it is regarded to the change of absorption cross section. Thus, this feature shows the unique place of photorefractive characteristics among other ones in order to characterize the spectral, photoconductive, photorefractive and dynamic properties of the optical materials.

It should be mentioned that promising nanoobjects, such as the fullerenes, the carbon nanotubes (CNTs), the quantum dots (QDs), the shungites, and the graphenes permit to found different area of applications of these nanoobjects [1-6]. The main reason to use the fullerenes, shungites, and quantum dots is connected with their unique energy levels and high value of electron affinity energy. The basic features of carbon nanotubes and graphenes are regarded to their high conductivity, strong hardness of their C-C bonds as well as complicated and unique mechanisms of charge carrier moving.

These peculiarities of carbon nanoobjects and their possible optoelectronics, solar energy, gas storage, medicine, display and biology applications connecting with dramatic improvement of photorefractive, spectral, photoconductive and dynamic parameters will be under consideration in this paper. In comparison with other effective nanoobjects the mail accent will be given namely on carbon nanotubes (CNTs) and their unique features to modify the properties of the optical materials.

2. Experiment

The different experimental techniques have been used to study the properties of nanostructured materials.

To reveal the change of the photorefractive properties, as the systems under study the organic thin films based on conjugated monomer, polymer and liquid crystals sensitized with carbon nanotubes, fullerenes, shungites, graphenes oxides, or quantum dots have been chosen. Polyimides (PI), 2-cyclooctylamino-5-nitropyridine (COANP), *N*-(4-nitrophenyl)-(*L*)-prolinol (NPP), 2-(*N*-prolinol)-5-nitropyridine (PNP), nematic liquid crystals (NLCs) have been considered as organic matrixes. These conjugated systems are the good model with effective intramolecular charge transfer process which can be easy modify via sensitization by nanoobjects. Carbon nanotubes, fullerenes, shungites, graphenes oxides, quantum dots content was varied in the range of 0.003-5.0 wt.%. The solid thin films have been developed using centrifuge deposition. The general view of these films is shown in Fig.1. The thickness of the films was 2-5 micrometers. The LC cell thickness was 5-10 micrometers.

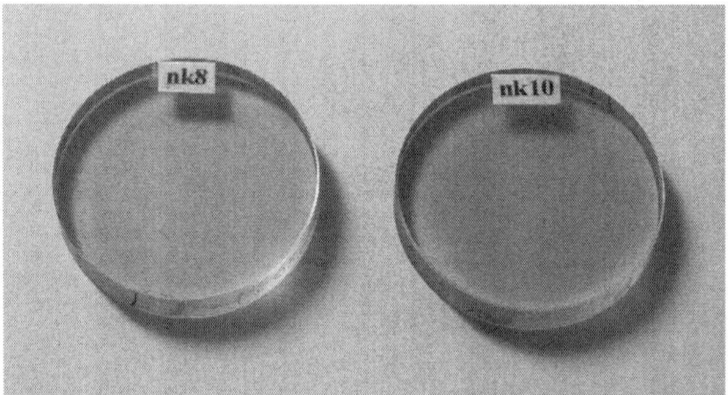

Figure 1. Photographs of samples of pure (nk8) and nanoobjects-containing (nk10) PI films.

The nanostructured LC films have been placed onto glass substrates covered with transparent conducting layers based on ITO contacts. The nanostructured monomer or polymer sol-

http://dx.doi.org/10.5772/50843

id films have been deposited on the substrate with ITO contact. For the electric measurements of volt-ampere parameters, gold contact has been put to the solid thin films upper side. The picture which can interpret the placement of the conducting contacts on the solid conjugated organic thin films is shown in Fig.2.

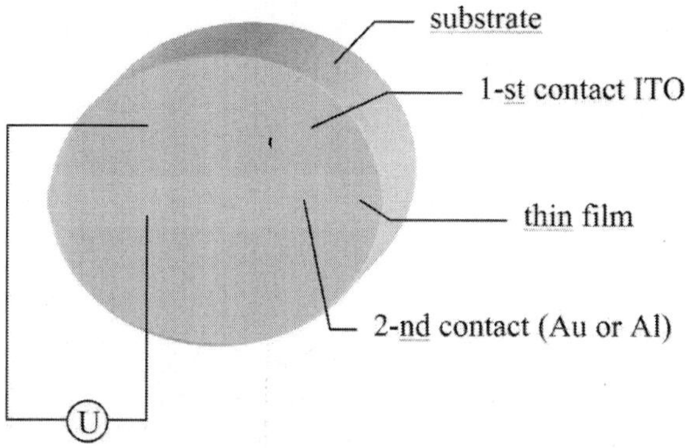

Figure 2. Interpretation of the solid thin films with the conducting layers

The bias voltage applied to the photosensitive polymer layers has been varied from 0 to 50 V. The current–voltage characteristics have been measured under the illumination conditions from dark to light. Voltmeter-electrometer B7—30 and Characteriscope—Z, type TR-4805 has been used for these photoconductive experiments.

The photorefractive characteristics have been studied using four-wave mixing technique analogous to paper [7]. The experimental scheme is shown in Fig.3.

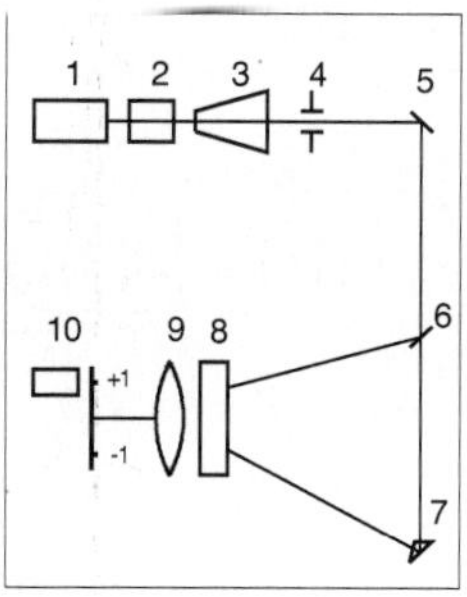

Figure 3. An experimental scheme: 1 – Nd-laser; 2 – second harmonic convertor; 3 – telescope; 4 – diaphragm; 5 – rotating mirror; 6 – beam-splitting mirror; 7 – prism; 8 – sample; 9 – lens; 10 - photodetector.

The second harmonic of pulsed Nd-laser at wave length of 532 nm has been used. The laser energy density has been chosen in the range of 0.005-0.9 J×cm^{-2}. The nanosecond laser regime with the pulse width of 10-20 ns has been applied. The amplitude-phase thin gratings have been recorded under Raman-Nath diffraction conditions according to which $\Lambda^{-1} \geq d$, where Λ^{-1} is the inverse spatial frequency of recording (i.e., the period of the recorded grating) and *d* is the film thickness. In the experiments the spatial frequency was in the range of 90-150 mm^{-1}.

The spectral characteristics have been tested using Perkin Elmer lambda 9 spectrophotometer. Dynamic features of nanoobjects-doped LC films have been studied via the four-wave mixing technique and the Frederick's scheme one. Atomic force microscopy (AFM) method using equipment of "NT-MDT" firm, "Bio47-Smena" in the "share-force" regime has been applied to analyze the diffraction relief into the solid conjugated nanostructured thin film.

3. Results and discussion

It should be noticed that previously, we demonstrated [5,6,8] the formation of barrier-free charge transfer pathways, increased dipole moment, and increased specific (per unit volume) local polarizability in some organic matrices doped with fullerenes, carbon nanotubes and quantum dots, where the formation of intermolecular complexes predominated over the intramolecular donor–acceptor interaction. The possible schemes of charge transfer between matrix organic molecule donor fragment and different efficient nanosensitizers including the additional graphene and shungites nanostructures are schematically shown in Fig. 4.

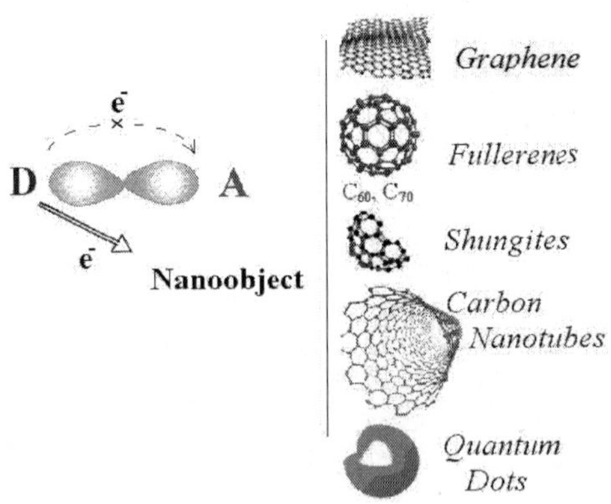

Figure 4. Schematic diagram of possible intermolecular charge transfer domination under intramolecular ones.

Analyzing the Fig.4, one can say that it is necessary to take into account that the charge transfer between matrix organic molecule donor fragment and nanasensitizers can be organized due to their high electron affinity energy (for example, electron affinity energy is close to 2 eV for shungites [9], to 2.65 eV for fullerenes [5,8] and to 3.8-4.2 eV for quantum dots [10]) that is more than the ones for intramolecular acceptor fragments (for example, electron affinity energy of COANP acceptor fragment is close to 0.54 eV [11] and to 1.14-1.4 eV for polyimide one [12]). Regarding graphenes it is necessary to take into account the high surface energy and planarity of the graphenes plane which can provoke to organize the charge transfer complex (CTC) with good advantage too. Regarding the CNTs it should be drawn the attention on the variety of charge transfer pathways, including those along and across a CNT, between CNTs, inside a multiwall CNT, between organic matrix molecules and CNTs, and between the donor and acceptor fragments of an organic matrix molecule.

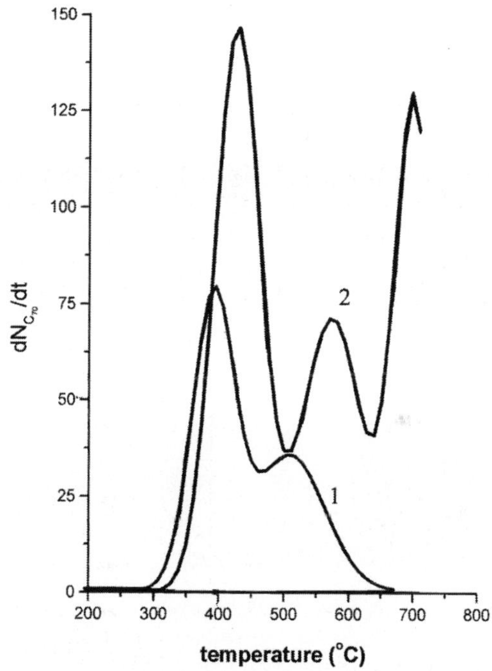

Figure 5. The rate of release of C_{70} molecules on heating of systems: (*1*) COANP with 5 wt % of C_{70} and (*2*) polyimide with 0.5 wt % of C_{70}

It should be noticed that some supporting CTC results for PIs and COANP systems sensitized with nanoobjects can be presented via mass-spectrometry experiments. It is easy to show the organization of CTC using fullerenes acceptor. Really, the mass spectroscopy data point to the effective CTC formation between fullerene and donor part of PI (triphenylamine) and between fullerene and the HN group of COANP systems, respectively. For the 5 wt.% C_{70}-COANP film, mass spectrometry curve contains two peaks. The first one at 400 C corresponds to the release rate of fragments with free fullerene masses. The second one is shifted to the temperature range of 520 °C and associated with the decomposition temperature of the fullerene-HN group complex. For the 0.5 wt.% C_{70}-PI film, curve contains three peaks. The first one is observed also close to 400 C. The second peak is located at 560 °C and associated with the decomposition of fullerene- triphenylamine complex. It should be noticed that the melting temperature of these PIs is 700-1000 °C, thus, the third peak at the temperature higher 700 °C corresponds to the total decomposition of PI. Figure 5 presents the mass-spectrometry data.

By monitoring the diffraction response manifested in the laser scheme (see Fig.6); it is possible to study the dynamics of a photo-induced change in the refractive index of a sample and to calculate via [13] the nonlinear refraction and nonlinear third order optical susceptibility (cubic nonlinearity). An increase in the latter parameter characterizes a change in the specific (per unit volume) local polarizability and, hence, in the macroscopic polarization of the entire system.

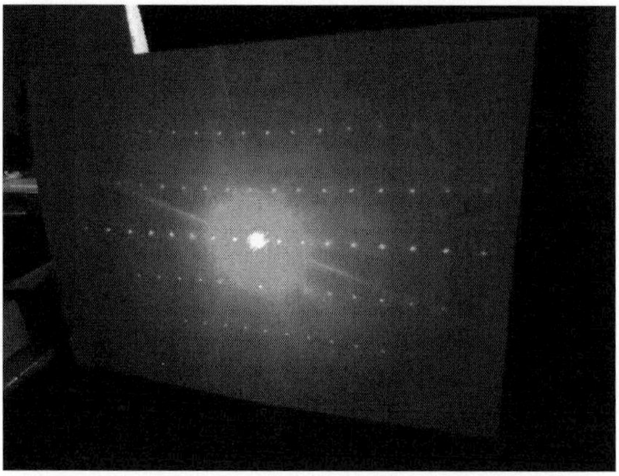

Figure 6. The visualization of the diffraction response in the organic films doped with nanoobjects.

The main results of this study are summarized in the Table 1 (Ref. 5,6,9,14-18) in comparison to the data of some previous investigations. An analysis of data presented in the Table 1 for various organic systems shows that the introduction of nanoobjects as active acceptors of electrons significantly influences the charge transfer under conditions where the intermolecular interaction predominates over the intramolecular donor–acceptor ones. Moreover, redistribution of the electron density during the recording of gratings in nanostructured materials changes the refractive index by at least one order of magnitude in comparision to that in the initial matrix. The diffusion of carriers from the bright to dark region during the laser recording of the interference pattern proceeds in three (rather than two) dimensions, which is manifested by a difference in the distribution of diffraction orders along the horizontal and vertical axes (see Fig.6). Thus, the grating displacement takes place in a three dimensional (3D) medium formed as a result of the nanostructirization (rather than in a 2D medium).

Some atomic force microscopy data are supported the realization of 3D-media via development of complicated diffraction relief into the solid thin conjugated films after transfer from the reversible regime to the irreversible one. Figure 7 demonstrates this fact. Two types of diffraction replica, namely due to interference of laser beams onto the thin films surface and due to the diffraction of these beams inside the body of the nanostructured media have been presented.

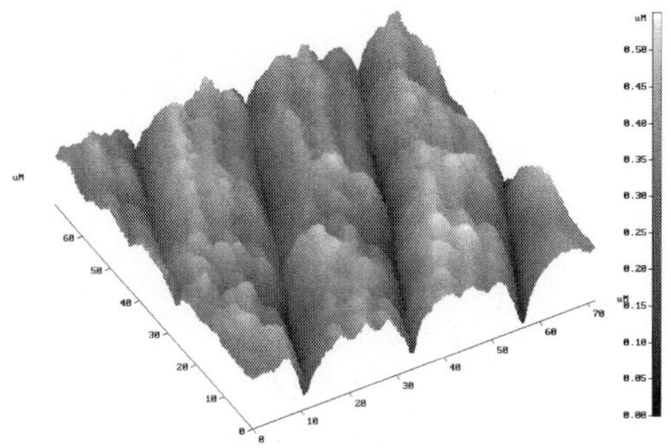

Figure 7. Demonstration of AFM evidence of new 3D-media development.

Using the obtained results, the nonlinear refraction n_2 and nonlinear third order optical susceptibility (cubic nonlinearity) $\chi^{(3)}$ for all systems have been calculated using a method described in [13,18]. In the current experiments using four-wave mixing technique, the nonlinear refraction coefficient and cubic nonlinearity (third order susceptibility) have been estimated via equations (1) and (2):

$$n_2 = \Delta n_i / I \tag{1}$$

$$\chi^{(3)} = n_2 n_0 c / 16\pi^2 \tag{2}$$

where I – is the irradiation intensity, n_o – is the linear refractive index of the media, c – is the speed of the light.

It was found that these parameters fall within $n_2 = 10^{-10}$–10^{-9} cm^2/W and $\chi^{(3)} = 10^{-10}$–10^{-9} cm^3/erg.

Moreover, it should be remained, that optical susceptibility $\chi^{(n)}$, from fundamental point of view, directly connected with the dipole system polarizability $^{(n)}$ via equation (3) written in the paper [19]:

$$\chi^{(n)} = \alpha^{(n)} / \upsilon \quad (3)$$

where $\alpha^{(n)}$ – dipole polarizability and υ – local volume.

Therefore, using the fact that polarizability of all structures can be accumulated from local volumes, it can be found that increased micropolarization of system (see eq.4) will predict the dynamic properties improvement and high electro-optical response speed.

$$P^{\circ} = {}^{\circ}\chi^{(1)}E^{\circ} + {}^{\circ}\chi^{(2)}E^{2\circ} + \chi^{(3)}E^{3\circ} + \ldots + {}^{\circ}\chi^{(n)}E^{n\circ} + \ldots \quad (4)$$

It should be mentioned (see Table 1) that the larger nonlinear optical parameters have been found for CNTs-doped organic systems or CNTs-nanofibers-doped ones. It is natural to suggest (see Fig.8) that variations of the length, of the surface energy, of the angle of nanoobject orientation relative to the intramolecular donor can significantly change the pathway of charge carrier transfer, which will lead to changes in the electric field gradient, dipole moment (proportional to the product of charge and distance), and mobility of charge carriers.

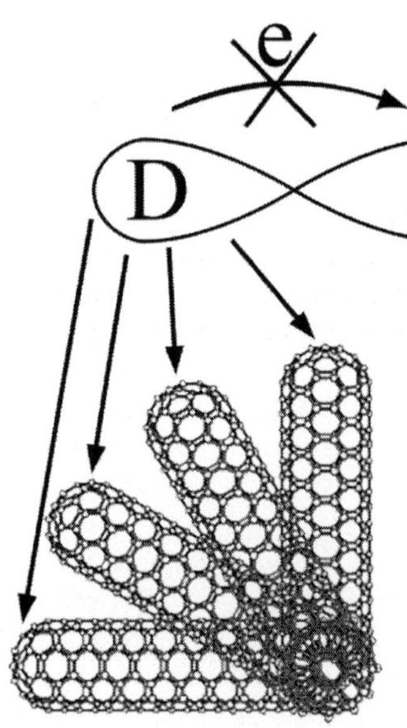

Figure 8. Schematic diagram of possible charge transfer pathways depending on the arrangement of introduced intermolecular acceptor relative to the intramolecular donor

In addition, the barrier free charge transfer will be influenced by competition between the diffusion and drift of carriers during the creation of diffraction patterns with various periods and, hence, differing charge localization at the grating nodes and antinodes. Indeed, in the case of a nanocomposite irradiated at small spatial frequencies (large periods of recorded grating), a drift mechanism of the carrier spreading in the electric field of an intense radiation field will moist probably predominate, while at large spatial frequencies (short periods of recorded grating) the dominating process is diffusion. This also naturally accounts for the aforementioned discrepancy of published data on photoinduced changes in the refractive index of nanocomposites, greater values of which were observed (see, e.g., data presented in the table for systems doped with CNTs and MIG nanofibers) at smaller spatial frequencies. Lower values of photoinduced changes in the refractive index of nanocomposites were observed at high spatial frequencies. This evidence predicts the strong correlations between photorefractive and photoconductive parameters.

To support the evidence on correlation between photorefractive and photoconductive features of the materials studied, the volt-current characteristics for nanoobjects-doped solid thin films and pure ones has been measured. After that charge carrier mobility has been estimated using the Child–Langmuir current–voltage relationship [20] following the formula (5) shown below:

$$\mu = 10^{13} I d^{3} \times \varepsilon^{-1} V^{-1} \quad (5)$$

For example, one can calculate the absolute values of the charge carrier mobility in pure and fullerene-modified PI samples. The results of these calculations show that the introduction of fullerenes leads to a tenfold increase in the mobility. The absolute values were estimated for a bias voltage of 10 V, a film thickness of $d = 2$ μm, a dielectric constant of $\varepsilon \sim 3.3$, a fullerene content of about 0.2 wt % C_{70}, and an upper electrode contact area with a diameter of 2 mm. Under these conditions, the carrier mobility in a fullerene-modified polyimide PI film is $\sim 0.3 \times 10^{-4}$ $cm^2/(V\ s)$, while the analogous value for pure PI is $\sim 0.17 \times 10^{-5}$ $cm^2/(V\ s)$. These values well agree with the data reported in [21], where it was demonstrated that the carrier mobility in pure PI films ranges in the interval from 10^{-7} to 0.5×10^{-5} $cm^2/(V\ s)$. Relationship (5) used for the estimation of charge carrier mobility is valid in the case of currents limited by the space charge. This situation is characteristic of most of the conjugated organic structures (in particular, PIs) in which the charge transfer processes are additionally determined by traps, although formula (5) contains no terms dependent on the illumination intensity. However, taking into account the aforementioned equality of the activation energies of conductivity and mobility in PIs, the results of calculations of the relative changes in the carrier mobility probably adequately reflect the general trends in mobility variations. This behavior does not contradict the pattern of changes in the mobility observed for the other conjugated organic systems, for example, for the fullerene–carbazole one [22].

We have also calculated the relative values of the charge carrier mobility μ and estimated that two orders of magnitude differences of charge carrier mobility for the pure and nanoobjects-doped films has been found. Moreover, the following relation for the charge carrier mobility has been proposed:

$$\mu_{\text{pure organic systems}} < \cdot\, \mu_{C70,C60} < \cdot\, \mu_{CNT,QD} \tag{6}$$

The observation of the increase of charge carrier mobility, high refractive index and high value of cubic nonlinearities predicts that the nonlinear optical and the dynamic feature of the nanostructured conjugated materials can be optimized via nanostructurization with good advantage. It should be noted that classical inorganic nonlinear volume media (including BSO, $LiNbO_3$, etc.) exhibit significantly lower nonlinearity, while bulk silicon based materials have nonlinear characteristics analogous to those of the organic thin film nanoobjects-doped materials under consideration.

Structure	Content of dopants, wt.%	Wavelength, nm	Energy density, Jcm^{-2}	Spatial frequency, mm^{-1}	Laser pulse duration, ns	Laser-induced change in the refractive index, (n	References
NPP	0	532	0.3	100	20	0.65x10^{-3}	[14]
NPP+C_{60}	1	532	0.3	100	20	1.65x10^{-3}	[14]
NPP+C_{70}	1	532	0.3	100	20	1.2x10^{-3}	[14]
PNP*	0	532	0.3	100	20	*	
PNP+C_{60}	1	532	0.3	100	20	0.8x10^{-3}	[14]
PI	0	532	0.6	90-100	10-20	10^{-4}-10^{-5}	[5]
PI+malachite green dye	0.2	532	0.5-0.6	90-100	10-20	2.87x10^{-4}	[5]
PI+graphene oxides	0.1	532	0.2	100	10	3.4x10^{-3}	present
PI+graphene oxides	0.2	532	0.28-0.3	100	10	3.65x10^{-3}	present
PI+shungites	0.1	532	0.6	150	10	3.1x10^{-3}	present
PI+shungites	0.2	532	0.063-0.1	150	10	5.3x10^{-3}	[9,15]
PI+C_{60}	0.2	532	0.5-0.6	90-100	10-20	4.2x10^{-3}	[5]
PI+C_{70}	0.2	532	0.6	90-100	10-20	4.68x10^{-3}	[5]
PI+C_{70}	0.5	532	0.6	90-100	10-20	4.87x10^{-3}	[5]
PI+C_{70}	0.1-0.5	1315	0.2–0.8	100	50	~10^{-3}	[14]

PI+quantum dots based on CdSe(ZnS)	0.003	532	0.2-0.3	100	10	$2.0x10^{-3}$	[10]
PI+CNTs	0.1	532	0.5-0.8	90-100	10-20	$5.7x10^{-3}$	[6]
PI+ CNTs	0.05	532	0.3	150	20	$4.5x10^{-3}$	[6,14]
PI+ CNTs	0.07	532	0.3	150	20	$5.0x10^{-3}$	[6,14]
PI+ CNTs	0.1	532	0.3	150	20	$5.5x10^{-3}$	[6,14]
PI + double-walled carbon nanotube powder	0.1	532	0.1	100	10	$9.4x10^{-3}$	[15]
PI + double-walled carbon nanotube powder	0.1	532	0.1	150	10	$7.0x10^{-3}$	[15]
PI+ mixture of CNT and nanofibers (type MIG)	0.1	532	0.3-0.6	90-100	10	$11.7x10^{-3}$	[15]
PI+ mixture of CNT and nanofibers (type MIG)	0.1	532	0.3-0.6	150	10	$11.2x10^{-3}$	[15]
Polymer-dispersed LC based on PI–C_{70} complex	0.2	532	0.1	90-100	10	$1.2x 10^{-3}$	[10]
COANP	0	532	0.9	90-100	10-20	10^{-5}	[14]
COANP+ TCNQ**	0.1	676	2.2 Wcm^{-2}			$2x10^{-5}$	[16]
COANP+C_{60}	5	532	0.9	90-100	10-20	$6.21x10^{-3}$	[14]
COANP+C_{70}	0.5	532	0.6	100	10	$5.1x10^{-3}$	present
COANP+C_{70}	5	532	0.9	90-100	10-20	$6.89x10^{-3}$	[14]
Polymer-dispersed LC	0.5	532	$30x10^{-3}$	100	10	$1.2x10^{-3}$	present

based on COANP–C_{70} complex							
Polymer-dispersed LC based on COANP–C_{70} complex	5	532	17.5×10^{-3}	100	20	1.4×10^{-3}	[17]
Polymer-dispersed LC based on COANP–nanotubes	0.1	532	30×10^{-3}	100	10	2.8×10^{-3}	present
Polymer-dispersed LC based on COANP–nanotubes	0.5	532	18.0×10^{-3}	90-100	10-20	3.2×10^{-3}	[18]

Table 1. Laser-induced change in the refractive index in some organic structures doped with nanoobjects. * The diffraction efficiency has not detected for these systems at this energy density. ** Dye TCNQ - 7,7,8,8,-tetracyanoquinodimethane – has been used in the paper [16].

4. Conclusion

Thus, analysis of the obtained results leads to the following conclusions:

- Doping with nanoobjects significantly influences the photorefractive properties of nanobjects-doped organic matrices. An increase in the electron affinity (cf. shungite, fullerenes, QDs) and specific area (cf. QDs, CNTs, nanofilers) implies a dominant role of the intermolecular processes leading to an increase in the dipole moment, local polarizability (per unit volume) of medium, and mobility of charge carriers.
- A change in the distance between an intramolecular donor and intermolecular acceptor as a result of variation of the arrangement (rotation) of the introduced nanosensitizer leads to changes on the charge transfer pathway in the nanocomposite.
- The variations of the length, of the surface energy, of the angle of nanoobject orientation relative to the intramolecular donor fragment of matrix organics can significantly change the pathway of charge carrier transfer, which will lead to changes in the electric field gradient and dipole moment.
- Different values of nonlinear optical characteristics in systems with the same sensitizer type and concentration can be related to a competition between carrier drift and diffusion processes in a nanocomposite under the action of laser radiation.

http://dx.doi.org/10.5772/50843

- Special role of the dipole moment as a macroscopic parameter of a medium accounts for a relationship between the photorefraction and the photoconductivity characteristics.
- The photorefractive parameters change can be considered as the indicator of following dynamic and photoconductive characteristics change.

As the result of this discussion and investigation, new area of applications of the nanostructured materials can be found in the optoelectronics and laser optics, medicine, biology, telecommunications, display, microscopy technique, etc. Moreover, the nanostructured materials can be used for example, for development of 3D media with high density of recording information, as sensor in the gas storage and impurity testing, as photosensitive layer in the spatial light modulators, convertors, limiters, etc. devices.

Acknowledgements

The author would like to thank their Russian colleagues: Prof. N. M. Shmidt (Ioffe Physical-Technical Institute, St.-Petersburg, Russia), Prof. E.F.Sheka (University of Peoples' Friendship, Moscow, Russia), Dr.N.N.Rozhkova (Institute of Geology, Karelian Research Centre, RAS), Dr.A.I.Plekhanov (Institute of Automation and Electrometry SB RAS, Novosibirsk, Russia), Dr.V.I.Studeonov and Dr.P.Ya.Vasilyev (Vavilov State Optical Institute, St.-Petersburg, Russia), as well as foreign colleagues: Prof. Francois Kajzar (Université d'Angers, Angers, France), Prof. D.P. Uskokovic (Institute of Technical Sciences of the Serbian Academy of Sciences and Arts, Belgrade, Serbia), Prof. I.Kityk (Politechnica Czestochowska, Czestochowa, Poland), Dr. R.Ferritto (Nanoinnova Technologies SL, Madrid, Spain) for their help in discussion and study at different their steps. The presented results are correlated with the work supported by Russian Foundation for Basic Researches (grant 10-03-00916, 2010-2012).

Author details

Natalia V. Kamanina[1*]

Address all correspondence to: nvkamanina@mail.ru

1 Vavilov State Optical Institute, 12, Birzhevaya Line, St. Petersburg, , Russia

References

[1] Couris, S., Koudoumas, E., Ruth, A. A., & Leach, S. (1995). Concentration and wavelength dependence of the effective third-order susceptibility and optical limiting of C_{60} in toluene solution. *J. Phys. B: At. Mol. Opt. Phys.*, 8, 4537-4554.

[2] Robertson, J. (2004). Realistic applications of CNTs. *.Mater. Today*, 7, 46-52.

[3] Lee, Wei, & Chen, Hsu-Chih. (2003). Diffraction efficiency of a holographic grating in a liquid-crystal cell composed of asymmetrically patterned electrodes. *Nanotechnology*, 14, 987-990.

[4] Buchnev, Oleksandr, Dyadyusha, Andriy, Kaczmarek, Malgosia, Reshetnyak, Victor, & Reznikov, Yuriy. (2007). Enhanced two-beam coupling in colloids of ferroelectric nanoparticles in liquid crystals. J. Opt. Soc. Am. B/ , 24(7), 1512-1516.

[5] Kamanina, N. V., Emandi, A., Kajzar, F., & Attias, Andre´-Jean. (2008). Laser-Induced Change in the Refractive Index in the Systems Based on Nanostructured Polyimide: Comparative Study with Other Photosensitive Structures. *Mol. Cryst. Liq. Cryst.*, 486, 1-11.

[6] Kamanina, N. V., Serov, S. V., Savinov, V. P., & Usković, D. P. (2010). Photorefractive and photoconductive features of the nanostructured materials. *International Journal of Modern Physics B (IJMPB)*, 24(6-7), 695-702.

[7] Kamanina, N. V., & Vasilenko, N. A. (1997). Influence of operating conditions and of interface properties on dynamic characteristics of liquid-crystal spatial light modulators. *Opt. Quantum Electron.*, 29, 1-9.

[8] Kamanina, N. V. (2005). Fullerene-dispersed liquid crystal structure: dynamic characteristics and self-organization processes. *Physics-Uspekhi*, 48, 419-427.

[9] Kamanina, N. V., Serov, S. V., Shurpo, N. A., & Rozhkova, N. N. (2011). Photoinduced Changes in Refractive Index of Nanostructured Shungite-Containing Polyimide Systems. *Technical Physics Letters*, 37(10), 10(10), 949-951.

[10] Kamanina, N. V., Shurpo, N. A., Likhomanova, S. V., Serov, S. V., Ya, P., Vasilyev, V. G., Pogareva, V. I., Studenov, D. P., & Uskokovic, . (2011). Influence of the Nanostructures on the Surface and Bulk Physical Properties of Materials. *ACTA PHYSICA POLONICA A*, 119(2), 2(2), 256-259.

[11] Kamanina, N. V., & Plekhanov, A. I. (2002). Mechanisms of optical limiting in fullerene-doped conjugated organic structures demonstrated with polyimide and COANP molecules. *Optics and Spectroscopy*, 93(3), 408-415.

[12] Cherkasov, Y. A., Kamanina, N. V., Alexandrova, E. L., Berendyaev, V. I., Vasilenko, N. A., & Kotov, B. V. (1998). Polyimides: New properties of xerographic, thermoplastic, and liquid-crystal structures. (SPIE International Symposium on Optical Science, Engineering and Instrumentation, San Diego, CA, USA, 1998) Proceed. of SPIE , 3471, 254-260.

[13] Akhmanov, S. A., & Nikitin, S. F. (1998). Physical Optics. (Izdat. Mos. Gos. Univ., Moscow, 1998) [in Russian].

[14] Kamanina, N. V., Vasilyev Ya, P., Serov, S. V., Savinov, V. P., Bogdanov. Yu, K., & Uskokovic, D. P. (2010). Nanostructured Materials for Optoelectronic Applications. *Acta Physica Polonica A*, 117(5), 786-790.

[15] Kamanina, N. V., Serov, S. V., Shurpo, N. A., Likhomanova, S. V., Timonin, D. N., Kuzhakov, P. V., Rozhkova, N. N., Kityk, I. V., Plucinski, K. J., & Uskokovic, D. P. (2012). Polyimide-fullerene nanostructured materials for nonlinear optics and solar energy applications. J Mater Sci: Mater Electron DOI: 10.1007/s10854-012-0625-9, published on-line 26 January 2012.

[16] Ch Bosshard. Bosshard., K., Sutter, P., & Chapuis. Günter, G. (1989). Linear- and non-linear-optical properties of 2 -cyclooctylamino-5-nitropyridine. *J. Opt. Soc. Am.*, B6, 721-725.

[17] Kamanina, N. V. (2002). Optical investigations of a C_{70}-doped 2-cyclooctylamino-5-nitropyridine-liquid crystal system. *Journal of Optics A: Pure and Applied Optics,*, 4(4), 4(4), 571-574.

[18] Kamanina, N. V., & Uskokovic, D. P. (2008). Refractive Index of Organic Systems Doped with Nano-Objects. *Materials and Manufacturing Processes*, 23, 552-556.

[19] Chemla, D. S., & Zyss, J. (1987). Nonlinear Optical Properties of Organic Molecules and Crystals . (Orlando Academic Press, 1987), Translated into Russian, Moscow: Mir, 1989., 2

[20] Gutman, F., & Lyons, L. E. (1967). Organic Semiconductors. Wiley, New York.

[21] Mylnikov, V. S. (1994). in Advances in Polymer Science. Photoconducting Polymers/ Metal-Containing Polymers Springer-Verlag Berlin , 115, 3-88.

[22] Wang, Y., & Suna, A. (1997). Fullerenes in Photoconductive Polymers. Charge Generation and Charge Transport. *J. Phys. Chem.*, B 101, 5627-5638.

Chapter 18

Interconnecting Carbon Nanotubes for a Sustainable Economy

Steve F. A. Acquah, Darryl N. Ventura,
Samuel E. Rustan and Harold W. Kroto

Additional information is available at the end of the chapter

http://dx.doi.org/10.5772/51781

1. Introduction

Concerns about depleting natural resources have been circulating for decades with alarming predictions that have turned out to be less than accurate. What has become clear, however, is the need for a decrease in the utility of a fossil based economy and a focus on a more sustainable one. This chapter reviews some of the recent progress made in the use of interconnected carbon nanotubes (CNTs) in the hydrogen, photovoltaics and thermoelectric alternative energy based economies.

The move towards a hydrogen economy is a concept that has gained traction over the last 5 years with advances in hydrogen fuel cells that are economically viable. It is envisaged that the automotive industry will begin to implement measures for the development of vehicles with hydrogen fuel cells as the economy begins to recover. However, such a move will also require a substantial investment in the infrastructure to support these vehicles. Key to the development of such technologies is the need to continuously improve the efficiency, while monitoring the safety. CNTs have been used as frameworks for a number of key areas in the hydrogen economy [1]. The most notable area is that of fuel cell integration, where the tubes are mixed with platinum or palladium to aid in the process of catalysis.

CNTs with palladium attached to their surface have also been used for the construction of hydrogen sensors, expanding the research field from the consumption to the detection of hydrogen. The recent advances in cross-linked CNT papers are stimulating the development of new materials, such as flexible palladium embedded CNT sensors [2] (Fig. 1.). This section of the chapter will explore some of the latest results from the use of interconnected CNTs in hydrogen fuel cells and sensor development.

Figure 1. A cross-linked CNT paper with embedded Pd nanoparticles that can be used to construct a hydrogen sensor.

The field of photovoltaics is regarded as a major contributor to a sustainable economy. However; purveyors of large scale solar panels have been experiencing a degree of volatility in the market due in part to the decreasing price of the technology, increased competition and a dependence on government subsidies. At the opposite end of the scale, there is a surge in small solar powered gadgets such as pocket LED torches and mobile device chargers, which adorn many airport convenience outlets. The demand for pocket sized solar powered devices is helping to stimulate research into making the energy conversion process more efficient. There were three major advances in photovoltaics, the development of photovoltaic devices from crystalline silicon, which dominate the commercial market, cadmium telluride (CdTe) and dye sensitized solar cells (DSSCs). CNTs are currently being investigated as a way to enhance electron transfer and replace the standard platinum based counter electrodes, especially with DSSCs. CNT thin films and mats are currently being tested as components of these photovoltaic devices. This section of the chapter will explore how the CNTs have been used to enhance dye-sensitized [3], CdTe [4] and silicon [5] based solar cells, and address some the concerns about the race to produce novel photovoltaic devices and the toxic warnings from the past that may ultimately define the balance between safety and efficiency.

The last section of this chapter will focus on the development of CNT based thermoelectric devices which may bridge the gap between conventional and sustainable economies. Energy loss in the form of heat is clearly an important concept to address, and capturing the heat from combustion engines is one avenue being pursued by research. Around 75% of the energy produced from fuel with internal combustion engines is lost to the environment, so it may be possible to recapture some of this energy using a thermoelectric device between the engine coolant system to the exhaust manifold [6]. However, problems have been encountered with low efficiency so CNTs have been investigated as a suitable component of thermoelectric devices due to a number of characteristics, such as their low dimensional structure, their electrical conductivity, and their axial thermal conductivity [7, 8].

http://dx.doi.org/10.5772/51781

2. The Hydrogen Economy

Many nations are looking into alternative sources of energy to address issues of environmental responsibility and energy independence. Some of these energy sources include solar power, wind energy, natural gas, and hydrogen. As society explores hydrogen as an alternative energy source, the question is how effective can CNTs be in helping to solve some of the problems in the structure, function and safety of this emerging industry?

2.1. Fuel Cells & Hydrogen Storage

In the simplest case, a hydrogen fuel cell is comprised of a permeable membrane placed between an anode and a cathode. There are various types of fuel cells: polymer electrolyte membrane, direct methanol, alkaline, phosphoric acid, molten carbonate, and solid oxide. Hydrogen fuel cells fall under the polymer electrolyte membrane fuel cell (PEMFC) category and are sometimes also referred to as a proton exchange membrane fuel cell. In a typical PEMFC, the permeable membrane consists of a proton-conductive polymer such as perfluorosulphonic acid, also known commercially as Nafion. The fuel cell works by using a catalyst to oxidize hydrogen at the anode, converting it into a positively charged proton and a negatively charged electron. The electrons travel through a wire creating an electrical current to power a device while the protons travel through the permeable membrane to the cathode. At the cathode, the protons recombine with the electrons and react with oxygen to form water which is eventually drained from the system.

Despite recent advances in research, there are still a few obstacles that need to be overcome in order for fuel cells to become mainstream technology. In order to integrate with existing technologies, fuel cells need to become considerably cheaper. Currently, they are expensive to construct, mainly due to the use of platinum catalysts. According to the United States Department of Energy, the cost per kilowatt would need to decrease in order for fuel cells to be competitive and economically viable. In order to compete commercially with the combustion engine, it is estimated that the fuel cell cost would need to be cut to approximately $25–$35/kW. Another aspect of fuel cells that needs improvement is the operational lifetime. The permeable membrane is made of a synthetic polymer which is susceptible to chemical degradation. Reliability in automotive applications, can be defined by the lifetime of a car engine, approximately 150,000 miles, so research has focused both on improving the efficiency of the catalytic process and the durability of the components. CNTs have been proposed as a substitute to the carbon powder currently used in PEMFCs. (Fig. 2.) CNTs have excellent conductive properties, a low mass density, and robust physical properties making them an ideal and durable material for fuel cell electrodes. Furthermore, nanotubes assembled into such macrostructures have a high surface area making them a suitable substrate for Pt catalysts and hydrogen adsorption [9].

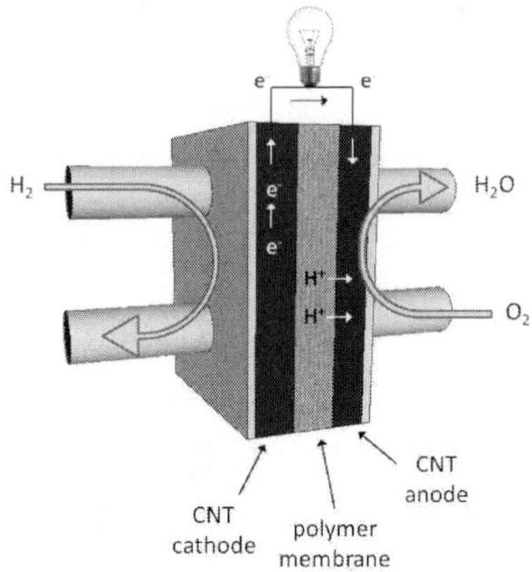

Figure 2. Schematic of a CNT composite hydrogen fuel cell.

In 2003, researchers from the University of California, Riverside explored the use of MWCNTs as a carbon support for platinum catalysts in an attempt to maximize Pt interfacing between all the components in a fuel cell. The problem in conventional fuel cells is that the addition of the polymer tends to isolate the carbon particles reducing electron transport, resulting in the requirement of additional Pt particles to increase the power output. To resolve this issue and improve conductivity, Wang et al. grew nanotubes directly on carbon paper and electrodeposited Pt particles onto the CNTs [10]. Although their experiments produced promising results, their CNT based fuel cell still had a lower performance compared to conventional PEMFCs. Despite this low performance, this proof of concept was important to other researchers using CNTs in fuel cells. The following year in 2004, Girishkumar et al. investigated ways to improve the electrodes in direct methanol fuel cells (DMFCs) [11]. Their team developed a way to synthesize SWCNT thin films onto optically transparent electrodes using electrophoretic deposition techniques. It was determined that there was an improvement in catalytic activity mainly due to a larger surface area provided by the CNTs. This high surface area and porosity maximizes interactions between the fuel, electrode, and catalyst interface thereby enhancing Pt utilization and potentially reducing fuel cell manufacturing costs. Li et al. (2006) also explored the use of CNTs in PEMFCs. They developed a facile and cost-effective method for the synthesis of an aligned Pt/CNT film [12]. They were interested in producing oriented CNT films due to enhanced conductivity. It was also suggested that there would be higher gas permeability and better water removal with aligned nanotubes. The aligned CNTs did show an improvement in Pt utilization as 60% of the metal particles were being used during catalysis [11].

Using covalently cross-linked CNTs is another promising avenue for fuel cell electrodes [13]. Our work at Florida State University focused on the covalent cross-linking of multi-walled carbon nanotubes via a Michael addition reaction mechanism to form thin, flexible mats

http://dx.doi.org/10.5772/51781

[14]. We then explored an alternative cross-linking system to avoid the use of thiols and embedded palladium nanocrystals into the cross-linked network [2].

Research into hydrogen storage with interconnected CNT networks started by looking into SWCNTs using a procedure called temperature programmed desorption. Experiments on MWCNTs followed with work focusing on metal doped tubes. However, problems began to arise when increasing values of CNT storage capacities, up to 21 wt%, were reported. A detailed review of the findings can be found by Yunjin Yao and serves as an interesting footnote towards the role of CNTs and the need for a better understanding of their chemistry in materials [15]. In summary, the main concerns were that elevated hydrogen storage percentages may have be due to a number of factors including the insufficient characterization of CNT composites due to the presence of SWCNTs, DWCNTs and MWCNTs with a variety of open and closed ended tubes. Contamination of the CNTs during the process of ultrasonic probe treatments was a concern, because in one example the value for SWCNTs were reported to have a hydrogen storage capacity of around 4.5% at 30 kPa and 70 K, but the ultrasonic probe was made from a titanium alloy that was known to act as a hydrogen storage material.

2.2. Water Splitting

The research field based on water splitting has, not surprisingly, found a niche in the development of the hydrogen economy due to the clean production of hydrogen and oxygen. However this integration has a far more significant impact when combined with hydrogen fuel cells. The waste product of hydrogen fuel cells is water, and it is formed during the reaction with oxygen, so the water could fuel the process of splitting and this in turn can fuel hydrogen cells.

CNTs have been used to enhance the water splitting performance of titania photocatalysts [16] but an alternative use for CNTs has been found in membranes. Nafion, a sulfonated tetrafluoroethylene based fluoropolymer-copolymer, is a membrane that has had commercial success in the fuel cell industry. Research groups are looking into enhancing the properties of the film with the addition of CNTs. Nafion/CNT composites with low concentrations of CNTs have been shown to have an effect on solvent permeation and mechanical stability. At high concentrations of CNTs the membranes have the ability to separate proton and electron conduction pathways in the membrane. Using this concept, many applications can be envisaged for these membranes with one example being that of using sunlight to produce hydrogen from water splitting. Current research has focused on the measurements of the electron and proton transport characteristics of Nafion and MWCNT composite films.[17] These films can be assembled by the addition of Nafion solution to MWCNTs, followed by the dispersion of the MWCNTs in an ultrasonic bath. Various concentrations of MWCNTs were investigated to a maximum of 5% MWCNTs by dry weight of Nafion. After the addition of isopropyl alcohol, to further aid the dispersion of the MWCNTs, the slurry was poured into petri-dishes and left to undergo solvent evaporation for 3 hours. The dishes with various

CNT concentrations were placed in an oven set to 40 °C for a few hours before being washed with deionized water and removed from the petri-dishes.

To test the membrane, an artificial leaf system was constructed. (Fig. 3.). The membrane separated the anode, which was exposed to sunlight where water droplets were present, and the cathode.

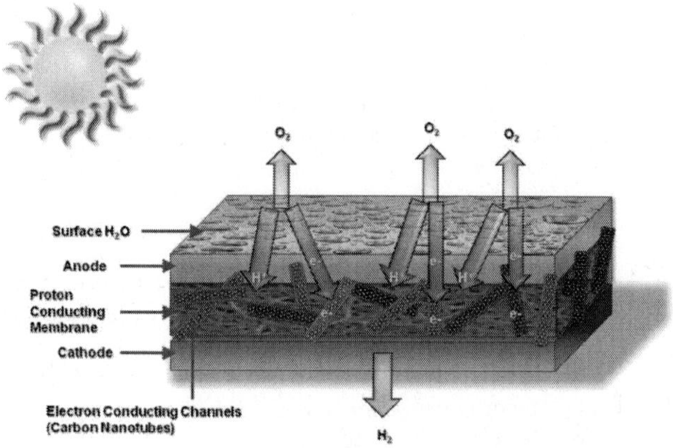

Figure 3. Schematic of the water-splitting device. The anode contained a chromophore and an oxygen evolving complex. The cathode contained a proton reducing catalyst. Image adapted from V. Ijeri *et al.* (2010).

The results highlighted a few points. Firstly pure Nafion exhibited insulating behavior and with increasing MWCNT percentage, a non linear behavior is observed with I–V curves, which is an indication of non-ohmic conductivity. The membranes were tested in both wet (1% H_2SO_4) and dry conditions to evaluate the electron conductivity. Before and after wetting the conductivity values increase with increasing filler content, but again without a linear relationship, which meant a critical concentration at which the membrane changed from insulating to conducting/semiconducting had to be established. This was done by looking at the values higher than 10^{-1} mS/cm which were obtained when MWCNTs > 3%. The next task was to investigate proton conductivity, and with standard conditions, this was generally low. However, with an increasing MWCNT percentage there was a subsequent increase. Although the effects of MWCNTs on proton conductivity is still not fully understood, most researchers will fall back on the semi-empirical quantum mechanical calculations too at least provide an insight into the possible conduction pathways.

When the membranes were subjected to 1% H_2SO_4 they did show an increase in proton conductivity, which was due to the various proton transfer mechanisms. The hydrogen bonding of the $-SO_3$ groups with an H_3O^+ ion and water molecules results in a change in the side chains of Nafion. It was difficult to determine the contribution of MWCNTs in the process of electron transfer because of the amount of water molecules. However membranes with no MWCNTs demonstrated the best proton conductivity, while the others have slightly lower conductivities. The answer could be as simple as a decrease in the amount of Nafion. Either way, this study has shown great potential for the integration of CNTs for membrane applications.

2.3. CNT Hydrogen Sensors

Another application of great interest in the field of CNTs is hydrogen sensing. Advancements in the development of fuel cell design and technology means that a variety of sensors would be required to maintain a safe operational environment. CNTs are an ideal material for components of sensors due to their durability, and electronic properties.

One of the first breakthroughs in CNT sensor technology occurred in 2001 when Kong et al. constructed hydrogen sensors by decorating SWCNTs with Pd nanoparticles [18]. Their H_2 sensor exhibited significant changes in conductivity when exposed to small amounts of H_2 and was able to operate at room temperature. Kong et al. were able to achieve this by depositing Pd particles on CVD grown SWCNTs via electron beam evaporation methods. When they placed this in a hydrogen atmosphere, a decrease in the CNT conductivity was observed. It has been proposed that this lower work function promotes electron transport from the Pd NPs into the CNTs resulting in a decreased amount of hole-carriers and conductivity. The reaction is also reversible. Under a hydrogen atmosphere, Pd reacts with H_2 to become palladium hydride. The dissolved hydrogen in Pd metal combines with oxygen in air and results in H_2O, recovering the electrical characteristics of the sensor. Kong et al. reported that their detector had a limit at 400 ppm, a response time of 5-10 s, and a recovery time of approximately 400 s [18].

One design principle of CNT composites that has defined the nature of efficiency is that of aligned CNTs. From aligned thin films of Buckypaper to forests of vertically grown CNTs on substrates, control over the direction of individual tubes and connected bundles is essential for unlocking the full potential of the tubes. An investigation was made into the development of aligned CNT sensors using a method involving nanoplating and firing to produce cracks in a CNT composite film, exposing horizontally aligned carbon nanotubes (HACNTs) [19]. This research used arc produced MWCNTs as the basis for the composite film, which was rather enlightening in a field geared towards chemical vapor deposition (CVD) produced tubes. Research with arc produced MWCNTs has almost become a relic of the early years of CNT research. They were made by using a 150 mm long graphite rod for the anode and a graphite disc on a copper block for the cathode. After purification steps the CNTs were acid-oxidized using the standard 3:1 ratio of nitric acid (HNO_3) to sulfuric acid (H_2SO_4) and washed several times in DI water before being dried in air at 120 °C. The acid treatment was required to increase the interfacial adhesion between the CNTs and metals. To produce the sensor, a sample of the purified CNTs was dispersed in DI water with a polyvinylpyrrolidone surfactant (PVP K30), which produced a CNT suspension. The CNT/Ni composite was produced by the addition of nickel sulfate solution containing sodium phosphinate, maleic acid disodium salt hydrate, citric acid monohydrate, lead(II) acetate trihydrate and sodium acetate trihydrate. The composite film was produced on a glass substrate by the immersion of the glass, with palladium particles on the surface, into the CNT/Ni solution for 60 seconds before drying the substrate at 100 °C to induce cracks in the film (Fig. 4.) exposing horizontally aligned CNTs. 18 Finger platinum electrodes were then deposited by DC sputtering to complete the sensor.

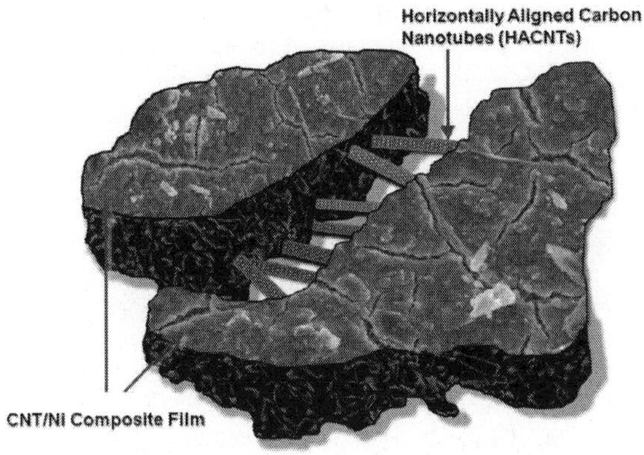

Figure 4. Schematic of the cracked composite film exposing horizontally aligned CNTs.

These results are described in two papers and although the idea of horizontal alignment is important, it is difficult to accurately quantify the results of the papers since in both cases there is an abundance of nanoparticle palladium in both the CNT/Ni system (Pd deposited on the glass) [19] and the Pd/CNT/Ni (Pd deposited on the CNT/Ni film) system [20]. Fig. 5. shows the process of assembly for the sensors, which use a similar procedure in both of the research papers.

The HACNT-based sensors were also shown to have a sensitivity response to carbon dioxide, methane and ethene with a gas concentration of 200 ppm, with the highest sensitivity for H_2. One of the points raised in this research, that was fundamental to the mechanism of sensing, was the role of atomized hydrogen. These atoms, produced by the metal particles, migrated to the sidewalls and the defects of CNTs, diffusing into the lattice of nanoparticles. It was stated that a dipole layer formed at that interface and affected the charge-carrier concentration, and the hydrogen atoms donated their electrons to the CNTs, which resulted in a decrease in conductivity.

http://dx.doi.org/10.5772/51781

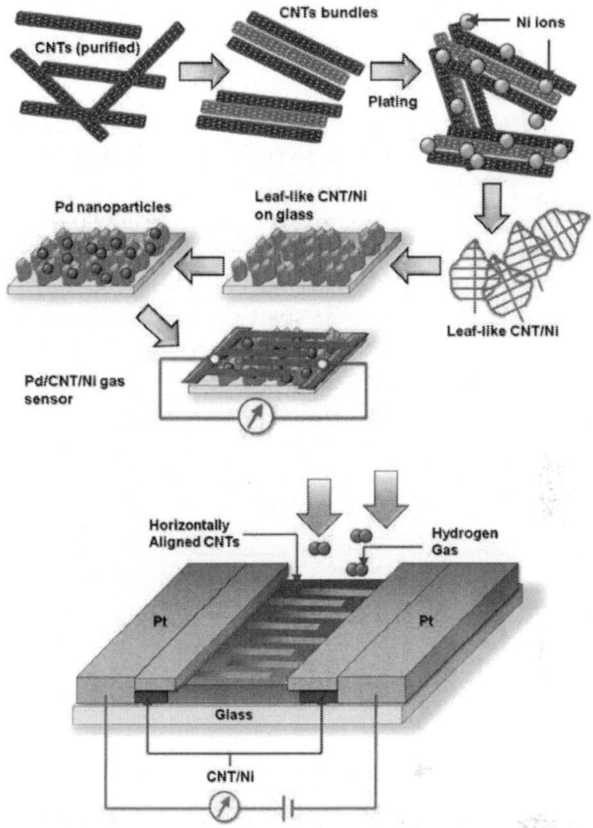

Figure 5. Schematic of the steps involved in the construction of hydrogen sensors on glass substrates with the use of the Pd nanoparticle functionalized CNT/Ni composite film. Image adapted from Lin *et al.* (2012). Schematic illustration of a HACNT-based gas sensor on glass substrate. Image adapted from B-R Huang *et al.* (2012).

In another example, a hydrogen sensor was constructed using SWCNTs and chitosan (CHIT).[21] The CHIT which covered the SWCNTs was able to filter out polar molecules and allow hydrogen to flow to the SWCNTs. The CHIT conjugate which is porous is insulating by nature, but can be made water soluble in an acidic environment which is then useful for making a film. Additional benefits can be found in the many functional hydroxyl (–OH) and amino (–NH_2) groups that react with analytes, so the effect Of a CHIT conjugate with SWCNTs for the development of a hydrogen sensor was investigated. The CHIT film was prepared by making a 2 wt% solution dissolving CHIT in a 5% acetic acid solution. This was used to coat a glass substrate or SWCNTs depending on the sensor preparation and followed by the removal of solvent to form the films. To evaluate the sensor performance three different types were made (Fig. 6.). The Type I sensor was assembled simply by depositing SWCNTs onto the glass substrate with Pt electrodes placed by sputter deposition. The Type II sensor was assembled by casting the glass slide with a film of CHIT before being placed into an arc-discharge chamber to deposit SWCNTs. The Pt electrodes were added in a similar method. The Type III sensor was assembled using the initial preparation for a Type I sensor followed by CHIT film coating and Pt electrode deposition. There were slight differences in the interaction of the CHIT film with the SWCNTs. In the Type II sensor, there was some mixing of the CNTs with CHIT but only at the interface. With the Type III sensor, the CNTs were immersed in the CHIT matrix.

Resistance measurements of the films were made between the electrodes, and the values were around 100 Ω for Type I and II films and around 10^6 Ω for the Type III film. The high resistance could be accounted for by the contact of the electrode with chitosan, although it was noted by the authors that ohmic contacts were present.

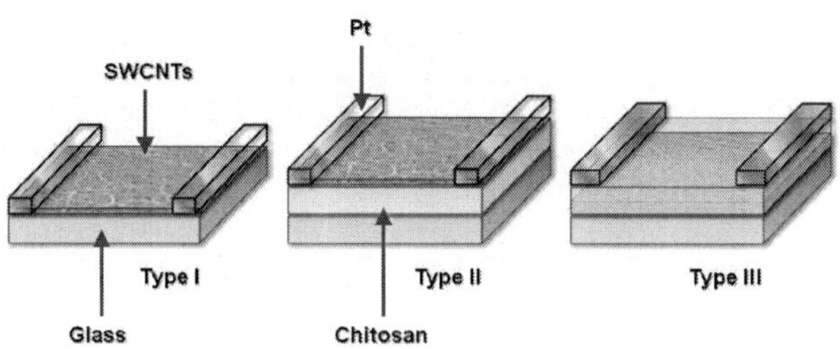

Figure 6. Diagram of the 3 types of sensors. Image adapted from Li *et al.* (2010).

The response of the sensors was measured at room temperature and the results showed 15, 33, and 520% for Type I, Type II, and Type III sensors, respectively. One interesting point made by the authors was that although the Pd decoration of SWCNTs is typically used to enhance hydrogen sensing, the response can be less than the effect of chitosan at 4% H_2 gas. This research provided an important step towards the use of CNTs in sensors without the requirement of Pd.

In summary, the use of CNTs in the hydrogen economy has highlighted some interesting points. Is the race to develop more efficient hydrogen powered devices really producing a sustainable economy? And has the focus on reducing the utility of some of the rare raw ma-

http://dx.doi.org/10.5772/51781

terials been lost? It is well known that platinum and palladium are extremely important to the fuel cell and sensor industries, with CNTs enhancing their properties, but an increase in alternative energy devices based on these metals, whatever the concentration, may cause issues of sustainability in the future.

3. Photovoltaics

The research field of photovoltaics has certainly become a hot topic over the last few years with a lot of attention based on increasing the efficiency of dye sensitized solar cells (DSSCs) in the hope that they will one day be as prevalent as the silicon based alternative. CNTs are an important addition to the field of photovoltaics with the focus on the nanotubes acting as p-type materials or enhancing/replacing the counter electrodes.

3.1. Dye Sensitized Solar Cells

If there were an enclave for truly beautiful chemistry, then the research behind dye sensitized solar cells (DSSCs) would clearly be the centerpiece. The chemistry behind the operation of these devices is inspiring a generation of researchers to address the concerns of renewable energy with a different approach to the well established silicon based solar cells. Generally, the DSSCs are comprised of an anode, electrolyte and cathode. The anode is usually assembled from nano-crystalline titania particles (TiO_2) and a dye attached to the particles. The cathode, also known as the counter electrode (CE), is where the catalysis must occur and typically contains platinum. The iodide electrolyte facilitates the iodide/triiodide redox couple where after the excitation of the dye and loss of an electron, it regains one from iodide, oxidizing it to triiodide. The best reported efficiency for DSSCs is 11.4% as documented by the National Institute for Material Science (NIMS).

CNTs have been used as a potential replacement for the platinum based CE. In a study by Jo et al. (2012), interconnected ordered mesoporous carbon–carbon nanotube nanocomposites were used to demonstrate Pt-like CE behavior in a dye-sensitized solar cell [22]. CNT fibers have been used as a conductive material to support the dye-impregnated TiO_2 particles. The CNTs were first spun from an array synthesized by chemical vapor deposition and resulted in highly aligned macroscopic fibers [23]. The research was novel in the application of these fibers as both the working electrode and the counter electrode.

The CNT/TiO_2 composite fiber was produced by submersing the pure CNT fiber in a TiO_2 colloid solution which was followed by sintering at 500 °C for 60 min. The thickness of TiO_2 layer was determined to be between 4 and 30 μm, depending on the submersion time. The dye used for the cell was cis-diisothiocyanato-bis (2,2′-bipyridyl-4,4′-dicarboxylato) ruthenium(II) bis (tetrabutylammonium) which is better known as N719. For DSSCs with a metal CE the I^-/I_3^- couple does eventually cause corrosion, but the CNT fibers exhibit a high stability and are relatively cheap. Fig. 7. shows the schematic of the working device with the two fibers in an electrolytic solution.

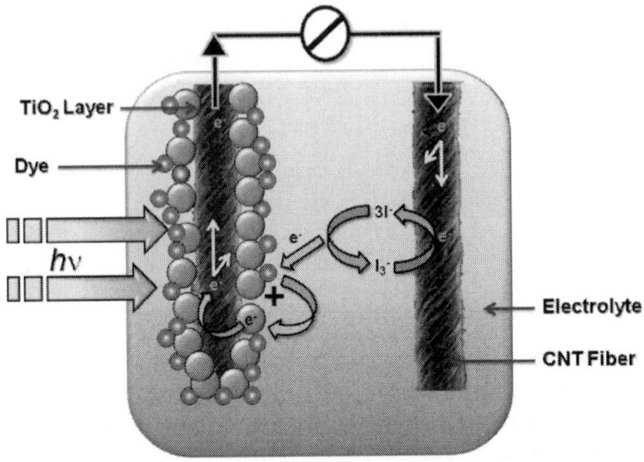

Figure 7. Schematic illustration of a wire-shaped DSSC made from two CNT fibers. Figure adapted from Chen *et al.* (2012).

The mechanical properties of the fiber are quite good with tensile strength measurements that exceed 700 MPa. The enhanced electrical conductivity also ranges from 100 to 1000 S/cm. The fiber-shaped DSSC demonstrated an efficiency of 2.94% which was a significant accomplishment. The fibrous nature of the material would make large-scale composites easy to fabricate. One of the more exciting applications is that of woven fabrics that may be used for the development of smart textiles for consumers, or extended use for space based electronics.

3.2. Quantum Dot Solar Cells

Cadmium telluride (CdTe) has been shown to be a promising low-cost component photovoltaic material, however the incorporation of quantum dot (QD) based technologies will likely raise fears about the toxicity of cadmium and cadmium based compounds. Significant progress has been made during the past several years with the highest efficiency reported for CdTe based photovoltaic devices at 17.3% produced by the company First Solar.

Although research is shifting towards CdTe/graphene composites [24], there is still room for CNT based devices. SWCNT/polyelectrolyte/QD nanohybrids have been produced that take advantage of the negatively charged thioglycolic acid capped CdTe QDs and SWCNTs coated with a positively charged polyelectrolyte facilitating electrostatic interactions [25]. In this work, SWCNTs coated with a positively charged polyelectrolyte showed typical transitions and emission attributes in the visible and near-infrared spectrum. The application of steady state absorption spectra was useful in outlining the superimposition of QD and SWCNT characteristics. The results of the study also confirmed charge transfer between SWCNTs and QDs, underlined by femtosecond transient absorption spectroscopy. Microscopic studies suggested that statically formed SWCNT/polyelectrolyte/QD nanohybrids with individually immobilized QDs were generated. It is clear that this study focuses on the importance of the interactions between the components of the nanohybrids and creates a pathway for looking at the development of the layer-by-layer coating of SWNTs and recruitment of photoactive particles for photovoltaic applications.

3.3. Silicon Based Solar Cells

With the exception of multi-junction cells and gallium arsenide (GaAs) based devices, crystalline silicon based cells are still the best choice with efficiencies at 20.4% for multicrystalline structures to 27.6% for single crystal based cells. However, there is clearly room for improvement as the increase in efficiency has generally reached a plateau over the last few years. What may be required is a different approach to the design and chemistry of these photovoltaic devices. CNTs have again been applied on the strength of their p-type conduction. In one recent example, polyaniline (PANI) and CNTs were used to construct heterojunction diode devices on n-Type silicon [26]. If was found that both PANI and SWCNTs could act as photovoltaic materials in a bilayer configuration with n-type Silicon: n-Si/PANI and n-Si/SWCNT. Four devices were tested (Fig. 8.) and it was determined that the short circuit current density increased from 4.91 mA/cm^2 for n-Si/PANI (Fig. 8a) to 12.41 mA/cm^2 n-Si/PANI/SWCNT (Fig. 8c). The n-Si/SWCNT/PANI device (Fig. 8d) and its control n-Si/SWCNT (Fig. 8b) exhibited a decrease in the short-circuit current density.

PANI was synthesized using the MacDiarmid method [27] before being spin-coated at 600 rpm to form a film. The SWCNTs were dispersed in DMF by sonication over a period of 12 h in 3 hour intervals, with the any solids removed by centrifugation. The supernatant was then removed and sonicated for an additional 6 hours before being used to make the devices. The devices were assembled by spraying SWCNTs using an airbrush deposition technique at 150 °C. It was found that the characteristics of the devices were affected by their design structure with better hole transport from PANI to SWCNTs and less efficient transport of holes from PANI to SWCNTs in the multilayer devices.

Other examples of CNT-Silicon hybrid photovoltaic devices include the investigation of the optimal thickness of SWCNT films on n-type silicon in order to maximize photovoltaic conversion [28] giving percentage efficiencies between 0.4 and 2.4%, and the effect of the number of walls of MWCNTs on the photon to electron conversion [29].

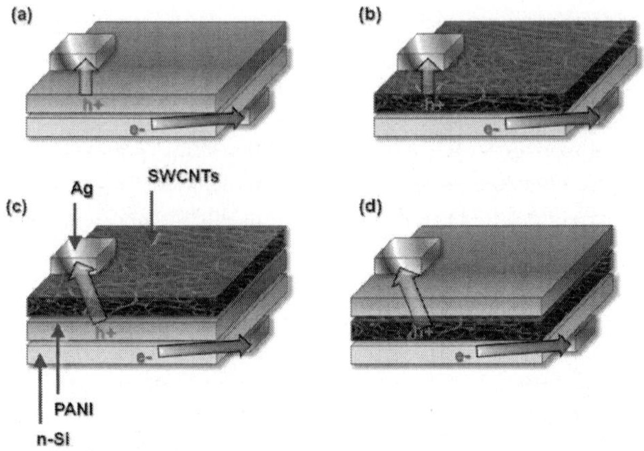

Figure 8. Schematics for (a) n-Si/PANI, (b) n-Si/SWCNTs, (c) n-Si/PANI/SWCNTs, and (d) n-Si/SWCNT/PANI devices. Image adapted from Bourdo *et al.* (2012).

In summary, photovoltaics have been shown to be very popular within the scientific field and the commercial market. Consumer electronics have been marketed with solar power chargers as a way to promote sustainability and environmental responsibility. The research into ruthenium based DSSCs is very popular but again there are concerns about the use of ruthenium for a sustainable economy. Fortunately, there are many photosensitive dyes that don't contain ruthenium which are currently being explored, but it is clear that the integration of interconnected CNTs can play an important role in the development of novel photovoltaic devices.

4. Thermoelectrics

In 1821 Thomas Johann Seebeck made the first discovery in the series of thermoelectric effects. The Seebeck effect described the electromotive force (emf) produced by heating the junction between two different metals. In essence, the kinetic energy of the electrons in the warmer part of a metal would facilitate the transfer of the electrons to the cooler metal faster than electron transfer from the cooler to the warmer metal, essentially creating an electronic potential where the cooler metal obtains a net negative charge. Harnessing the heat lost from a system and converting it to electricity will help to reduce the strain on electricity providers, but the difficulties surrounding the efficiency of the conversion process need to be addressed.

4.1. Thermoelectric Fabrics

One of the more futuristic ideas is that of wearable electronics, and this has been envisaged for many in the field of photovoltaics, but an Interesting alternative can be found in the field of thermoelectrics. Recent advancements in research have shown that composite films of MWCNT and polyvinylidene fluoride (PVDF) assembled in a layered structure can be designed to have the effect of felt-like fabric.[30] A thermoelectric voltage can be generated by these fabrics as a result of the individual layers increasing the amount of power produced. More importantly, these fabrics would be more economical to produce clearing the way for a new generation of energy harvesting devices that could power portable electronics. Fig. 9. shows a schematic of a fabric with every alternate conduction layer made with p-type CNTs (B) followed by n-type CNTs (D). The insulating layers allow for alternating p/n junctions when all the layers are stacked, pressed and heated to melt the polymer. It was noted that layers A–D could be repeated to reach a desired number of conduction layers N, and when the film is exposed to a change in temperature ($\Delta T = T_h - T_c$), the charge carriers which can be holes (h) or electrons (e) migrate from T_h to T_c generating a thermoelectric current I.

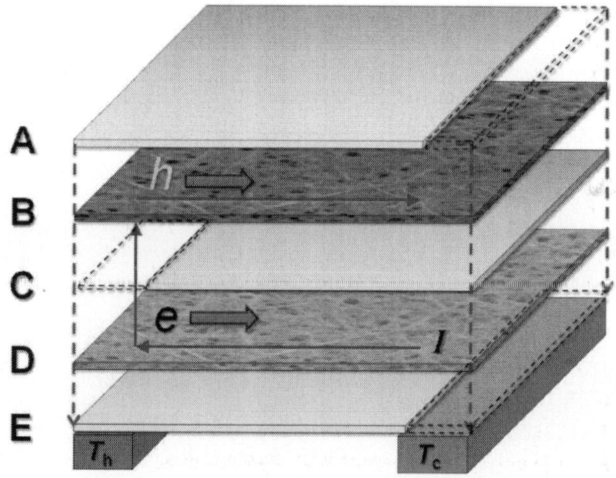

Figure 9. A Layered arrangement for the multilayered fabric. The CNT/PVDF conduction layers (B,D) are alternated between the PVDF insulation layers (A,C,E). Figure adapted from Hewitt *et al.* (2012).

When more power is required, ΔT would have to be increased. Subsequently, if the heat source were sufficiently large enough, the number of conduction layers could be increased. This would be a huge benefit for manufacturing industries that use high temperature equipment. In terms of energy output, a fabric composed of 300 layers with a ΔT = 100 K, may produce up to 5 μW. This is certainly a promising material that could potentially be integrated into many thermal systems and help with waste heat recovery.

4.2. Micro-Thermal Electrics

The addition of CNTs to microelectrical mechanical systems (MEMS) typically proceeds by either a bottom-up approach which focuses on the deposition of catalytic nanoparticles to control the location of CNT growth or a top-down which concerns the manipulation of the CNTs to the correct position. A top-down method was use to make a CNT thin film on a microelectrical mechanical system which was then characterized in terms of the thermoelectric coefficients of the aligned SWCNTs [8]. Using the process of 'super-growth' which incorporates water-assisted chemical vapor deposition, a CNT film was made and patterned by electron beam lithography into the required dimensions. By patterning a formed array of gold–SWCNT thermocouples it was found that under standard room temperature the Seebeck coefficient of the aligned SWCNT film was between 18 and 20 $\mu V\ C^{-1}$. The Seebeck effect of the SWCNT film was documented using thermocouples made of gold–SWCNT (Fig. 10.). Electrodes, a hot end and cold end temperature sensor, and a heater were produced by photolithography, and with a gold lift-off process on top of a silicon substrate that was covered by an insulating layer of Si_3N_4. The SWCNT film was then constructed on the gold surface using the process of top-down assembly.

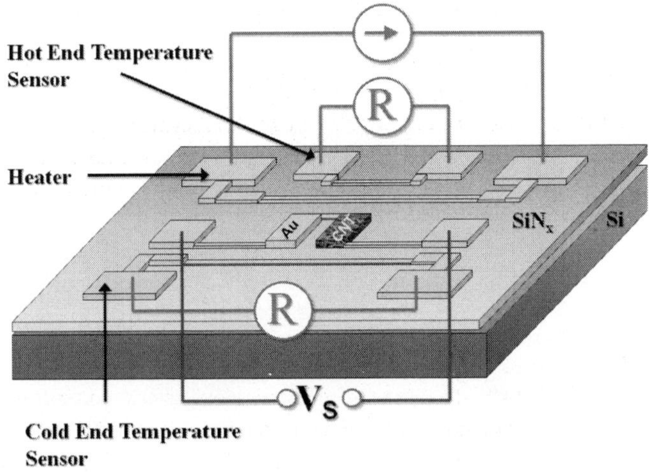

Figure 10. Schematic of a device for measuring the Seebeck effect in a CNT film. Figure Adapted from Dau *et al.* (2010).

When the device was used, an output voltage of 54 μV was recorded with a temperature difference of 3.07 °C. This gave a Seebeck voltage of 19.38 $\mu V\ K^{-1}$ which on average re-

http://dx.doi.org/10.5772/51781

mained constant. Aligned CNT bundles may have smaller Seebeck coefficients (thermoelectric sensitivity) than randomly oriented CNTs. The authors suggested that the difference may be a result of the contribution of inter-tube barriers, relative to ΔT, although more work is required to fully understand the effect of CNT films for the integration of them into thermoelectric devices.

5. Conclusion

CNTs have seldom been just another material for novel composites, but their true potential has yet to be transferred from the nano- to macro-scale. More than a two decades after their discovery, their influence has reached almost every aspect of scientific research from engineering to medicine. Faced with concerns about sustainably and climate change, the use of CNTs have helped to transform our approach to renewable energy. Advances in hydrogen fuel cells with CNT composite electrodes or membranes are helping to reduce and eliminate the need for rare and expensive catalysts. Safety is also another issue for the hydrogen based economy. Many different types of sensors will be required to promote a safe operational environment especially when the ignition concentration of hydrogen can be as low as 4%. The same technology that is used in the catalysis process in hydrogen fuel cells can be used to create hydrogen sensors, and work with interconnected CNTs has provided sensitivity values that contend with conventional sensors.

The role of interconnected CNTs in the photovoltaic research field is popular because of the potential to make novel hybrid solar cells, whilst increasing the overall efficiency of the device. While the early results look promising, there are still some difficult questions to address, like how does the presence of defects on the CNT surface affect the chemistry and ultimately the efficiency of a DSSC?

The integration of CNTs into thermoelectric devices currently does not have the same level of development as the other alternative energy resources, possibly because the field is more geared towards cost saving on an industrial scale and the development of component systems for vehicles rather than consumer gadgets or devices, but research into waste heat recovery is substantial. It is likely that thermoelectric devices will conform more to a silent revolution with an uptake in industries that work with high temperature equipment looking at converting some of the heat produced back to electricity. However, the research into thermoelectric fabrics has shown the potential for consumer products that may find a market in the future.

In summary, we are beginning to see a shift towards alternative fuel sources, with a focus on hybrid technologies like those found in the automotive industries, but we need to address the impact of our current economy as we transition to a more sustainable one.

Author details

Steve F. A. Acquah*, Darryl N. Ventura, Samuel E. Rustan and Harold W. Kroto

*Address all correspondence to: acquah51@hotmail.com

Florida State University, United States

References

[1] Dillon, A. C., Jones, K. M., Bekkedahl, T. A., Kiang, C. H., Bethune, D. S., & Heben, M. J. (1997). Storage of hydrogen in single-walled carbon nanotubes. *Nature.*, 386(6623), 377-379.

[2] Ventura, D. N., Li, S., Baker, C. A., Breshike, C. J., Spann, A. L., Strouse, G. F., Kroto, H. W., & Acquah, S. F. A. (2012). A flexible cross-linked multi-walled carbon nanotube paper for sensing hydrogen. *Carbon.*, 50(7), 2672-2674.

[3] Velten, J., Mozer, A. J., Li, D., Officer, D., Wallace, G., Baughman, R., & Zakhidov, A. (2012). Carbon nanotube/graphene nanocomposite as efficient counter electrodes in dye-sensitized solar cells. *Nanotechnology.*, 23(8), 6.

[4] Barnes, T. M., Wu, X., Zhou, J., Duda, A., van de Lagemaat, J., Coutts, T. J., Weeks, C. L., Britz, D. A., & Glatkowski, P. (2007). Single-wall carbon nanotube networks as a transparent back contact in CdTe solar cells. *Applied Physics Letters*, 90(24).

[5] Jia, Y., Cao, A. Y., Bai, X., Li, Z, Zhang, L. H., Guo, N., Wei, J. Q., Wang, K. L., Zhu, H. W., et al. (2011). Achieving High Efficiency Silicon-Carbon Nanotube Heterojunction Solar Cells by Acid Doping. *Nano Lett.*, 11(5), 1901-1905.

[6] Hatzikraniotis, E. (2012). On the Recovery of Wasted Heat Using a Commercial Thermoelectric Device. *Acta Phys Pol A.*, 121(1), 287-289.

[7] Kunadian, I., Andrews, R., Menguc, M. P., & Qian, D. (2009). Thermoelectric power generation using doped MWCNTs. *Carbon.*, 47(3), 589-601.

[8] Van Thanh, D., Dzung, Viet. D., Takeo, Y., Bui, Thanh. T., Kenji, H., & Susumu, S. (2010). Integration of SWNT film into MEMS for a micro-thermoelectric device. Smart Materials and Structures. , 19(7), 075003.

[9] Oh, S. H., Kim, K., & Kim, H. (2011). Polypyrrole-modified hydrophobic carbon nanotubes as promising electrocatalyst supports in polymer electrolyte membrane fuel cells. *International Journal of Hydrogen Energy*, 36(18), 11564-11571.

[10] Wang, C., Waje, M., Wang, X., Tang, J. M., Haddon, R. C., & Yan, Y. S. (2004). Proton exchange membrane fuel cells with carbon nanotube based electrodes. *Nano Lett.*, 4(2), 345-348.

[11] Girishkumar, G., Vinodgopal, K., & Kamat, P. V. (2004). Carbon nanostructures in portable fuel cells: Single-walled carbon nanotube electrodes for methanol oxidation and oxygen reduction. *J Phys Chem B.*, 108(52), 19960-19966.

[12] Li, L., Wu, G., & Xu, B. Q. (2006). Electro-catalytic oxidation of CO on Pt catalyst supported on carbon nanotubes pretreated with oxidative acids. *Carbon*, 44(14), 2973-2983.

[13] Acquah, S. F. A., Ventura, D. N., & Kroto, H. W. (2011). Strategies To Successfully Cross-link Carbon Nanotubes. *Electronic Properties of Carbon Nanotubes: InTech.*

[14] Ventura, D. N., Stone, R. A., Chen, K. S., Hariri, H. H., Riddle, K. A., Fellers, T. J., Yun, C. S., Strouse, G. F., Kroto, H. W., et al. (2010). Assembly of cross-linked multi-walled carbon nanotube mats. *Carbon.*, 48(4), 987-994.

[15] Yao, Y. (2012). Hydrogen Storage Using Carbon Nanotubes. In: Marulanda JM, ed., *Carbon Nanotubes*, 543-562.

[16] Li, N., Ma, Y. F., Wang, B., Huang, Y., Wu, Y. P., Yang, X., & Chen, Y. S. (2011). Synthesis of semiconducting SWNTs by arc discharge and their enhancement of water splitting performance with TiO2 photocatalyst. *Carbon*, 49(15), 5132-5141.

[17] Ljeri, V., Cappelletto, L., Bianco, S., Tortello, M., Spinelli, P., & Tresso, E. (2010). Nafion and carbon nanotube nanocomposites for mixed proton and electron conduction. *Journal of Membrane Science*, 363(1-2), 265-270.

[18] Kong, J., Chapline, M. G., & Dai, H. J. (2001). Functionalized carbon nanotubes for molecular hydrogen sensors. *Advanced Materials.*, 13(18), 1384-1386.

[19] Huang, B. R., & Lin, T. C. (2011). A novel technique to fabricate horizontally aligned CNT nanostructure film for hydrogen gas sensing. *International Journal of Hydrogen Energy.*, 36(24), 15919-15926.

[20] Lin, T. C., & Huang, B. R. (2012). Palladium nanoparticles modified carbon nanotube/nickel composite rods (Pd/CNT/Ni) for hydrogen sensing. *Sensors and Actuators B-Chemical.*, 162(1), 108-113.

[21] Li, W., Hoa, N. D., & Kim, D. (2010). High-performance carbon nanotube hydrogen sensor. *Sensors and Actuators B-Chemical.*, 149(1), 184-188.

[22] Jo, Y., Cheon, J. Y., Yu, J., Jeong, H. Y., Han-H, C., Jun, Y., & Joo, S. H. (2012). Highly interconnected ordered mesoporous carbon-carbon nanotube nanocomposites: Pt-free, highly efficient, and durable counter electrodes for dye-sensitized solar cells. *Chemical Communications.*

[23] Chen, T., Qiu, L., Cai, Z., Gong, F., Yang, Z., Wang, Z., & Peng, H. (2012). Intertwined Aligned Carbon Nanotube Fiber Based Dye-Sensitized Solar Cells. *Nano Lett.*, 12(5), 2568-2572.

[24] Bi, H., Huang, F. Q., Liang, J., Tang, Y. F., Lu, X. J., Xie, X. M., & Jiang, M. H. (2011). Large-scale preparation of highly conductive three dimensional graphene and its applications in CdTe solar cells. *J Mater Chem.*, 21(43), 17366-17370.

[25] Leubner, S., Katsukis, G., & Guldi, D. M. (2012). Decorating polyelectrolyte wrapped SWNTs with CdTe quantum dots for solar energy conversion. *Faraday Discuss.*, 155, 253-265.

[26] Bourdo, S. E., Saini, V., Piron, J., Al-Brahim, I., Boyer, C., Rioux, J., Bairi, V., Biris, A. S., & Viswanathan, T. (2012). Photovoltaic Device Performance of Single-Walled Carbon Nanotube and Polyaniline Films on n-Si: Device Structure Analysis. *ACS applied materials & interfaces.*, 4(1), 363-368.

[27] Mattoso, L. H. C., Manohar, S. K., Macdiarmid, A. G., & Epstein, A. J. (1995). Studies on the Chemical Syntheses and on the Characteristics of Polyaniline Derivatives. *Journal of Polymer Science Part a-Polymer Chemistry.*, 33(8), 1227-1234.

[28] Kozawa, D., Hiraoka, K., Miyauchi, Y., Mouri, S., & Matsuda, K. (2012). Analysis of the Photovoltaic Properties of Single-Walled Carbon Nanotube/Silicon Heterojunction Solar Cells. *Appl Phys Express*, 5(4).

[29] Castrucci, P., Del Gobbo, S., Camilli, L., Scarselli, M., Casciardi, S., Tombolini, F., Convertino, A., Fortunato, G., & De Crescenzi, M. (2011). Photovoltaic Response of Carbon Nanotube-Silicon Heterojunctions: Effect of Nanotube Film Thickness and Number of Walls. *J Nanosci Nanotechnol.*, 11(10), 9202-9207.

[30] Hewitt, C. A., Kaiser, A. B., Roth, S., Craps, M., Czerw, R., & Carroll, D. L. (2012). Multilayered Carbon Nanotube/Polymer Composite Based Thermoelectric Fabrics. *Nano Lett.*, 12(3), 1307-1310.

Chapter 19

Carbon Nanotube-Enzyme Biohybrids in a Green Hydrogen Economy

Anne De Poulpiquet, Alexandre Ciaccafava,
Saïda Benomar, Marie-Thérèse Giudici-Orticoni and
Elisabeth Lojou

Additional information is available at the end of the chapter

http://dx.doi.org/10.5772/51782

1. Introduction

Alternative energy pathways to replace depleting oil reserves and to limit the effects of global warming by reducing the atmospheric emissions of carbon dioxide are nowadays required. Dihydrogen appears as an attractive candidate because it represents the highest energy output relative to the molecular weight (120 MJ kg^{-1} against 50 MJ kg^{-1} for natural gas), and because its combustion delivers only water and heat. Whereas the main renewable sources of energy available in nature (solar, wind, geothermal…) need to be transformed, dihydrogen is able to transport and store energy. Dihydrogen can be produced from renewable energies, indirectly from photosynthesis *via* biomass transformation, or directly by bacteria. It can be converted into electricity using fuel cell technology. From all these properties and because it does not compete with food and water resources, dihydrogen has been defined as third generation biofuel. It thus emerges as a new fully friendly environmental energy vector. The use of dihydrogen as an energy carrier is not a new idea. Let us simply remember that Jules Verne, a famous French visionary novelist, wrote early in 1874: "I believe that O_2 and H_2 will be in the future our energy and heat sources" [1]. His prediction simply relied on the discovery a few years before of the fuel cell concept by C. Schönbein, then W. Groove, who demonstrated that when stopping water electrolysis, a current flow occurred in the reverse way [2]. However in order to implement the dihydrogen economy and replace fossil fuels, there are significant technical challenges that need to be overcome in each of the following domains:

1. dihydrogen production and generation,
2. dihydrogen storage and transportation,
3. dihydrogen conversion to electrical energy.

As opposed to widespread opinions, natural dihydrogen sources exist alone on the earth's surface. Local and continuous emanations of dihydrogen can be observed in cratonic zones, ophiolitic rocks or oceanic ridges [3]. Dihydrogen is effectively produced in the upper mantle of the earth through natural oxidation of iron (II)-rich minerals, like ferromagnesians, by water of the hydrosphere. The ferrous iron is oxidized in ferric iron and water is concurrently reduced in dihydrogen, as given by following equation: $2Fe^{2+}$ (mineral) + 2 H^+ (water) → 2 Fe^{3+} (mineral) + H_2. The same reaction can occur with other ions like Mn^{2+}. Exploitation of these sources remains however difficult so far as dihydrogen does not accumulate on the earth subsurface, especially for two reasons. First because as a powerful energy source dihydrogen is quickly consumed (biologically or abiotically), and second because as the lightest and most mobile gas it is not much retained by Earth's attraction and escapes in the atmosphere.

Combined with water and hydrocarbons dihydrogen is nevertheless the most abundant element on earth. Green means to ecologically convert H containers into dihydrogen still remain however challenging. The energetic volume density of dihydrogen is low (10.8 MJ m^{-3} against 40 MJ m^{-3} for natural gas) so that storage and transportation appear as bottlenecks for large scale development in transportation for example. Conversion of dihydrogen to electricity in fuel cells presents high electrical efficiency (more than 50% against less than 30% for gas engines), but requires the use of catalysts both for H_2 oxidation and O_2 reduction. These are mainly based on platinum catalysts, which are highly expensive, weakly available on earth, and non biodegradable. Extensive researches thus aim to decrease the amount of platinum catalysts in fuel cells. Following the discovery of carbon nanotubes (CNTs) [4, 5], their large scale availability opened a new avenue in these three domains. Due to their intrinsic properties, such as high stability, high electrical and thermal conductivities [6] and high developed surface areas, carbon nanotubes constitute attractive materials, able to enhance the credibility of an hydrogen economy.

Besides, platinum catalysts are inhibited by very low amount of CO and S (0.1% of CO is sufficient to decrease one hundred fold the catalytic activity of Pt in ten minutes!), thus requiring strong steps of H_2 purification [7]. They are not specific to either O_2 or H_2 catalysis, thus requiring the use of a membrane to separate the anodic and cathodic compartments. Nafion® perfluoronated membrane is currently the only really performing polymer [8], increasing its cost. Replacement of platinum-based catalysts is thus highly needed. In that way, a new concept appeared less than five years ago, when looking at the pathways microorganisms use for the production of ATP, their own energetic source [9, 10]. As an example, the hyperthermophilic, microaerophilic bacterium *Aquifex aeolicus*, couples H_2 oxidation to O_2 reduction *via* a membrane quinone pool (Figure 1). The redox coupling generates a proton gradient through the cell membrane for ATP synthesis. Clearly, this pathway can be considered as an "*in vivo* biofuel cell". The question rises if we could take benefit of bacterial energetic pathways for our own energetic needs. The idea thus emerged that microorganisms or enzymes could be used instead of chemical catalysts for the development of efficient electricity producing devices. These innovative batteries called biofuel cells rely on enzymes highly specific for various fuels and oxidants [11]. A mandatory condition is that these enzymes have to be immobilized onto electrodes. One of the most common biofuel cell uses glucose oxidase and laccase, two enzymes specific for glucose oxidation and oxygen reduction, respectively. A few years ago, a new concept of biofuel cells appeared based on enzymes specific of dihydrogen oxidation. This biohydrogen economy relies on the opportunity to use low-cost materials for efficient conversion of solar energy to dihydrogen and of dihydrogen to electricity. Many microorganisms biosynthesize hydrogenase, the metalloenzyme that catalyzes the dihydrogen conversion. At least two modes of application of dihydrogen-metabolizing protein catalysts are nowadays considered within dihydrogen as a future energy carrier. Hydrogenases may be used as catalysts in dihydrogen production by coupling oxygenic photosynthesis to biological dihydrogen production [12]. Hydrogenases can also be used directly as anode catalysts in biofuel cells instead of chemical catalysts [13]. The improved knowledge of hydrogenase structure and of catalytic mechanisms allows nowadays to design the development of biofuel cells functioning as Proton Exchange Membrane (PEM) fuel cells.

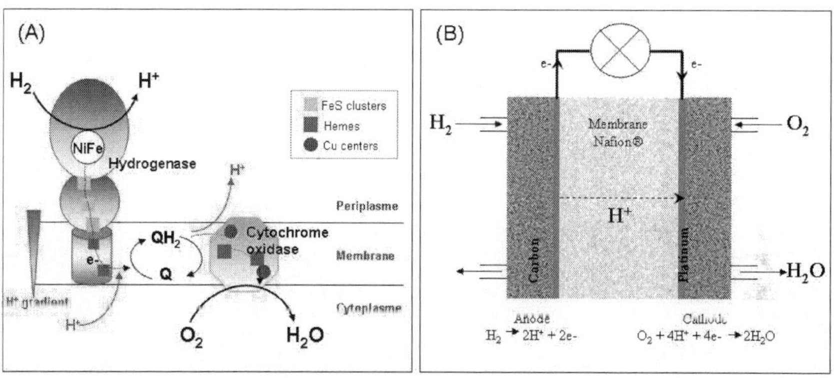

Figure 1. A) Energetic metabolism of the bacterium *Aquifex aeolicus*: H_2 oxidation in the periplasm is coupled to O_2 reduction in the cytoplasm *via* a membrane quinone pool to generate a trans-membrane proton gradient for ATP synthesis; (B) General view of a chemical PEM fuel cell.

For all these innovative concepts, one of the key points is the increase in power density, thus in the current density furnished by a redox couple displaying a large as possible potential difference. Apart from the improvement in enzyme stability, the increase in the current densities supposes an optimization of both the interfacial electron transfer rate and the amount of connected enzymes at the electrode. Carbon nanotubes which develop large surface areas and can be functionalized constitute an attractive platform for such enzyme immobilization. CNTs are described as graphene sheets rolled into tubes. They exist under various structural configurations (single-walled (SWCNTs), multi-walled (MWCNTs)) differing in electrical properties, thus tuning the platform properties for enzyme immobilization. The end of the tubes is capped by a fullerene-type hemisphere that yields selective functionalization of the CNTs [14].

With the objective of dihydrogen as a future green energy vector, this review focuses on the last developments in the fuel -and more especially biofuel- cell field thanks to the advantageous use of carbon nanotubes. In a first part, carbon nanotubes for H_2 storage enhancement are discussed. Then fuel cells in which carbon nanotubes help to decrease the amount of high cost noble metal catalysts are described. Green H_2 economy is then emphasized considering the key role of hydrogenase, the enzyme responsible for dihydrogen conversion. This requires the functional immobilization of the biocatalysts onto electrodes. The use of carbon nanotubes in this immobilization step is underlined, including the modes of carbon nanotube functionalization and enzyme or microbes grafting. Then the advantages of developing biofuel cells in which chemical catalysts are replaced by enzymes or microbes are described. A short review of the sugar/O_2 biofuel cells, the most widely investigated biofuel cell, is given with a particular attention on the devices based on carbon nanotube-modified bioelectrodes. The last developments based on carbon nanotube networks for hydrogenase immobilization, or mimicking synthetic complex immobilization, in view of efficient dihydrogen catalytic oxidation are finally described in order to allow the design of a future H_2/O_2 biofuel cell.

2. Carbon nanotubes: an attractive carbon material

The discovery of carbon nanotubes (CNTs) has induced breakthroughs in many scientific domains, including H_2 economy, biosensors, bioelectrochemistry…This is due to their remarkable properties, such as good electronic, mechanical and thermal properties. Their nanometric size compares with that of proteins and enzymes, offering the possibility of electrical connection. Their large developed surface area allows the development of devices in smaller volumes. SWCNTs are sp2 hybridized carbon in a hexagonal honeycomb structure that is rolled into hollow tube morphology [15]. MWCNTs are multiple concentric tubes encircling each other [5]. Depending on the chirality, CNTs can be metallic or semiconducting. The distinction between metallic and semiconducting is very important for application, but the physical separation of allotropes is one of the most difficult challenges to overcome. In MWCNTs, a single metallic layer results in the entire nanotubes metallic behavior. Most often mixtures of these two forms are present in CNTs preparation. More information on the physical and electronic structures can be found in many published reviews [16]. CNTs are produced by various methods such as arc discharge, laser ablation, and chemical vapor deposition (CVD). Commercially CNTs are generally produced by CVD during the pyrolysis of hydrocarbon gases at high temperature. The control of synthesis parameters (reagent gas, T °, metal catalysts) allows for the control of CNT properties. Metal impurities may remain in the CNTs sample, thus requiring purification steps. CNTs may be treated to functionalize the surface.

http://dx.doi.org/10.5772/51782

3. Carbon nanotubes for safe and efficient H_2 storage

The use of H_2 in fuel cells to generate electricity has been proved early in the middle of the nineteenth century. Surprisingly this discovery by C. Schönbein in 1839 of current generation by use of H_2 and O_2 in sulphuric acid was applied by NASA only late in 1960. Despite intensive studies over the last two decades, fuel cells still suffer from high cost and low durability. The first difficulty responsible for this slow large scale development lies on dihydrogen storage and transportation, both regarded as bottlenecks considering dihydrogen specific volumic density as a gas. For convenience the gas must be intensely pressurized to several hundred atmospheres and stored in a pressure vessel. The ways to store dihydrogen with minimum hazard are under liquid state under cryogenic temperatures (at a temperature of -253 °C), or more efficiently in a solid state. Storage of dihydrogen in hydride form uses an alloy that can absorb and hold large amounts of dihydrogen by bonding with hydrogen and forming hydrides. A dihydrogen storage alloy is capable of absorbing and releasing dihydrogen without compromising its own structure, according to the reaction: $M + H_2 \leftrightarrow MH_2$, where M represents the metal and H, hydrogen. Qualities that make these alloys useful include their ability to absorb and release large amounts of dihydrogen gas many times without deteriorating, and their selectivity toward dihydrogen only. In addition, their absorption and release rates can be controlled by adjusting temperature or pressure. The dihydrogen storage alloys in common use occurs in four different forms: AB_5 (e.g., $LaNi_5$), AB (e.g., FeTi), A_2B (e.g., Mg_2Ni) and AB_2 (e.g., ZrV_2). Metal hydrides, such as MgH_2, Mg_2NiH_4 or $LiBH_4$, constitute secure reserves of dihydrogen [17-19]. Dihydrogen is released from MH_2 upon increase in temperature and/or decrease in pressure.

Material	H_2 gas, 200 bar	H_2 liquid, -253 C	MgH_2	Mg_2NiH_4	$FeTiH_2$	$LaNi_5H_6$
H-atom per cm^3 (x10^{22})	0.99	4.2	6.5	5.9	6.0	5.5

Table 1. H density as a function of storage method.

Much progress has been made during the last years in that domain, including the highlight of the advantages offered by using CNTs. An efficient approach appears to be the formulation of new carbon/transition metal catalyst composites of specific composition and molecular structure, which can greatly stimulate and improve the chemical reactions involving dihydrogen relocation in alkali-metal aluminium materials. Absorption kinetics and dihydrogen storage capacity were shown to be enhanced by mixing MH_2 with SWCNTs as a result of an increase in interfacial area, decrease in MH_2 particle agglomeration and nanoplatform for efficient H_2 diffusion [20, 21]. The hydriding and dehydriding kinetics of SWCNT/catalyzed sodium aluminium composite were found to be much better than those of the material ground without carbon additives. Temperature of H_2 desorption was lowered [22]. The presence of carbon creates new dihydrogen transition sites and the high dihydrogen diffusivity of the nanotubes facilitates hydrogen atom transition. Faster thermal energy transfer through the nanotubes may also help reduce hydriding and dehydriding times.

Dihydrogen can be stored through physisorption on CNTs, based on Van der Waals interaction. Based on the surface area of a single graphene sheet, the maximum value for the storage of dihydrogen capacity is around 3 wt%. Dihydrogen can also be stored through chemisorption in CNTs matrix. If the π-bonding between carbon atoms were fully utilized, every carbon atom could be a site for chemisorption of one hydrogen atom. Dillon et al. first reported in 1997 dihydrogen storage in SWCNT networks [23]. Both SWCNTs and MWCNTs store dihydrogen in microscopic pores on the tubes [24, 25]. Similar to metal hydrides in their mechanism for storing and releasing dihydrogen, the carbon nanotubes hold the potential to store a significant volume of dihydrogen. The storage capacity is dependent on many parameters of the CNTs, including their structure, structure defects, pretreatment, purification, geometry (surface area, tube diameter, length), arrangement of tubes in bundles, storage pressure, temperature,...Dihydrogen uptake varies linearly with tube diameter, because the uptake is proportional to the surface area, *i.e.* the number of carbon atoms. The adsorption sites exist inside and outside the tube, between tubes in bundles, between the shells in MWCNTs. For dihydrogen storage into the tube dihydrogen must pass through the CNT wall or the tube must be opened. Hydrogen forms stable C-H bonds on SWCNT surface at room temperature that can dissociate above 200°C. According to SWCNT diameter 100% hydrogenation can be obtained, thus more than 7 wt % dihydrogen storage capacity, which is above the target fixed by the US Department of Energy's Office of Energy Efficiency and Renewable Energy [26].

4. Carbon nanotubes for a decrease in the amount of noble metal catalysts in fuel cells

Among the different types of fuel cells, PEM fuel cell operates at low temperatures around 100°C. For small portable application requiring less than 10 kW, they are more suitable than higher powering solid oxide fuel cells (functioning at 700°C) due to the possible use of usual materials for electronic connectors (mainly based on carbon) and membrane. However the necessary use of platinum-based catalysts on electronic connectors to accelerate the rate of dihydrogen oxidation and oxygen reduction is a real brake towards the fuel cell development. Platinum is scarce enough on earth to be a limiting factor in case of large scale development of fuel cells. Consequently platinum currently accounts for 25% in the total cost of a fuel cell. Over the past five years, the price of platinum has ranged from just below $800 to more than $2,200 an ounce. Carbon black particles offer a high surface area support, able to decrease the amount of platinum particles. But they suffer from mass transfer limitations and strong carbon corrosion.

Among the low-cost alternatives to platinum, carbon appears to be the most promising. Due to their nano-structure and unique chemical and physical properties, CNTs have appeared as ideal supporting materials to improve both catalytic activity and electrode stability. The enhancement of fuel cell performances by using CNT/Pt or Pt-alloy catalysts may arise from:

i. higher dispersion of Pt nanoparticles,

ii. increased electron transfer rates,

iii. porous structure of CNT layers.

Various CNT-Pt composites were used to reduce the platinum amount while preserving high catalytic activity in PEM fuel cells. Platinum nanodots sputter-deposited on a CNT-grown carbon paper [27], or deposited on functionalized MWCNTs [28] exhibited great improvement in cell performance compared to platinum on carbon black. This was primarily attributed to high porosity and high surface area developed by the CNT layer. Compared to a commercial Pt/carbon black catalyst, Pt/SWCNT films cast on a rotating disk electrode was shown to exhibit a lower onset potential and a higher electron-transfer rate constant for oxygen reduction. Improved stability of the SWCNT support was also confirmed from the minimal change in the oxygen reduction current during repeated cycling over a period of 36 h [29]. Platinum particles deposited on MWCNT encapsulated in micellar surfactant were also explored as efficient catalysts for fuel cells [30, 31]. An in situ synthetic method was reported for preparing and decorating metal nanoparticles at sidewalls of sodium dodecyl sulfate micelle functionalized SWCNTs/MWCNTs. Accelerated durability evaluation was carried out by conducting 1500 potential cycles between 0.1 and 1.2 V at 80°C. These nanocomposites were demonstrated to yield a high fuel cell performance with enhanced durability. The membrane electrode assembly with Pt/MWCNTs showed superior performance stability with a power density degradation of only 30% compared to commercial Pt/C (70%) after potential cycles. Identically electrocatalytically active platinum nanoparticles on CNTs with enhanced nucleation and stability have been demonstrated through introduction of electron-conducting polyaniline (PANI) [32]. A bridge between the Pt nanoparticles and MWCNTs walls was demonstrated with the presence of platinum nitride bonding and π-π bonding. The synthesized PANI was found to wrap around the CNT as a result of π-π bonding, and highly dispersed Pt nanoparticles were loaded onto the CNT with narrowly distributed particle sizes ranging from 2.0 to 4.0 nm. The Pt-PANI/CNT catalysts were electroactive and exhibited excellent electrochemical stability, therefore constitute promising potential applications in proton exchange membrane fuel cells. Strong evidence thus emerges that CNTs/Pt composites are efficient as catalysts for fuel cells. Although platinum content has been dramatically decreased, industrials consider that further optimization is mandatory for a large scale fuel cell production. In addition Nafion® membrane between the cathodic and anodic compartment delays the large scale application of fuel cells, due to cost and problem of mass transfer. Breakthrough research towards these two bottlenecks could surely enforce a hydrogen economy.

5. Towards a green H_2 economy: carbon nanotubes for enzyme and microbe immobilization

Replacement of chemical catalysts is thus nowadays highly needed in view of the development of a green energy economy. Microorganisms contain many biocatalysts, namely enzymes, which are highly efficient and specific towards various substrate conversions. Given they are produced in large enough quantities, these enzymes could be used as catalysts in biotechnological devices. A mandatory condition to develop heterogeneous catalysis is to succeed in the functional immobilization and in the stabilization of the enzymes on solid supports. The redox active site of enzymes is indeed buried inside the protein moiety so that the enzymatic property can be maintained under environmental stresses. Specific channels are often involved to allow the substrate to reach the active site. Complex but highly organized electron transfer chains occur for energetic metabolism. Electron transfer between two physiological partners associated with transformation of the substrate involves specific recognition site. The game for a bioelectrochemist that aims to get the highest electron transfer rate for heterogeneous catalysis is to reproduce at the electrode interface the physiological electron transfer recognition process. Given the usual size of an enzyme (5-10 nm), electron transfer cannot occur *via* electron tunneling from the active site to the surface of the enzyme. In some enzymes, electron relays, one being located at the protein surface, act as a conductive line for electron shuttling. If the electrode interface is built so that it fits the surface electron relay environment, one can expect to favor a direct electrical connection of the enzyme onto the electrode. In case of direct electron transfer failure, an artificial redox mediator that acts as a fast redox system and shuttles electrons between the enzyme and the electrode can be used (Figure 2) [13, 33].

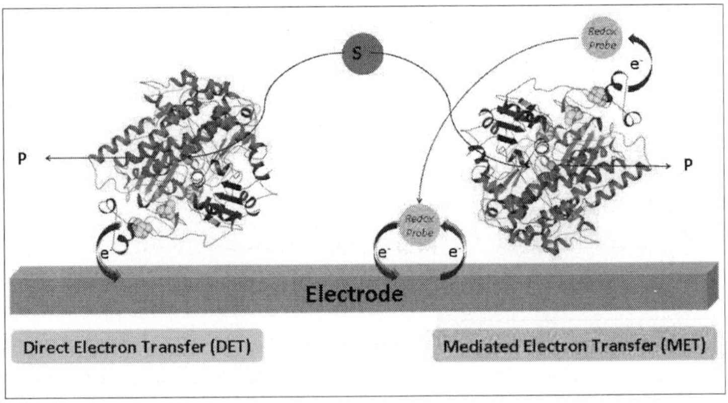

Figure 2. Interfacial electron transfer between an enzyme and an electrode can be achieved by direct (left) or mediated (right) electron transfer process.

Direct electron transfer process is preferred to mediated one, because it is not limited by the affinity between the enzyme and the redox mediator, and because it avoids the co-immobilization of enzyme and mediator. It is furthermore expected to yield the highest power density because enzymes, as biocatalysts, transform their substrate into products with very low overvoltages. However it requires the knowledge of the protein structure and the construction of a tuned electrochemical interface that fits the electron transfer site.

There are many strategies for efficient enzyme immobilization onto electrochemical interfaces, including simple physical adsorption, covalent attachment, cross-linking or entrapment in polymers. The objectives are to optimize the immobilization procedure so that the efficiency of the enzyme and its stability are preserved. Moreover, due to the size of enzymes compared to chemical catalysts, large surface area interfaces baring many anchorage sites are required to obtain high catalytic currents.

To reach these goals, 3D structures are preferred, and CNT-based electrodes are very popular, both SWCNTs and MWCNTs. CNTs can be directly grown onto electrode surface, or adsorbed on it, or imbedded in polymer coating. In most cases, higher activity was reported for enzymes physically adsorbed onto CNTs [34]. Hydrophobic interactions between the enzyme and the CNT walls and π-π interactions between side walls of CNTs and aromatic rings of the enzyme are thought to be the driving force for direct adsorption of enzymes on CNTs [35]. Electrostatic interaction between the defect sites of CNTs and protonated amino residues of the enzyme plays also a role in the adsorption process [35]. CNTs are quite easily functionalized, allowing covalent, thus stable specific attachment of enzymes. The oxidation in strong acidic solutions at high temperature was demonstrated to remove the end caps and shorten the lengths of the CNTs. The length of the CNTs was shown to be a function of the oxidation duration [36]. Acid treatment also adds oxide groups, primarily carboxylic acids, to the tube ends and defect sites [37]. The control of reactants and/or oxidation conditions may control the locations and density of the functional groups on the CNTs, which can be used to control the location and density of the attached enzymes [37]. Covalent immobilization is induced by carbodiimide reaction between the free amine groups on the enzyme surface and carboxylic groups generated by side wall oxidation of CNTs.

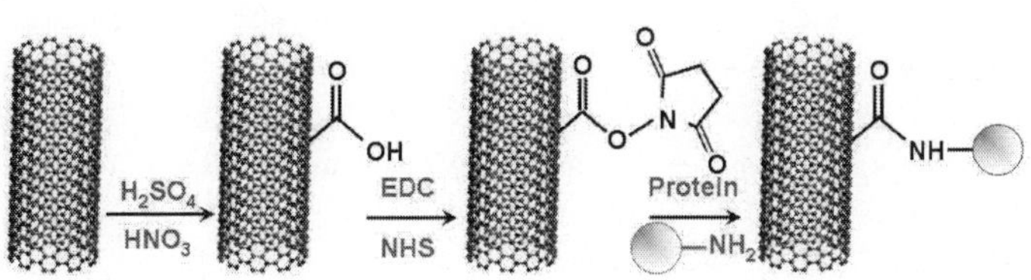

Further chemical reactions can be performed at the oxide groups generated on the oxidized CNTs to functionalize with groups such as amides, thiols, etc...From an electrochemical point of view, the side walls of CNTs were suggested to behave as basal plane of pyrolytic graphite, while their open ends resemble the edge planes [38, 39]. But recent work has demonstrated that the side wall may be responsible for electrochemical activity [40]. It has been furthermore suggested that the uncovered surface of CNTs promotes the accessibility of the substrate to the enzyme [41]. It is also interesting to note that the open spaces between CNTs are accessible to large species such as entire bacteria [42], opening the way for the development of fuel cells using whole microorganisms instead of purified enzymes. The cost and complexity of CNT manufacturing seem to be still clogging issues in that field.

Abundant literature exists on the ways CNTs are architectured for efficient enzyme immobilization, including those specific for development of enzymatic fuel cells. Enzymes and proteins as various as glucose oxidase and dehydrogenase, tyrosinase, laccase and bilirubin oxidase, peroxidase, haemoglobin and myoglobin, *i.e.* flavin, copper or heme containing active sites, have been studied. Whereas direct electron transfer between protein or enzyme and an electrochemical interface has been for long time supposed to be restricted to small proteins (<15kDa) possessing active sites exposed to the surface (it is the case for many cytochromes as example [43]), the use of CNT-modified electrodes has greatly enhanced the number and kinds of enzymes able to be directly connected to an electrode. Enzymes as large as one hundred kDa, with many cofactors are now considered for direct electron transfer. Consequently, recent works during the last years focus and report on direct communication between enzymes and electrode interface through CNT network. The induced porosity of the film depends on the type of CNTs used. But generally the nanometric size of the CNTs compared to the size of enzymes favors a direct electronic connection of the enzyme whatever its orientation [44]. The physical properties of CNTs, including high electrical conductivity, explain why CNT layers can be built up on electrodes most often yielding high rate direct electron transfer for enzymatic product transformation. Many researches report on the increase in electroactive surface area by use of CNT coatings that contribute to an increase in the direct electron transfer process [45-52]. CNTs are usually deposited on electrodes as thick films. Alternatively, layer-by-layer (LBL) process induces a quite stable protein film with nice electrocatalytic properties [53-55]. LBL is based on electrostatic interaction between oppositely charged monolayers in an alternating assembling. Although CNTs greatly amplify the current response, layer-by-layer architecture suffers from weak stability of the build-up and decrease in electron transfer for the upper layers. Besides vertically aligned CNTs were suggested to act as molecular wires that ensure the electrical communication between enzyme and electrode [56-58]. The carboxylic functions induced by acidic treatment of CNTs can be used for further chemical modifications. Amine- [59-61], thionin- [62, 63], diazonium salts [64, 65], pyrene [66, 67] (Figure 3) or other π-π stacking interactions [68] were used to functionalize CNTs. These modifications were demonstrated to be efficient platforms for enzyme immobilization.

Mixing CNTs with surfactant [69-71] was claimed to assist in the dispersion of CNTs while avoiding oxidative functionalization which may disrupt their π-network. Polymer modified CNTs [72, 73] and sol-gel-CNT nanocomposite films [74] were proved to behave as friendly platforms for enzyme encapsulation.

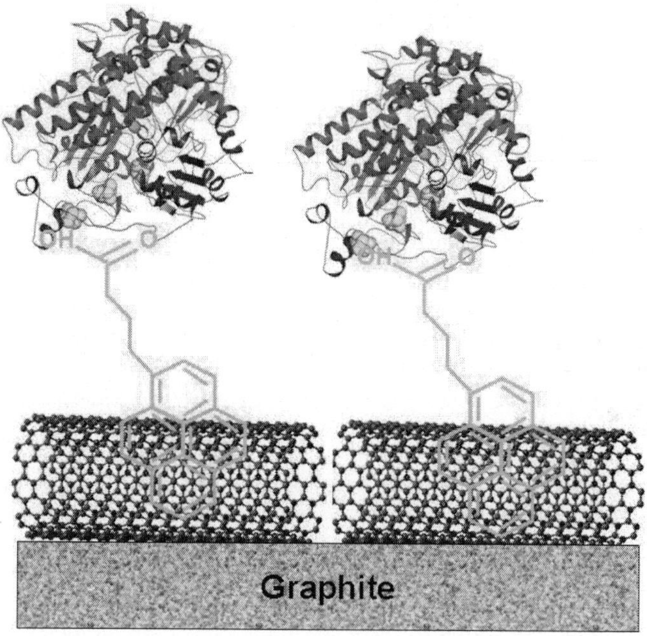

Figure 3. Schematic drawing of the build-up of enzyme on SWCNTs *via* π-π interactions.

Many enzymes however cannot be electrically connected to the electrode interface and require redox mediator to electrochemically follow substrate conversion. In that case, electrode kinetics is mainly dependant on mediator kinetics, so that the choice of the redox mediator mainly impedes the power density. Another issue is that the mediator can be co-immobilized with the enzyme at the electrode, while still being capable of efficient interaction with the enzyme. CNTs have also been used for building networks enabling co-immobilization of enzymes and redox mediators. In that way, one of the most popular redox entities is osmium polymer which forms hydrogels with enzymes allowing both charge transfer reaction between enzyme and mediators and diffusion of substrate and product [75]. Composite CNT/osmium films were used To immobilize bacteria [76], or enzymes [77]. By optimizing the CNT and polymer amounts, enhanced current responses were obtained linked to a promotion of the electron transfer within the composite. Various phenothiazine derivatives were also used to form nanohybrids with CNTs acting as efficient redox mediator platforms [78-80]. Phenothiazine derivatives strongly adsorb onto CNTs leading to great enhancement of redox dye loading onto the electrode, but also to improved electrochemical sensing devices. Another strategy involves the use of a redox polymer as redox mediator platform. Electropolymerization of the redox conducting polymer onto CNTs enhances the amount of redox units and the electrical conductivity of the coating [81]. An interesting construction has also been obtained by immobilization of physiological cofactor onto CNT layers *via* π-π interactions, then immobilization of the enzyme [82]. The covalent coupling between the enzyme and its natural cofactor which was immobilized onto CNTs was proved to be efficient towards mediated substrate catalysis. This overview of multiple architectures involving enzymes and CNTs highlights the deep efforts engaged in the last years for efficient biocatalyst immobilization that open avenues towards biotechnological devices.

6. Carbon nanotubes for biological production of dihydrogen

Apart from replacement of noble metal catalysts in fuel cells, a new green technology for production of dihydrogen is required. It currently relies on steam reforming of hydrocarbons under high temperature and pressure conditions, which starts from fossil fuels, thus producing greenhouse gases. Dihydrogen production *via* water electrolysis appears as a renewable solution given that the energy input comes from a renewable source, ideally solar energy. Many bacteria gain energy by the oxidation of dihydrogen assisted by a number of complex mechanisms. Various species evolve H_2 under anaerobic conditions. This is also a human being process since bacteria in our digestive tract produce H_2, though not detectable because immediately recycled by other bacteria. Photosynthetic organisms such as microalgae and cyanobacteria are very efficient in water splitting [83]. They possess photosensitizers for photon capture and charge separation, and enzymes for water oxidation to oxygen and water reduction to dihydrogen. This chemical activity relies on the expression of very efficient enzymes, called hydrogenases [84], which catalyze with high turn-over (one molecule of hydrogenase produces up to 9000 molecules of H_2 per second at neutral pH and 37°C) and low overvoltage the conversion of protons into dihydrogen and the oxidation of dihydrogen. The sequences of 450 hydrogenases are now available. Hydrogenases differ in size, structure, electrons donors. They also differ by their position in the cell (soluble in the periplasm, membrane-bound), and by their activity preferentially towards H_2 oxidation or protons reduction. Hydrogenase active site is composed of non noble metals such as iron and nickel, unlike platinum catalyst necessary for the chemical electrolysis of water. Three distinct classes can be split which differ from the type of metal content in the active site: [NiFe], [FeFe] and [Fe] hydrogenases. [NiFe] and [FeFe] hydrogenases possess dinuclear active centers which are connected through thiolate bridges. [NiFe] hydrogenase (Figure 4) is the most usual hydrogenase in microorganisms. It is composed of two subunits. The larger subunit harbors the [NiFe] active site. The small subunit contains FeS clusters. Electrons are transferred to the active site along these FeS clusters distant less than 10 Å that act as a conductive line. [FeFe] hydrogenases are monomeric. In addition to the active site they contain additional domains which accommodate FeS clusters.

In order to use these biocatalysts for green dihydrogen production, two main research domains are currently concerned: the understanding of the catalytic mechanisms of H_2 production, and the optimization of enzyme immobilization. Adsorption onto graphite electrodes [85, 86] was largely used to study the mechanisms by which hydrogenases produce H_2.

Grafting of hydrogenase onto gold electrode modified by thiolated Self-Assembled-Monolayer [87] allowed efficient proton reduction into dihydrogen in aqueous buffer solutions. Hydrogenase is also considered as a promising biocatalyst for photobiological production of dihydrogen when coupled to a photocatalyst [88]. Hybrid complexes of hydrogenases with TiO_2 nanoparticles [89, 90] were studied for H_2 production. The optimized system was shown to produce H_2 at a turnover frequency of approximately 50 (mol H_2) s^{-1} (mol total hydrogenase)$^{-1}$ at pH 7 and 25 °C, even under the typical solar irradiation of a northern European sky. Cd-based nanorods [91, 92] were recently studied. The CdS nanorod/hydrogenase complexes photocatalyzed reduction of protons to H_2 at a hydrogenase turnover frequency of 380-900 s^{-1} and photon conversion efficiencies of up to 20% under illumination at 405 nm. Cd-based complexes allowed photoproduction of dihydrogen for a couple of hours, but still suffer from quick inhibition of hydrogenase.

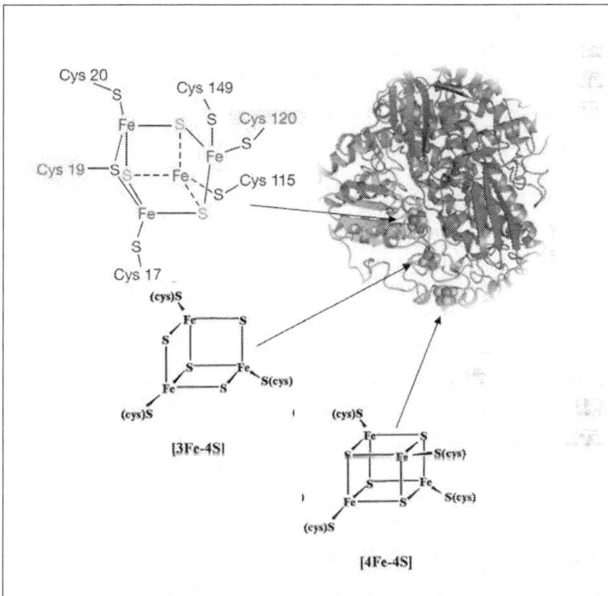

Figure 4. Structure of an oxygen-tolerant [NiFe] hydrogenase.

Although a very attractive way, little work has been done towards enhancement of green hydrogen production using CNTs. Three studies from the same group reported however catalytically active hydrogenase-SWCNT biohybrids [93, 94]. Surfactant-suspended SWCNTs were shown to spontaneously self-assemble with hydrogenase. Photoluminescence excitation and Raman spectroscopy showed that SWCNTs act as molecular wires to make electrical contact with at least one of the FeS electron relay. Hydrogenase was demonstrated to be strongly attached to the SWCNTs and to mediate electron injection into nanotubes. The displacement of the surfactant by hydrogenase to gain access to the SWCNTs was strongly suggested by photoluminescence studies. Furthermore, Raman studies of charge transfer complexes between hydrogenase and either metallic (m) or semiconducting (s) SWCNTs revealed a difference in oxygen deactivation of hydrogenase according to the SWCNT species. m-SWCNTs most probably interact with hydrogenase to produce a more oxygen-tolerant species. The study further suggested that purified m-SWCNTs or s-SWCNTs, rather than mixed preparation, would be more suitable for hydrogenase-SWCNTs biohybrids. The formation of these catalytically active biohybrids in addition with the intrinsic properties developed by CNT networks on electrodes certainly accounts for the improved dihydrogen production observed in the following studies. Kihara *et al.* immobilized hydrogenase on a SWCNT-forest with a unique dense structure of vertically aligned millimetre-scale height SWCNTs [95]. Hydrogenase was demonstrated to spontaneously assemble between adjacent nanotubes. The maximum rate of dihydrogen production was reported to be 720 nmol/min/(mg hydrogenase) and the electron transfer efficiency was estimated to be 32%. It is two thousand fold higher than reported before using the same hydrogenase on Langmuir-Blodgett film [96]. Nevertheless, one key point in the development of biotechnological devices is the long term stability of enzymes. If these biological catalysts are very efficient *in vivo*, they often suffer from weak stability when extracted from their physiological environment. Enzyme encapsulation in silica-derived sol-gel materials has been demonstrated to stabilize many enzymes. This procedure was applied to hydrogenase [97]. The majority of hydrogenase was shown to be entrapped in the gel and protected against proteolysis. Hydrogenase/sol-gel pellets retained 60% of the specific mediated activity for H_2 production displayed by hydrogenase in solution. The gel-encapsulated enzyme retained its activity for long periods, *i.e.* 80% of the activity after four weeks at room temperature. Notably, by doping the hydrogenase-containing sol-gel materials with MWCNTs Zadvorny *et al.* demonstrated a 50% increase in dihydrogen production [98]. Furthermore stabilization of hydrogenase was proved through encapsulation process.

One alternative for green hydrogen production is to synthesize metal complexes that mimic the active site of enzymes. Huge work has been done in that field in order to obtain bioinspired models that could produce H_2 as efficiently as hydrogenase, while being much more stable [99]. The most performing complex involves mononuclear nickel diphosphine complex. This complex is inspired from the active sites of both [NiFe] and [FeFe] hydrogenases and displays remarkable catalytic proton reduction in organic solvent [100]. Le Goff *et al.* took benefice from this complex and from the results obtained by immobilization of hydrogenase on CNT networks [44]. The authors successfully immobilized the nickel complex onto carbon nanotube networks by covalent coupling [101]. Such construction was demonstrated to be very efficient for dihydrogen production in aqueous solution, evolving dihydrogen with overvoltage less than 20 mV and exceptional stability.

7. Carbon nanotubes for biofuel cells: an attractive green alternative

Beside researches towards decrease in chemical catalyst amount and discovery of less expensive catalysts (as alloys for example), a new concept emerged early in 1964 by Yahiro *et al.* [102]. A fuel cell was constructed using usual O_2 reduction at platinum modified electrode in the cell cathodic compartment, but using glucose as a fuel in the anodic compartment. The innovative idea was the use of an enzyme specific for fuel oxidation instead of platinum. For glucose oxidation, glucose oxidase was tested as the anodic catalyst. The fuel cell delivered 30 nA cm^{-2} at 330 mV…a very low power density indeed but the proof of concept of biofuel cell was born. Generally speaking these biofuel cells function as fuel cells but used enzymes instead of noble metals as catalysts (Figure 5). They are referred as enzymatic biofuel cells. Microorganisms can also be used as catalysts, defining microbial fuel cells. Microbial biofuel cells use the metabolism of microorganisms under anaerobic conditions to oxidize fuel [103-104]. Although a promising concept, little is known yet about the mechanisms by which fuel is oxidized at the anode. The involvement of nanowires, electron transfer mediators, either membrane-bound or excreted, is supposed to be responsible for the cell current. Enzymatic biofuel cells are however more efficient because no mass transfer limitations across the cell membrane exist.

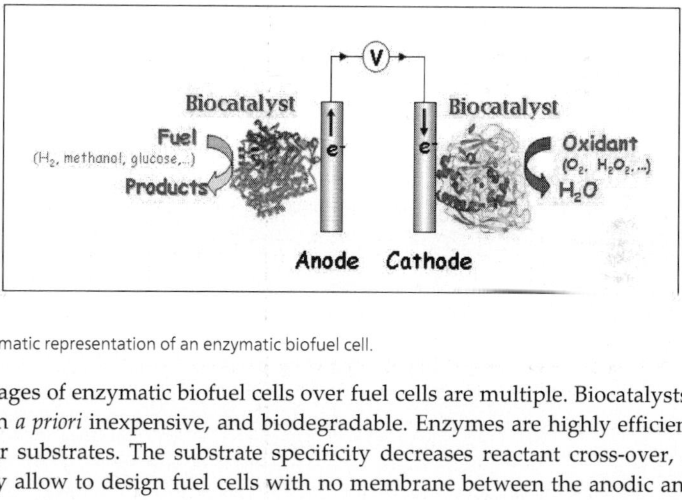

Figure 5. Schematic representation of an enzymatic biofuel cell.

The advantages of enzymatic biofuel cells over fuel cells are multiple. Biocatalysts are widespread, then *a priori* inexpensive, and biodegradable. Enzymes are highly efficient and specific to their substrates. The substrate specificity decreases reactant cross-over, and might theoretically allow to design fuel cells with no membrane between the anodic and cathodic compartments. Both costs are reduced and the design is simplified. A large variety of fuels and oxidants can be used to feed the biofuel cells, as opposed to the poor available fuels and oxidant in classical fuel cells (dihydrogen, methanol, oxygen). Indeed, many enzymes are nowadays characterized which differ by their natural abundant substrates. Dihydrogen, but also various inexpensive sugars can thus be used as efficient fuels at the anode. Furthermore, the involvement of cascades of enzymes can enhance the cell performance because of the summation of the electrons from each enzymatic reaction [105]. Finally, biofuel cells can deliver power under soft working conditions, as enzymes usually perform their enzymatic reactions at mild pH and temperature. Nevertheless, some extremophilic enzymes operate in extreme acidic or basic pH, as well as at high temperatures (around 90°C) or high pressure, offering the possibility to develop biofuel cell devices for special applications requiring extreme working conditions [106]. The applications of biofuel cells are still in their infancy. They are mainly thought to power small portable devices. Remarkable progress has been reported for implantable biofuel cells during the last year to power drug pumps, glucose sensors, vision devices [107-109].

The most common redox couple that has been used in biofuel cells is sugar/O_2, essentially because of sugar and O_2 abundance in nature and their essential role in living metabolism. In particular, glucose is an important metabolite and a source of energy for many living organisms. In that field, CNTs have been widely used, both at the anode and cathode. Glucose/O_2 biofuel cell is thus a very pertinent investigation field to investigate the role of CNTs. A view of some typical results is presented in Table 2.

Enzymes **Anode / Cathode**	**Mediators** **Anode / Cathode**	**Power density** **μW cm^{-2}**	**Ref**
Gox / Laccase	Ferrocene / -	15	[111]
GDH / BOD	PQQ / -	23	[82]
Gox / Pt	Ferrocenecarboxaldehyde / -	51	[112]
GDH/ BOD	Poly(brilliant cresyl blue) / -	54	[113]
GDH / laccase	Azine dies / -	58	[114]
Gox / Pt	Benzoquinone / -	77	[52]
Gox / Laccase	Ferrocene / ABTS	100	[115]
Gox / BOD	Ferrocene methanol / ABTS	120	[116]
GDH / Laccase	- / -	131	[117]
CDH / Pt	Os complex / -	157	[118]
Gox / Laccase	- / -	1300	[119]

Gox: Glucose oxidase; GDH: Glucose dehydrogenase; BOD: Bilirubin oxidase; ABTS: 2, 2'-azino-bis(3-ethylbenzothiazoline-6-sulfonate) diammonium; CDH: cellobiose dehydrogenase.

Table 2. Performances of glucose/O_2 fuel cells.

Data highlight that kinetics of bioelectrochemical reactions, thus power density, largely depends on the experimental conditions, *i.e.* enzyme and mediators, T°, pH, concentration of substrate, electrolyte and type of electrode construction. Highest values are obtained with mediatorless fuel cells, reaching power densities upper than 1 mW cm^{-2} which is sufficient to power small electrical devices. It appears that direct connection of copper enzymes, namely laccase or BOD, for oxygen reduction at the cathode can be quite easily obtained with the help of CNT network. Direct connection of enzymes for glucose oxidation is conversely hardly observed, even on CNT coatings. From literature examination direct connection of

http://dx.doi.org/10.5772/51782

Gox at electrode interfaces is still controversial. Due to the peculiar structure of Gox, a dimer with flavin adenine dinucleotide active site buried within a thick and isolated protein shell, it is understandable that electrical connection of Gox could be unexpected. A recent work concluded that CNTs were capable to electrically connect Gox, but this connection was unfruitful for glucose catalytic oxidation [110].

8. Carbon nanotubes for bioelectrooxidation of H_2: towards H_2/O_2 biofuel cells

We already described above hydrogenases, the enzymes that convert with high specificity and efficiency protons into dihydrogen. Most of these biocatalysts are also efficient in the oxidation of dihydrogen into protons. Consequently this allows to imagine biofuel cells in which the fuel would be dihydrogen, exactly as in PEM fuel cells. As hydrogenases are able to oxidize dihydrogen with very low overvoltage, the open circuit voltage for the biofuel cell using oxygen at the cathode, is expected to be not far from the thermodynamic one, *i.e.* 1.23 V. Hence, high power densities are expected, provided that a strong and efficient electrical connection between hydrogenase and electrode can be achieved. Simple adsorption of hydrogenase was performed in a first step, because it allowed a direct oxidation of dihydrogen without any redox mediators [120]. Catalytic mechanisms associated with dihydrogen oxidation at the active site were largely studied. The effect of strong hydrogenase inhibitors such as oxygen and CO were explored by this mean, leading to nice developments in engineering of more tolerant hydrogenases [121] or use of naturally resistant hydrogenases [122, 123]. However, this immobilization procedure relies on a monolayer of enzyme, which furthermore suffers from quick desorption. Otherwise, multilayer enzymatic films require a redox mediator so that even the last layer far from the electrode could be connected. Other immobilization processes are thus needed, that can favor an enhancement in both the amount of connected hydrogenases as well as their stability, while preserving their functionality.

Carbon nanotube networks constituted technological breakthroughs in that way. All the recent developments using immobilization of hydrogenases onto carbon nanotubes point out improved catalytic currents essentially related to an increase in the active area of the electrode. The respective role of metallic-SWCNTs against semiconducting one was explored for dihydrogen oxidation by immobilized hydrogenase [124]. A higher oxidation process was revealed when the nanotube mixture was enriched in metallic SWCNT. The study furthermore suggested no need of oxygenated SWCNTs for efficient anchoring of hydrogenases. The catalytic current enhancement was claimed to be due to an increase in active electrode surface area and an improved electronic coupling between hydrogenase redox active sites and the electrode surface. In most cases, however, CNTs are used as a mixture of metallic and semi-conducting tubes. Oxidation of the mixture yields the defects and functionalities described above in this review. Advantage is gained due to these chemical functions quite easily generated on the surface of the carbon nanotubes. Electrodes modified by carbon nanotubes are thus expected to offer numerous anchoring sites for stable hydrogenase immobilization. The literature provides a few examples of efficient immobilization of hydrogenase on carbon nanotubes coatings bearing various functionalities. Both SWCNTs and MWCNTs are used. Notably, more and more articles are devoted nowadays to this domain in hydrogenase research. A bionanocomposite made of the hydrogenase, MWCNTs and a thiopyridine derivative was proved to form stable monolayers when transferred by Langmuir-Blodgett method on indium tin oxide electrode surfaces [125]. A greater amount of electroactive hydrogenase towards dihydrogen oxidation was demonstrated to be adsorbed on the Langmuir-Blodgett films. De Lacey and co-workers grew MWCNTs on electrode by chemical vapor deposition of acetylene [65]. A high density of vertically aligned carbon nanotubes was obtained, which were functionalized by electroreduction of a diazonium salt for covalent binding of hydrogenase. High coverage of electroactive enzyme was measured, suggesting that almost all the functionalized CNT surface was accessible to hydrogenase. Great stabilization of the catalytic current for H_2 oxidation was obtained, with no decrease in current density after one month. Another work by Heering and co-workers studied a gold electrode pre-treated by polymyxin then a multilayer of carbon nanotubes [126]. Polymyxin was shown to help in the stable attachment of hydrogenase on the gold electrode. Using adsorption of hydrogenase on a nanotube layer pretreated with polymyxin the current density for H_2 oxidation was an order of magnitude higher than at the gold electrode only modified by polymixin. This result was supposed to origin from greater surface area even though only the top of the nanotube layer was supposed to be accessible to the enzyme. The catalytic current was stable with time, at least for two hours under continuous cycling, and several days upon storage under ambient conditions. AFM visualization of hydrogenase immobilized onto polymyxin-treated SWCNT layer on SiO_2 revealed that hydrogenase was structurally intact and preferentially adsorbed on the sidewalls of the CNTs rather than on SiO_2 [126].

In our laboratory, we immobilized the [NiFe] hydrogenase from a mesophilic anaerobic bacterium (the sulfate reducing bacterium *Desulfovibrio fructosovorans* Df) by adsorption onto SWCNT films [44]. The current for direct H_2 oxidation was shown to increase with the amount of SWCNTs in the coating (Figure 6).

Because non-turnover signals were not detected for hydrogenase in these conditions, the increase in surface area was evaluated using a redox protein as a probe. It was shown that SWCNTs induced one order larger surface area. The same hydrogenase was entrapped in methylviologen functionalized polypyrrole films coated onto SWCNTs and MWCNTs [127]. Although no direct electrical hydrogenase connection was observed, an efficient dihydrogen oxidation through a mediated process occurred. It was concluded that the entrapment of hydrogenase into the redox polymer coated onto CNTs combined the electron carrier properties of redox probes, the flexibility of polypyrroles, and the high electroactive area developed by CNTs. The reason why no direct connection could be observed is however not clearly understood yet. In our group we handled immobilization of hydrogenase on a film obtained by electropolymerization of a phenothiazine dye on a SWCNT coating [81]. The phenothiazine dye was shown to be able to mediate dihydrogen oxidation but also to serve as an anchor for the enzyme when adsorbed or when electropolymerized. Higher current density than in the absence of SWCNT was observed. In addition, a wider potential window for dihydrogen oxidation was reached as well as very stable electrochemical signals with

http://dx.doi.org/10.5772/51782

time. We postulated that the conductive polymer which was electropolymerized onto CNTs could play a multiple role: enhancement of the electroactive surface area, enhancement of redox mediator units due to phenothiazine monomers entrapped in the polymer matrix, enhancement of hydrogenase anchorage sites. We have already mentioned in this review the advantages of a direct electron transfer over a mediated one for H_2 oxidation, including gain in over-potential values, less interferences due to enzyme specificity, absence of redox mediators that could be difficult to co-immobilize with the enzyme… Functionalized carbon nanotube films were evaluated in our group as platforms for various hydrogenases, that present a very different environment of FeS cluster electron relay. Dihydrogen oxidation was studied at gold electrodes modified with functionalized self-assembled-monolayers [128]. As expected, dihydrogen oxidation process was demonstrated to be driven by electrostatic or hydrophobic interactions according to the specific environment of the surface electron relay. Interestingly, at CNT coatings, although CNTs were negatively charged, direct electrical connection of hydrogenases that present a negatively charged patch around the FeS surface electron relay was observed [44, 123]. In other words, despite unfavourable electrostatic interactions, direct electron transfer process for dihydrogen oxidation was achieved. One important conclusion was that on such CNT films, the nanometric size of the CNTs allows a population of hydrogenases to be directly connected to a neighbouring nanotube, hence allowing direct electron transfer for H_2 oxidation, whatever the orientation of the enzyme.

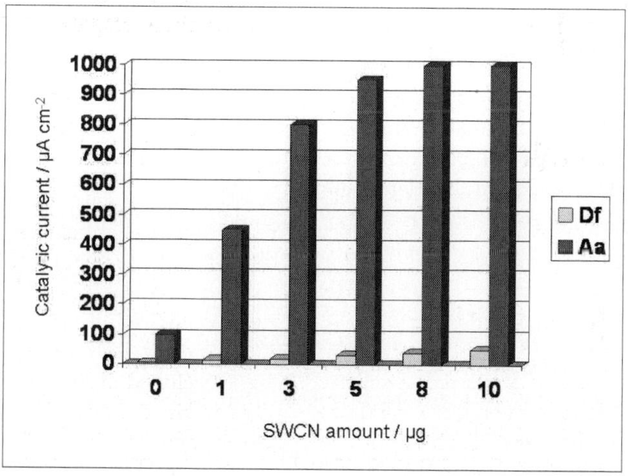

Figure 6. Comparative evolution of the catalytic current for dihydrogen oxidation with the amount of SWCNTs deposited at a graphite electrode in the case of hydrogenases from *Aquifex aeolicus* (Aa) or *Desulfovibrio fructosovorans* (Df). Catalytic currents are measured using voltammetry under H_2 at 60 and 25°C for Aa and Df respectively.

However, the extreme oxygen sensitivity of hydrogenases used in the former studies yielded an intensive research towards more resistant enzymes. During the last years, four [NiFe] membrane-bound hydrogenases have been discovered from aerobic or extremophilic organisms [128, 129-132]. They have been demonstrated to oxidize H_2 in the presence of oxygen and CO. The crystallographic structure of three of them has been resolved, showing that an uncommon [4Fe-3S] cluster proximal to the active site prevents deleterious oxygen attack. Of course, the sensitivity to oxygen, and also to CO, of most hydrogenases known before was a strong limitation for their potential use in biotechnological devices. Therefore these resistant biocatalysts open new avenues towards a biohydrogen economy. No doubt that these researches will increase in the next future. To date, two main studies report the immobilization of resistant hydrogenase on CNT-modified electrodes. Krishnan *et al.* very recently modified MWCNTs by pyrenebutyric acid, and demonstrated it was an efficient platform for stable O_2-resistant hydrogenase linkage [133]. In our group, original use of a hyperthermophilic O_2- and CO-resistant hydrogenase allowed the increase in the catalytic current for direct H_2 oxidation on a large range of temperature up to 70°C. Attempts to enhance the number of electrically connected hydrogenase succeeded by use of coatings of chemically oxidized SWCNTs [123]. Values as high as 1 mA cm^{-2} were reached depending on the amount of SWCNTs used in the coating (Figure 6). For the lowest amounts of SWCNTs, the increase in the catalytic current was demonstrated to be essentially due to the increase in surface area. However the catalytic current rapidly reached a plateau, although the peak current for the redox probe still increased, suggesting rapid saturation of the surface.

9. Design of a H_2/O_2 biofuel cell based on carbon nanotubes-modified electrodes

H_2/O_2 biofuel cells did not get much attention before O_2 and CO resistant hydrogenases were proved to be efficient for H_2 oxidation when immobilized onto electrode surfaces. Even though more and more efficient hydrogenase immobilization procedures are nowadays reported, few H_2/O_2 biofuel cells are described. An early study by Armstrong's group in 2006 [134] demonstrated that simple adsorption on graphite electrode of hydrogenase at the anode and laccase (a copper protein for O_2 reduction) at the cathode, allowed a wristwatch to run for 24h. Power density of around 5 μW cm^{-2} at 500 mV was delivered with no membrane between the two compartments providing hydrogenase was extracted from *Ralstonia metallireducens*. As this is an aerobic bacterium, the result underlined that the H_2/O_2 biofuel cell could operate only with O_2 resistant hydrogenase. In 2010, the same group improved the device by using another O_2 resistant hydrogenase from *Escherichia coli* and bilirubin oxidase (BOD), another copper protein more efficient than laccase towards oxygen reduction because being able to function at neutral pH [135]. The oxygen reductase was covalently linked to the graphite electrode which had been modified by diazonium salt reduction. The power density was enhanced compared to the former study reaching 63 μW cm^{-2}. But most of all, this work provided a nice understanding of the operating conditions of such H_2/O_2 fuel cells involving hydrogenase as anode catalyst.

Due to the understanding of how hydrogenases could be efficiently connected at CNT-coated electrodes, a huge step jumped over very recently. First, using covalent attachment of both O_2 resistant hydrogenase and BOD on pyrene derivative functionalized MWCNTs, a membrane-less biofuel cell was designed fed with a non-explosive 80/20 dihydrogen/air mixture [133]. This biofuel cell displayed quite a good stability with time and a much higher power density than reported before. Indeed, an average power density of 119 μW cm^{-2} was measured. Low solubility of oxygen and weak affinity of BOD for oxygen was shown to limit the cathodic current. Secondly in our group, a more performant H_2/O_2 mediatorless biofuel cell was constructed based on one step covalent attachment directly on SWCNTs of an hyperthermophilic O_2 resistant hydrogenase at the anode and BOD at the cathode [136] (Figure 7).

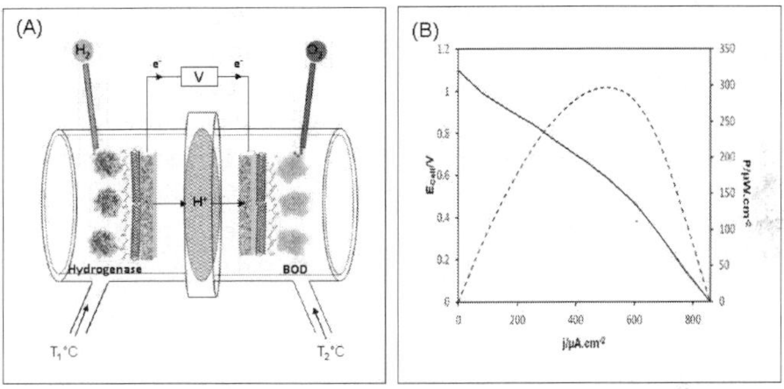

Figure 7. (A) Schematic representation of H_2/O_2 biofuel cell with O_2 resistant hydrogenase at the anode and bilirubin oxidase (BOD) at the cathode. Each half cell, separated by a Nafion® membrane, is independently thermoregulated with waterbaths. (B) Performance of the H_2/O_2 biofuel cell.

Taking advantage of temperature, the biofuel cell delivered power densities up to 300 μW cm^{-2} at 0.6V with an OCV of 1.1V, which is the highest performance ever reported. Furthermore, promising stability of the biofuel cell during 24h of continuous use lets us consider this device as an alternative power supply for small portable applications. The analysis of the fuel cell parameters during polarization, allows us to define the potential window in which the fuel cell fully operates. Interestingly, in Armstrong's group [135] and in our group, different approaches on the settings of biofuel cell working conditions, led to similar observations of an unexpected increasing anodic potential. This high oxidizing potential generates an inactive state of hydrogenase active site. It is worth noticing that this hydrogenase inactivation occurred under anaerobic conditions in our group while it was under aerobic conditions in Armstrong's group. Consequently, dramatic loss in power densities was observed. By applying negative potential to the anode, and thus providing electrons to the active site, we were unable to reactivate hydrogenase. Another protocol used by Armstrong, consisted to add a second hydrogenase coated anode, unconnected to the system but present in the anodic half-cell which was consequently unaffected by the oxidizing potential but still in presence of O_2. This second anode, under H_2 oxidation was used as an electron supplier and connected to the first electrode. This procedure reactivated hydrogenase and allowed full recovery of OCV. It is of relevant interest to overcome hydrogenase inactivation in H_2/O_2 biofuel cell.

10. Future directions

As reviewed in this chapter, many of the CNTs based technologies are promising for the development of a green hydrogen economy. Not only abiotic dihydrogen storage, but also microbial dihydrogen production and use of this green dihydrogen in biofuel cells can take advantages of the outstanding properties of CNTs. In all these applications, CNTs appear to play multiple roles including increase in surface area, increase in electron transfer rate, increase in directly connected enzymes. Possible protection against oxygen damage of enzymes has even been strongly suggested. Use of CNTs thus allows to architecture three-dimensional nanostructured interfaces which can be an alternative to strictly orientated proteins or enzymes for high direct electron transfer interfacial processes. The ease in obtaining tuned surface functionalizations is one of the very attracting points in view of the development of efficient bioelectrodes.

This is in particular the case for biofuel cells using dihydrogen as a fuel. During the last years, tremendous research on hydrogenase, the key enzyme for dihydrogen conversion, has led to the discovery, then control of some hydrogenases presenting properties that allow their use in biotechnological devices. During this year, based on these new resistant enzymes and on improved knowledge of how CNTs can enhance direct current densities, two H_2/O_2 biofuel cells have been reported. Although these biofuel cells constitute the first device using hydrogenases, they already deliver sufficient power density for small portable applications. No doubt that this research field will gain more and more interest in a next future.

However, various directions might be followed to further improve the biological system in such a way it could be commercially available. One is the enhancement of long-term stability of the device, which is obviously the critical point shared by (bio)fuel cells, yet. Search for more stable enzymes in the biodiversity or enzyme engineering has to be explored. Protection of enzymes by various encapsulation procedures could be another solution given efficient interfacial electron transfer can be reached. The use of whole microorganisms with controlled and driven metabolism, or at least immobilization of naturally encapsulated enzymes will be a next step. As an example, reconstitution of proteoliposomes with a membrane-bound hydrogenase was proved to enhance the stability of the enzyme [137]. This could be a novel route for preserving enzymes in their physiological environment, hence enhancing their stability. New enzymes, with outstanding properties (T°, pH, inhibitors, substrate affinity…) have to be discovered and studied. Notably, two very recent publications report on a new thermostable bilirubin oxidase and a tyrosinase which present outstanding resistances to serum constituents [138, 139]. These two new enzymes appear to be able to efficiently replace the currently used BOD for implantable applications of biofuel cells.

More sophisticated materials interfaces, constituted of mixtures of CNTs with other conducting materials could bring a hierarchical porosity necessary for both enzyme immobiliza-

http://dx.doi.org/10.5772/51782

tion and substrate diffusion. Carbon fibers, mesoporous carbon templates could be used to build very interesting new electrochemical interfaces. This diversity in potential carbon materials for efficient enzyme immobilization would be a key step to go through the difficulties linked to CNTs, *i.e.* effective cost for separation and purification as well as possible toxicity. Finally, to avoid the membrane between the cathodic and anodic compartments, and build a miniaturized biofuel cell, unusual cell designs, such as microfluidic or flow-through systems, are likely to open new avenues. All these future developments will certainly require a multidisciplinary approach, coupling electrochemists with biochemists and physicists, and coupling methods such as electrochemistry and spectrometry, electrochemistry and molecular genetics or electrochemistry and materials chemistry. This multidisciplinary willingness will help in the elucidation of the interactions between enzymes and nanostructured materials at the nanoscale and yield innovative nanobiotechnological approaches and applications.

Acknowledgements

We gratefully acknowledge the contribution of Marielle Bauzan (Fermentation Plant Unit, IMM, CNRS, Marseille, France) for growing the bacteria, Dr Marianne Guiral, Dr Marianne Ilbert and Pascale Infossi for fruitful discussions. This work was supported by research grants from CNRS, Région PACA and ANR.

Author details

Anne De Poulpiquet, Alexandre Ciaccafava, Saïda Benomar,
Marie-Thérèse Giudici-Orticoni and Elisabeth Lojou*

*Address all correspondence to: lojou@imm.cnrs.fr

Bioénergétique et Ingénierie des Protéines, CNRS - AMU - Institut de Microbiologie de la Méditerranée, France

References

[1] Verne, J. (1874). L'île mystérieuse.

[2] Grove, W. (1838). On a new voltaic combination. *Philosophical Magazine and Journal of Science*, 13, 430.

[3] Charlou, J. L., Donval, J. P., Konn, C., Ondréas, H., Fouquet, Y., Jean-Baptiste, P., & Fourré, E. (2010). High production and fluxes of H_2 and CH_4 and evidence of abiotic hydrocarbon synthesis by serpentinization on ultramafic-hosted hydrothermal systems on Mid-Atlantic Ridge. Rona P., Devey C., Dyment J. Murton B. Editors,. *"Di-*

versity of hydrothermal systems on slow spreading ocean ridges" Edited by AGU Geophysical monograph series, 188, 265-296.

[4] Oberlin, A., Endo, M., & Koyama, T. (1976). Filamentous growth of carbon through benzene decomposition. *Journal of Crystal Growth*, 32(3), 335-349.

[5] Iijima, S. (1991). Helical microtubules of graphitic carbon. *Nature*, 354-56.

[6] Saito, R., Dresselhaus, G., & Dresselhaus, M. (1998). Physical Properties of Carbon Nanotubes. *Imperial College Press, London.*

[7] Kirk, Othmer. (1996). Encyclopedia of Chemical Technology, (4th ed.). *Wiley and Sons, New York.*

[8] Heitner-Wirguin, C. (1996). Recent advances in perfluorinated ionomer membranes: structure, properties and applications. *Journal of Membrane Science*, 120(1), 1-33.

[9] Cracknell, J., Vincent, K., & Armstrong, F. (2008). Enzymes as working or inspirational electrocatalysts for fuel cells and electrolysis. *Chemical Review*, 108(7), 2439-2461.

[10] Guiral, M., Prunetti, L., Aussignargues, C., Ciaccafava, A., Infossi, P., Ilbert, M., Lojou, E., & Giudici-Orticoni, M. T. (2012). The hyperthermophilic bacterium *Aquifex aeolicus*: from respiratory pathways to extremely resistant enzymes and biotechnological applications. *Advances in Microbiological Physiology; to be edited.*

[11] Ivanov, I., Vidakovic-Koch, T., & Sundmacher, K. (2010). Recent Advances in Enzymatic Fuel Cells:. *Experiments and Modeling Energies*, 3(4), 803-846.

[12] Tran, P., Artero, V., & Fontecave, M. (2010). Water electrolysis and photoelectrolysis on electrodes engineered using biological and bio-inspired molecular systems. *Energy & Environmental Science*, 3(6), 727-747.

[13] Lojou, E. (2011). Hydrogenases as catalysts for fuel cells: Strategies for efficient immobilization at electrode interfaces. *Electrochimica Acta*, 56(28), 10385-10397.

[14] Schnorr, J. M., & Swager, T. M. (2011). Emerging Applications of Carbon Nanotubes. *Chemistry of Matererials*, 23(3), 646-657.

[15] Iijima, S., & Ichihashi, T. (1993). Single-shell carbon nanotubes of 1-nm diameter. *Nature*, 363, 603-605.

[16] Dresselhaus, M., Dresselhaus, G., & Jorio, A. (2004). Unusual properties and structure of carbon nanotubes. *Annual Review of Material Research*, 34-247.

[17] Botzung, M., Chaudourne, S., Gillia, O., Perret, C., Latroche, M., Percheron-Guegan, A., & Marty, P. (2008). Marty Simulation and experimental validation of a hydrogen storage tank with metal hydrides. *International Journal of Hydrogen Energy*, 33(1), 98-104.

[18] Vajo, J., Li, W., & Liu, P. (2010). Thermodynamic and kinetic destabilization in $LiBH_4/Mg_2NiH_4$: promise for borohydride-based hydrogen storage. *Chemical Communications*, 46(36), 6687-6689.

[19] Li, C., Peng, P., Zhou, D., & Wan, L. (2011). Research progress in $LiBH_4$ for hydrogen storage: A review. *International Journal of Hydrogen Energy*, 36(22), 14512-14526.

[20] Yao, X., Wu, C., Du, A., Lu, G., Cheng, H., Smith, S., Zou, J., & He, Y. (2006). Mg-based nanocomposites with high capacity and fast kinetics for hydrogen storage. *Journal of Physical Chemistry B*, 110(24), 11697-11703.

[21] Luo, Y., Wang, P., Ma , L. P, & Cheng, M. (2007). Enhanced hydrogen storage properties of MgH_2 co-catalyzed with NbF_5 and single-walled carbon nanotubes. *Scripta Materialia*, 56(9), 765-768.

[22] Wu, C., Wang, P., Yao, X., Liu, C., Chen, D., Lu, G., & Cheng, H. (2006). Effect of carbon/noncarbon addition on hydrogen storage behaviour of magnesium hydride. *Journal of Alloys Compounds*, 414-259.

[23] Dillon, A., Jones, K., Bekkedahl, T., Kiang, C., Bethune, D., & Heben, M. (1997). Storage of hydrogen in single-walled carbon nanotubes. *Nature*, 386(6623), 377-379.

[24] Ding, F., & Yakobson, B. (2011). Challenges in hydrogen adsorptions: from physisorption to chemisorption. *Frontiers of Physics*, 6(2), 142-150.

[25] Nikitin, A., Li, X., Zhang, Z., Ogasawara, H., Dai, H., & Nilsson, A. (2008). Hydrogen storage in carbon nanotubes through the formation of stable C-H bonds. *Nano Letters*, 8(1), 162-167.

[26] Sahaym, U., & Norton, M. (2008). Advances in the application of nanotechnology in enabling a hydrogen economy. *Journal of Material Sciences*, 43(16), 5395-549.

[27] Tang, Z., Poh, C., Lee, K. K., Tian, Z., Chua, D., & Lin, J. (2010). Enhanced catalytic properties from platinum nanodots covered carbon nanotubes for proton-exchange membrane fuel cells. *Journal of Power Sources*, 195(1), 155-159.

[28] Lin, J., Kamavaram, V., & Kannan, A. (2010). Synthesis and characterization of carbon nanotubes supported platinum nanocatalyst for proton exchange membrane fuel cells. *Journal of Power Sources*, 195(2), 466-470.

[29] Kongkanand, A, Kuwabata, S, Girishkumar, G, & Kamat, P. (2006). Single-wall carbon nanotubes supported platinum nanoparticles with improved electrocatalytic activity for oxygen reduction reaction. *Langmuir*, 22(5), 2392-2396.

[30] Lin, J., Mason, C., Adame, A., Liu, X., Peng, X., & Kannan, A. (2010). Synthesis of Pt nanocatalyst with micelle-encapsulated multi-walled carbon nanotubes as support for proton exchange membrane fuel cells. *Electrochimica Acta*, 55(22), 6496-6500.

[31] Lee, C., Ju, Y. C., Chou, P. T., Huang, Y. C., Kuo, L. C., & Oung, J. C. (2005). Preparation of Pt nanoparticles on carbon nanotubes and graphite nanofibers via self-regulated reduction of surfactants and their application as electrochemical catalyst. *Electrochemistry Communications*, 7(4), 453-458.

[32] He, D., Zeng, C., Xu, C., Cheng, N., Li, H., & Mu, S. (2011). Polyaniline-Functionalized Carbon Nanotube Supported Platinum Catalysts. *Langmuir*, 27(9), 5582-5588.

[33] Ludwig, R., Harreither, W., Tasca, F., & Gorton, L. (2011). Cellobiose Dehydrogenase: A Versatile Catalyst for Electrochemical Applications. *ChemPhysChem.*, 11(13), 2674-2697.

[34] Cang-Rong, J., & Pastorin, G. (2009). The influence of carbon nanotubes on enzyme activity and structure: investigation of different immobilization procedures through enzyme kinetics and circular dichroism studies. *Nanotechnology*, 20(25), 255102.

[35] Matsuura, K., Saito, T., Okasaki, T., Oshima, S., Yumura, M., & Iijima, S. (2006). Selectivity of water-soluble proteins in single-walled carbon nanotube dispersions. *Chemical Physics Letters*, 429(4-6), 497-502.

[36] Patolsky, F., Weizmann, Y, & Willner, I. (2004). Long-range electrical contacting of redox enzymes by SWCNT connectors. *Angewandte Chemistry International Edition*, 43(16), 2113-2117.

[37] Liu, J., Chou, A., Rahmat, W., Paddon-Row, M., & Gooding, J. (2005). Achieving direct electrical connection to glucose oxidase using aligned single walled carbon nanotube arrays. *Electroanalysis*, 17(1), 38-46.

[38] Wang, J. (2005). Carbon-nanotube based electrochemical biosensors: A review. *Electroanalysis*, 17(1), 7-14.

[39] Wildgoose, G., Banks, C., Leventis, H., & Compton, R. (2006). Chemically modified carbon nanotubes for use in electroanalysis. *Microchimica Acta*, 152(3-4), 187-214.

[40] Dumitrescu, I., Unwin, P., & Macpherson, J. (2009). Electrochemistry at carbon nanotubes: perspective and issues. *Chemical Communications*, 45, 6886-4901.

[41] Ji, P., Tan, H., Xu, X., & Feng, W. (2010). Lipase Covalently Attached to Multiwalled Carbon Nanotubes as an Efficient Catalyst in Organic Solvent. *AIChE Journal*, 56(11), 3005-3011.

[42] Upadhyayula, V., & Gadhamshetty, V. (2010). Appreciating the role of carbon nanotube composites in preventing biofouling and promoting biofilms on material surfaces in environmental engineering: A review. *Biotechnological Advances*, 28(6), 802-816.

[43] Lojou, E., Luciano, P., Nitsche, S., & Bianco, P. (1999). Poly(ester-sulfonic acid):modified carbon electrodes for the electrochemical study of c-type cytochromes. *Electrochimica Acta*, 44(19), 3341-3352.

[44] Lojou, E., Luo, X., Brugna, M., Candoni, N., Dementin, S., & Giudici-Orticoni, M. T. (2008). Biocatalysts for fuel cells: efficient hydrogenase orientation for H_2 oxidation at electrodes modified with carbon nanotubes. *Journal of Biological Inorganic Chemistry*, 13(7), 1157-1167.

[45] Minteer, S., Atanassov, P., Luckarift, H., & Johnson, G. (2012). New materials for biological fuel cells. *Material Today*, 15(4), 166-173.

[46] Weigel, M., Tritscher, E., & Lisdat, F. (2007). Direct electrochemical conversion of bilirubin oxidase at carbon nanotube-modified glassy carbon electrodes. *Electrochemistry Communications*, 9(4), 689-693.

[47] Pumera, M., & Smid, B. (2007). Redox protein noncovalent functionalization of double-wall carbon nanotubes: Electrochemical binder-less glucose biosensor. *Journal of Nanosciences and Nanotechnology*, 7(10), 3590-3595.

[48] Willner, I., Yan, Y. M., Willner, B., & Tel-Vered, R. (2009). Integrated Enzyme-Based Biofuel Cells-A Review. *Fuel Cells*, 09(1), 7-24.

[49] Ueda, A., Kato, D., Kurita, R., Kamata, T., Inokuchi, H., Umemura, S., Hirono, S., & Niwa, O. (2011). Efficient Direct Electron Transfer with Enzyme on a Nanostructured Carbon Film Fabricated with a Maskless Top-Down UV/Ozone Process. *Journal of the American Chemical Society*, 133(13), 4840-4846.

[50] Tasca, F., Gorton, L., Harreither, W., Haltrich, D., Ludwig, R., & Nöll, G. (2088). Direct electron transfer at cellobiose dehydrogenase modified anodes for biofuel cells. *Journal of Physical Chemistry C*, 112(26), 9956-9961.

[51] Zheng, W., Zhao, H., Zhou, H., Xu, X., Ding, M., & Zheng, Y. (2010). Electrochemistry of bilirubin oxidase at carbon nanotubes. *Journal of Solid State Electrochemistry*, 14(2), 249-254.

[52] Zheng, W., Zhou, H., Zheng, Y., & Wang, N. (2008). A comparative study on electrochemistry of laccase at two kinds of carbon nanotubes and its application for biofuel cell. *Chemical Physics Letters*, 381-385.

[53] Zhao, L., Liu, H., & Hu, N. (2006). Assembly of layer-by-layer films of heme proteins and single-walled carbon nanotubes: electrochemistry and electrocatalysis. *Analytical Bioanalytical Chemistry*, 384(2), 414-422.

[54] Liu, G., & Lin, Y. (2006). Amperometric glucose biosensor based on self-assembling glucose oxidase on carbon nanotubes. *Electrochemistry Communications*, 8(2), 251-256.

[55] Iost, R., & Crespilho, F. (2012). Layer-by-layer self-assembly and electrochemistry: Applications in biosensing and bioelectronics. *Biosensors Bioelectronics*, 31(1), 1-10.

[56] Gooding, J., Wibowo, R., Liu, J., Yang, W., Losic, D., Orbons, S., Mearns, F., Shapter, J., & Hibbert, D. (2003). Protein electrochemistry using aligned carbon nanotube arrays. *Journal of the American Chemical Society*, 125(30), 9006-9007.

[57] Yu, X., Chattopadhyay, D., Galeska, I., Papadimitrakopoulos, F., & Rusling, J. (2003). Peroxidase activity of enzymes bound to the ends of single-wall carbon nanotube forest electrodes. *Electrochemistry Communications*, 5(5), 408-411.

[58] Esplandiu, M., Pacios, M., Cyganek, L., Bartroli, J., & Del Valle, M. (2009). Enhancing the electrochemical response of myoglobin with carbon nanotube electrodes. *Nanotechnology*.

[59] Santhosh, P., Gopalan, A., & Lee, K. (2006). Gold nanoparticles dispersed polyaniline grafted multiwall carbon nanotubes as newer electrocatalysts: Preparation and performances for methanol oxidation. *Journal of Catalysis*, 238(1), 177-185.

[60] Nazaruk, E., Karaskiewicz, M., Zelechowska, K., Biernat, J., Rogalski, J., & Bilewicz, R. (2012). Powerful connection of laccase and carbon nanotubes Material for mediator-free electron transport on the enzymatic cathode of the biobattery. *Electrochemistry Communications*, 14(1), 67-70.

[61] Sadowska, K., Stolarczyk, K., Biernat, J., Roberts, K., Rogalski, J., & Bilewicz, R. (2010). Derivatization of single-walled carbon nanotubes with redox mediator for biocatalytic oxygen electrodes. *Bioelectrochemistry*, 80(1), 73-80.

[62] Jeykumari, D., & Narayanan, S. (2008). Fabrication of bienzyme nanobiocomposite electrode using functionalized carbon nanotubes for biosensing applications. *Biosensors Bioelectronics*, 23(11), 1686-1693.

[63] Wang, Z., Li, M., Su, P., Zhang, Y., Shen, Y., Han, D., Ivaska, A., & Niu, L. (2008). Direct electron transfer of horseradish peroxidase and its electrocatalysis based on carbon nanotube/thionine/gold composites. *Electrochemistry Communications*, 10(2), 306-310.

[64] Le Floch, F., Thuaire, A., Bidan, G., & Simonato, J. P. (2009). The electrochemical signature of functionalized single-walled carbon nanotubes bearing electroactive groups. *Nanotechnology*, 20(14), 45705.

[65] Alonso-Lomillo, M., Rüdiger, O., Maroto-Valiente, A., Velez, M., Rodriguez-Ramos, I., Munoz, F., Fernandez, V., & De Lacey, A. (2007). Hydrogenase-coated carbon nanotubes for efficient H_2 oxidation. *Nano Letters*, 7(6), 1603-1608.

[66] Jönsson-Niedziolka, M., Kaminska, A., & Opallo, M. (2010). Pyrene-functionalised single-walled carbon nanotubes for mediatorless dioxygen bioelectrocatalysis. *Electrochimica Acta*, 55(28), 8744-8750.

[67] Karachevtsev, V., Stepanian, S., Glamazda, A., Karachevtsev, M., Eremenko, V., Lytvyn, O., & Adamowicz, L. (2011). Noncovalent Interaction of Single-Walled Carbon Nanotubes with 1-Pyrenebutanoic Acid Succinimide Ester and Glucose oxidase. *Journal of Physical Chemistry C*, 115(43), 21072-21082.

[68] Lau, C, Adkins, E, Ramasamy, R, Lackarift, H, Johnson, G, & Atanassov, P. (2012). Design of Carbon Nanotube-Based Gas-Diffusion Cathode for O_2 Reduction by Multicopper Oxidases. *Advanced Energy Materials*, 2(1), 162-168.

[69] Xu, H., Xiong-Y, H., Zeng-X, Q., Jia, L., Wang, Y., & Wang-F, S. (2009). Direct electrochemistry and electrocatalysis of heme proteins immobilized in single-wall carbon nanotubes-surfactant films in room temperature ionic liquids. *Electrochemistry Communications*, 11(2), 286-289.

[70] Yan, Y., Zheng, W., Zhang, M., Wang, L., Su, L., & Mao, L. (2005). Bioelectrochemically functional nanohybrids through co-assembling of proteins and surfactants onto

carbon nanotubes: Facilitated electron transfer of assembled proteins with enhanced faradic response. *Langmuir*, 21(14), 65606566.

[71] Cosnier, S., Ionescu, R., & Holzinger, M. (2008). Aqueous dispersions of SWCNTs using pyrrolic surfactants for the electro-generation of homogeneous nanotube composites. Application to the design of an amperometric biosensor. *Journal of Materials Chemistry*, 18(42), 5129-5133.

[72] Gao, M., Dai, L., & Wallace, G. (2003). Biosensors based on aligned carbon nanotubes coated with inherently conducting polymers. *Electroanalysis*, 15(13), 1089-1094.

[73] Tsai, C. Y., Li, C. S., & Liao, W. S. (2006). Electrodeposition of polypyrrole-multiwalled carbon nanotube-glucose oxidase nanobiocomposite film for the detection of glucose. *Biosensors Bioelectronics*, 22(4), 495-500.

[74] Chen, H, & Dong, S. (2007). Direct electrochemistry and electrocatalysis of horseradish peroxidase immobilized in sol-gel-derived ceramic-carbon nanotube nanocomposite film. *Biosensors Bioelectronics*, 22(8), 1811-1815.

[75] Heller, A. (2006). Electron-conducting redox hydrogels: design, characteristics and synthesis. *Current Opinion in Chemical Biology*, 10(6), 664-672.

[76] Timur, S, Anik, U, Odaci, D, & Gorton, L. (2007). Development of a microbial biosensor based on carbon nanotube (CNT) modified electrodes. *Electrochemistry Communications*, 9(7), 1810-1815.

[77] Song, J., Shin, H., & Kang, C. (2011). A Carbon Nanotube Layered Electrode for the Construction of the Wired Bilirubin Oxidase Oxygen Cathode. *Electroanalysis*, 23(12), 2941-2948.

[78] Tiwari, I., & Singh, M. (2011). Preparation and characterization of methylene blue-SDS- multiwalled carbon nanotubes nanocomposite for the detection of hydrogen peroxide. *Microchimica Acta*, 174(3-4), 223-230.

[79] Pakapongpan, S., Palangsuntikul, R., & Surareungchai, W. (2011). Electrochemical sensors for hemoglobin and myoglobin detection based on methylene blue-multiwalled carbon nanotubes nanohybrid-modified glassy carbon electrode. *Electrochimica Acta*, 56(19), 6831-6836.

[80] Hoshino, T., Sekiguchi, S., & Muguruma, H. (2012). Amperometric biosensor based on multilayer containing carbon nanotube, plasma-polymerized film, electron transfer mediator phenothiazine, and glucose dehydrogenase. *Bioelectrochemistry*, 84-1.

[81] Ciaccafava, A., Infossi, P., Giudici-Orticoni, M. T., & Lojou, E. (2010). Stabilization role of a phenothiazine derivative on the electrocatalytic oxidation of hydrogen via *Aquifex aeolicus* hydrogenase at graphite membrane electrodes. *Langmuir*, 26(23), 18534-18541.

[82] Tanne, C., Göbel, G., & Lisdat, F. (2010). Development of a (PQQ)-GDH-anode based on MWCNT-modified gold and its application in a glucose/O_2-biofuel cell. *Biosensors Bioelectronics*, 26(2), 530-535.

[83] Reisner, E. (2011). Solar Hydrogen Evolution with Hydrogenases: From Natural to Hybrid Systems European. *Journal of Inorganic Chemistry* [7], 1005-1016.

[84] Vignais, P., & Billoud, B. (2007). Occurrence, classification, and biological function of hydrogenases: An overview. *Chemical Reviews*, 107(10), 4206-4272.

[85] Parkin, A., Cavazza, C., Fontecilla-Camp, J., & Armstrong, F. (2006). Electrochemical investigations of the interconversions between catalytic and inhibited states of the [FeFe]-hydrogenase from *Desulfovibrio desulfuricans*. *Journal of the American Chemical Society*, 128(51), 16808-16815.

[86] Dementin, S., Belle, V., Bertrand, P., Guigliarelli, B., Adryanczyk-Perrier, G., De Lacey, A., Fernandez, V., Rousset, M., & Léger, C. (2006). Changing the ligation of the distal [4Fe4S] cluster in NiFe hydrogenase impairs inter- and intramolecular electron transfers. *Journal of the American Chemical Society*, 128(15), 5209-5218.

[87] Krassen, H., Stripp, S., von, Abendroth. G., Ataka, K., Happe, T., & Heberle, J. (2009). Immobilization of the [FeFe]-hydrogenase CrHydA1 on a gold electrode: Design of a catalytic surface for the production of molecular hydrogen. *Journal of Biotechnology*, 142(1), 3-9.

[88] Zadvornyy, O., Lucon, J., Gerlach, R., Zorin, N., Douglas, T., Elgren, T., & Peters, J. (2012). Photo-induced H_2 production by [NiFe]-hydrogenase from *T. roseopersicina* covalently linked to a Ru(II) photosensitizer. *Journal of Inorganic Biochemistry*, 106(1), 151-155.

[89] Reisner, E., Powell, D., Cavazza, C., Fontecilla-Camps, J., & Armstrong, F. (2009). Visible Light-Driven H_2 Production by Hydrogenases Attached to Dye-Sensitized TiO_2 Nanoparticles. *J. Am. Chem. Soc.*, 131(51), 18457-18466.

[90] Morra, S., Valetti, F., Sadeghi, S., King, P., Meyer, T., & Gilardi, G. (2011). Direct electrochemistry of an [FeFe]-hydrogenase on a TiO_2 Electrode. *Chemical Communications*, 47(38), 10566-10568.

[91] Brown, K., Wilker, M., Boehm, M., Dukovic, G., & King, P. (2012). Characterization of Photochemical Processes for H_2 Production by CdS Nanorod-[FeFe] Hydrogenase Complexes. *Journal of the American Chemical Society*, 143(12), 5627-5636.

[92] Brown, K., Dayal, S., Ai, X., Rumbles, G., & King, P. (2010). Controlled Assembly of Hydrogenase-CdTe Nanocrystal Hybrids for Solar Hydrogen Production. *Journal of the American Chemical Society*, 132(28), 9672-9680.

[93] Mc Donald, T., Svedruzic, D., Kim-H, Y., Blackburn, J., Zhang, S., King, P., & Heben, M. (2007). Wiring-up hydrogenase with single-walled carbon nanotubes. *Nano Letters*, 7(11), 3528-3534.

[94] Blackburn, J., Svedruzic, D., Mc Donald, T., Kim-H, Y., King, P., & Heben, M. (2008). Raman spectroscoipy of charge transfer interaction between single wall carbon nanotubes and [FeFe] hydrogenase. *Dalton Transactions*, 5454-5461.

[95] Kihara, T., Liu-Y, X., Nakamura, C., Park-M, K., Yasuda-W, S., Qian-J, D., Kawasaki, K., Zorin, N., Yasuda, S., Hata, K., Wakayama, T., & Miyake, J. (2001). Direct electron transfer to hydrogenase for catalytic hydrogen production using a single-walled carbon nanotubes forest. *International Journal of Hydrogen Energy*, 36(13), 7523-7529.

[96] Noda, K., Zorin, N., Nakamura, C., Miyake, M., Gogotov, I., Asada, Y., Akutsu, H., & Miyake, J. (1998). Langmuir-Blodgett film of hydrogenase for electrochemical hydrogen production. *Thin Solid Films*, 327-329, 639-642.

[97] Elgren, T, Zadvorny, O, Brecht, E, Douglas, T, Zorin, N, Maroney, M, & Peters, J. (2005). Immobilization of active hydrogenases by encapsulation in polymeric porous gels. *Nano Letters*, 5(10), 2085-2087.

[98] Zadvorny, O., Barrows, A., Zorin, N., Peters, J., & Elgren, T. (2010). High level oh hydrogen production activity achieved for hydrogenase encapsulated in sol-gel material doped with carbon nanotubes. *Journal of Materials Chemistry*, 20-1065.

[99] Fontecave, M., & Artero, V. (2011). Bioinspired catalysis at the crossroads between biology and chemistry: A remarkable example of an electrocatalytic material mimicking hydrogenases. *Compte Rendu Chimie*, 14(4), 362-371.

[100] Dubois, M., & Dubois, D. (2009). The roles of the first and second coordination spheres in the design of molecular catalysts for H_2 production and oxidation. *Chemical Society Reviews*, 38(1), 62-72.

[101] Le Goff, A., Artero, V., Jousselme, B., Tran, P. D., Guillet, N., Metaye, R., Fihri, A., Palacin, S., & Fontecave, M. (2009). From Hydrogenases to Noble Metal-Free Catalytic Nanomaterials for H_2 Production and Uptake. *Science*, 326(5958), 1384-1387.

[102] Yahiro, A., Lee, S., & Kimble, D. (1964). Bioelectrochemistry I. Enzyme utilizing Biofuel cell studies. *Biochimica Biophysica Acta*, 88, 375-383.

[103] Franks, Ashley. E., & Nevin, Kelly. P. (2010). Microbial Fuel Cells. *A Current Review. Energies*, 3(5), 899-919.

[104] Zhou, M., Chi, M., Luo, J., He, H., & Jin, T. (2011). An overview of electrode materials in microbial fuel cells. *Journal of Power Sources*, 196(10), 4427-4435.

[105] Sokic-Lazic, D., & Minteer, S. (2008). Citric acid cycle biomimic on a carbon electrode. *Biosensors Bioelectronics*, 24(4), 939-944.

[106] Egorova, K., & Antranikian, G. (2005). Industrial relevance of thermophilic Archaea. *Current Opinion in Microbiology*, 8(6), 649-655.

[107] Halamkova, L., Halamek, J., Bocharova, V., Szczupak, A., Alfonta, L., & Katz, E. (2012). Implanted Biofuel Cell Operating in a Living Snail. *Journal of the American Chemical Society*, 134(11), 5040-5043.

[108] Cinquin, P., Gondran, C., Giroud, F., Mazabrard, S., Pellissier, A., Boucher, F., Alcaraz, J. P., Gorgy, K., Lenouvel, F., Mathé, S., Porcu, P., & Cosnier, S. (2010). A Glucose BioFuel Cell Implanted in Rats. *PLoS ONE*, 5, e10476.

[109] Falk, M., Andoralov, V., Blum, Z., Sotres, J., Suyatin, D., Ruzgas, T., Arnebrant, T., & Shleev, S. (2012). Biofuel cell as a power source for electronic contact lenses. *Biosensors Bioelectronics*, 37(1), 38-45.

[110] Zebda, A., Gondran, C., Le Goff, A., Holzinger, M., Cinquin, P., & Cosnier, S. (2011). Mediatorless high-power glucose biofuel cells based on compressed carbon nanotube-enzyme electrodes. *Nature Com.*, 2-370.

[111] Zhao, H., Zhou, H., Zhang, J., Zheng, W., & Zheng, Y. (2009). Carbon nanotube-hydroxyapatite nanocomposite: A novel platform for glucose/O_2 biofuel cell. *Biosensors Bioelectronics*, 25(2), 463-468.

[112] Liu, J., Zhang, X., Pang, H., Liu, B., Zou, Q., & Chen, J. (2012). High-performance bioanode based on the composite of CNTs-immobilized mediator and silk film-immobilized glucose oxidase for glucose/O_2 biofuel cells. *Biosensors Bioelectronics*, 31(1), 170-175.

[113] Gao, F., Yan, Y., Su, L., Wang, L., & Mao, L. (2007). An enzymatic glucose/O_2 biofuel cell: Preparation, characterization and performance in serum. *Electrochemistry Communications*, 9(5), 989-996.

[114] Li, X., Zhou, H., Yu, P., Su, L., Ohsaka, T., & Mao, L. (2008). A miniature glucose/O_2 biofuel cell with single-walled carbon nanotubes-modified carbon fiber microelectrodes as the substrate. *Electrochemistry Communications*, 10(6), 851-854.

[115] Nazaruk, E., Sadowska, K., Biernat, J., Rogalski, J., Ginalska, G., & Bilewicz, R. (2010). Enzymatic electrodes nanostructured with functionalized carbon nanotubes for biofuel cell applications. *Analytical and Bioanalytical Chemistry*, 398(4), 1651-1660.

[116] Lim, J., Malati, P., Bonet, F., & Dunn, B. (2007). Nanostructured sol-gel electrodes for biofuel cells. *Journal of the Electrochemical Society*, 154(2), A140-A145.

[117] Karaskiewicz, M., Nazaruk, E., Zelzchowska, K., Biernat, J., Rogalski, J., & Bilewicz, R. (2012). Fully enzymatic mediatorless fuel cell with efficient naphthylated carbon nanotube-laccase composite cathodes. *Electrochemistry Communications*, 20-124.

[118] Tasca, F., Gorton, L., Harreither, W., Haltrich, D., Ludwig, R., & Nöll, G. (2008). Highly efficient and versatile anodes for biofuel cells based on cellobiose dehydrogenase from *Myriococcum thermophilum*. *Journal of Physical Chemistry C*, 112(35), 13668-13673.

[119] Wang, Y., & Yao, Y. (2012). Direct electron transfer of glucose oxidase promoted by carbon nanotubes is without value in certain mediator-free applications. *Microchimica Acta*, 176(3-4), 271-277.

[120] Vincent, K., Parkin, A., & Armstrong, F. (2007). Investigating and exploiting the electrocatalytic properties of hydrogenases. *Chemical Reviews*, 107(10), 4366-4413.

[121] Liebgott, P. P., de Lacey, A., Burlat, B., Cournac, L., Richaud, P., Brugna, M., Fernandez, V., Guigliarelli, B., Rousset, M., Léger, C., & Dementin, S. (2011). Original Design of an Oxygen-Tolerant [NiFe] Hydrogenase: Major Effect of a Valine-to-Cysteine Mutation near the Active Site. *Journal of the American Chemical Society*, 133(4), 986-997.

[122] Cracknell, J. A., Vincent, K. A., Ludwig, M., Lenz, O., Friedrich, B., & Armstrong, F. A. (2008). Enzymatic Oxidation of H_2 in Atmospheric O_2: The Electrochemistry of Energy Generation from Trace H_2 by Aerobic Microorganisms. *Journal of the American Chemical Society*, 130(2), 424-425.

[123] Luo, X. J., Brugna, M., Infossi, P., Giudici-Orticoni, M. T., & Lojou, E. (2009). Immobilization of the hyperthermophilic hydrogenase from *Aquifex aeolicus* bacterium onto gold and carbon nanotube electrodes for efficient H_2 oxidation. *Journal of Biological Inorganic Chemistry*, 14(8), 1275-1288.

[124] Svedruzic, D., Blackburn, J., Tenent, R., Rocha, D. J., Vinzant, T., Heben, M., & King, P. (2011). High-performance hydrogen production and oxidation electrodes with hydrogenase supported on metallic single-wall carbon nanotubes networks. *Journal of the American Chemical Society*, 133(12), 4299-4306.

[125] Sun, Q., Zorin, N. A., Chen, D., Chen, M., Liu, X. T., Miyake, J., & Qian, J. D. (2010). Langmuir-Blodgett films of pyridyldithio-modified multiwalled carbon nanotubes as a support to immobilize hydrogenase. *Langmuir*, 26(12), 10259-10265.

[126] Hoeben, F. J. M., Heller, I., Albracht, S. P. J., Dekker, C., Lemay, S. G., & Heering, H. A. (2008). Polymyxin-coated Au and Carbon nanotubes electrodes for stable [NiFe]-hydrogenase film voltammetry. *Langmuir*, 24(11), 5925-5931.

[127] Baur, J., Le Goff, A., Dementin, S., Holzinger, M., Rousset, M., & Cosnier, S. (2011). Three-dimensional carbon nanotube-polypyrrole-[NiFe] hydrogenase electrodes for the efficient electrocatalytic oxidation of H_2. *International Journal of Hydrogen Energy*, 36(19), 12096-12101.

[128] Ciaccafava, A., Infossi, P., Ilbert, M., Guiral, M., Lecomte, S., Giudici-Orticoni, M. T., & Lojou, E. (2012). Electrochemistry, AFM, and PM-IRRA Spectroscopy of Immobilized Hydrogenase: Role of a Hydrophobic Helix in Enzyme Orientation for Efficient H_2 Oxidation. *Angewandte Chemistry International Edition*, 51(4), 953-956.

[129] Volbeda, A., Amara, P., Darnault, C., Mouesca, J. M., Parkin, A., Roessler, M. M., Armstrong, F. A., & Fontecilla-Camps, J. C. (2012). X-ray crystallographic and computational studies of the O_2-tolerant [NiFe]-hydrogenase 1 from *Escherichia coli*. *Proceeding of the National Academic Sciences*, 10(14), 5305-5310.

[130] Shomura, Y., Yoon, S. K., Nishihara, H., & Higuchi, Y. (2011). Structural basis for a [4Fe-3S] cluster in the oxygen-tolerant membrane-bound [NiFe]-hydrogenase. Nature NIL_143., 479(7372), 253.

[131] Pandelia, M., Fourmond, V., Tron, P., Lojou, E., Bertrand, P., Léger, C., Giudici-Orticoni, M. T., & Lubitz, W. (2010). Membrane-Bound Hydrogenase I from the Hyperthermophilic Bacterium *Aquifex aeolicus:* Enzyme Activation, Redox Intermediates and Oxygen Tolerance. *Journal of the American Chemical Society*, 132(20), 6991-7004.

[132] Fritsch, J., Scheerer, P., Frielingsdorf, S., Kroschinsky, S., Friedrich, B., Lenz, O., & Spahn, C. M. T. (2011). The crystal structure of an oxygen-tolerant hydrogenase uncovers a novel iron-sulphur centre. Nature NIL_134., 479(7372), 249.

[133] Krishnan, S., & Armstrong, F. A. (2012). Order-of-magnitude enhancement of an enzymatic hydrogen-air fuel cell based on pyrenyl carbon nanostructures. *Chemical Science*, 3(4), 1015-1023.

[134] Vincent, K., Cracknell, J., Clark, J., Ludwig, M., Lenz, O., Friedrich, B., & Armstrong, F. (2006). Electricity from low-level H_2 in still air- an ultimate test for an oxygen tolerant hydrogenase. *Chemistry Communications*, 5033-5035.

[135] Wait, A., Parkin, A., Morley, G., dos, Santos. L., & Armstrong, F. (2010). Characteristics of enzyme-based hydrogen fuel cells using an oxygen-tolerant hydrogenase as the anodic catalyst. *Journal of Physical Chemistry C*, 114(27), 12003-12009.

[136] Ciaccafava, A., de Poulpiquet, A., Techer, V., Giudici-Orticoni, M. T., Tingry, S., Innocent, C., & Lojou, E. (2012). An innovative powerful and mediatorless H_2/O_2 biofuel cell based on an outstanding bioanode. *Electrochemistry Communications*, 23, 25-28.

[137] Infossi, P., Lojou, E., Chauvin, J. P., Herbette, G., Brugna, M., & Giudici-Orticoni, M. T. (2010). *Aquifex aeolicus* membrane hydrogenase for hydrogen bioxidation: role of lipids and physiological partners in enzyme stability and activity. *International Journal of Hydrogen Energy*, 35(19), 10778-10789.

[138] Reuillard, B., Le Goff, A., Agnès, A., Zebda, A., Holzinger, M., & Cosnier, S. (2012). Direct electron transfer between tyrosinase and multi-walled carbon nanotubes for bioelectrocatalytic oxygen reduction. *Electrochemistry Communications*, 20, 19-22.

[139] Durand, F., Kjaergaard, C., Suraniti, E., Gounel, S., Hadt, R., & Solomon, E. (2012). Mano N Bilirubin oxidase from *Bacillus pumilus*: A promising enzyme for the elaboration of efficient cathodes in biofuel cells. *Biosensors Bioelectronics*, 35(1), 140-146.

Chapter 20

Adsorption of Methylene Blue on Multi-Walled Carbon Nanotubes in Sodium Alginate Gel Beads

Fang-Chang Tsai, Ning Ma, Lung-Chang Tsai, Chi-Min Shu, Tao Jiang, Hung-Chen Chang, Sheng Wen, Chi Zhang, Tai-Chin Chiang, Yung-Chuan Chu, Wei-Ting Chen, Shih-Hsin Chen, Han-Wen Xiao, Yao-Chi Shu and Gang Chang

Additional information is available at the end of the chapter

http://dx.doi.org/10.5772/50714

1. Introduction

Surface water contamination by pollutants is common in highly industrialized countries due to direct discharge of industrial effluents into bodies of water or precipitation of air-borne pollutants into surface water [Murakamia et. al., 2008]. Dyes from the pollutants released along with industrial effluents are easily detected because of their inherently high visibility, meaning that concentrations as low as 0.005 mg/L can easily be detected and capture the attention of the public and the authorities [Ray et. al., 2003, Ray et. al., 2002]. Apart from the aesthetic problems caused by dyes, the greatest environmental concern with dyes is their absorption and reflection of sunlight entering the water, which interferes with the growth of bacteria, such that bacteria levels are insufficient to biologically degrade impurities in the water [Pierce, 1994. Ledakowicz et. al., 2001]. Methylene blue (MB) and methyl violet are two common dyes that have been shown to induce harmful effects on living organisms during short periods of exposure [Hameed et. al., 2009]. Oral ingestion of MB results in a burning sensation and may cause nausea, vomiting, diarrhea, and gastritis. The accidental consumption of large dose induces abdominal and chest pain, severe headache, profuse sweating, mental confusion, painful micturation, and methemoglobinemia [Yasemin et. al., 2006]. Inhalation of methyl violet may cause irritation to the respiratory tract, vomiting, diarrhea, pain, headaches, and dizziness; long-term exposure may cause damage to the mucous membranes and gastrointestinal tract [Allen & Koumanova, 2005]. The majority of dyes in this class are synthetic and usually composed of aromatic rings, which makes them carcinogenic and mutagenic [Ghosh & Bhattacharyya, 2002. Chen et. al., 2003]; they are inert and non-biodegradable when discharged into waste streams [Mittal & Gupta, 1996, Seshadri et. al., 1994]. With the social and economic development, the environmental consciousness of citizens and governing agencies was enhanced. Environmental pollution issues have garnered a considerable amount of attention throughout the world [Renmin et. al., 2005]. MB is a good representative of organic dyes that are difficult to degrade and substantially damage the environment due to their toxicity and dark color [Ho et. al., 2005].

Since carbon nanotube (CNT) was first discovered by S. Ijima in 1991, it has become an academic research subject of great interest [Olson, 1994]. CNT is the thinnest tubular structure humans can presently fabricate. It is lightweight and has high strength, high toughness, flexibility, high surface area, high thermal conductivity, and good electric conductivity and is chemically stable [Baughman et. al., 2002, Thostenson et. al., 2001, Banerjee et. al., 2005]. To fully exploit the superior mechanical, electrical and optical properties of multi-walled carbon nanotube (MWCNT), dispersion and adhesion to a polymeric matrix is a key issue [Iijima, 1991]. Both the dispersibility and matrix adhesion of MWCNT can be improved either by covalent or noncovalent functionalization. For covalent functionalization, several approaches studied, each having its advantages and drawbacks; examples of such methods include wet chemical methods with typical treatment times of up to 24 h [Sahoo et. al., 2010, Liu et. al., 1998, Chen et. al., 1998], treatment in air at elevated temperatures [Tsai et. al., 2010, Ajayan et. al., 1993], by ozone oxidation [Ago et. al., 1999, Mawhinney et. al., 2000] and treatment with low-pressure plasmas [Simmons et. al., 2006, Tseng et. al., 2006, Chen et. al., 2010, Zschoerper et. al., 2009].

Alginate is a collective term for a family of exopolysaccharides produced mainly by brown seaweeds. It has been widely used in the food, biomedical, pharmaceutical, and sewage-treatment industries, preferentially as sodium alginate due to its solubility in cold water. In molecular terms, alginate is composed of (1–4)-linked b-D-mannuronic acid (M) unbranched binary copolymer and a-L-guluronic acid (G) monomer residue, constituting M-, G-, and MG sequential block structures [Chen et. al., 2009]. Most applications that use alginate are based on its gel-forming ability through cation binding: the transition from water-soluble sodium alginate (SA) to water insoluble calcium alginate, for example. Divalent cations preferentially bind toward the G-block rather than the M-block [Moe et. al., 1995, Braccini et. al., 1999]. The composition of monomers and their sequential character (i.e., blackness) affects the gelatin behavior of alginate. In the presence of Ca^{2+}, G-rich samples generally form hard and brittle gels while M-rich samples from soft and elastic gels [Braccini & Perez, 2001, Courtois et. al., 1993, Thakur et. al., 1997, Pe´rez et. al., 1996]. The "egg-box" model has been accepted as a general model to describe gel formation [Morris et. al., 1978, Thom et. al., 1985]. Alginate is an excellent polymer flocculant and has been widely used in wastewater treatment.

This study reports for the first time the effect of the carboxylation method on CNT structure and property. The results can be used as reference for selecting the carboxylation method. Furthermore, the applicability of a new adsorbent, SA and MWCNT and the SA/MWCNT composite, for the sorption of MB dyes from an aqueous solution were investigated.

http://dx.doi.org/10.5772/50714

2. Experimental

2.1. Materials and methods

The SA, MB, and MWCNT were used as received from Fuchen Chemical Reagents Factory, Tianjin, China and Nanotechnologies Port Co., Ltd., Shenzhen, China. The MWCNTs was treated with a mixture of sulfuric and nitric acid under ultrasonic vibration, as seen in Table 1. According to the series reaction time in Table 1, the optimized ratio of the MWCNT to acid mixture is 3:1 by volume. Ultrasonic treatment was applied for the duration of varying reaction times. Filtration was conducted with a micropore filter and sand core filter. Pure de–ionized water (pure DI water) was used to rinse the filtrate until the pH of the aqueous solution was neutral. The compositions of the SA, MWCNT, and SA/MWCNT series specimens prepared in this study are summarized in Table 2. Ten milliliters of an aqueous solution of SA/MWCNT was added drop-wise to 50 mL of calcium chloride (10%, w/v) aqueous solution for 20 min, followed by the sampling of supernatant at the specified time intervals. The gel particles were pre-consolidated under a pressure of 8–30 kPa in a consolidation cell with an inner diameter of 2.0–3.0 mm to produce a packed gel bed to determine their expression characteristics. The schematic evolution of the SA and MWCNT in the microsphere, as a function of the calcium chloride, is shown in Figure 1. Other supplementary agents were of analytical grade (purity > 99.8 mass%) and all solutions were prepared with double distilled water.

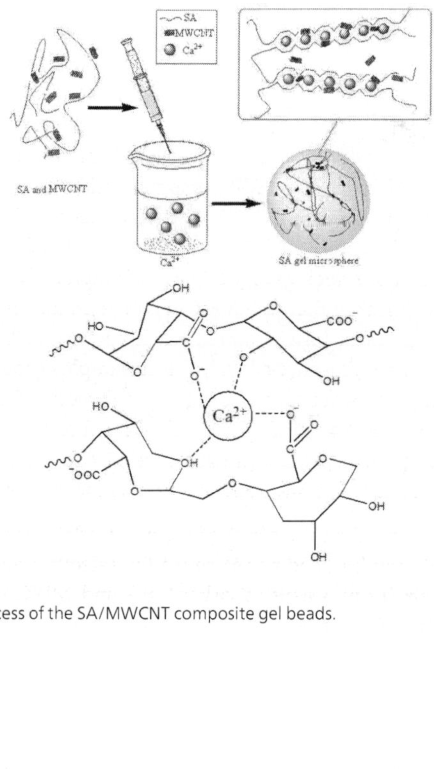

Figure 1. The preparation process of the SA/MWCNT composite gel beads.

A–series reaction group	Mixed-acid treatment time (h)	Hydrogen peroxide treatment time (h)	B–series reaction group	Mixed-acid treatment time (h)	Hydrogen peroxide treatment time (h)
A0	0	0	B0	0	0
A1	1	0	B1	1	0.5
A2	2	0	B2	2	1
A3	4	0	B3	4	2
A4	6	0	B4	6	3
A5	8	0	B5	8	4

Table 1. Formulation for CNT carboxylation.

Sample		SA (%, w/v)	MWCNT (%, w/v)	$CaCl_2$ (%, w/w)
0#	SA_0MWCNT_0	0	0	0
1#	SA_2MWCNT_0	2	0	10
2#	$SA_2MWCNT_{0.03}$	2	0.03	10
3#	$SA_2MWCNT_{0.06}$	2	0.06	10
4#	$SA_2MWCNT_{0.09}$	2	0.09	10
5#	$SA_2MWCNT_{0.12}$	2	0.12	10
6#	$SA_2MWCNT_{0.15}$	2	0.15	10

Table 2. The composites of SA/MWCNT series samples.

2.2. MWCNT dispersed polarity

Six small reagent bottles were filled with 6 mL pure DI water, 4 mL toluene and a small amount of MWCNT derived as shown in Table 1. They were ultrasonically treated for 0.5 h, and then, after the solution was stored for 12 h, they were recovered and observed.

2.3. Particle size analysis

The particle size analysis measurements of MWCNT and modified MWCNT series specimens were recorded using a Dandong Bettersize Instruments Ltd. BT-9300H at 25°C and 50% relative humidity, wherein six scans with a size range of 0.1–340 μm were collected during each data measurement. Particle size analysis samples of powder specimens were collected using approximately 15 mL pure DI water and a small amount of MWCNT derived as shown in Table 1.

http://dx.doi.org/10.5772/50714

2.4. Adsorption property

All sorption measurements were performed by batch type with 50 mL of MB solution in a shaking incubator to form a final concentration of 50 mg/L ($A_{665\ nm}$ = 2.9966) at room temperature for 3 h. The equilibrium MB concentration was measured by means of double beam ultraviolet–visible spectroscopy (Shanghai Precision & Scientific Instrument Co., Ltd, UV762, China), and the pH values of the solution were measured using a pH meter (Shanghai Yulong Instrument Co., Ltd., PHS-3 C, China) with a calomel and glass electrode (E201-9). The dye decolorization percentage was defined as follows:

$$\text{Decoloration percentage}(\%) = (A_0 - A) / A_0 \times 100\% \tag{1}$$

where A_0 is the dye absorbance of the control specimen, A is the dye absorbance of the reacted sample.

2.5. Electrical conductivity

To understand the electrical conductivity properties of MWCNT in SA specimens dispersed in MB solution, the electrical conductivity of the SA and SA/MWCNT solutions were measured at 25°C and 50% relative humidity using a conductivity meter (LIDA Instrument Factory, DDS-12A, China).

3. Results and discussion

3.1. Carbon nanotube dispersed polarity

A typical photograph of the polarity of MWCNT and modified MWCNT specimens is shown in Fig. 2. Fig. 2 shows the dispersion of the modified MWCNT in aqueous and organic solvent solutions after being exposed to the treatments highlighted in Table 1 and then left undisturbed for 12 h. The figure shows that in the six groups of MWCNT, except the unmodified carbon nanotube, there always exists an interface of two phases that cannot be dissolved in one another. All five of the other groups show different extents of dispersion. $MWCNT^{b3}$ shows the most stable dispersion in aqueous phase; even after being aged for a week, it still maintained the state seen in Fig. 2.

Figure 2. Photograph depicting the polarity of pure MWCNT specimens.

3.2. Particle Size Analysis

Particle size analysis was conducted on A–series and B–series of carboxylated MWCNT. Figs. 3 and 4 show that as carboxylation reaction time increases the extent of carbon nanotube shortening is increased; this is particularly true for the B–series, where the mixed-acid, hydrogen peroxide, and ultrasonic treatment times were all shortened. The B–series samples are much shorter than the A–series samples of MWCNT treated only with mixed-acid and ultrasonic treatment. This finding further corroborates the FT-IR results. With longer carboxylation reaction times the MWCNT is more severely damaged, inducing greater rupture on the C–C bond of the CNT. The higher activity at MWCNT ends facilitates bonding with free O and H from water or solution and the formation of carboxyl groups on the fracture site, increasing the functionalized carboxyl groups and the extent of MWCNT carboxylation.

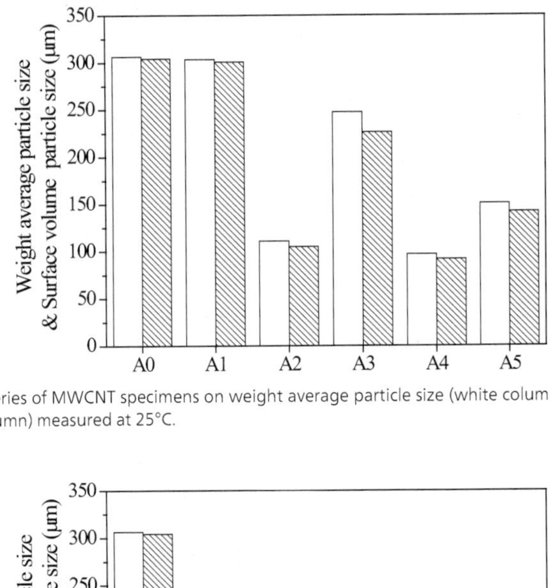

Figure 3. The A–series of MWCNT specimens on weight average particle size (white column) and surface volume particle size (slash column) measured at 25°C.

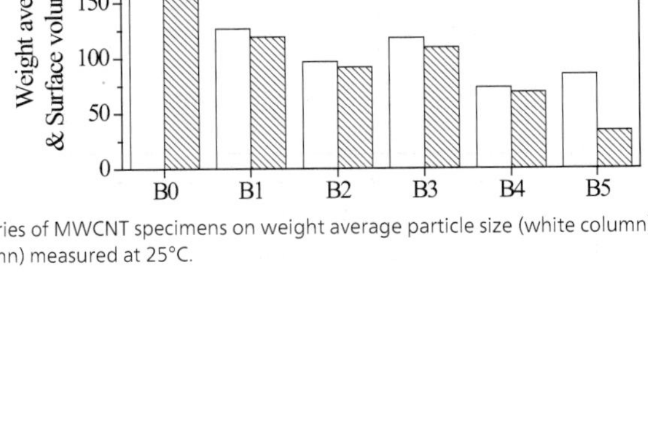

Figure 4. The B–series of MWCNT specimens on weight average particle size (white column) and surface volume particle size (slash column) measured at 25°C.

http://dx.doi.org/10.5772/50714

3.3. Adsorption property

As shown in Figs. 5, 6, and 7, the absorption of MB increased with time. The decolorization of sample 1 (SA_2MWCNT_0) reached 40.16% after 120 h, meaning that SA itself has the ability to absorb MB. Compared with sample #1, the decolorizations of 2#, 3#, 4#, 5#, and 6# ordered by increasing content of MWCNT, were 55.78, 66.62, 76.9, 82.06, and 83.46%, respectively, when tested under the same conditions. This trend of increased decolorization with increased MWCNT concentration is attributed to the fact that the surface of MWCNT has substantial amounts of carbonyl that reacted with MB (see Fig. 8). Another reason for this decolorization may be due to the large specific surface area of MWCNT that greatly affects adsorption ability. Voids present in the MWCNT may also favor of the adsorption of MB.

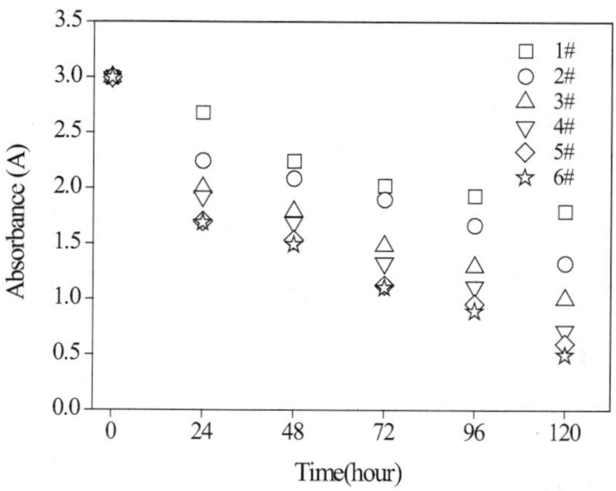

Figure 5. Absorbance of MB by SA/MWCNT microsphere, determined at 25°C.

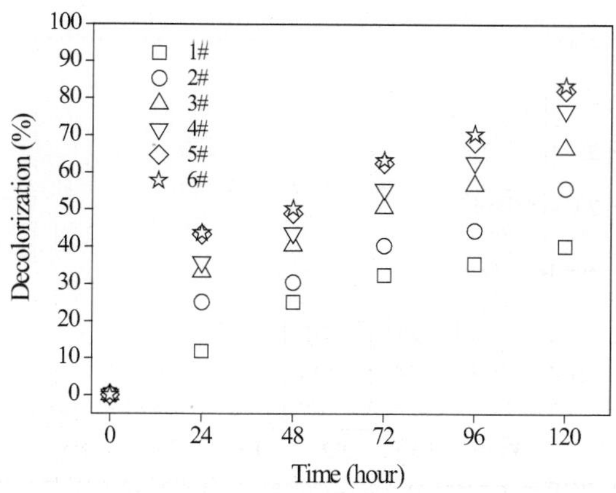

Figure 6. Decolorization of MB by SA/MWCNT microsphere determined at 25°C.

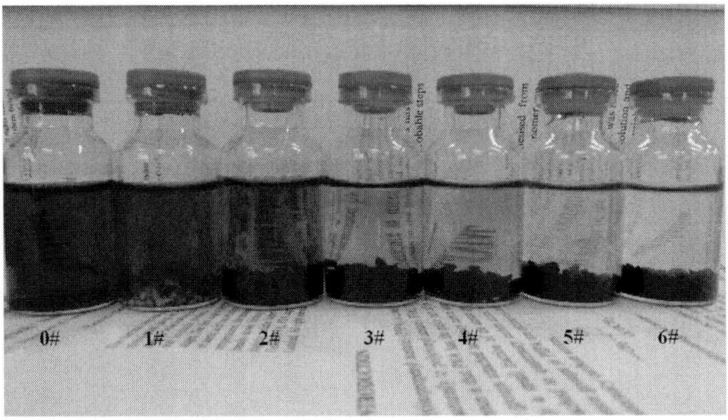

Figure 7. The photograph of MB absorbtion by different amounts of MWCNT.

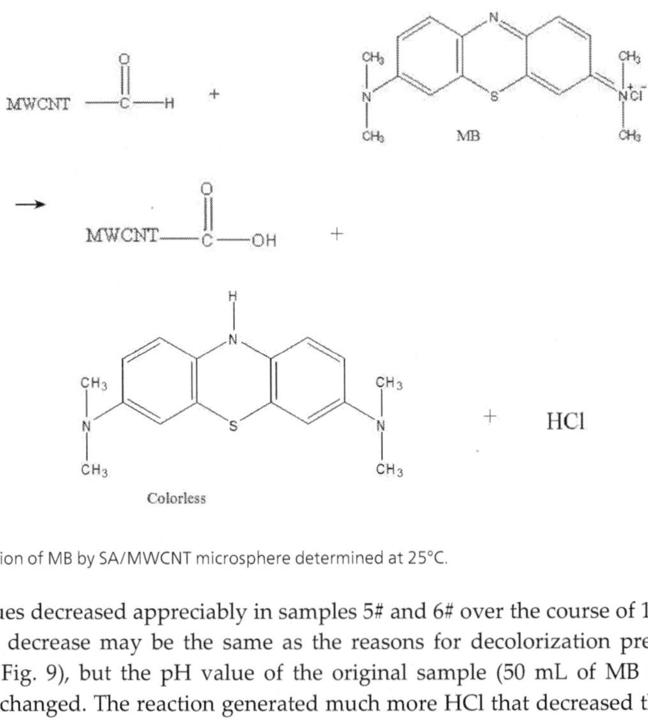

Figure 8. Reaction of MB by SA/MWCNT microsphere determined at 25°C.

The pH values decreased appreciably in samples 5# and 6# over the course of 120 h. The reason for this decrease may be the same as the reasons for decolorization previously mentioned (see Fig. 9), but the pH value of the original sample (50 mL of MB solution) was virtually unchanged. The reaction generated much more HCl that decreased the pH values, but the reaction rate eventually diminished after 120 h because there was less HCl generated and the adsorption of MWCNT surfaces was also nearly complete.

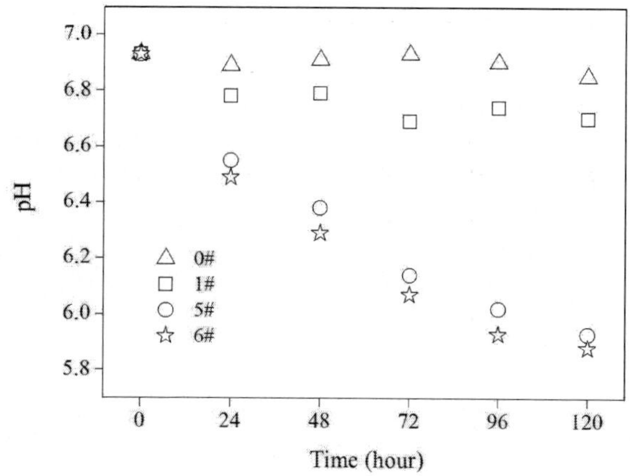

Figure 9. pH values of MB by SA/MWCNT microsphere determined at 25°C.

3.4. Electrical conductivity

The electrical conductivity was initially fixed at 79.3 μS/cm for the original sample, but increased considerably with increasing reaction time. The electrical conductivity of samples 5#, 6# increased sharply over the course of 120 h, but the electrical conductivity of the original sample (50 mL of MB solution) remained virtually unchanged, indicating that the original sample was stable.

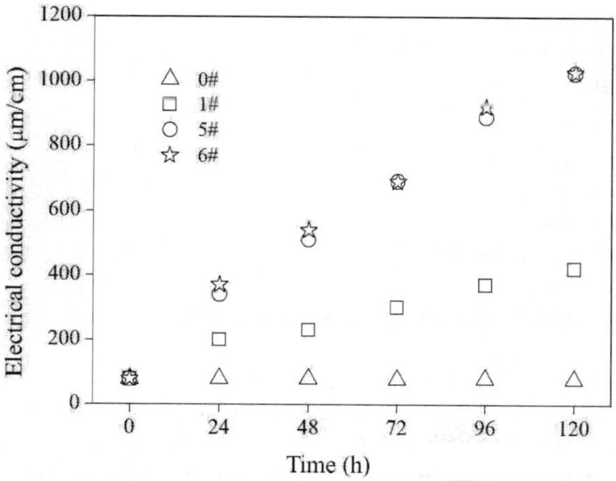

Figure 10. Electrical conductivity of MB by SA/MWCNT microsphere determined at 25°C.

4. Conclusions

This study effectively analyzed the adsorption of MB using gel beads prepared by sol–gel with SA and MWCNT. The formation conditions and mechanism of adsorption of the gel beads were also discussed. The decolorization of MB showed that the stability and reusability of SA/MWCNT could prove potentially advantageous in wastewater treatment.

Author details

Fang-Chang Tsai[1*], Ning Ma[1], Lung-Chang Tsai[2], Chi-Min Shu[2], Tao Jiang[1], Hung-Chen Chang[3], Sheng Wen[4], Chi Zhang[1], Tai-Chin Chiang[2], Yung-Chuan Chu[2], Wei-Ting Chen[2], Shih-Hsin Chen[5], Han-Wen Xiao[1], Yao-Chi Shu[6] and Gang Chang[1]

*Address all correspondence to: tfc0323@gmail.com

1 Ministry of Education Key Laboratory for the Green Preparation and Application of Functional Materials, Faculty of Materials Science and Engineering, Hubei University, P. R. China

2 Department of Safety, Health, and Environmental Engineering, National Yunlin University of Science and Technology, Douliou, Yunlin, Taiwan ROC

3 Department of Chemical and Materials Engineering, National Chin-yi University of Technology, Taichung, Taiwan ROC

4 Faculty of Chemistry and Materials Science, Hubei Engineering University, P. R. China

5 Department of Food Science, National I-Lan University, I-Lan, Taiwan ROC

6 Department of Cosmetic Applications & Management, Lee Ming Institute of Technology, Taipei, Taiwan ROC

References

[1] Ago, H., Kugler, T., Cacialli, F., Salaneck, W. R., Shaffer, M. S. P., & Windle, A. H. (1999). *J Phys Chem B*, 103, 8116-8121.

[2] Ajayan, P. M., Ebbesen, T. W., Ichihashi, T., Iijima, S., Tanagaki, K., & Hiura, H. (1993). *Nature*, 362, 522-523.

[3] Allen, S. J., & Koumanova, B. (2005). *J Univ Chem Technol Metall*, 40(3), 175-192.

[4] Banerjee, S., Hemraj-Benny, T., & Wong, S. S. (2005). *Adv. Mater*, 17, 17-29.

[5] Baughman, R. H., Zakhidov, A. A., & de Heer, W. A. (2002). *Science*, 297, 787-792.

[6] Braccini, I., & Perez, S. (2001). *Biomacromolecules*, 2, 1089-1096.

[7] Braccini, I., Grasso, R. P., & Perez, S. (1999). *Carbohydr Res*, 317, 199-130.

[8] Chen, C., Liang, B., Lu, D., Ogino, A., Wang, X., & Nagatsu, M. (2010). *Carbon*, 48, 939-948.

[9] Chen, C., Liang, B., Ogino, A., Wang, X., & Nagatsu, M. J. (2009). *J Phys Chem C*, 113, 7659-7665.

[10] Chen, J., Hamon, M. A., Chen, H. H. Y., Rao, A. M., Eklund, P. C., & Haddon, R. C. (1998). *Science*, 282, 95-98.

[11] Chen, K. C., Wu, J. Y., Huang, C. C., Liang, Y. M., & Hwang, S. C. J. J. (2003). *Biotechnol*, 101, 241-252.

[12] Courtois, J., Courtois, B., Heyraud, A., Colin-Morel, P., Dantas, L., Stadler, T., & David, P. (1993). *Agro-Food-Industry Hi- Tech.*, 6, 31-34.

[13] Ghosh, D., & Bhattacharyya, K. G. (2002). *Appl Clay Sci.*, 20, 295-300.

[14] Hameed, B. H., & Ahmad, A. A. (2009). *J. Hazard. Mater.*, 164, 870-875.

[15] Ho, Y. S., Chiang, T. H., & Hsueh, Y. M. (2005). *Proc Biochem*, 40(1), 119-124.

[16] Iijima, S. (1991). *Nature*, 354, 56-58.

[17] Ledakowicz, S., Solecka, M., & Zylla, R. J. (2001). *Biotechnol.*, 89, 175-184.

[18] Liu, J., Rinzler, A. G., Dai, H., Hafner, J. H., Bradley, R. K., & Boul, P. J. (1998). *Science*, 280, 1253-1256.

[19] Mawhinney, D. B., Naumenko, V., Kuznetsova, A., Yates, J. T., Jr., Liu, J., & Smalley, R. E. (2000). *J Am Chem Soc*, 122, 2383-2384.

[20] Mittal, A. K., & Gupta, S. K. (1996). *Water Sci Technol*, 34(10), 81-87.

[21] Moe, S. T., Draget, K. I., Skja, K. B. G., & Smidsrod, O. (1995). Marcel Dekker, New York, USA, 245-286.

[22] Morris, E. R., Rees, D. A., Thom, D., & Boyd, J. (1978). *Carbohydr Res*, 66, 145-154.

[23] Murakamia, M., Satob, N., Anegawac, A., Nakadad, N., Haradad, A., Komatsue, T., Takadaa, H., Tanakaf, H., Onoc, Y., & Furumaig, H. (2008). *Water Research*, 42, 2745-2755.

[24] Olson, T. M. (1994). *Water Research*, 6, 1383-1391.

[25] Pe´rez, S., Kouwijzer, M., Mazeau, K., & Engelsen, S. B. J. (1996). *Mol. Graph.*, 14, 307-321.

[26] Pierce, J. (1994). *J Soc Dyers Color*, 110, 131-134.

[27] Ray, S. S., & Okamoto, M. (2003). *Prog Polym Sci*, 28, 1539-1641.

[28] Ray, S. S., Yamada, K., Okamoto, M., & Ueda, K. (2002). *Polymer*, 44, 857-866.

[29] Renmin, G., Mei, L., Chao, Y., Yingzhi, S., & Jian, C. J. (2005). *Hazard. Mater*, 121, 247-250.

[30] Sahoo, N. G., Rana, S., Cho, J. W., Li, L., & Chan, S. H. (2010). *Prog Polym Sci*, 35, 837-867.

[31] Seshadri, S., Bishop, P. L., & Agha, A. M. (1994). *Waste Manage*, 15, 127-137.

[32] Simmons, J. M., Nichols, B. M., Baker, S. E., & Marcus, M. S. (2006). *J Phys Chem B*, 110, 7113-7118.

[33] Thakur, B. R., Singh, R. K., & Handa, A. K. (1997). *Crit. Rev. Food Sci. Nutr.*, 37, 47-73.

[34] Thom, D., Grant, G. T., Morris, E. R., & Rees, D. A. (1982). *Carbohydr Res*, 100, 29-42.

[35] Thostenson, E. T., Ren, Z., & Chou, T. W. (2001). *Compos. Sci.Technol*, 61, 1899-1912.

[36] Tsai, F. C., Shu, C. M., Tsai, L. C., Ma, N., Wen, Y., & Wen, S. (2011). Carbon nanotube industrial applications. In: Jose Mauricio Marulanda (Editor), *Carbon Nanotubes/ Book 2*, Rijeka, InTech-Open Access Publisher, 387-404.

[37] Tseng, C. H., Wang, C. C., & Chen, C. Y. (2006). *Nanotechnology*, 17, 5602-5606.

[38] Yasemin, B., & Haluk, A. (2006). *Desalination*, 194, 259-267.

[39] Zschoerper, N. P., Katzenmaier, V., Vohrer, U., Haupt, M., Oehr, C., & Hirth, T. (2009). *Carbon*, 47, 2174-2185.

Chapter 21

The Role of Carbon Nanotubes in Enhancement of Photocatalysis

Tawfik A. Saleh

Additional information is available at the end of the chapter

http://dx.doi.org/10.5772/51050

1. Introduction

The chemical, physical and mechanical properties of carbon nanotubes (CNTs) have stimulated extensive investigation since their discovery in the early 1990s (Iijima, 1991). CNTs, which are considered quasi-one dimensional nanostructures, are graphite sheets rolled up into cylinders with diameters of the order of a few nanometers and up to some millimeters in length. Types of nanotubes are the single-walled nanotubes (SWCNTs), double-walled nanotubes (DWCNTs) and the multi-walled nanotubes (MWCNTs). The MWCNTs consist of multiple layers of graphite arranged in concentric cylinders.

During the early stage, the primary research interests include the synthesis or growth of CNTs to prepare enough amounts of CNTs with desired dimension and purity. Several methods like arc discharge, laser ablation of graphite, the more productive chemical vapor deposition (CVD) and plasma enhanced CVD method, have been used to prepare high purity CNTs with controllable wall-thickness and length and acceptable price (Meyyappan, 2004). CNTs attract considerable attention due to their special structure and high mechanical strength which makes them to be good candidates for advanced composites. They can be either semiconducting, semimetallic or metallic, depending on the helicity and the diameter of the tube (Ebbesen et al., 1996; Yang et al., 2003). Based on the structure and shape, CNTs conduct electricity due to delocalization of the pi bond electrons. On the other side, researchers found that CNTs are efficient adsorbents due to their large specific surface area, hollow and layered structures and the presence of pi bond electrons on the surface. Besides that, more active sites can be created on the nanotubes. Thus, CNTs can be used as a promising material in environmental cleaning.

Photocatalytic oxidation using a semiconductor such as TiO_2, ZnO and WO_3 as photocatalyst is one of the advanced oxidation processes used for degradation of various pollutants in industrial wastewaters. As the semiconductor is illuminated with photons having energy content equal to or higher than the band gap, the photons excite valence band (VB) electrons across the band gap into the conduction band(CB), leaving holes behind in the valence band.Thus, there must be at least two reactions occurring simultaneously: oxidation from photogenerated holes, and reduction from photogenerated electrons.

The holes react with water molecules or hydroxide ions (OH^-) producing hydroxyl radicals ($^{\bullet}OH$). The generation of such radicals depends on the pH of the media. Targeted pollutants which are adsorbed on the surface of the catalyst will then be oxidized by $^{\bullet}OH$. On the other hand, the excited electrons (e^-) to the conduction band (CB) can generate hydroxyl radical ($^{\bullet}OH$) and can also react with O_2 and trigger the formation of very reactive superoxide radical ion ($O_2^{-\bullet}$) that can oxidize the target.

The band gap is characteristic for the electronic structure of a semiconductor and is defined as the energy interval (ΔE_g) between the VB and CB (Koci et al.,2011). VB is defined as the highest energy band in which all energy levels are occupied by electrons, whereas CB is the lowest energy band without electrons. The rate of a photo catalytic reaction depends on several parameters. First and most important is the type of the photo catalytic semiconductor. The second factor is the light radiation used or the stream of photons, as over supply of light accelerates electron–hole recombination (Koci et al.,2008). Third factor is pH of the medium with which the semiconductor surface is in contact with the targeted molecules. Fourth factor is the concentration of the substrate influencing the reaction kinetics. Fifth parameter is the temperature of the media where higher temperatures cause frequent collision between the semiconductor and the substrate (Koci et al.,2010).

The degradation rate can be enhanced by reducing the electron-hole recombination rate; preventing the particles agglomeration; and increasing the adsorption capacity, as it is a key process in the photocatalysis. In order to improve the photocatalytic efficiency, several methods have been investigated. This includes:

1. increasing the surface area of the metal oxide by synthesizing nano-size materials;
2. generation of defect structures to induce space-charge separation and thus reduce the recombination;
3. modification of the semiconductors with metal or other semiconductor; and
4. adding a co-sorbent such as silica, alumina, zeolite or clay (Yu et al. 2002; Rusu and Yates, 1997)

CNTs based composites have attracted considerable attentions due to the intrinsic properties that have been created owing to the addition of CNTs into the composite. Functionalization of CNTs, or attachment of individual atoms, molecules or their aggregates to CNTs, further extend the field of application of these nanosystems in different fields like in photocatalysis process (Dresselhaus & Dresselhaus, 2001; Burghard, 2005; Saleh, 2011). CNT/Metal oxide composites have been recently reported to be used for the treatment of contaminated water. In this chapter, therefore, the application of CNTs to enhance the photocatalytic activity of TiO_2, ZnO and WO_3 will be discussed.

http://dx.doi.org/10.5772/51050

2. Synthesis of carbon nanotube/catalyst composites

There are two main steps for the synthesis of CNT/catalyst nanocomposites. The first step is the grafting of oxygen-containing groups on the surface of the nanotubes and the second step is the attachment of the metal oxides on the active surface of the nanotubes.

2.1. Grafting of oxygen-containing groups on CNTs

Grafting of oxygen-containing groups on the surface of the nanotubes or activation of CNTs can be achieved by oxidation treatment. It can be performed using oxidizing agents such as nitric acid, sulfuric acid, or a mixture of both. For example, oxygen-containing groups can be grafted on the surface of the nanotubes by the following procedure. Initially, CNTs are dispersed by sonication in concentrated acidic media. Then, the mixture is treated by reflux while stirring vigorously at temperature of 100-120°C. After refluxing process, the mixture is allowed to cool at room temperature. The oxidized CNTs are purified by extraction from the residual acids by repeated cycles of dilution with distilled water, centrifugation and decanting the solutions until the pH is approximately 5-6. After the purification process, the oxidized CNTs are dried overnight in an oven at 100°C. After that, the dry oxidized CNTs are pulverized in a ball-mill.

The presence of oxygen containing groups on the surface of the oxidized nanotubes are characterized by the means of Fourier transform infrared spectroscopy (FT-IR), X-ray powder diffraction (XRD), field emission scanning electron microscopy (FESEM) and the transmission electron microscopy (TEM).

As an example, IR spectra, in the range of 400-4000 cm^{-1}, were recorded in KBr pellets using a Thermo Nicolet FT-IR spectrophotometer at room temperature. Samples were prepared by gently mixing 10 mg of each sample with 300 mg of KBr powder and compressed into discs at a force of 17 kN for 5 min using a manual tablet presser. Figure 1 depicts IR spectrum of oxidized MWCNTs. In the spectrum, a characteristic peak at 1580 cm^{-1} can be assigned to C=C bond in MWCNTs. The band at about 1160cm^{-1} is assigned to C–C bonds. Also, the spectrum shows the carbonyl characteristic peak at 1650 cm^{-1}, which is assigned to the carbonyl group from quinine or ring structure. More characteristic peak to the carboxylic group is the peak at 1720 cm^{-1} (Ros et al., 2002; Yang et al., 2005; Xia et al., 2007). The observation of IR spectra corresponding to the oxidized MWCNTs suggests the presence of carboxylic and hydroxylic groups on the nanotube surface.

Figure 2 depicts the typical XRD pattern of the oxidized MWCNTs. The strongest diffraction peak at the angle (2θ) of 25.5° can be indexed as the C(002) reflection of the hexagonal graphite structure (Rosca et al., 2005; Saleh et al., 2011; Lu et al., 2008). The sharpness of the peak at the angle (2θ) of 25.5° indicates that the graphite structure of the MWCNTs was acid-oxidized without significant damage since any decrease in the order of crystallinity in

MWCNTs will make the XRD peaks broader and shift the peak diffraction towards lower angles. The other characteristic diffraction peaks of graphite at 2θ of about 43°, 53° and 77° are associated with C(100), C(004) and C(110) diffractions of graphite, respectively.

Energy dispersive X-ray spectroscope (EDX) measurement is also used as a quantitative analysis for the presence of the oxygen containing groups on the surface of the nanotubes. Figure 4 represents the results of the oxidized MWCNTs. The results shows the presence of oxygen in the sample in addition to carbon element. SEM and TEM are used to characterize the morphology of the nanotube and to ensure that the structure of the nanotube has not been destroyed by the acid treatment. As an example, SEM image and the inset TEM image in Figure 3 confirm that there is no damage effect on the nanotubes using mixtures of nitric acid sulfuric acid for the treatment of the nanotubes.

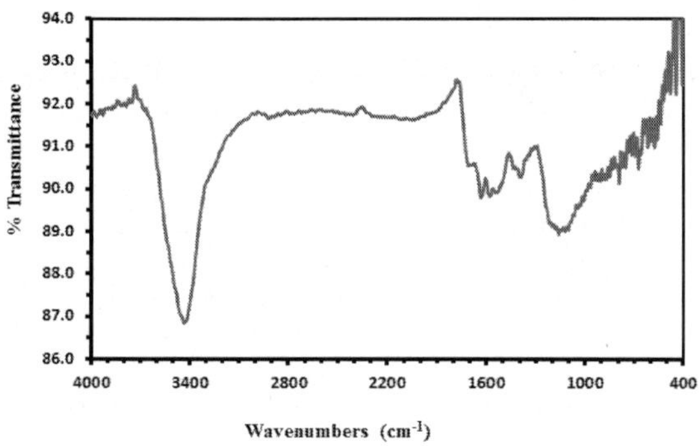

Figure 1. FTIR spectrum of MWCNT oxidized with H_2SO_4/HNO_3 mixture for 6h at 100°C.

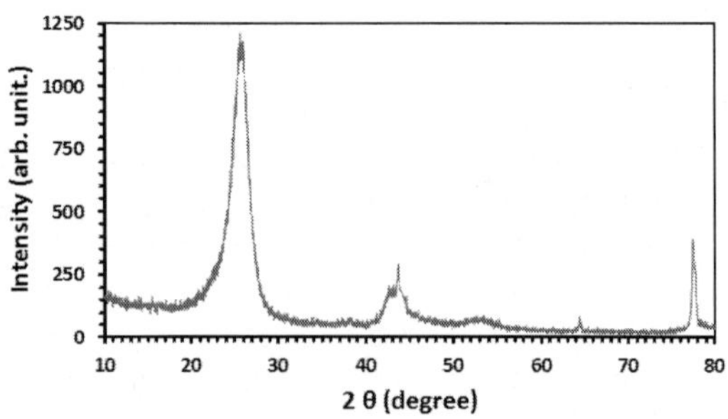

Figure 2. XRD patterns of MWCNT oxidized with H_2SO_4/HNO_3 mixture for 6h at 100°C.

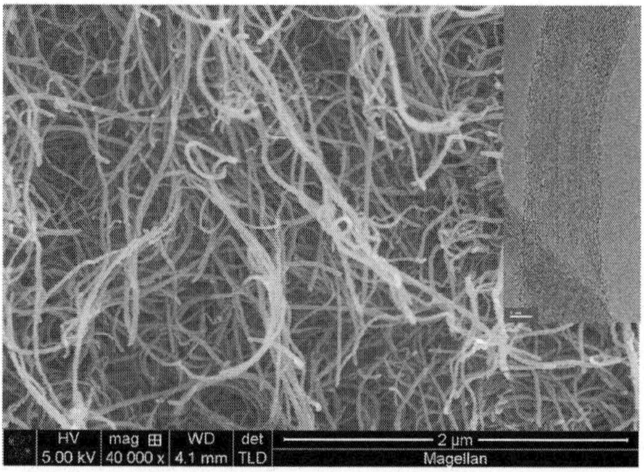

Figure 3. Field emission scanning electron microscopy (FESEM) image of the MWCNTs oxidized with H_2SO_4/HNO_3 mixture for 6h at 100°C; Inset is the transmission electron microscopy (TEM) image of the same.

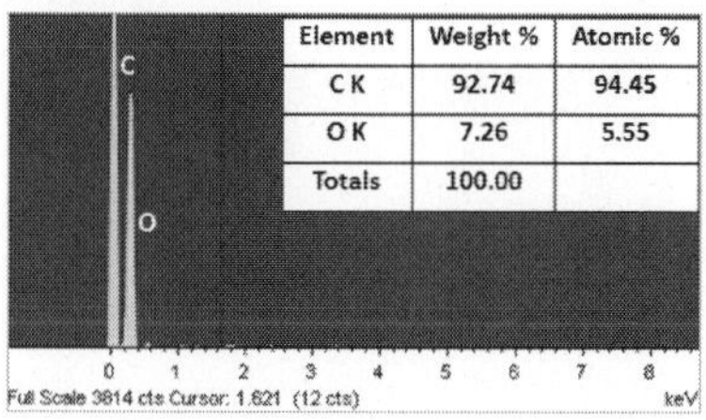

Figure 4. EDX spectrum of the MWCNTs oxidized with H_2SO_4/HNO_3 mixture for 6h at 100°C; inset is the table showing the percentage of each component in the nanotubes.

2.2. Synthesis of CNT/catalysts nanocomposites

CNT/metal oxide nanocomposites can be synthesized by different methods which fall into two basic classes. The first class involves the prior synthesis of nanoparticles that subsequently connected to surface functionalized CNTs by either covalent or noncovalent interactions (Eder, 2010; Peng et al., 2010; Hu et al., 2010). The second class is the one step method which involves direct deposition of nanoparticles onto MWCNT surface, in situ formation of nanoparticles through redox reactions or electrochemical deposition on CNTs (Chen et al., 2006; Gavalas et al., 2001; Yang et al., 2010; Sahoo et al., 2001; Lee et al., 2008). The second class has the advantages where uniform nanomaterials can be prepared due to the presence of active sites on oxidized CNT surfaces.

As an example, CNT/ZnO nanocomposites are prepared by the following procedure (Saleh et al., 2010). Zinc precursor like $Zn(NO_3)_2.6H_2O$, is dissolved in doubly deionized water. Then, ammonia is added drop-wise under continuous stirring into the solution to form a clear solution. The oxidized MWCNTs is added into the solution. The mixture is refluxed at 100°C. The composite are separated and dried at 80°C prior to the calcination in vacuum at 300°C.

Different techniques can be applied for the characterization of the nanocomposite. For example XRD is employed to determine crystalline phases and average crystalline size. FT-IR is used for qualitative analysis of the binding of the metal oxide into the nanotube surface. The morphology of the nanotubes and particle size are examined by the field emission scanning electron microscope (FESEM) and high resolution transmission electron microscopy (HRTEM). EDX measurement is also used as a quantitative analysis for the presence of the oxygen containing groups on the surface of the nanotubes. As an example, Figure 5 depicts the EDX data of the CNT/ZnO nanocomposite. The table shows the percentage of each component in the composite. Figure 6, SEM image and the inset HRSEM image, confirm the presence of zinc oxide particles on the surface of the nanotubes.

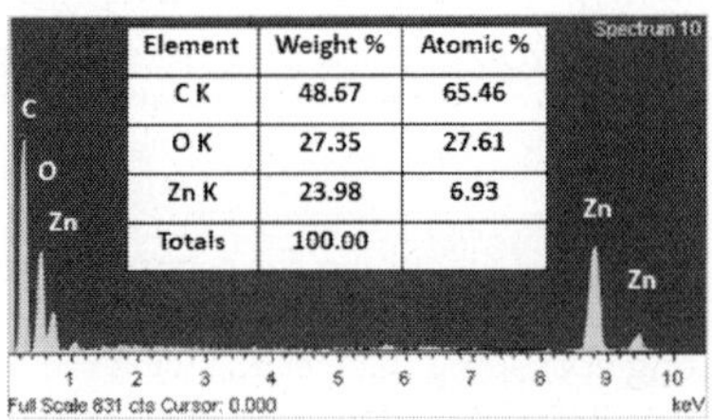

Figure 5. EDX spectrum of the MWCNT/ZnOnanocomposites; inset is the table showing the percentage of each component in the nanotubes.

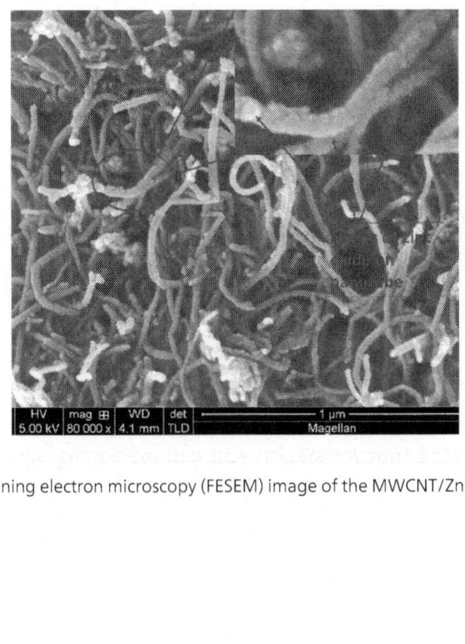

Figure 6. Field emission scanning electron microscopy (FESEM) image of the MWCNT/ZnO; Inset is the HRSEM image.

3. Applications of CNT/Catalyst nanocomposites

CNTs are considered to be good support material for catalysts, because they provide large surface area support and also stabilize charge separation by trapping electrons transferred from metal oxides, thereby hindering charge recombination.

A significant number of papers have been published on the application of CNTs in conjunction with TiO_2, reflecting the focus of recent research (Jitianu et al., 2004; Huang and Gao, 2003; Woan et al., 2009; Feng et al., 2005). One of the most important applications of such composite is to act as photocatalyst for some chemical reactions, especially for the decontamination of organic pollutants in waste waters. The photocatalytic activity of MWCNT/TiO_2 composite toward the degradation of acetone under irradiation of UV light was investigated by the detection of the hydroxyl radical (•OH) signals using electron paramagnetic resonance. It has been reported that the agglomerated morphology and the particle size of TiO_2 in the composites change in the presence of CNTs, which provide a large surface area resulting in more hydroxyl group on the surface of the composite with no effect on the mesoporous nature of the TiO_2. The composite have been reported to be of higher photocatalytic activity than commercial photocatalyst (P25) and TiO_2/activated carbon (AC) composite (Yu et al., 2005a,b).

The photocatalytic activities of MWCNT/TiO_2under visible light have also been reported using the decolorization of dyes like methylene blue, methyl orange, azo dye and other dyes in model aqueous solutions (Cong et al., 2011; Gao et al., 2009; Hu et al., 2007; Saleh and Gupta, 2012; Yu et al., 2005; Kuo, 2009). TiO_2 loading of 12% was found to result in the highest photoactivity in comparison with 6% and 15% loadings. Little TiO_2 or excessive nanotubes addition shielded the TiO_2 and reduced the UV intensity, due to photon scattering by the nanotubes. However, a high TiO_2 content was found to be ineffective in suppressing exciton recombination because of the large distance between the titania and the nanotubes (Li et al., 2012). Optimum ratio of titania and nanotubes provides a large surface area support and stabilize charge separation by trapping electrons transferred from TiO_2, thereby hindering charge recombination with minimum photon scattering. The composite provides high surface area which is beneficial for photocatalytic activity, as it provides high concentration of target organic substances around sites activated by ultraviolet (UV) radiation.

Also, the activity of MWCNT/TiO_2 composites has been investigated in photodegradation of phenol and photocatalytic oxidation of methanol under irradiation of visible light (Wang et al., 2005; An et al., 2007; Yao et al., 2008; Dechakiatkrai et al., 2007). The catalysts exhibited enhanced photocatalytic activity for degradation of toluene in gas phase under both visible and simulated solar light irradiation compared with that of commercial Degussa P25 (Wu et al., 2009). It exhibited high activity for the photoreduction of Cr(VI) in water (Xu et al., 2008). Its efficiency was higher compared to a mechanical mixture of TiO_2 and MWCNTs. A probable synergistic effect of TiO_2 and MWCNTs in a composite MWCNT/TiO_2 on the enhancement of visible light performance, have been proposed where MWCNTs act as support, absorbent, photo-generated transfer station and carbon-doping source to narrow the band gap of TiO_2.

The composite has been reported for photoinactivation of E. coli in visible light irradiation (Akhavan et al., 2009). The efficiency of the nanocomposite was high toward photocatalytic hydrogen generation and for the reduction of CO_2 with H_2O (Dai et al., 2009; Xia et al., 2007).

Zinc oxide, a direct wide band gap (3.37 eV) semiconductor with a large excitation binding energy (60 meV), has been investigated as a potential non-toxic photocatalyst used to successfully degrade organic pollutants. Recently, ZnO nanoparticles have received much attention due to its high photoactivity in several photochemical, UV light response, photoelectron-chemical processes and its low cost production possibility (Wu et al., 2008; Neudeck et al., 2011; Gondal et al., 2010; Drmosh et al., 2010).

Experimental results proved that CNT/ ZnO nanocomposites display relatively higher photocatalytic activity than ZnO nanoparticles for the degradation of some dyes like rhodamine B, azo-dyes, methylene blue, methylene orange (Dai et al., 2012; Zhu et al., 2009). The complete removal of azo-dyes such as acid orange, acid bright red, acid light yellow, after selection of optimum operation parameters such as the illumination intensity, catalyst amount, initial dye concentration and the different structures of the dye on the photocatalytic process, can be achieved in relatively short time by using CNT/ZnO composites.

The MWCNT/ZnO nanocomposites exhibits excellent photocatalytic activity toward other pollutants such as acetaldehyde and cyanide in model solutions (Saleh et al., 2011; Saleh et al., 2010). CNTs act as a photogenerated electron acceptor and retard the recombination of photoinduced electron and hole. The adsorption and photocatalytic activity tests indicate that the CNTs serve as both an adsorbent and a visible light photocatalyst. The experimental results show that the photocatalytic activity of the ZnO/MWCNTs nanocomposites strongly depends on the synthetic route, which is probably due to the difference of surface states resulted from the different preparation processes (Zhang, 2006; Kim and Sigmund, 2002; Jiang and Gao, 2005;Agnihotri et al., 2006).

CNT/WO_3 nanocomposites have been synthesized via different routs (Pietruszka et al., 2005; Wang et al., 2008; Saleh and Gupta, 2011). The utilization of carbon nanotubes to enhance photocatalytic activity of tungsten trioxide has also been investigated. The photocatalytic activities are greatly improved when CNT/WO_3 nanocomposite has been used for the degradation of pollutants such as rhodamine B under ultraviolet lamp or under sunlight. The results showed that photocatalytic activity of the MWCNT/WO_3 composites prepared by chemical process is higher than that prepared by mechanical mixing. The photocatalytic activity is enhanced when WO_3 nanoparticles are loaded on the surface of CNTs. The enhanced photocatalytic activity may be ascribed to the effective electron transfer between the nanotubes and the metal nanoparticles.

A possible synergistic effect between the semiconductor nanoparticles and CNTs on the enhancement of photocatalytic activity is proposed in Figure 7. The mechanism is based on the results of the structure characterizations and the enhancement in photocatalytic activity of the prepared composite.

When the catalyst is irradiated by photons, electrons (e^-) are excited from the valence band (VB) to the conduction band (CB) of catalysts or the metal oxide nanoparticles (NP) creating a charge vacancy or holes (h^+), in the VB. Some of the charges quickly recombine without creating efficient photodecomposition of the pollutant. In the case where the composite is applied, the strong interaction between the nanotube and the metal oxide results in a close contact to form a barrier junction which offers an effective route of reducing electron-hole recombination by improving the injection of electrons into the nanotube. Therefore, CNTs acts as a photo-generated electron acceptor to promote interfacial electron transfer process since CNTs are relatively good electron acceptor while the semiconductor is an electron donor under irradiation (Saleh and Gupta, 2011; Riggs et al., 2000; Subramanian et al., 2004; Geng et al., 2008). The adsorbed oxygen molecules on the nanotubes react with the electrons forming very reactive superoxide radical ion ($O2^{\bullet -}$) which oxidize the target. On the other side, the hole (h^+) oxidize hydroxyl groups to form hydroxyl radical ($^{\bullet}OH$) which can decompose the target.

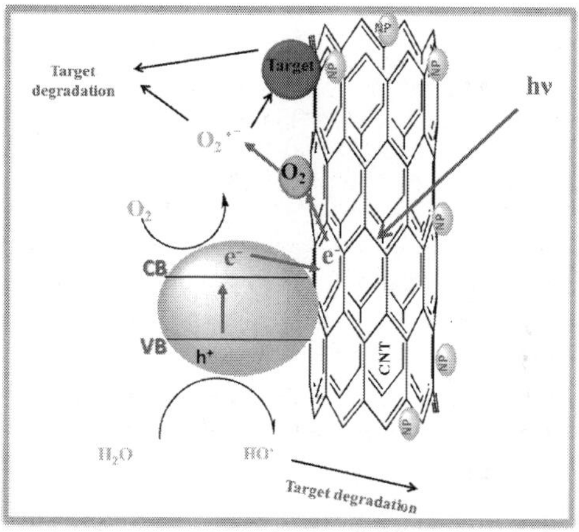

Figure 7. Schematic diagram of the proposed mechanism of photodegradation over CNT/MO composite.

Some important points of such process can be highlighted as:

- Stronger adsorption on photocatalyst for the targeted moleculs of pollutant is achieved by the incorporation of the nanotubes, due to their large specific surface area and high quality active sites.
- The nanotubes can act as effective electron transfer unitbecause of their high electrical conductivity and high electron storage capacity.
- The nanotubes manifest higher capture electron ability and can prompt electron transfer from the conduction band of the metal oxide or semiconductor nanoparticles (NP) towards the nanotube surface due to their lower Fermi level (Cong et al., 2011).
- Schottky barrier forms at the interface between the CNTs and the semiconductor. The photo-generated electrons may move freely towards the CNT surface, thus the left holes may move to the valence band (Woan et al., 2009; Chen et al., 2005).

- The presence of the nanotubes in the composite can inhibit the recombination of photo-generated electrons and holes, thus, improving the photocatalytic activity.
- The transmission stability of promoted electron between the nanotubes and the conduction band is enhanced by the strong interaction and intimate contact between the nanoparticles and the surface of the nanotubes.

4. Conclusion

The chapter discusses the preparation of the nanocomposites consisting of carbon nanotubes and metal oxides like titania, zinc oxide and tungsten trioxide. For the preparation of such composite, the oxygen-containing groups are grafted on the surface of the nanotubes by acid treatment. This is followed by the attachment of the metal oxides nanoparticles on the nanotubes surface. The chapter also highlights the means by which the composite is characterized. These include Fourier transform infrared spectroscope, X-ray powder diffraction, field emission scanning electron microscope, energy dispersive X-ray spectroscope and transmission electron microscope.

The UV, visible light and sunlight photocatalytic activity of the CNT-based nanocomposites is higher than that of the metal oxide or mechanical mixture of the metal oxide and CNTs. CNTs are considered to be good support materials for semiconductors like TiO_2, ZnO and WO_3 because nanotubes provide a large surface area support with high quality active sites. Also they stabilize charge separation by trapping electrons, thereby hindering electron–hole recombination by modification of band-gap and sensitization.

Acknowledgements

The author would like to acknowledge the support of Chemistry Departments, Center of Research Excellence in Nanotechnology & King Fahd University of Petroleum and Minerals, (KFUPM) Dhahran, Saudi Arabia, for this work.

Author details

Tawfik A. Saleh*

Address all correspondence to: tawfik@kfupm.edu.sa

Chemistry Department, Center of Excellence in Nanotechnology, King Fahd University of Petroleum & Minerals, Saudi Arabia

References

[1] Agnihotri, S., Mota, J. P. B., Rostam-Abadi, M., & Rood, M. J. (2006). Adsorption site analysis of impurity embedded single-walled carbon nanotube bundles. *Carbon*, 44(1), 2376-2383.

[2] Akhavan, O., Abdolahad, M., Abdi, Y., & Mohajerzadeh, S. (2009). Synthesis of titania/carbon nanotube heterojunction arrays for photoinactivation of E. coli in visible light irradiation. *Carbon*, 47(14), 3280-3287.

[3] An, G., Ma, W., Sun, Z., Liu, Z., Han, B., Miao, S., Miao, Z., & Ding, K. (2007). Preparation of titania/carbon nanotube composites using supercritical ethanol and their photocatalytic activity for phenol degradation under visible light irradiation. *Carbon*, 45, 1795-1801.

[4] Burghard, M. (2005). Electronic and vibrational properties of chemically modified singlewall carbon nanotubes. *Surf. Sci. Rep.*, 0167-5729, 58(1-4), 1-109.

[5] Chen, W., Pan, X., Willinger, M. G., Su, D. S., & Bao, X. (2006). Facile autoreduction of iron oxide/carbon nanotube encapsulates. *J. Am. Chem. Soc.*, 128, 3136-3137.

[6] Chen, Y., Crittenden, J. C., Hackney, S., Sutter, L., & Hand, H. W. (2005). Preparation of a Novel TiO2-Based p–n Junction Nanotube Photocatalyst. *Environ. Sci. Technol.*, 39(5), 1201-1208.

[7] Cong, Y., Li, X., Qin, Y., Dong, Z., Yuan, G., Cui, Z., & Lai, X. (2011). Carbon-doped TiO_2 coating on multiwalled carbon nanotubes with higher visible light photocatalytic activity. *Applied Catalysis B: Environmental*, 107, 128-134.

[8] Dai, K., Dawson, G., Yang, S., Chen, Z., & Lu, L. (2012). Large scale preparing carbon nanotube/zinc oxide hybrid and its application for highly reusable photocatalyst. *Chemical Engineering Journal*, 191, 571-578.

[9] Dai, K., Peng, T., Ke, D., & Wei, B. (2009). Photocatalytic hydrogen generation using a nanocomposite of multi-walled carbon nanotubes and TiO_2 nanoparticles under visible light irradiation. *Nanotechnology*, 20(12), 125603.

[10] Dechakiatkrai, C., Chen, J., Lynam, C., Phanichphant, S., & Wallace, G. G. (2007). Photocatalytic oxidation of methanol using titanium dioxide/single-walled carbon nanotube composite. *J Electrochem Soc.*, 154(5), A407-411.

[11] Dresselhaus, M. S., & Dresselhaus, G. (2001). Carbon Nanotubes: Synthesis, Structure, Properties and Applications: Topics in Applied Physics, Springer-Verlag. ISBN 3-54041-086-4, Berlin.

[12] Drmosh, Q. A., Gondal, M. A., Yamani, Z. H., & Saleh, T. A. (2010). Spectroscopic Characterization Approach to Study Surfactants Effect On ZnO Nanoparticles Synthesis by Laser Ablation Process. *Applied Surface Science*, 256, 4661-4666.

[13] Ebbesen, T. W., Lezec, H. J., Hiura, H., Bennett, J. W., Ghaemi, H. F., & Thio, T. (2010). Electrical conductivity of individual carbon nanotubes. *Nature*, 382, 6586 (1996) 54-56.

[14] Eder, D. Carbon nanotube-inorganic hybrids. *Chem. Rev.*, 110, 1348-1385.

[15] Feng, W., Feng, Y., Wu, Z., Fujii, A., Ozaki, M., & Yoshino, K. (2005). Optical and electrical characterizations of nanocomposite film of titania adsorbed onto oxidized multiwalled carbon nanotubes. *J Phys Condens Matter*, 17(27), 4361-4368.

[16] Gao, B., Chen, G. Z., & Li, P. G. (2009). Carbon nanotubes/titanium dioxide (CNTs/TiO_2) nanocomposites prepared by conventional and novel surfactant wrapping sol-gel methods exhibiting enhanced photocatalytic activity. *Appl Catal*, B89(3-4), 503-509.

[17] Gavalas, V. G., Andrews, R., Bhattacharyya, D., & Bachas, L. G. (2001). Carbon nanotube sol-gel composite materials. *Nano Lett.*, 1, 719-721.

[18] Geng, Q., Guo, Q., Cao, C., & Wang, L. (2008). Investigation into NanoTiO_2/ACSPCR for Decomposition of Aqueous Hydroquinone. *Ind. Eng. Chem. Res.*, 47, 2561-2568.

[19] Gondal, M. A., Drmosh, Q. A., Yamani, Z. H., & Saleh, T. A. (2010). Effect of post-annealing temperature on structural and optical properties of nano-ZnO synthesized from ZnO_2 by Laser Ablation Method. *International Journal of NanoParticles*, 3(3), 257-266.

[20] Hu, G., Meng, X., Feng, X., Ding, Y., Zhang, S., & Yang, M. (2007). Anatase TiO_2 nanoparticles/carbon nanotubes nanofibers: preparation, characterization and photocatalytic properties. *J Mater Sci*, 42(17), 7162-7170.

[21] Hu, L., Hecht, D. S., & Gruner, G. (2010). Carbon nanotube thin films: fabrication, properties, and applications. *Chem. Rev.*, 110, 5790-5844.

[22] Huang, Q., & Gao, L. (2003). Immobilization of rutile TiO_2 on multiwalled carbon nanotubes. *J Mater Chem*, 13(7), 1517-9.

[23] Iijima, S. (1991). Helical Microtubules of Graphitic Carbon. *Nature*, 354, 56-58.

[24] Jiang, L., & Gao, L. (2005). Fabrication and characterization of ZnO-coated multi-walled carbon nanotubes with enhanced photocatalytic activity. *Mater. Chem. Phys.*, 91, 313-316.

[25] Jitianu, A., Cacciaguerra, T., Benoit, R., Delpeux, S., Beguin, F., & Bonnamy, S. (2004). Synthesis and characterization of carbon nanotubes-TiO_2 nanocomposites. *Carbon*, 42(5-6), 1147-1151.

[26] Kim, H., & Sigmund, W. (2002). Zinc oxide nanowires on carbon nanotubes. *Appl. Phys. Lett.*, 81, 2085-2088.

[27] Koci, K., Mateju, K., Obalova, L., Krejcikova, S., Lacny, Z., Placha, D., Capek, L., Hospodkova, A., & Solcova, O. (2010). Effect of silver doping on the TiO_2 for Photocatalytic reduction of CO_2 . *Applied Catalysis B: Environmental*, 96, 239-244.

[28] Koci, K., Obalová, L., & Lacný, Z. (2008). Photocatalytic reduction of CO_2 over TiO_2 based catalysts. *Chemical Papers*, 62, 1-9.

[29] Koci, K., Reli, M., Kozák, O., Lacny, Z., Plachá, D., Praus, P., & Obalov, L. (2011). Influence of reactor geometry on the yield of CO2 Photocatalytic reduction. *Catalysis Today*, 176(1), 212-214.

[30] Kuo, C-Y. (2009). Prevenient dye-degradation mechanisms using UV/TiO_2/carbon nanotubes process. *J Hazard Mater*, 163(1), 239-244.

[31] Lee, D. H., Park, J. G., Choi, K. J., Choi, H. J., & Kim, D. W. (2008). Preparation of brookite-Type TiO_2/carbon nanocomposite electrodes for application to Li ion batteries. *Eur. J. Inorg. Chem.*, 6, 878-882.

[32] Li, Y., Leiyong, Li., Chenwan, Li., Chen, W., & Zeng, M. (2012). Carbon nanotube/titania composites prepared by a micro-emulsion method exhibiting improved photocatalytic activity. *Applied Catalysis A: General*, 427(428), 1-7.

[33] Lu, C., Su, F., & Hu, S. (2008). Surface modification of carbon nanotubes for enhancing BTEX adsorption from aqueous solutions. *Applied Surface Science*, 254, 7035-7041.

[34] Meyyappan, M. (2004). (Ed.), *Carbon Nanotubes: Science and Applications*, CRC Press, 0-84932-111-5.

[35] Neudeck, C., Kim, Y. Y., Ogasawara, W., Shida, Y., Meldrum, F., & Walsh, D. (2011). General route to functional metal oxide nanosuspensions, enzymatically deshelled nanoparticles, and their application in photocatalytic water splitting. *Small*, 7, 869-873.

[36] Peng, X., Sfeir, M. Y., Zhang, F., Misewich, J. A., & Wong, S. S. (2010). Covalent synthesis and optical characterization of double-walled carbon nanotube-nanocrystal heterostructures. *J. Phys. Chem. C*, 114, 8766-8773.

[37] Pietruszka, B., Gregorio, F. D., Keller, N., & Keller, V. (2005). High-efficiency WO3/carbon nanotubes for olefin skeletal isomerization. *Catal. Today*, 102-103, 94-100.

[38] Riggs, J. E., Guo, Z., Carroll, D. L., & Sun, Y. P. (2000). Strong Luminescence of Solubilized Carbon Nanotubes. *J. Am. Chem. Soc.*, 122, 5879-5880.

[39] Ros, T. G., van Dillen, A. J., Geus, J. W., & Koningsberger, D. C. (2002). Surface oxidation of carbon nanofibres. *Chem Eur J*, 8, 1151-1162.

[40] Rosca, I. D., Watari, F., Uo, M., & Akasaka, T. (2005). Oxidation of multiwalled carbon nanotubes by nitric acid. *Carbon*, 43, 3124-31.

[41] Rusu, C. N., & Yates Jr, J. T. (1997). Defect sites on TiO2(110). Detection by O2 photodesorption. *Langmuir*, 13(16), 4311-4316.

[42] Sahoo, S., Husale, S., Karna, S., Nayak, S. K., & Ajayan, P. M. (2011). Controlled assembly of Ag nanoparticles and carbon nanotube hybrid structures for biosensing. *J. Am. Chem. Soc.*, 133, 4005-4009.

[43] Saleh, T. A. (2011). The influence of treatment temperature on the acidity of MWCNT oxidized by HNO3 or a mixture of HNO3/H2SO4. *Applied Surface Science*, 257, 17 June, 7746-7751.

[44] Saleh, T. A., & Gupta, V. K. (2011). Functionalization of tungsten oxide into MWCNT and its application for sunlight-induced degradation of rhodamine B. *Journal of Colloid and Interface Science*, 362(2), 337-344.

[45] Saleh, T. A., & Gupta, Vinod. K. (2012). Photo-catalyzed degradation of hazardous dye methyl orange by use of a composite catalyst consisting of multi-walled carbon nanotubes and titanium dioxide. *Journal of Colloid and Interface Science*, 371(1), 101-106.

[46] Saleh, T. A., Gondal, M. A., Drmosh, Q. A., Z Yamani, H. A., & AL-yamani, A. (2011). Enhancement in photocatalytic activity for acetaldehyde removal by embedding ZnO nano particles on multiwall carbon nanotubes. *Chemical Engineering Journal*, 166(1), 407-412.

[47] Saleh, T. A., Gondal, M. A., & Drmosh, Q. A. (2010). Preparation of a MWCNT/ZnO nanocomposite and its photocatalytic activity for the removal of cyanide from water using a laser. *Nanotechnology*, 21(49), 8, doi:10.1088/0957-4484/21/49/495705.

[48] Subramanian, V., Wolf, E., & Kamat, P. V. (2004). Catalysis with TiO2/Gold Nanocomposites. Effect of Metal Particle Size on the Fermi Level Equilibration. *J. AM. CHEM. SOC.*, 126, 4943-4950.

[49] Wang, S., Xiaoliang, Shi., Gangqin, Shao., Xinglong, Duan., Hua, Yang., & Tianguo, Wang. (2008). Preparation, characterization and photocatalytic activity of multi-walled carbon nanotube-supported tungsten trioxide composites. *Journal of Physics and Chemistry of Solids*, 69, 2396-2400.

[50] Wang, W. D., Serp, P., Kalck, P., & Faria, J. L. (2005). Visible light photodegradation of phenol on MWCNT-TiO_2 composite catalysts prepared by a modified sol-gel method. *J Mol Catal A: Chem*, 235(1-2), 194-9.

[51] Woan, K., Pyrgiotakis, Georgios., & Sigmund, Wolfgang. (2009). Photocatalytic Carbon-Nanotube-TiO2 Composites. *Advanced Materials*, 21(21), 2233-2239.

[52] Wu, X., Jiang, P., Cai, W., Bai, X. D., Gao, P., & Xie, S. S. (2008). Hierarchical ZnO micro-/nanostructure film. *Adv. Eng. Mater.*, 10, 476-481.

[53] Wu, Z., Fan, Dong., Weirong, Zhao., Haiqiang, Wang., Yue, Liu., & Baohong, Guan. (2009). The fabrication and characterization of novel carbon doped TiO_2 nanotubes, nanowires and nanorods with high visible light photocatalytic activity. *Nanotechnology*, 20(23), 235701.

[54] Xia, W., Wang, Y., Bergstraberr, R., Kundu, S., & Muhler, M. (2007). Surface characterization of oxygen-functionalized multi-walled carbon nanotubes by high-resolution X-ray photoelectron spectroscopy and temperature-programmed desorption. *Applied Surface Science*, 254, 247-250.

[55] Xia, X-H., Jia, Z-J., Yu, Y., Liang, Y., Wang, Z., & Ma, L-L. (2007). Preparation of multi-walled carbon nanotube supported TiO_2 and its photocatalytic activity in the reduction of CO_2 with H_2O. *Carbon*, 45(4), 717-21.

[56] Xu, Z., Long, Y., Kang, S-Z., & Mu, J. (2008). Application of the composite of TiO_2 nanoparticles and carbon nanotubes to the photoreduction of Cr(VI) in water. *J Dispersion Sci Technol*, 29(8), 1150-2.

[57] Yang, C., Wohlgenannt, M., Vardeny, Z. V., Blau, W. J., Dalton, A. B., Baughman, R., & Zakhidov, A. A. (2003). Photoinduced charge transfer in poly(p-phenylene vinylene) derivatives and carbon nanotube/C60 composites. *Physica B: Condensed Matter*, 338(1-4), 366-369.

[58] Yang, D.-Q., Rochette, J-F., & Sacher, E. (2005). Functionalization of multiwalled carbon nanotubes by mild aqueous sonication. *J Phys Chem B*, 109, 7788-7794.

[59] Yang, J., Jiang, L. C., Zhang, W. D., & Gunasekaran, S. (2010). Highly sensitive nonenzymatic glucose sensor based on a simple two-step electrodeposition of cupric oxide (CuO) nanoparticles onto multi-walled carbon nanotube arrays. *Talanta*, 82, 25-33.

[60] Yao, Y., Li, G., Ciston, S., Lueptow, R. M., & Gray, K. A. (2008). Photoreactive TiO_2/carbon nanotube composites: synthesis and reactivity. *Environ Sci Technol*, 42(13), 4952-7.

[61] Yu, J. C., Zhang, L., & Yu, J. (2002). Rapid synthesis of mesoporous TiO_2 with high photocatalytic activity by ultrasound-induced agglomeration. *New Journal of Chemistry*, 26(4), 416-420.

[62] Yu, Y., Yu, J. C., Chan, C-Y., Che, Y-K., Zhao, J-C., & Ding, L. (2005a). Enhancement of adsorption and photocatalytic activity of TiO2 by using carbon nanotubes for the treatment of azo dye. *Appl Catal B*, 61(1-2), 1-11.

[63] Yu, Y., Yu, Jimmy C., Yu, Jia-Guo., Kwok, Yuk-Chun., Che, Yan-Ke., Zhao, Jin-Cai., Ding, Lu, Ge, Wei-Kun, & Wong, Po-Keung. (2005b). Enhancement of photocatalytic activity of mesoporous TiO2 by using carbon nanotubes. *Applied Catalysis A: General*, 289, 186-196.

[64] Zhang, W. D. (2006). Growth of ZnO nanowires on modified well-aligned carbon nanotube arrays. *Nanotechnology*, 17, 1036-1040.

[65] Zhu, L.-P., Liao, Gui-Hong, Huang, Wen-Ya, Ma, Li-Li, Yang, Yang, Yu, Ying, & Fu, Shao-Yun. (2009). Preparation, characterization and photocatalytic properties of ZnO-coated multi-walled carbon nanotubes. *Materials Science and Engineering B*, 163, 194-198.

Chapter 22

Carbon Nanotubes for Energy Applications

Dennis Antiohos, Mark Romano, Jun Chen and
Joselito M. Razal

Additional information is available at the end of the chapter

http://dx.doi.org/10.5772/51784

1. Introduction

1.1. The energy problem

The energy crisis during the 1970s sparked the development of renewable energy sources and energy conservation measures. As supply eventually met demand, these programs were scaled back. Ten years later, the hazards of pollution led to work on minimisation and reversal of the environmental impact of fossil fuel extraction, transport and consumption [1]. The United States Department of Energy predicts that 20 years from now, the world's energy consumption will increase by 20% (Figure 1). The growing concerns over the constant use of fossil fuels and its effect on climate change [2], has once again spurred research on sustainable energy development and on enhancement in renewable energy systems. Advances in energy storage and conversion systems that will make our energy usage more efficient are essential if we are to meet the challenge of global warming and the finite nature of fossil fuels [2, 3].

The need for the development of efficient energy storage systems is paramount in meeting the world's future energy targets, especially when energy costs are on the increase and more people need access to electricity [4, 5]. Energy storage technologies can improve efficiencies in supply systems by storing the energy when it is in excess, and then release it at a time of high demand [4]. Further material progression in research and development fundamentals, as well as engineering improvements need to be continued in order to create energy storage systems that will help alleviate humanities energy storage and conversion dilemmas.

Low grade heat (around 130°C) is a by-product of almost all human activity, especially when energy conversion is involved. It is also known as "waste heat" because the dissipated heat into the environment is unutilised. Progress in the field of thermal energy conversion can lead to effective use of limited fossil fuels and provide supplemental power to current energy conversion systems [6].

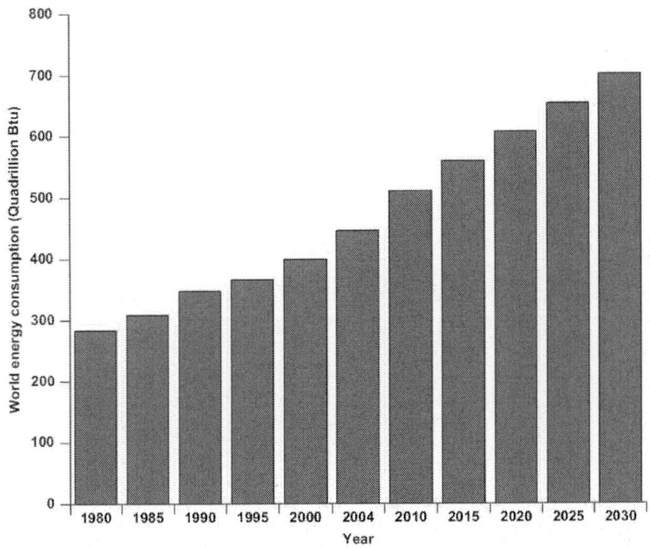

Figure 1. The United States Department of Energy values and forecasts for energy utilisation in the period from 1980 to 2030 [5]. Reproduced with permission from Elsevier.

1.2. How and why carbon nanotubes can address the issues of energy storage and conversion

Nanostructured materials are of great interest in the energy storage and conversion field due to their favourable mechanical, and electrical properties [3, 7]. Carbon nanotubes (CNTs) are one type of nanostructured material that possess these favourable electrical and mechanical properties due to the confinement of one dimension, combined with the surface properties that contribute to the enhanced overall behaviour. The potential of nanostructured materials is not only limited to energy storage and conversion devices; but also to nanotransistors [8, 9], actuators [8, 9], electron field emission [8, 9], and biological sensing devices [10, 11].

The use of carbon-based nanomaterials as electrode materials is practical and economically viable because cheap carbon pre-cursor materials are abundant [12]. As the research into carbon nanotubes (CNTs) has increased over the last 20 years, the cost of these materials has significantly reduced alongside improvements in processability and scalability [13].

The advantage of incorporating carbon materials and specifically CNTs as part of the electrode material is the excellent mechanical and electrical properties. They provide mechanical support to the substrate while enhancing the conductive and electrochemical properties. The low cost of the carbon precursor material used to synthesise CNTs makes device fabrication scalable and economically viable [14]. CNT assemblies can have extremely high specific surface areas, which are extremely important in capacitor design. CNT electrode materials can be confined to a smaller area increasing the electrode-electrolyte contact and decreasing the weight of the device therefore maximising the overall gravimetric performance of the device [15]. CNTs are also chemically stable, which enhances the resistance to degradation of the electrode surface [16].

2. Supercapacitors

2.1. Background information

Electrical energy can be stored in two different forms and can best be described when considering a battery and a capacitor. In a battery, it is the available chemical energy through the release of charges that performs work when two electroactive species undergo oxidation and reduction [17]; this is termed a Faradaic reaction. In a capacitor, electrostatic forces between two oppositely charged plates will separate charge. The generated potential is due to an excess and deficiency of electron charges between the two plates without charge transfer taking place [17]. The current that is observed can be considered as a displacement current due to the rearrangement of charges [2]; this effect is termed as non-Faradaic in nature.

2.1.1. Supercapacitor operation and types

There are two types of electrochemical capacitors that are referred to as 1) electric double layer capacitors (EDLC) and 2) pseudocapacitors. The construction of these devices can vary, with electrodes being fabricated from porous carbon materials including activated carbons, graphene, carbon nanotubes, templated carbons, metal oxides and conducting polymers [18, 19]. EDLC or supercapacitors have two electrodes immersed in an electrolyte solution, separated by a semi-permeable dielectric that allows the movement of ions to complete the circuit but prevents a short circuit from being formed. EDLCs are advantageous as they are able to provide relatively large power densities and larger energy densities than conventional capacitors, and long life cycles compared to that of a battery and ordinary capacitor [20]. The performance of supercapacitors is affected by the power density requirements, high electrochemical stability, fast charge/discharge phenomena, and low self-discharging [21]. Table 1 below shows a comparison between the three types of devices.

Parameters	Capacitor	Supercapacitor	Battery
Charge time	$10^{-6} - 10^{-3}$ sec	1- 30 sec	0.3 – 3 hrs
Discharge time	$10^{-6} - 10^{-3}$ sec	1 – 30 sec	1 – 5 hrs
Energy Density (Wh/kg)	<0.1	1 - 10	20 - 100
Power Density (W/kg)	>10 000	1000 - 2000	50 - 200
Cycle life	>500 000	> 100 000	500 - 2000
Charge/discharge efficiency.	≈ 1	0.90 – 0.95	0.7 – 0.85

Table 1. Comparison of key parameters for a capacitor, supercapacitor and battery [22].

Energy storage is achieved by the build-up and separation of electrical charge that is accumulated on two oppositely charged electrodes as shown in Figure 2 [12]. As stated previously, no charge transfer takes place across the electrode-electrolyte interface and the current that is measured is due to a rearrangement of charges. The electrons involved in the non-Faradaic electrical double layer charging are the conduction band electrons of the electrode. These electrons leave or enter the conduction band state depending on the energy of the least tightly bound electrons or the Fermi level of the system [2]. Supercapacitors exhibit very high energy storage efficiencies exceeding 95 % and are relatively stable for up to10^4-10^5 cycles [4, 5]. The energy given by the equation, $E = 0.5CV^2$, means that the operating voltage is the key in determining the energy characteristics of a supercapacitor. The choice of electrolyte when designing and fabricating a supercapacitor device dictates the operating voltage [23]. Operating voltages are approximately 1.2V, 2.7V, and 3.5V respectively for aqueous, organic and ionic liquid with all of them having associated advantages and disadvantages [4, 5].

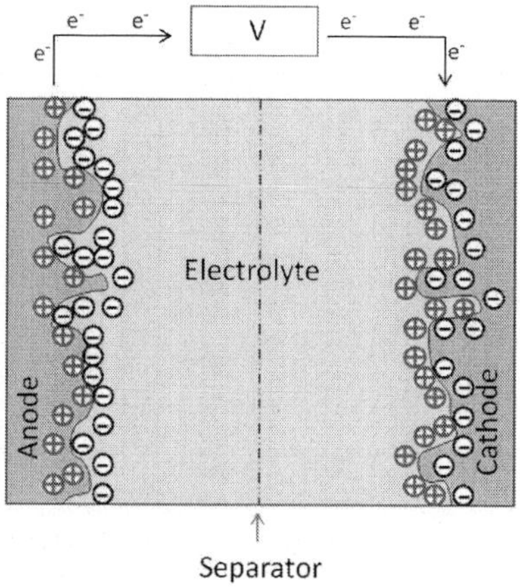

Figure 2. Schematic diagram of an EDLC supercapacitor with a positive and negative electrode, separator and porous carbon.

Like EDLC, a pseudocapacitor consists of two porous electrodes with a separator between them all immersed in an electrolyte solution [24]. However, the difference is that the charge is accumulated during Faradaic reactions near to or at the surface of the electrodes [18], hence non-Faradaic double-layer charging and Faradaic surface processes occur simultaneously [25]. The pseudocapacitance arises from a Faradaic reaction when some of the charge (q) passed in an electrode process is related to the electrode potential (V) via thermodynamical considerations [26]. The two principal cases are adsorption pseudocapacitance arising in underpotential deposition processes [26], and homogeneous redox pseudocapacitance where the reaction is reversible [26, 27]. Pseudocapacitors thus combine features of both capacitors and batteries [18, 28]. A comparison of energy density and power density for various electrical energy storage systems is depicted in Figure 3. Current commercial uses of supercapacitors include personal electronics, mobile telecommunications, back-up power storage, and industrial power and energy management [29, 30]. A recent application is the use of supercapacitors in emergency doors on the Airbus A380, highlighting their safe and reliable performance [30].

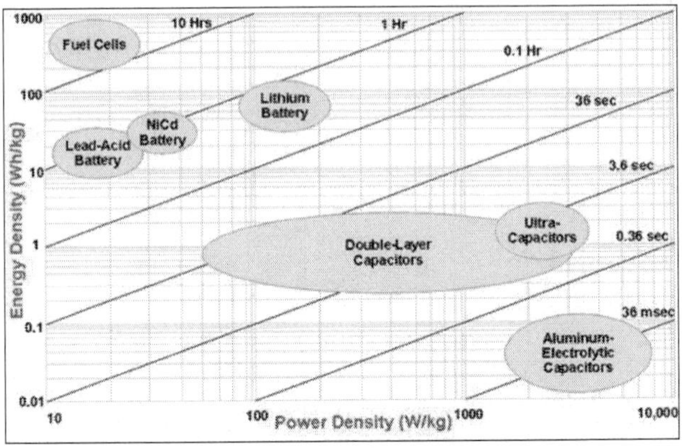

Figure 3. Ragone plot showing specific power against specific energy for various electrical energy storage systems. The times shown are the time constants of the device, which are obtained by dividing the energy density and power density [31].

2.2. Different electrolytes used and their advantages and disadvantages

The choice in electrolyte is extremely important for supercapacitor design as it influences the performance for energy storage and delivery. The extremely large surface area can allow for enhanced energy and power density as long as the micro-porosity and meso-porosity is tailored to suit the type of electrolyte being used. The electrolyte can be designed to enhance the cyclability, to sustain target power densities during operation, and to have an excellent rate capability (i.e. excellent charge/ discharge behaviour) [32].

2.2.1. Aqueous electrolytes

For aqueous electrolytes, the maximum operating voltage is theoretically limited by the reduction potential of water (1.23 V at room temperature) [32]. Most aqueous electrolyte systems tend to have an electrochemical window of approximately 1 V [32]. Electrolyte conductivity has a significant effect on the equivalent series resistance (esr) of the cell, which determines the power output [4]. Concentrated electrolytes are required to minimise the esr and maximise power capability [32]. In general, strong solutions of acidic electrolytes are much more corrosive than strong basic electrolytes meaning that the electrolyte has to be carefully selected for the particular electrode material.

Aqueous electrolytes tend to have very good kinetic behaviour of the electrolyte ions leading to very efficient charge/discharge rates. This behaviour is due to the relatively high conductivity and low viscosity of the concentrated solutions [4, 33]. For example, the conductivity of 1M H_2SO_4 is 730 mS cm^{-1} compared to the much lower value of 10-20 mS cm^{-1} for organic solutions of lithium salts [34]. Time constants of symmetrical carbon supercapacitors using H_2SO_4 were reported to be 0.1s [34].

2.2.2. *Organic electrolytes*

The use of non-aqueous electrolytes in supercapacitors has the main advantage of higher operating voltages compared to aqueous systems. Voltage windows can range up to 2.5 V and since the stored energy increases as V^2, it is possible to attain large energy and power densities [32, 34]. It must be noted that to operate at these higher voltages, non-aqueous electrolytes must be free of water and oxygen which will ensure no evolution of O_2 and/or H_2O at potential differences above 1.23 V [35]. Salts are added to the system to provide mobile ion movement at the electrode/electrolyte interface. The most common salt used generally consists of lithium ions as these ions move very well under an electric field and the effective ion diameter is very small [17].

The major disadvantages of non-aqueous systems are the lower conductivity and the higher viscosity resulting in higher equivalent series resistance (ESR) and reduced wettability if the electrode is hydrophilic. A decrease in wettability will effectively reduce the surface area used by the electrolyte, reducing the energy and power density. Most commercial systems that use organic electrolytes are manufactured in inert atmospheres and are costly to be produced [4].

2.2.3. *Ionic liquids*

Ionic liquids are another class of electrolytes that is proving a great area of research for electrolytes in supercapacitors. These electrolytes can be considered as molten salts with melting temperatures usually below room temperature where the ionic conductivity is no more than 20 mS cm^{-1} [34].

Common ions include BF_4^-, PF_6^- $(CF_3SO_2)_2N^-$, $CF_3SO_3^-$ as well as imidazolium, pyridinium, and quaternary ammonium salts [36]. The physical properties depend on the type of anion and cation and the alkyl chain length [37]. The main advantages are the good solvating properties, relatively high conductivity, non-volatility, low toxicity, large potential window, negligible vapour pressure and good electrochemical stability [37, 38]. Disadvantages include high viscosities and low conductivities compared to that of aqueous electrolytes; while some ionic liquid mixtures yield a potential window that is not much greater than that of aqueous systems. Capacitances approaching 100 F/g for activated carbon (AC) electrodes have been reported by Frackowiak et. al. by using $(CF_3SO_2)_2N^-$ anions and phosphoniumcations. Balducciet et. al. reported capacitance values of 115 F/g for asymmetric poly(3-methylthiophene)/AC electrodes using 1-buytl-3-methyl-imidazolium ionic liquids [36, 37].

2.3. Carbon nanotube powders

Carbon nanotubes (CNTs) were first discovered in 1953 through research in the Soviet Union, but the first accessible results were by Sumio Iijima [39], in 1991 as a result of research into buckminster fullerenes. CNTs have a cylindrical shape that can be considered as a graphene sheet rolled up; either individually as a single-walled carbon nanotube (SWNT), or concentrically as a multi-walled carbon nanotube (MWNT) as depicted in Figure 4and Figure 5. However, these sheets can have varying degrees of twist along its length that can lead to the nanotubes to be either metallic or semi-conducting as the change in chiralities induces different orbital overlaps [9]. They exhibit remarkable electrical transport and mechanical properties [7], which is why interest and research into this material has increased over the last two decades. CNT powders have the potential to be tailored to specific energy storage and conversion applications with there being an added advantage that they can be used in all electrolyte environments that encompass aqueous, organic and ionic liquids [40].

2.3.1. *CNT synthesis overview*

There are a variety of different methods for making SWNTs and MWNTs that have been developed since CNTs were first discovered. These include laser ablation, arc discharge, chemical vapour deposition (CVD), and high pressure carbon monoxide disproportionation (HiPCO). Recently, work by Harris et. al. has successfully scaled-up the synthesis of CNT using a fluidised bed reactor [41]. All growing conditions for synthesising CNTs require a catalyst to achieve high yields, where the size of the catalyst nanoparticles will determine the diameter and chirality of the CNT [42]. The CNTs that are formed are generally in a mixture with other carbonaceous product including amorphous carbon and graphitic nanoparticles.

2.3.2. *Main synthesis methods for CNT growth*

Both Laser ablation and arc-discharge methods for the growth of CNTs involve the condensation of carbon atoms generated from the evaporation of carbon sources. High temperature is involved, ranging from 3000 °C - 4000° C [43]. In arc discharge, various gases such as Helium or Hydrogen are induced into plasma by large currents generated at a carbon anode and cathode. This process leads to the evaporation of carbon atoms which produces very high quality MWNTs and SWNTs [44, 45]. Laser ablation also produces very high quality CNTs with a high degree of graphitisation by focusing a CO_2 laser (in pulsed or in continuous wave mode) for a period of time onto a rotating carbon target [46]. The HiPCO process utilises clusters of Fe particles as catalysts to create very high quality SWNTs [47]. Catalyst is formed *in situ* by thermal decomposition of iron pentacarbonyl, which is delivered intact within a cold CO flow and then rapidly mixed with hot CO in the reaction zone. Upon heating, the $Fe(CO)_5$ decomposes into atoms that condense into larger clusters. SWNTs nucleate and grow on these particles in the gas phase [48, 49].

The CVD method usually consists of a furnace, catalyst material, carbon source, a carrier gas, a conditioning gas, and a collection device (usually a substrate). The carrier gas is responsible for taking the reacting material onto the substrate where CNT growth occurs at catalyst sites [43]. The components mentioned are essential; however, different groups and researchers have alternative experimental conditions which can contain multiple types of furnaces, and a variety of catalyst and carbon sources. The key advantage of this technique is its capability to directly deposit the CNTs onto the substrate, unlike arc discharge and laser ablation that produces a soot / powder. Recent developments by Harris et. al. [41] has led to the development of a large scale batch process for fabricating MWNTs. Here, a furnace like system called a fluidised bed reactor continuously flows a carrier gas over a porous alumina powder that is impregnated with the catalyst material, leading a to continuous creation of MWNTs where tens of grams can be synthesised in one run.

2.3.3. *Single walled nanotubes*

SWNTs have been studied extensively as a supercapacitor and hybrid energy material [4, 50, 51]. The structure of a SWNTis illustrated in Figure 4 with a cylindrical nature apparent as previously stated. Its advantage is that it has very good thermal and conductive properties where the thermal conductivity can exceed 6000 $Wm^{-1}K^{-1}$ and a potential current carrying capacity of 10^9 A/cm^2 [52, 53].

The maximum reported gravimetric capacitance for SWNT fabricated electrodes (PVA / PVC binder; pressed into pellet) is 180 F/g with an energy density of 7 Wh/kg and a power density of 20 kW/kg using KOH electrolyte [54, 55]. Hu et. al. [56] have recently reported a solid state paper based SWNT supercapacitor, which has a specific capacitance of 115 F/g, energy density of 48 Wh/kg and a large operating voltage of 3V. The electrode preparation involved pre-processing where cotton sheets were immersed in the SWNT dispersion, annealed then immersed in an PVA/ H_3PO_4 electrolyte.

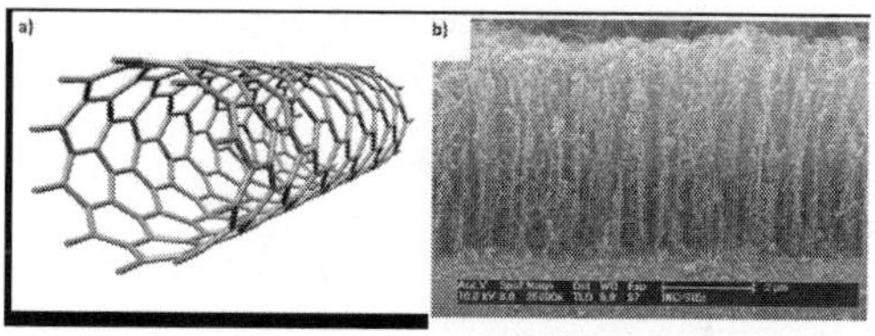

Figure 4. (a) Schematic representation of a SWNT[57].(b) FESEM of SWNTs grown onto a Si wafer substrate[58]. Reproduced with permission from Elsevier.

2.3.4. *Multi-walled nanotubes*

Like their SWNT counter parts, MWNTs have also been studied extensively as electrode materials for supercapacitors [4, 50, 51]. The advantages over SWNTs are their ability to be more easily synthesised on much larger scales, making them more suitable for commercial application. The concentric nature of MWNTs can be observed in the SEM image of Figure 5. The maximum gravimetric capacitance attained for electrodes constructed from MWNTs range between 4-140 F/g with the best available commercial result at 130 F/g from Maxwell's Boost capacitor [59]. Wang et. al. [50] have recently reported partially exfoliated MWNTs on carbon cloth that gave a specific capacitance in the range of 130-165 F/g with a coulombic efficiency of 98 %.

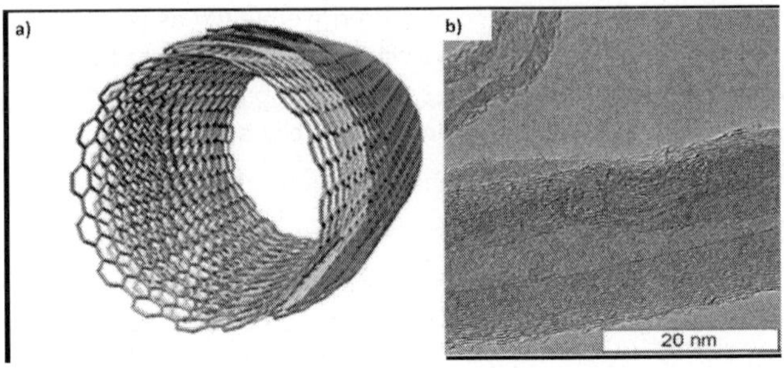

Figure 5. (a) Schematic representation of a SWNT [57]. (b) TEM images of pristine MWNTs [59]. Reproduced with permission from Elsevier

2.3.5. Surface functionalities

The presence of surface functionalities such as oxygen, nitrogen, hydrogen, boron and catalyst nanoparticles (dependent on the synthesis environment and pre-cursor materials) can affect the capacitative behaviour of the electrode through the introduction of Faradaic reactions [60], changes in electric and ionic conductivity [23], and influencing wettability [61]. A schematic representation of an sp^2 hybridised carbon lattice with various dopants is show in Figure 6.

Oxygen

Carbon materials will have functional groups present on their surface as a result of the precursors and preparation conditions [23]. Most of these functional groups are in the form of –COOH, =CO as well as phenol, quinone and lactone groups [4, 23]. Activation procedures such as post treatment with H_2SO_4 and / or HNO_3 also leads to acid oxygen functionalities [4].

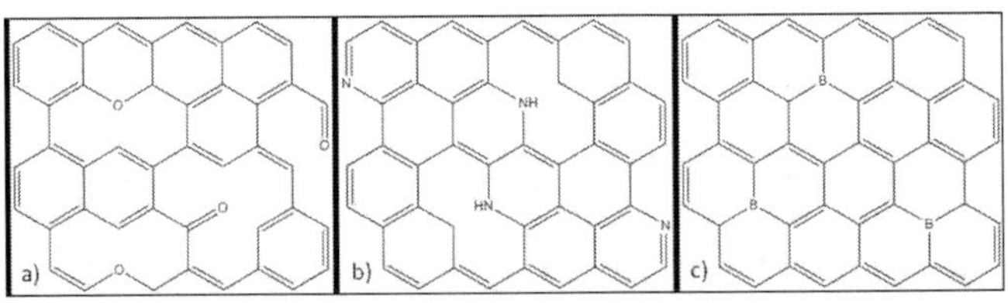

Figure 6. Schematic representation of a sp^2 hybridised carbon lattice that has been doped with; (a) oxygen functional groups, (b) nitrogen functional groups, (c) boron.

Most of these groups are bonded with carbon atoms at the edge of hexagonal carbon layers where Faradaic reactions (via interactions with the electrolyte) lead to pseudocapacitance such as those developed with transition metal oxides RuO_2 and MnO_2 [23, 62]. These functional groups can also be purposely added onto the surface of carbons via oxidation with O_2 or acid treatment with HNO_3 or H_2SO_4 [63]. In aqueous systems, the presence of oxygen containing functional groups can lead to an enhanced wettability as well as pseudocapacitance as mentioned above, which maximises the electroactive surface area leading to larger energy densities [8, 23]. It has been proposed that the pseudocapacitative reactions for oxygen functionalised CNTsinvolve carboxyl groups undergoing electron transfer[64]:

$$C{-}OH \;--\; C{=}O + H^{+} + e^{-}$$

$$C{=}O + e^{-} \;--\; C - O^{-}$$

In non-aqueous systems, however, oxygen functionalities are detrimental to device performance. Parasitic redox reactions can lead to a degradation of the electrode, as well as adverse effects relating to voltage proofing and increased leakage current [4, 65]. These redox reactions will reduce the cycle life of a device, as well as lowering the operating voltage.

Shenet. al. [66] reported in 2011 the effects of changing the carboxylic group concentration on SWNTs. The specific capacitance, power density and energy density 0.5 M H_2SO_4 electrolyte increased with carboxylic group density reaching a maximum of 149.1 F/g, 304.8 kW/kg, and 20.71 Wh/kg, respectively. The 10 μm film electrodes were fabricated using vacuum filtration to create "bucky papers" onto a mixed cellulose estate membrane.

Nitrogen

Nitrogen containing carbons have recently attracted interest due to its n-type behaviour that promotes large pseudocapacitance, which can be obtained even if the surface area of the material is decreased [67]. In some instances, up to 3-fold increase in capacitance have been reported [68]. Typical examples of redox reactions involving nitrogen are described below [69]:

$$C + NH + 2e^{-} + 2H^{+} \leftrightarrow CH_2{-}NH_2$$

$$CH{-}NHOH + 2e^{-} + 2H^{+} \leftrightarrow CH_2{-}NH_2 + H_2O$$

The chosen precursor material affects the types of functionalities that are attached to the carbon backbone. Nitrogen-containing groups may be added via various methods with compounds containing nitrogen including treatment with urea, melamine, aldehyde resins and polyacylonitrile [4, 70-73]. Surface areas for nitrogen-doped carbon materials are thought to be in excess of 400 m^2/g [23]. This is much lower than pure SWNTs and pure MWNTs that have been reported to attain a surface area greater than 1315 m^2/g and 830 m^2/g respectively; suggesting that pseudocapacitance makes up a substantial portion of the total capacitance [74, 75]. Y. Zhang et. al. [76] have showed that N-doped MWNTs synthesised via CVD growth exhibited a capacitance of 44.3 F/g, which was more than twice the value obtained than that of the un-doped MWNTs in a 6M KOH electrolyte. K. Lee et. al. [77] have shown that the nitrogen content on vertically aligned CNTs increases the capacitance until a certain point due to an increased donation of an electron by the N (N acts as an n-type dopant) and an enhanced wettability in aqueous systems. Excessive N-doping significantly reduced the conductivity and inhibited charge storage and delivery [77]. The doped and un-doped CNTs were directly grown onto a stainless steel substrate using CVD [77].

Boron

Boron is another interesting material for doping CNTs due to its p-type nature which promotes CNT growth and increases the oxidation temperature of the nanotubes [78]. However, the development of boron doped CNTs for the use as electrodes in supercapacitor devices is not well established [23]. Work by Shiraishi et.al. showed that boron doping MWNTs, increased the capacitance per surface area from 6.5 $\mu F/cm^2$ to 6.8 $\mu F/cm^2$ in 0.5M $LiBF_4$/PC [79]. These electrodes were once again synthesised using CVD [79]. Wang et. al. reported in 2008 that interfacial capacitance was increased by 1.5-1.6 times in boron-doped carbon than that in boron-free carbon with alkaline electrolyte (6 M KOH) and/or acid electrolyte (1 M H2SO4) [80]. The carbon material was made into a slurry using carbon black and PTFE binder and pasted onto a Ni mesh current collector [80].

2.3.6. *Advantages, limitations and comparison*

It can be seen that CNTs can be tailored different ways in order to tune (to a degree) the performance of the electrode material. This control has been demonstrated by firstly, varying the chiralitiy of the nanotube to produce the single-walled or the multi-walled variety. Both CNT types have associated advantages and disadvantages with SWNTs being able to be synthesised with a high degree of purity; while MWNTs can be synthesised on a larger scale. CNTs can also have functionalities (through addition of oxygen or nitrogen containing groups) added to their structure through treatment in order to change the surface properties and hence wettability of the material. These functionalities enable enhanced compatibility to an electrolyte to maximise electroactive surface area usage and hence performance. Further doping with specific elements such boron and nitrogen can introduce a p-type/n-type behaviour where electrons contribute a Faradaic response to the system and enhance capacitance and energy density. However, it must be noted that when faradaic processes occur at the electrode/electrolyte interface, irreversible processes increase degradation of the electrode over time. Specific capacitance of CNTs (three electrode and device testing) is in the order of 5 – 165 F/g with an increase thereafter as a result of doping (i.e due to Faradaic contribution). It must be pointed out that with electrical energy devices, there is always a trade-off between energy and power density. Therefore the electrode material has to be tailored to meet the requirements of the specific application.

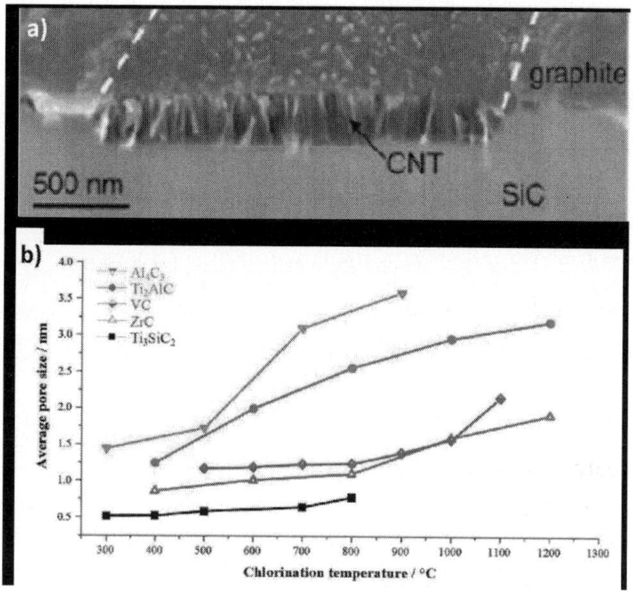

Figure 7. (a) SEM image depicting the growth of templated porous CNTs. (b) The tunability of average pore size distribution of binary and ternary carbides Al_4C_3, Ti_2AlC,VC,ZrC, Ti_3SiC_2 by varying the chlorination temperature [29]. Reproduced with permission from John Wiley & Sons

2.4. Templated porous carbons

Templated porous carbons are of recent great interest in the field of energy storage due to the tunability in porosity, which is necessary to meet the materials application requirements [29, 30]. These templated carbons are commonly known as carbide derived carbons (CDC) as the carbon materials are derived from carbon precursors through physical and/or chemical processes [29]. Briefly, the synthesis involves halogenations (usually chlorination) where the carbon is formed by selective extraction of the metal and metalloid atoms, transforming the carbide structure into pure carbon. The carbon layer is formed by inward growth, with retention of the original shape and volume of the precursor [29]. If any metal chlorides are trapped, they can be usually removed by hydrogenation or vacuum annealing [29]. The general reaction scheme is as follows where M = Si, Ti, Zr[23, 29];

$$MC(s) + 2Cl_2(g) \rightarrow MCl_4(g) + C(s)$$

The advantage of forming carbon structures this way is the ability to form a tailored and narrow pore size distribution with a large surface area as can be seen in Figure 7 [29].

Inagaki et. al. in their very comprehensive review of carbon materials for electrochemical capacitors reported a maximum surface area of S_{BET} of 2000 m^2/g for CDC, which gives rise to possible electrode materials with extremely large energy densities and power densities [23].

Gao et. al. have recently reported flexible CDC electrodes fabricated into a device which obtained a specific capacitance of 135 F/g in 1 M H_2SO_4 and 120 F/g in 1.5 M tetraethyl ammonium tetrafluoroborate ($TEABF_4$) [81]. Ordered mesoporous carbon spheres with impregnated NiO, and a maximum surface area of 944 m^2/g yielded a specific capacitance of 205.3 F/g in 2 M KOH [82]. Reported also by Y. Korenblit [83] was a high surface area CDC (2430 m^2/g) with aligned mesopores, which yielded a specific capacitance of 170 F/g and an extremely high capacity retention of 85% at high current densities of 20 A/g.

2.5. Composite electrode materials

Typical carbonaceous electrode materials (activated carbon, CNTs, graphene, CDC) with high surface areas used in supercapacitors have somewhat reached a limit when it comes energy storage capacity, thus restricting their possible applications [84]. Pseudocapacitor materials that are able to meet the needs of higher energy density are currently being developed and combined with carbonaceous materials in order to create composites that when designed into hybrid supercapacitors have advantages of fast rate capability, high storage capacity, and long cyclability[84, 85].

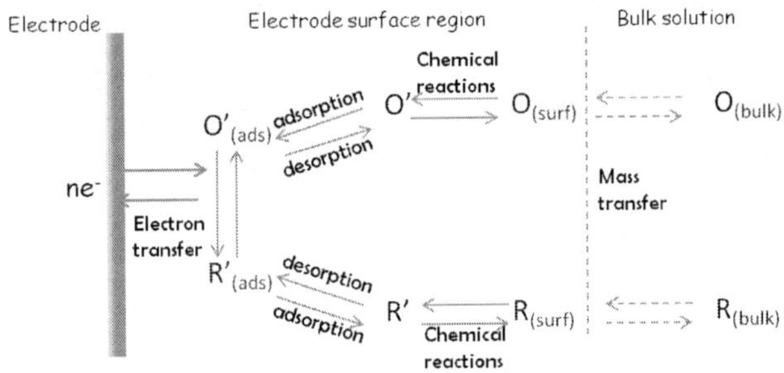

Figure 8. Schematic diagram of a reversible redox reaction, as well as EDLC occurring at the electrode/electrolyte interface leading to pseudocapacitance.

2.5.1. CNT / polymer

Electrode materials comprised of inherently conducting polymers (ICPs) and CNTs is a promising area of research. The conductive polymer matrix, combined with the network like structure of the CNTs provides an enhanced electronic and ionic conductivity that can considerably improve charge storage and delivery [86-88].

Antiohos et. al. reported a SWNT / Pedot-PSS composite electrode material that was fabricated into a device which had a specific capacitance of 120 F/g (1 M $NaNO_3$ / H_2O), coupled with an excellent stability (~90% capacity retention) over 1000 cycles using galvanostatic charge / discharge [89]. The SWNT / Pedot-PSS composite is depicted in Figure 9 where SWNTs are thoroughly dispersed throughout the Pedot-PSS conducting polymer matrix. Kim et. al. recently fabricated a ternary composite material consisting of MWNTs, graphene, and PANI where a specific capacitance of 1118 F/g was achieved. This electrode was stable with 85% capacity retention after 500 cycles using galvanostatic charge / discharge [90]. Hu et. al. [91] have recently reported a composite electrode materials containing MWNTs coated with polypyrrole that achieved a high capacitance of 587 F/g in a 0.1 M $NaClO_4$ / acetonitrile electrolyte.

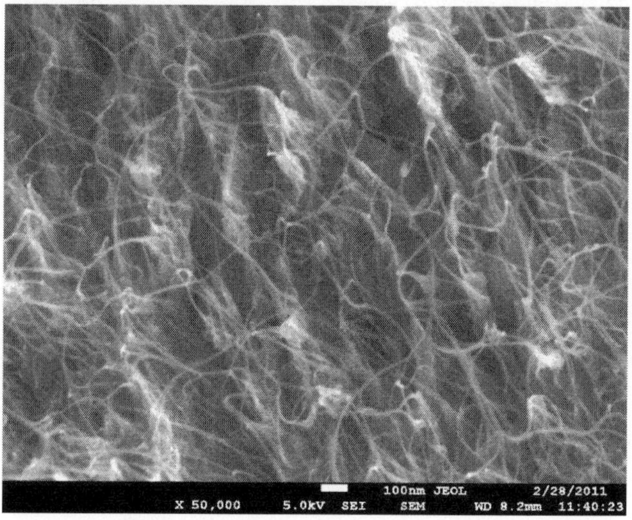

Figure 9. SEM image of PEDOT/PSS-SWNT composite showing PEDOT/PSS polymer to be integrated with the SWNT [89]. Reproduced with permission from The Royal Society of Chemistry.

2.5.2. CNT / metal oxide

Metal oxides exhibit pseudocapacitative behaviour over small rages of potentials, through redox processes which contribute electron transfer between the electrode / electrolyte interface (Figure 8). Common materials used in the construction of such devices are oxides of Mn, Ru, Ir, Pt, Rh, Pd, Au, Co and W [22, 26]. By combining metal oxides with CNTs, composites can be formed that combine both Faradaic and non-Faradic effects enabling a larger energy density to be obtained while still holding reasonable power density. Figure 10 shows MnO_2 particles that have been formed (*insitu*) in the presence of MWNTs.

Very recent work on carbon / metal oxide composites can be found in the review by Wang et. al. [50]. Myoungkiet. al. reported recently in their a RuO_2 / MWNT, electrode material which achieved a specific capacitance of 628 F/g[92]. The electrode was fabricated by dispersing the mixture in ethanol and casting onto carbon paper [92]. Li. et. al. reported that when MWNT were coated with MnO_2, a capacitance of 350 F/g was achieved [93]. More novel materials have been created by incorporating MWNTs and Co_3O_4,which yielded specific capacitances of 200 F/g [94] (acetylene black / PVDF slurry on Ni gauze); while Jayalakshmi et. al. reported in 2007 $V_2O_5.xH_2O$ / CNT film with a specific capacitance of 910 F/

gwith the material being ground into a paste with paraffin and spread onto a graphite electrode, and tested in 0.1 M KCl [95].

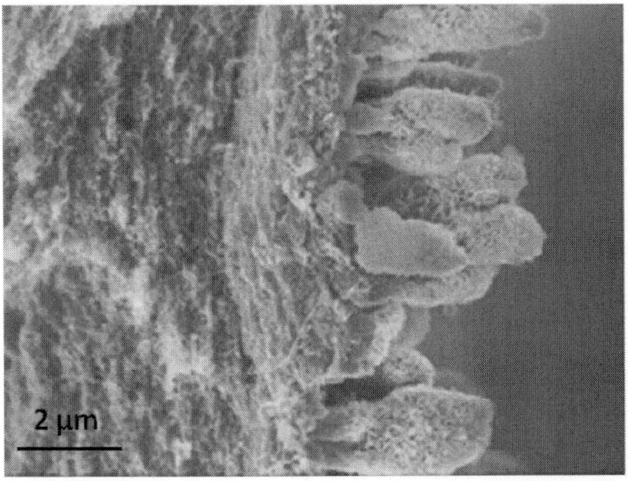

Figure 10. Surface cross-section morphology of MnO_2 particles being grown (insitu) onto MWNTs[96].Reproduced with permission from Elsevier.

2.5.3. *CNT / carbons*

Creating composite materials from CNT and different forms of carbon such as graphene or carbide derived carbons (CDC) can be advantageous due to the fact that the CNTs provide microporosity (large surface area to maximise capacitance and hence energy density); while graphene and CDC can be used to tailor the mesoporosity which improve ions kinetics, enhancing the power density [21]. In Figure 11, a composite of reduced graphene oxide coated with SWNTs is depicted that has been formed into a porous film. The edges of the graphene oxide protrude out with a uniform coating of SWNTs. Recently, Li et. al. fabricated different mass loadings of graphene and CNT composite electrodes by solution casting onto glass, annealing then peeling off [97]. They reported capacitance ranges of 70-110 F/g at a scan rate of 1 mV/s in 1M H_2SO_4 [97]. Luet. al. have reported a CNT / graphene composite which was bound together with polypyrrole (through a filtration process) that achieved a specific capacitance of 361 F/g at a current density of 0.2 A/g in 1 M KCl. The electrode exhibited excellent stability with only a 4% capacity loss over 2000 cycles [97]. Dong et. al. have shown that is it possible to form SWNT/graphene oxide core shell structures and spray coat the subsequent material onto a current collector [98]. The performance of these core structures yielded a material with a specific capacitance of 194 F/g using galvanostatic charge / discharge at a high current density of 0.8 A/g in 1 M KOH [98].

http://dx.doi.org/10.5772/51784

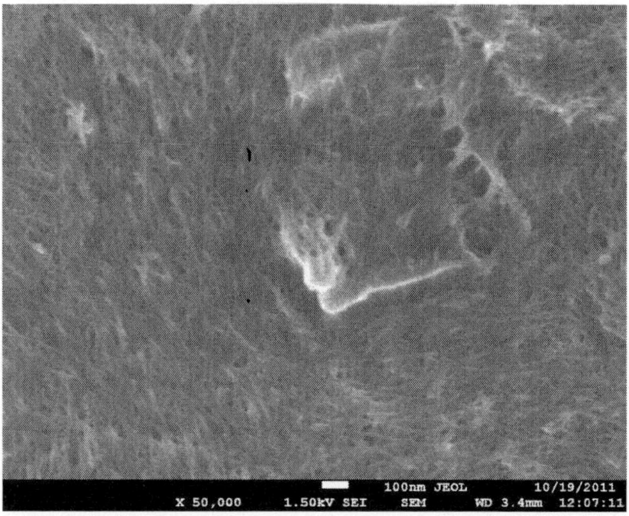

Figure 11. SEM image of a reduced graphene oxide / SWNT composite formed into a film.

2.6. Conclusions

It can be seen that there has been extensive research and development in the use of CNT as electrode materials for energy storage applications. Currently, they provide an excellent platform for devices that require high power density due to the very high surface areas and fast rate capability. Further studies need to be implemented in order to better understand the relationship between electrode porosity and electrolyte. An enhanced understanding of the role of micro and meso-porosity and its effect on system performance is critical. Electrolyte selection is also critical to device performance as it is proportional to the square of the voltage. The main classes of electrolytes are aqueous-based, organic-based and room temperature ionic liquids. Evolving work has focused on using CNT materials in conjunction with doping of various functional groups such as carboxyls, amines and elements such as boron and nitrogen in order to enhance the electrode performance through increased usage of electrode surface area and / or Faradaic contributions. The most recent work has focused on the creation of composite materials via the combination of CNTs with conducting polymers or metal oxides. CNT composites have amassed into a prevalent area of research through the search for the discovery of hybrid energy storage devices that are able to have high energy and high power density which are beneficial for creating more energy efficient systems and providing a greater range of applications.

3. Thermal Energy Harvesting

3.1. Introduction to thermogalvanic cells

Studies on the conversion of heat to electricity have been conducted as early as the 1960s [99]. Since then, several thermal converters have been developed: thermocouples, thermionic converters, thermally recharged cells, thermogalvanic cells, etc. [100]. The discussion in the subsequent sections will be limited to thermogalvanic cells, also known as thermocells. These are electrochemical systems that are able to directly transduce thermal energy to electrical energy [101]. The simple design of these systems allows them to function without the need for moving components. Their stability allows operation for extended periods without regular maintenance. Thermocells also have zero carbon emission hence it will not contribute to the environmental impact of electrical power generation.

3.1.1. Low grade heat sources and conversion through thermogalvanic cells

Various unharnessed low grade heat sources

The second law of thermodynamics dictates that a heat engine can never have perfect efficiency and will always produce surplus heat (usually around 100 °C). This waste heat (or low grade heat) is one of the world's most ubiquitous sources of untapped energy. (i.e. waste heat is produced by simply turning on an automobile). Roughly 70 % of the energy generated by an automobile motor is wasted; part of it ends up as a hot exhaust pipe and warm brakes. The Wartsila-Sulzer RTA96-C turbocharged two-stroke diesel engine, one of the most efficient engines in the world, is only able to convert around 50 % of the energy in the fuel to useful motion. The rest of that energy gets dissipated as waste heat [102]. Waste heat also exists in factories, particularly in the steel and glass production industry. Pipes that carry hot liquids are also low grade heat sources. Other scenarios wherein heat simply dissipates into the environment are power plants, household appliances, and various electronic gadgets. Research done to convert waste heat into electrical power by the use of ferromagnetic materials, thermocouples and thermionic converters has resulted in low efficiencies [103-105]. Advances in thermoelectric systems have been hampered by its high initial cost and material limitations; as these systems operate in the temperature range much higher than low grade heat [106].

Description of how a thermogalvanic cell works

A thermogalvanic cell, also known as a thermocell, is a thermal energy converter that utilises electrochemical reactions to attain conversion of low grade heat to electrical power. The two half cells of the system are held at different temperatures causing a difference in redox potentials of the mediator at the anode and cathode [107]. This reaction can drive electrons through an external circuit that allows generation of current and power. A schematic of a thermocell with a ferri/ferrocyanide redox couple is shown in Figure 12.

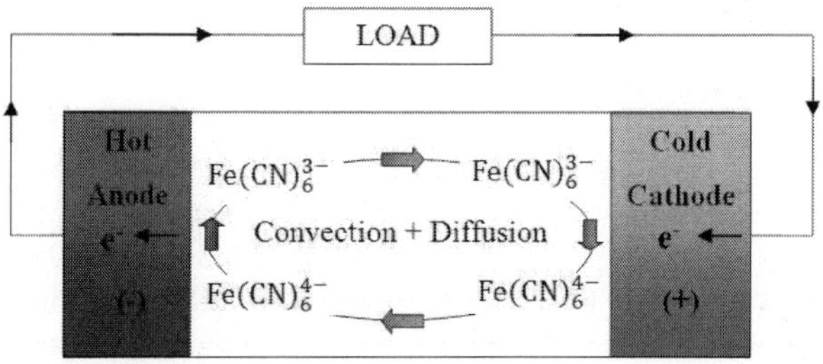

Figure 12. Ferri/Ferro Cyanide redox thermogalvanic cell [108]Reproduced with permission from Springer Science +Business Media

Ferrocyanide is oxidized at the hot anode, the electron generated then travels through an external circuit and returns to the cell via the cold cathode where it is consumed in the reduction of ferricyanide[109]. The accumulation of reaction products at either half-cell is prevented by the diffusion and convection of the electrolyte that occurs naturally, thus eliminating the need for moving mechanical components.

3.1.2. Desirable material properties for thermal conversion cells

Power conversion efficiency and how it is affected by electrode material properties

The power conversion efficieny (Φ) of a thermocell is defined as follows:

$$\Phi = \frac{\text{Electrical output power}}{\text{Thermal power flowing through the cell}} \tag{1}$$

The thermal power flowing through the cell is largely controlled by cell design and electrolyte selection. When a reversible redox couple is used, no net consumption of the electrolyte occurs and the thermal power is given by:

$$\textbf{Thermal power flowing through cell} = \mathbf{KA}\frac{\Delta T}{d} \tag{2}$$

Where K is the thermal conductivity of the electrolyte, A is the electrode cross sectional area, ΔT is the thermal gradient and d is the distance between the two electrodes [110].

Qualitative behaviour of the current and voltage dependencies are shown in Figure 13; it depicts that the maximum electrical output power (P_{max}) is obtained when the external and internal loads are equal and is given by:

$$P_{max} = 0.25 V_{OC} I_{SC} \tag{3}$$

Where V_{oc} is the open circuit voltage and I_{sc} is the short circuit current. V_{oc} is highly dependent on the reaction entropy of the redox couple and the thermal gradient at which the electrodes are exposed to as shown in Equation 4:

$$V_{oc} = \frac{\Delta S_{B,A} \Delta T}{nF} \tag{4}$$

where $\Delta S_{B,A}$ is the reaction entropy for a hypothetical redox couple $A \leftrightarrow ne^{-} B$, n is the number of reactions involved in the redox reaction and F is Faradays constant [100].

Combining Equation 2 and Equation 3 allows the power conversion efficiency to be expressed as:

$$\Phi = \frac{0.25 V_{oc} I_{sc}}{KA\left[{}^{\Delta T}/_{d}\right]} \tag{5}$$

Ohmic, mass transport and activation overpotentials are losses that need to be minimised in order to realise an improvement in thermocell conversion efficiency. At large electrode separations, ohmic overpotential is dictated by the electrolyte resistance; and mass transport overpotential is maximized. By decreasing the inter-electrode separation, an increase in generated power will be observed as both ohmic and mass transport overpotentials will decrease. However, the power conversion efficiency will be lowered as it will be harder to maintain the thermal gradient in the cell [112]. It has been shown that changes in electrolyte concentration affect its thermal conductivity [108]. Optimization of electrolyte concentration coupled with appropriate cell design is necessary to mitigate both overpotentials while maintaining large power conversion efficiency.

Activation overpotential is associated with the activation barrier needed to transfer an electrode to an analyte. For the same activation overpotential, larger current densities are realised when the exchange current density is increased. This increase is attained when the concentration of the redox couple in the electrolyte is maximised, the thermal gradient is increased and the number of possible reaction sites is augmented [113]. Porous electrodes have the advantage of increased electroactive surface area and will directly amplify the short circuit current density [114]. It must be noted that for porous electrodes, short circuit current density does not increase indefinitely with electrode thickness as mass transfer overpotential will become limiting. The reaction products formed within the pores of the anode will not be able to diffuse fast enough to the cathode and vice versa, generating concentration gradients around both electrodes. Another way to decrease the activation overpotential in thermocells is by using catalytic electrodes attained by doping [115].

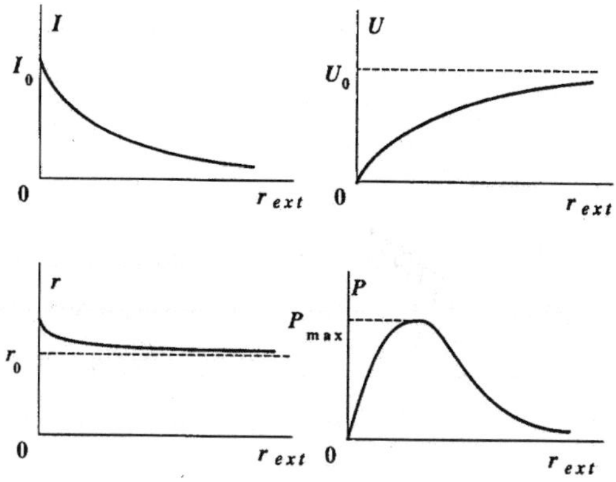

Fig. 2. The typical dependences of the current *I*, the effective voltage *U*, the internal resistance *r* and the useful power from the external resistance r_{ext}.

Figure 13. The typical dependencies of the current (*I*) on the effective voltage (*U*), internal resistance (*r*) and the useful power on the external resistance (r_{ext}) [111]Reprinted with permission from Elsevier.

3.1.3. CNTs vs flat electrodes - why CNTs can improve thermal harvesting

Power conversion efficiency achieved by using flat electrodes and why recent developments in CNTs can augment thermogalvanic cell performance

The chemical stability of platinum led to its extensive investigation in thermogalvanic cells. In fact, a study on the effects of platinum electrode cleaning was performed and it was deduced that this affects the power delivery characteristics of thermogalvanic cells [116].

Comparison of thermoelectric converters operating at different conditions (i.e. thermal gradient, electrolyte, electrode separation, etc.) can be done by measuring their power conversion efficiency relative to a Carnot engine operating at the same temperature (Φ_r).

$$\Phi_r = \frac{\Phi_{\text{thermogalvanic cell operating at } \Delta T}}{\Phi_{\text{Carnot engine operating at } \Delta T}} \tag{6}$$

If power inputs are ignored, such as mechanical stirring, thermocells with platinum electrodes are able to attain power conversion efficiency relative to a Carnot engine of 1.2%. However, if power inputs are strictly excluded, the efficiency drops to 0.5% [100].

As mentioned previously, the discovery of CNTs led to widespread research on this material to investigate its potential uses, one of them being electrochemical applications [39, 117-120]. CNT electrodes are known to exhibit Nernstian behaviour and more importantly, fast electron transfer kinetics with the redox couple ferri/ferrocyanide. Peak potential separation in cyclic voltammograms obtained using micron-sized MWNT electrodes and 5 mM potassium ferrocyanide is 59 mV, which is the expected theoretical value and implies that the highest electron transfer rate was attained [121]. Incidentally, the ferri/ferrocyanide redox couple has been studied intensively in thermocell applications owing to the large voltage that can be induced by a thermal gradient, also known as the Seebeck coefficient. The Seebeck coefficient 1.4 mV/K for the ferri/ferro cyanide redox couple implies that an open circuit potential of 84 mV is attainable at a thermal gradient of 60 °C (the usual limit for aqueous systems without significant cooling). The fast electron transfer of CNTs in ferri/ferro cyanide primarily justifies its use as electrodes in thermocells.

The nanometre diameter of CNTs gives rise to large gravimetric and volumetric specific surface areas (SSA). Their unique aspect ratios allow porous electrodes to be fabricated by a variety of methods. Theoretically the SSA of CNTs can range from 50-1315 m^2/g, the value dictated by the number of walls [74]. Theoretical predictions are in good agreement with experimental values obtained by the measurement of amount of gas (usually N_2) adsorbed at 77 K and calculations using the Brunauer-Emmett-Teller (BET) isotherm. Kaneko et al. [122] have reported that MWNTs are mesoporous while Rao et al. [123] have shown that SWNTs are microporous. MWNT buckypapers of the same geometric area compared with platinum foil are known to have three times larger charging current density during cyclic voltammetry in ferri/ferrocyanide aqueous electrolyte [114], evidence of the large accessible SSA of CNT electrodes.The large SSA of CNTs allows for a greater number of electroactive sites. When the CNT electrode porosity is controlled and the tortuousity is minimised in thermogalvanic cells (so that mass transfer is not limited within the electrode), the short circuit current generated can be significantly augmented.

3.2. Different types of CNTs investigated

3.2.1. SWNT and MWNT

CNTs were first used as thermocell electrodes in 2009 [114]. Baughman et al. tested 0.5 cm^2 MWNT buckypaper electrodes (with less than 1 % catalyst and with MWNT diameter of around 10 nm) in a U-Cell with electrode separation of 5 cm, a temperature gradient of 60 °C wherein T_{cold}= 5 °C. A schematic of the cell they used is shown in Figure 14. A specific power density of 1.36 W/m^2 was obtained [114]. Platinum electrodes tested under the same conditions generated a specific power density of 1.02 W/m^2, proving that CNTs are viable materials for thermocell electrodes.

SWNT powders produced by arc discharge (ASA-100F, Hanwha Nanotech) with an average diameter of 1.3 nm, and composition of 20-30 wt. % CNTs, 40 wt. % carbon nanoparticles, 20 wt. % catalyst material, 10 wt% amorphous carbon and graphite, was tested by Kang et al. in thermal harvesting [113]. A vertical test cell with a "hot above cold" orientation (Figure 15), glass frit separator and electrode separation of 4 cm was employed with T_{hot}= 46.4 °C and T_{cold} = 26.4 °C. SWNT electrodes with an area of 0.25 cm^2 were immersed in 0.2M $K_3Fe(CN)_6/K_4Fe(CN)_6$ electrolyte. The specific power density obtained was around 5.15 W/kg. Commercially available purified SWNT powders (P-SWNT) sourced from Hanwha Nanotech (ASP-100F), refined by thermal and acid treatment (60-70 wt. % nanotubes, 10 wt. % catalyst material, 20 wt. % graphite impurities) was tested by the same group. The specific power density improved by 32 %, generating 6.8 W/kg. Using the same test conditions, commercially available purified MWNT having 3-6 walls with a median diameter of 6.6 nm (SMW100, Southwest Nanotechnologies, Inc) and approximately 98 wt. % carbon yielded a specific power density of 6.13 W/kg. It must be noted that these tests were not done to maximise the power generation capability of the thermocell but to gain further insight into CNT electrodes for thermal harvesting. Hence the small electrode separation, low electrolyte concentration and small thermal gradient.

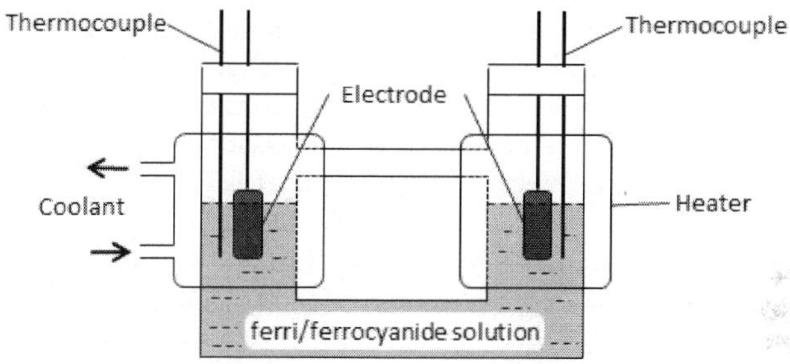

Figure 14. Schematic of the U-Cell used by Baughman et. al for thermal harvesting

Electrical impedance spectroscopy (EIS) of the various carbon nanomaterials tested by Kang et al revealed that the P-SWNT electrode has a marginally lesser ohmicoverpotential (21 Ω) than the pristine SWNT (22 Ω). This finding explains the increased specific power density generated when the P-SWNT electrodes are used.

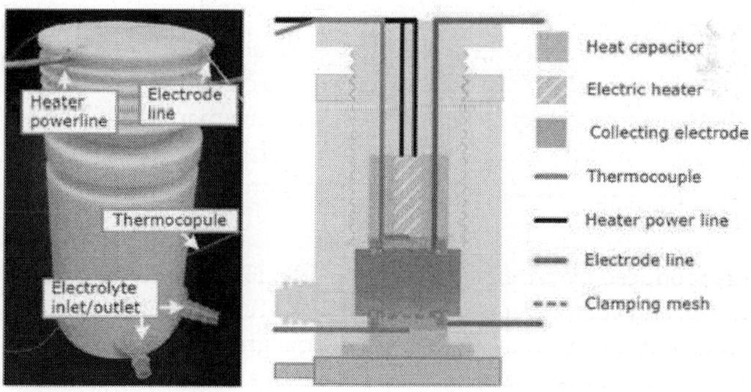

Figure 15. Vertical "hot above cold" thermocell[113]Reprinted with permission from John Wiley & Sons, Inc.

It has been proven that the catalytic nature of MWNTs is due to the edges or sites where the tube terminates, regions that are more numerous in MWNTs than in SWNTs [124]. Due to this and the fact that P-MWNTs had lower ohmic resistance (18 Ω) compared to P-SWNTs, it was expected that P-MWNTs would perform better in thermal harvesting. The authors attributed the enhanced performance of P-SWNT to the larger specific surface area, which compensated for the decreased electroactive sites and higher ohmic resistance.

3.2.2. *Functionalized CNTs*

Functionalising or doping (using Nitrogen or Boron atoms) may be used to fine tune the physical and chemical properties of CNTs [125-127]. With advances in technology, CNT functionalization is a reasonably simple process [128]. Dai et. al have shown that nitrogen-doped carbon nanotubes (NCNTs) have high electrocatalytic activity in oxygen reduction reactions as compared to undoped CNTs [129]. The increase in performance was brought about by a four electron pathway for oxygen reduction reactions that was attained by aligning the nanotubes and integrating nitrogen into the carbon lattice. The additional electrons contributed by nitrogen atoms can enhance electronic conductivity by providing electron carriers for the conduction band [130]. The many active defects and hydrophilic properties of NCNTs allow for enhanced electrolyte interaction in aqueous solutions [131]. Boron doped CNTs (BCNTs) are also attractive for electrochemical applications owing to the increased edge plane sites on the CNT surface; proven to be the predominant region for electron transfer [132]. Examples of the electrocatalytic performance of BCNTs are the improved detection of L-cysteine, enhanced electroanalysis of NADH and enhancement of field emission [126, 127,133].

The possibility of using nitrogen-doped CNT and boron-doped CNT electrodes in thermocells was investigated by Cola et.al [115]. Doping was attained by using a plasma-enhanced chemical vapour deposition process. Tests were run using a U-cell configuration (Figure 14), T_{cold} = 20 °C, thermal gradient of up to 40 °C, and electrolyte concentration of 0.1 M potassium ferri/ferrocyanide. The electrodes were sized to 0.178 cm^2 and were set up in a symmetric and asymmetric (N(hot)-B(cold) and B(hot)-N(cold)) fashion.

Results (Figure 16a) indicate inferior thermocell performance for both NCNT and BCNT as compared to Pt and pristine CNTs. The poor performance of the doped CNTs was brought about by the sluggish kinetics, evidenced by the large peak separations in the cyclic voltammograms taken at a scan rate of 100 mV/s (Figure 16b). It was theorized that the slow-moving kinetics for the doped electrodes was caused by the electrostatic effects at the electrode-electrolyte interface [134]. The positively charged BCNTs repulsed the similarly charged potassium (K^+) counter ion which decreased the electrolyte concentration in the vicinity of the electrode. The negatively charged NCNTs led to a strong electrostatic attraction with K^+, an effect which at low concentrations can improve electron transfer kinetics. However, the large bulk concentration (needed to achieve significant short circuit currents in thermocells) led to a high density of the K^+ ions in the vicinity of the electrode. This effectively acted as a barrier to the redox reactions that were supposed to occur at the electrode.

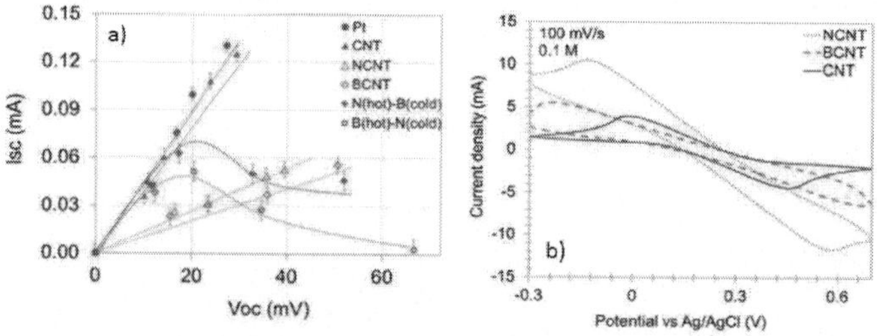

Figure 16. a) Thermal energy conversion response of various electrodes used by Cola et. Al b.Cyclicvoltammograms of various electrodes in 0.1M $K_3Fe(CN)_6/K_4Fe(CN)_6$ at 100 mV/s vs Ag/AgCl reference electrode [115]. Reproduced with permission from J. Electrochem. Soc.

Using BCNT and NCNT in an asymmetric configurations resulted in increased currents at small thermal gradients as compared to the symmetric arrangements. This current then decreased non-linearly as the temperature difference was increased. At small thermal gradients, with the BCNT at the cold side of the cell, the slower kinetics induced an accumulation of reactants at its surface. The faster kinetics at the NCNT, brought about by the increased temperature, kept the electrolyte concentration in its vicinity low. Both factors allow the redox reactions to occur rapidly until a threshold thermal gradient is reached. At this point, the ion concentration in the vicinity of the NCNT is sufficiently large to slow the kinetics and reduce the currents generated. The threshold temperature gradient is attained at lower temperatures when the NCNT is kept at the cold side of the cell because the slower kinetics at this temperature promotes accumulation of K^+ ions on the NCNT surface and leads to the "blocking" effect discussed previously.

3.2.3. *Composites*

The recent discovery of graphene through micromechanical exfoliation has sparked a flurry of research into its possible applications [135]. Graphene consists of a single layer of carbon atoms bonded in a hexagonal lattice. Like CNTs, its remarkable properties (charge carrier mobility of 200000 cm^2/V-s and specific surface area of 2630 m^2/g) make it an ideal candidate for electrochemical applications [136, 137]. In order to scale up graphene production, graphite is normally exfoliated in the liquid state through surfactant/solvent stablilization [138] or chemical conversion resulting in a graphene like structure known as reduced graphene oxide (RGO), shown in Figure 17 [139]. Being of the same composition as CNTs, investigation of the possibility of synthesizing composites of these two carbon materials and exploring their performance as electrode materials has been done by several research groups [83, 91, 140].

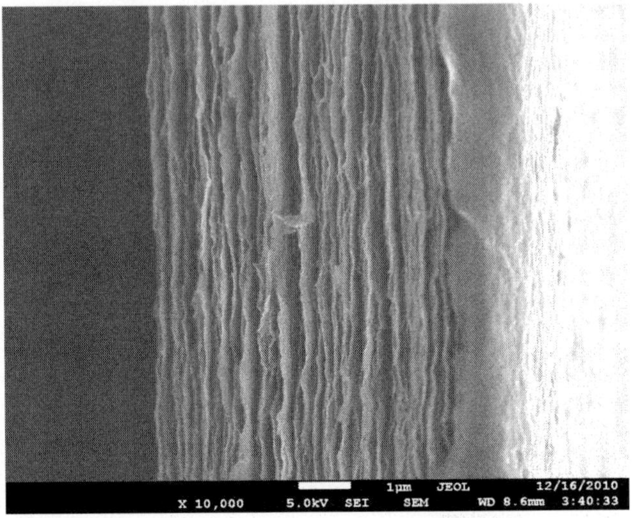

Figure 17. SEM image of the cross section of a reduced graphene oxide film

Kang et al. have shown that when composites composed of 1:1 weight RGO and P-SWNT are used in thermocells, the specific power generated (5.3 W/kg) is comparable to that of SWNTs (the experimental conditions used were discussed in section 3.2.1) [113]. It must be noted that when RGO alone is used, the specific power generated is 3.87 W/kg. However, the composite electrode produces a specific power that is only 78 % of the P-SWNT. This decrease in performance can be attributed to the large ohmic resistance observed in the RGO electrode (35.6Ω) that is 55 % higher than the P-SWNT electrode. The chemical conversion of graphite to RGO involves oxidising graphite, exfoliation and then a subsequent reduction. It is surmised that the incomplete removal of the oxygen containing functional groups is the cause of the pronounced ohmic resistance. Another reason for the poor performance of the composite electrodes is the restacking of the RGO sheets during electrode preparation, which impedes electrolyte diffusion and results in sluggish kinetics.

Optimisation of the RGO-SWNT composition for thermocell electrodes was done by Chen et. al [141]. The amount of RGO added ranged from 1 to 20 % by weight. A U-cell was used with 0.75 cm^2 electrodes separated by 10 cm, a thermal gradient of 60 °C with T_{cold}= 20 °C and 0.4 M ferri/ferrocyanide electrolyte. The optimised composite 99 % SWNT-1 % RGO generated a specific current density of 26.78 W/kg. By using large amounts of SWNTs the RGO sheets were prevented from restacking, which resulted in the appropriate nanoporosity that promoted redox mediator diffusion. The sheet like structure of the RGO provided increased pathways for electrons in the composite thus contributing to its enhanced performance. The interaction between SWNTs and RGO is clearly seen in Figure 18.

http://dx.doi.org/10.5772/51784

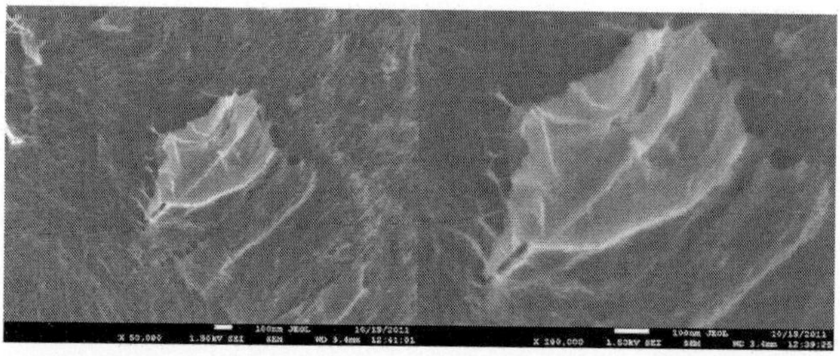

Figure 18. SEM images of reduced graphene oxide-SWNT composites

3.3. Developments in processing and fabrication of CNTs for better cell design

3.3.1. Current processing techniques for CNTs relevant to thermocells

Solvent/surfactant exfoliation

One of the major obstacles to research on CNT characterisation and application is their spontaneous aggregation brought about by attractive van der Waals interactions in both aqueous and organic solutions. The resulting aggregates or "bundles" can reach lengths of several microns and diameters of tens of nanometers. Debundling of these aggregates is essential as they have mediocre properties as compared to individual tubes. Reproducibility of results also becomes an issue when CNTs are not dispersed adequately.

Liquid phase separation is one of the simplest methods wherein stable CNT dispersions are attainable. Stable well-exfoliated CNT dispersions is achieved by appropriate selection of the solvent as forced dispersion via ultrasonication will result in agglomeration of the CNTs in a very short span of time. Selection of solvents can be based on the enthalpy of mixing per solvent volume ($\Delta \bar{H}_{mix}$) [138]:

$$\Delta \bar{H}_{mix} = \frac{2}{R_{bun}}(\delta_{NT} - \delta_{sol})^2 \varnothing \tag{7}$$

Where R_{bun} is the radius of the dispersed nanotube bundles, δ_{NT} and δ_{sol} are square roots of the nanotube and solvent surface energies and $\varnothing$ is the nanotube volume fraction. The solubility theory states that a negative free energy of mixing ($\Delta \bar{G}_{mix}$) is indicative of a stable dispersion.

$$\Delta \bar{G}_{mix} = \Delta \bar{H}_{mix} - T \Delta \bar{S}_{mix} \tag{8}$$

The entropy of mixing per unit volume ($\Delta \bar{S}_{mix}$) of nanotubes is generally small owing to their size and rigidity [142]. In order to realize a minimisation of $\Delta \bar{G}_{mix}$ then solvents which result in small values of ΔH_{mix} are necessary. Based on Equation 7, the most effective solvents at dispersing CNTs would be those that have a surface energy close to the nanotube surface energy (~70 mJ/m^2); i.e. solvents with surface tension around 40 mJ/m^2 [143].

Another method to attain stable dispersions of CNTs is through the use of surfactants. Its inherent advantage over solvent dispersion was the fact that it was carried out in aqueous media, lessening its hazards and environmental impact. It relies on the principle wherein colloids are stabilized by surface charges [144]; i.e. Coulomb repulsion. Adsorption of the amphiphilic surfactant molecules onto CNTs is attained through their hydrophobic tails. This introduces a removable surface charge that creates an electric double layer around the nanotube; of which the magnitude and sign is proportional to its zeta potential [145]. This double layer provides repulsive forces that counteract the attractive van der Waals forces [146]. Selection of surfactants for CNTs dispersion depends on the size of their molecules. Low molecular weight surfactants will be able to pack tightly around the nanotube surface resulting in better stabilization [147].

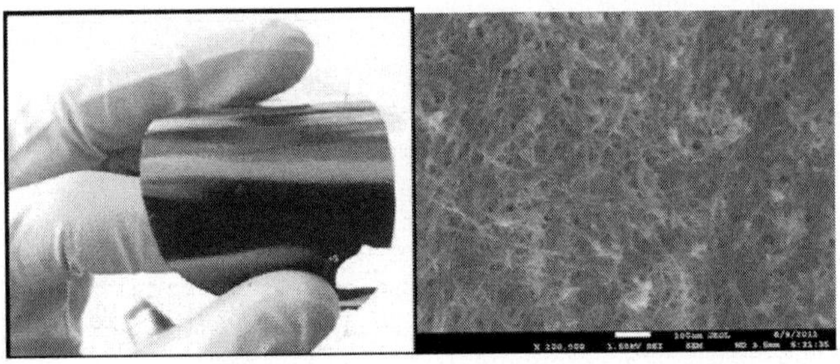

Figure 19. CNT bucky paper

Filtration of CNT dispersions results in a planar mat of randomly arranged tubes [148] that can be up to several hundred microns thick. These mats or "bucky papers" (Figure 19) have been instrumental in CNT evaluation (owing to their simple processing) as electrode materials not only for thermocells but for other electrochemical applications as well. Post-treatment of buckypapers via annealing or acid wash is essential to ensure complete removal of the solvent or surfactant used to attain the CNT dispersion [149].

Chemical vapour deposition

Chemical vapour deposition (CVD), a process that involves deposition of solids from a gas phase, has proven to be a viable method for attaining highly oriented CNTs on planar substrates (Figure 20) [150]. One of the theories behind the large degree of alignment is the reduction in free energy brought about by the van der waals interactions along the tube length inducing coordinated growth by holding the tubes together [151]. The ability to tailor the growth of CNTs in three dimensional configurations is highly advantageous in thermocell applications. This configuration promotes enhanced ion accessibility with the CNT matrix allowing larger current to be generated. The alignment of the tubes also minimises the tortuosity of the CNT electrodes which decreases the probability of forming concentration gradients within the electrode itself, leading to a decrease in the mass transfer overpotential. The wide range of substrates that can be used for CNT synthesis (metallic, carbon, etc) via CVD allows the fabrication of electrodes with a high degree of flexibility; materials that are highly desirable in thermocells [152].

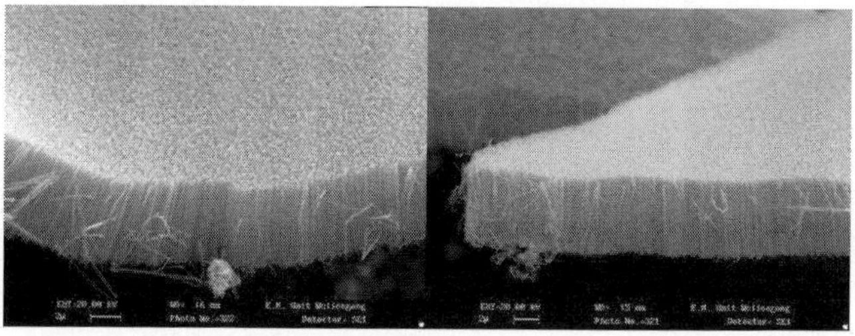

Figure 20. Aligned CNTs produced by CVD process

3.3.2. Cell design breakthroughs attained using CNTs

The development of flexible electrodes for electrochemical applications has paved the way for innovative cell designs for thermal energy conversion. CNT forests grown directly on thermocell casings, scroll electrodes, and thermocells that can be wrapped around cooling/heating pipes have been attained through flexible CNT electrodes [114].

Mark II thermocell

When inter electrode separation is decreased, larger specific power is generated as mass transfer is enhanced over shorter distances. However, this results in decreased power conversion efficiency as larger thermal energy is required to maintain a similar thermal gradient [112]. Scroll electrodes (Figure 21) can be employed to mitigate this problem. Using scrolled MWNT buck papers, each with a diameter of 0.3 cm and mass of 0.5 mg, aligned along their rolling axis inside a glass tube containing 0.4 M ferri/ferrocyanide, an electrode separation of 5 cm, T_{cold} of 5 °C and thermal gradient of 60 °C a specific power density of 1.8 W/m^2 was obtained. The power conversions efficiency of 0.24 % is an order of magnitude higher than thermocells using Pt electrodes tested under similar conditions [114]. The relative efficiency of the Mark II thermocell is 17 % higher than that obtained when using platinum electrodes, giving Φ_r of 1.4 %.

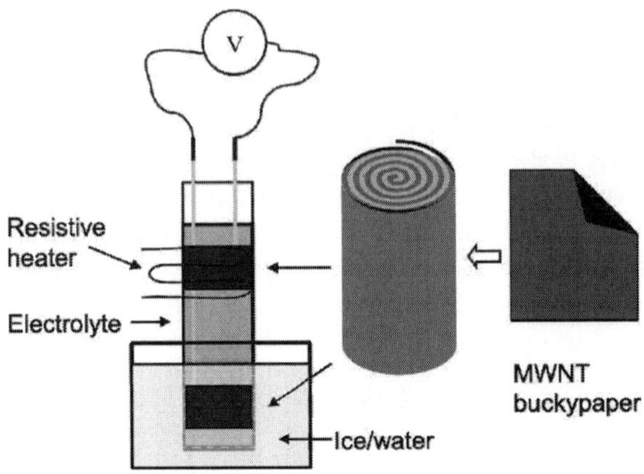

Figure 21. Mark II thermocell[114] Reprinted with permission from American Chemical Society

Coin cell

Thin coin type thermocells which could be powered by extremely low thermal gradients were developed using MWNTs and 0.4M ferri/ferrocyanide as the electrolyte (Figure 22). Coin cells fabricated using MWNT bucky paper electrodes and exposed to a thermal gradient of 45°C generated a specific power of 0.389 W/m^2 (equivalent to a normalised power density $P_{max}/\Delta T^2$ of 1.92 x 10^{-4} W/m^2K). Coin cells with electrodes made of MWNT forests around 100 μm tall, grown directly on the internal stainless steel surface of the packaging substrate using a trilayer catalyst (30 nm Ti, 10 nm Al, 2 nm Fe) plasma enhanced CVD method was also developed. The specific power generated at a thermal gradient of 60°C was 0.980 W/m^2,giving $P_{max}/\Delta T^2$ =2.72 x 10^{-4} W/m^2K. The larger normalised power density of the coin cell with MWNT forest electrodes is due to its nanotube alignment, which promotes electrolyte diffusion, and the lower thermal (0.01 cm^2K/W) and electrical resistance at the MWNT forest/substrate junction [153]. The thermal resistance for bucky papers is around 0.05 cm^2 K/W which leads to larger loss of thermal energy at the electrode/substrate junction and 30% less power conversion efficiency [154].

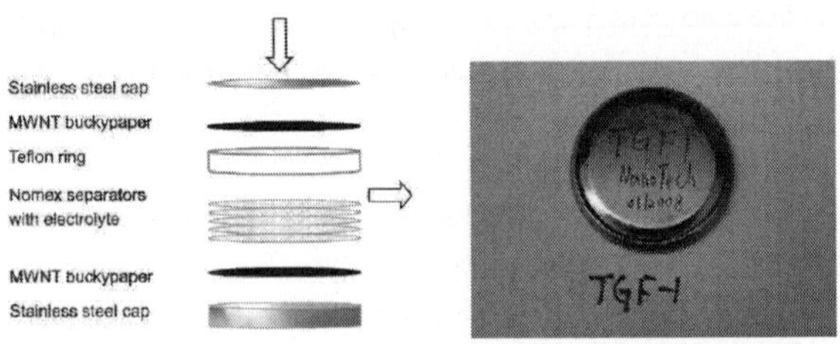

Figure 22. Coin cell for thermal energy conversion [114] Reprinted with permission from American Chemical Society

Flexible thermocell

One of the main applications of thermocells is to harvest thermal energy from automobile exhaust pipes and cooling or heating lines in industrial facilities. Flexible thermocells can be wrapped around these pipes and convert them to sources of electrical power. A flexible thermocells consisting of two MWNTbucky paper electrodes kept apart by 2 layers of NomexHT 4848 impregnated with 0.4M ferri/ferrocyanide and wrapped in a stainless steel sheetis shown in Figure 23. The cell was wrapped around a cooling pipe and a thermal gradient of 15°C was applied using a resistive heater. A specific power 0.39 W/m^2 was generated proving that flexible thermocells are now a possibility [114].

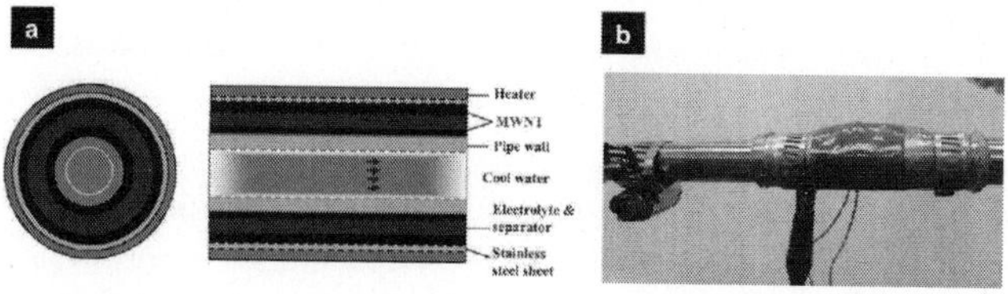

Figure 23. a) schematic b) photo of a flexible thermocell that can be wrapped around cooling/heating pipes [114] Reprinted with permission from American Chemical Society

3.4. Conclusion

Research on CNTs as electrode materials for thermogalvanic cells is still in its early stages. However, these initial results indicate that these nanocarbons are capable of generating significant amounts of power; much larger than when conventional electrodes are used. Without excluding the energy input from mechanical stirring, thermocells with platinum electrodes are able to attain a power conversion efficiency relative to a Carnot engine of 1.2% [100]. This value was surpassed with the use of CNT electrodes, reaching Φ_r = 1.4%, in a thermocell that did not utilise any mechanical stirring and relied only on convection and diffusion to cause mass transfer of reaction products. The most important breakthrough is in the area of cell design as the robustness of CNT electrodes allows them to be conformed into a variety of shapes in order to mitigate heat flow from the hot to cold side of the thermocell. Flexible thermocells are also possible; devices that can harvest heat from heating or cooling pipes. Optimisation of the porosity of these CNT electrodes is essential in order to minimise the tortuosity and to reduce the mass transport overpotential in these systems. Doping CNT electrodes can alter their electroactive surface area by up to 4-fold; this feature can be exploited by selecting the right electrolyte. The use of CNT-RGO composites has demonstrated the synergistic effect of these two materials, augmenting the power conversion efficieny of thermocells. Further developments in the field of CNT synthesis and processing will decrease the cost of these materials such that commercialisation of thermogalvanic cells may one day be realised.

4. Perspective and future developments

With energy consumption as a whole on the increase, coupled with the rapid economic development of countries such as Brazil, China, India, and Russia there will be a concerted effort to improve how energy is utilised. This expansion in industrialisation has already and will continue to lead to a further increase in the price of oil. Coupled to the rise in fossil fuel costs are drivers of an ageing energy infrastructure system and demand for a low-carbon emission economy through the use of renewable energy [155]. To help accommodate all these factors the supply and demand challenge may be addressed by tapping into otherwise wasted energy. Low grade heat, if effectively harvested can prove to be a viable source of power. Thermal converters have the potential to increase the efficiency of current energy conversion systems. Energy storage also plays a key role in providing a solution to the energy problem. Energy must be efficiently stored, when it is in excess, and released at a time of high demand. This is extremely important for renewables that are not load-following [156].

With these energy challenges and ongoing research and development, including those that have been conducted over the last decade, the awareness of the benefits of electrochemical capacitors is increasing. As the research and development into energy storage and conversion has increased, the applications of electrochemical capacitors has increased with the technology becoming more diverse meaning that systems can better be tailored /targeted for specific applications ranging from higher energy density to high power densities where fast charge / discharge efficiencies are needed [156]. The most commonly used material for supercapacitors has been activated carbon with new nanostructured materials such as carbon nanotubes and its derivatives coming to the forefront of the current fundamental research. It can be seen that the way forward in terms of trying to improve energy density and power density is in the use of CNT with composite materials such as other carbons, and conducting polymers or metal oxides in order to take advantage of the pseudocapacitative effects that these materials provide.

The research on thermogalvanic systems in the past has been generally limited to platinum electrodes [100]. This has enhanced the understanding of these electrochemical systems but has not advanced the research in terms of commercialisation due to its cost. The use of carbon nanomaterials has improved the performance of these devices immensely because of their fast transfer kinetics and large electroactive surface area and is also economically viable. A record threefold increase in power conversion efficiency (as compared to conventional systems wherein platinum is used) has been realised with the use of MWNT electrodes [114]. Flexible electrodes are now possible due to CNTs. These can be used as scroll electrodes or for thermocells that can be wrapped around pipes will make this system more versatile in terms of its possible applications. Further increase in thermocell performance may be realised with the use of CNTs-graphene composite materials.

Future development will most likely see supercapacitors and thermocells become a central part of hybrid energy storage and power delivery systems for large scale and domestic demand strategies. The integration of these two systems into one device will allow the converted waste heat to be stored then released when deemed necessary. These future advancements will not only enable better automotive and portable electronics, but they will revolutionise the fields of medicine, defence and consumer goods thus providing a step change in energy storage technology [5].

Author details

Dennis Antiohos*, Mark Romano, Jun Chen* and Joselito M. Razal

*Address all correspondence to: junc@uow.edu.au

Intelligent Polymer Research Institute, Australian Institute of Innovative Materials, Innovation Campus, University of Wollongong, Australia

References

[1] Turner, J. A. (1999). A Realizable Renewable Energy Future. *Science*, 285(5428), 687-689.

[2] Shukla, A. K. S. S., & Vijayamohanan, K. (2000). Electrochemical supercapacitors: Energy storage beyond batteries. *Current Science*, 79.

[3] Arico, A. S., et al. (2005). Nanostructured materials for advanced energy conversion and storage devices. *Nat Mater*, 4(5), 366-377.

[4] Hall, P. J., et al. (2010). Energy storage in electrochemical capacitors: designing functional materials to improve performance. *Energy & Environmental Science*.

[5] Hall, P. J., & Bain, E. J. (2008). Energy-storage technologies and electricity generation. *Energy Policy*, 36(12), 4352-4355.

[6] Murakami, T., et al. (2003). Thermoelectric Power of M-H Systems in Molten Salts and Application to M-H Thermogalvanic Cell. *Journal of The Electrochemical Society*, 150(7), A928-A932.

[7] Chung, J., et al. (2004). Toward Large-Scale Integration of Carbon Nanotubes. *Langmuir*, 20(8), 3011-3017.

[8] Frackowiak, E, et al. (2000). Supercapacitor electrodes from multiwalled carbon nanotubes. *Applied Physics Letters*, 77(15), 2421.

[9] Anantram, M.P, & Léonard, F. (2010). Physics of carbon nanotube electronic devices Rep. Prog. Phys. , 2006, 69.

[10] Pauliukaite, R., et al. Electrochemical impedance studies of chitosan-modified electrodes for application in electrochemical sensors and biosensors. *Electrochimica Acta*, 55(21), 6239.

[11] Dolatshahi-Pirouz, A., et al. (2008). Bovine serum albumin adsorption on nano-rough platinum surfaces studied by QCM-D. *Colloids and Surfaces B: Biointerfaces*, 66(1), 53.

[12] Aaron, Davies., & Yu, A. (2011). Material Advancements in Supercapacitors: From Activated Carbon to Carbon Nanotube and Graphene. *The Canadian Journal of Chemical Engineering*, 89, 1342-1357.

[13] Sherman, L. M. (2007). Carbon Nanotubes Lots of Potential--If the Price is Right. 01/05/12]; Available from:, www.ptonline.com/articles/carbon-nanotubes-lots-of-potentialif-the-price-is-right.

[14] Kierzek, K., et al. (2004). Electrochemical capacitors based on highly porous carbons prepared by KOH activation. *Electrochimica Acta*, 49(4), 515-523.

[15] Izadi-Najafabadi, A., et al. Extracting the Full Potential of Single-Walled Carbon Nanotubes as Durable Supercapacitor Electrodes Operable at 4 V with High Power and Energy Density. *Advanced Materials*.

[16] Kim, B. C., et al. (2009). Capacitive properties of RuO2 and Ru-Co mixed oxide deposited on single-walled carbon nanotubes for high-performance supercapacitors. *Synthetic Metals*, 159(13), 1389-1392.

[17] Bard, A. J., & Faulkner, L. R. (2001). Electrochemical Methods: Fundamentals and Applications. John Wiley & Sons, Inc.

[18] Yoon, S., et al. (2009). Preparation of mesoporous carbon/manganese oxide materials and its application to supercapacitor electrodes. *Journal of Non-Crystalline Solids*, 355(4-5), 252-256.

[19] Snook, G. A., Kao, P., & Best, A. S. (2011). Conducting-polymer-based supercapacitor devices and electrodes. *Journal of Power Sources*, 196(1), 1-12.

[20] Qu, Q. T., Wang, B., Yang, L. C., Shi, Y., Tian, S., & Wu, Y. P. (2008). Study on electrochemical performance of activated carbon in aqueous Li2SO4, Na2SO4 and K2SO4 electrolytes. *Electrochemistry Communications*, 10(10), 1652-1655.

[21] Bose, S., et al. (2012). Carbon-based nanostructured materials and their composites as supercapacitor electrodes. *Journal of Materials Chemistry*, 22(3), 767-784.

[22] Zhou, C. (2006). Carbon Nanotube Based Electrochemical Supercapacitors, in School of Polymer, Textile and Fiber Engineering. *Georgia Institute of Technology: Atlanta, Georgia*.

[23] Inagaki, M., Konno, H., & Tanaike, O. (2010). Carbon materials for electrochemical capacitors. *Journal of Power Sources*, 195(24), 7880-7903.

[24] Subramanian, V. R., Devan, S., & White, R. E. (2004). An approximate solution for a pseudocapacitor. *Journal of Power Sources*, 135(1-2), 361-367.

[25] Delahay, P., & Holub, K. (1968). Coupling of charging and faradaic processes: Electrode admittance for reversible processes. *Journal of Electroanalytical Chemistry*, 16(2), 131-136.

[26] Liu, T. C., Pell, W. G., & Conway, B. E. (1999). Stages in the development of thick cobalt oxide films exhibiting reversible redox behavior and pseudocapacitance. *Electrochimica Acta*, 44(17), 2829-2842.

[27] Pollak, E., Salitra, G., & Aurbach, D. (2007). Can conductivity measurements serve as a tool for assessing pseudocapacitance processes occurring on carbon electrodes? *Journal of Electroanalytical Chemistry*, 602(2), 195-202.

[28] Bradley, D. (2010). Ordered energy storage: Energy. *Materials Today*, 13(1-2), 9.

[29] Presser, V., Heon, M., & Gogotsi, Y. (2011). Carbide-Derived Carbons- From Porous Networks to Nanotubes and Graphene. *Advanced Functional Materials*, 21(5), 810-833.

[30] Zhang, L. L., & Zhao, X. S. (2009). Carbon-based materials as supercapacitor electrodes. *Chemical Society Reviews*, 38(9), 2520-2531.

[31] Battery and Energy Technologies. (2012). Available from:, http://www.mpower-uk.com/performance.htm.

[32] Conway, B. E. (1999). Electrochemical Supercapacitors: Scientific Fundamentals and Technological Applications. New York: Kluwer Acedemic / Plenum Publishers.

[33] Jong, H. J., et al. (2006). Supercapacitor Performance of Hydrous Ruthenium Oxide Electrodes Prepared by Electrophoretic Deposition. *Journal of The Electrochemical Society*, 153(2), A321-A328.

[34] Lewandowski, A., et al. (2010). Performance of carbon-carbon supercapacitors based on organic, aqueous and ionic liquid electrolytes. *Journal of Power Sources*, 195(17), 5814-5819.

[35] Kötz, R, & Carlen, M. (2000). Principles and applications of electrochemical capacitors. *Electrochimica Acta*, 45(15-16), 2483-2498.

[36] Balducci, A., et al. (2004). Ionic liquids for hybrid supercapacitors. *Electrochemistry Communications*, 6(6), 566-570.

[37] Frackowiak, E., Lota, G., & Pernak, J. (2005). Room-temperature phosphonium ionic liquids for supercapacitor application. *Applied Physics Letters*, 86(16), 164104-164103.

[38] Wei, D., & Ivaska, A. (2008). Applications of ionic liquids in electrochemical sensors. *Analytica Chimica Acta*, 607(2), 126-135.

[39] Iijima, S. (1991). Helical microtubules of graphitic carbon. *Nature*, 354(6348), 56-58.

[40] Yu, B, et al. (2006). The electrolyte switchable solubility of multi-walled carbon nanotube/ionic liquid (MWCNT/IL) hybrids. *Chemical Communications* [22], 2356-2358.

[41] Liu, J., et al. (2008). Postsynthesis microwave treatment to give high-purity multiwalled carbon nanotubes. *AIChE Journal*, 54(12), 3303-3307.

[42] Che, G., et al. (1999). Metal-Nanocluster-Filled Carbon Nanotubes:â€‰ Catalytic Properties and Possible Applications in Electrochemical Energy Storage and Production. *Langmuir*, 15(3), 750-758.

[43] Meyyappan, M. (2005). Carbon Nanotubes: Science & Applications. Boca Raton: CRC Press.

[44] Cai, X., Cong, H., & Liu, C. (2012). Synthesis of vertically-aligned carbon nanotubes without a catalyst by hydrogen arc discharge. *Carbon*, 50(8), 2726-2730.

[45] Liu, C, et al. (1999). Semi-continuous synthesis of single-walled carbon nanotubes by a hydrogen arc discharge method. *Carbon*, 37(11), 1865-1868.

[46] Yuge, R., et al. (2012). Characterization and field emission properties of multi-walled carbon nanotubes with fine crystallinity prepared by CO2 laser ablation. *Applied Surface Science.*

[47] Zhihua, P, et al. (2008). Investigation of the microwave absorbing mechanisms of HiPco carbon nanotubes. *Physica E: Low-dimensional Systems and Nanostructures*, 40(7), 2400-2405.

[48] Nikolaev, P. (2004). Gas-phase production of single-walled carbon nanotubes from carbon monoxide: A review of the HiPco process. *Journal of Nanoscience and Nanotechnology*, 307-316.

[49] Michael, J. Bronikowski, et al. (2001). Gas-phase production of carbon single-walled nanotubes from carbon monoxide via the HiPco process: A parametric study. *J. Vac. Sci. Technol. A.*, 19, 1800.

[50] Wang, G., Zhang, L., & Zhang, J. (2012). A review of electrode materials for electrochemical supercapacitors. *Chemical Society Reviews*, 797-828.

[51] Huang, C. W., et al. (2012). Electric double layer capacitors based on a composite electrode of activated mesophase pitch and carbon nanotubes. *Journal of Materials Chemistry*, 7314-7322.

[52] Dai, L. (2006). Carbon Nanotechnology. Amsterdam: Elsevier.

[53] Deng, F., & Zheng, Q. (2009). Interaction models for effective thermal and electric conductivities of carbon nanotube composites. *Acta Mechanica Solida Sinica*, 22(1), 1-17.

[54] Obreja, V. V. N. (2008). On the performance of supercapacitors with electrodes based on carbon nanotubes and carbon activated material--A review. *Physica E: Low-dimensional Systems and Nanostructures*, 40(7), 2596-2605.

[55] Kay, Hyeok An, et al. (2001). Characterization of Supercapacitors Using Singlewalled Carbon Nanotube Electrodes. *Journal of the Korean Physical Society*, 39, S511-S517.

[56] Hu, S., Rajamani, R., & Yu, X. (2012). Flexible solid-state paper based carbon nanotube supercapacitor. *Applied Physics Letters*, 100(10).

[57] School of Pharmacy, . Research Program: University of Waterloo. 20/16/12]; Available from:, http://science.uwaterloo.ca/~foldvari/images/SWNT-MWNT.jpg.

[58] Morant, C., et al. (2012). Mo-Co catalyst nanoparticles: Comparative study between TiN and Si surfaces for single-walled carbon nanotube growth. *Thin Solid Films*, 39(520), 16-5238.

[59] Candelaria, S. L., et al. (2012). Nanostructured carbon for energy storage and conversion. *Nano Energy*, 1(2), 195-220.

[60] Chen, P. J. C., Qiu, J., & Zhou, C. (2010). *Nano Research*, 3, 594-603.

[61] Chau, T. T., et al. (2009). A review of factors that affect contact angle and implications for flotation practice. *Advances in Colloid and Interface Science*, 150(2), 106-115.

[62] Kuratani, K., , T. K., & Kuriyama, N. (2009). *Journal of Power Sources*, 189, 1284-1291.

[63] Bakhmatyuk, B. P., , B. Y. V., Grygorchak, I. I., & Micov, M. M. (2008). *Journal of Power Sources*, 180, 890-895.

[64] Pan, H., Li, J., & Feng, Y. (2010). Carbon Nanotubes for Supercapacitor. *Nanoscale Research Letters*, 5(3), 654-668.

[65] Fuertes, A. B., et al. (2005). Templated mesoporous carbons for supercapacitor application. *Electrochimica Acta*, 50(14), 2799-2805.

[66] Shen, J., et al. (2011). How carboxylic groups improve the performance of single-walled carbon nanotube electrochemical capacitors? *Energy & Environmental Science*, 4(10), 4220-4229.

[67] Wu, Z. S., et al. (2011). Doped Graphene Sheets as Anode Materials with Superhigh Rate and Large Capacity for Lithium Ion Batteries. *ACS Nano*, 5463-5471.

[68] Babel, K., & Jurewicz, K. (2002). Electrical capacitance of fibrous carbon composites in supercapacitors. *Fuel Processing Technology*, 77-78, 181-189.

[69] Béguin, F, et al. (2005). A Self-Supporting Electrode for Supercapacitors Prepared by One-Step Pyrolysis of Carbon Nanotube/Polyacrylonitrile Blends. *Advanced Materials*, 17(19), 2380-2384.

[70] Drage, T. C., et al. (2007). Preparation of carbon dioxide adsorbents from the chemical activation of urea-formaldehyde and melamine-formaldehyde resins. *Fuel*, 86(1-2), 22-31.

[71] Inagaki, N., et al. (2007). Implantation of amino functionality into amorphous carbon sheet surfaces by NH3 plasma. *Carbon*, 45(4), 797-804.

[72] Li, W., et al. (2007). Nitrogen enriched mesoporous carbon spheres obtained by a facile method and its application for electrochemical capacitor. *Electrochemistry Communications*, 9(4), 569-573.

[73] Stein, A., Wang, Z., & Fierke, M. A. (2009). Functionalization of Porous Carbon Materials with Designed Pore Architecture. *Advanced Materials*, 21(3), 265-293.

[74] Peigney, A., et al. (2001). Specific surface area of carbon nanotubes and bundles of carbon nanotubes. *Carbon*, 39(4), 507-514.

[75] Niu, J. J., et al. (2007). An approach to carbon nanotubes with high surface area and large pore volume. *Microporous and Mesoporous Materials*, 100(1-3), 1-5.

[76] Zhang, Y., et al. (2010). Preparation and electrochemical properties of nitrogen-doped multi-walled carbon nanotubes. *Materials Letters*, 65(1), 49-52.

[77] Lee, K. Y., et al. (2010). Influence of the nitrogen content on the electrochemical capacitor characteristics of vertically aligned carbon nanotubes. *Physica E: Low-dimensional Systems and Nanostructures*, 42(10), 2799-2803.

[78] Antal, A. Koósa, et al. (2010). Comparison of structural changes in nitrogen and boron-doped multi-walled carbon nanotubes. *Carbon*, 48(11), 3033-3041.

[79] Shiraishi, S., et al. (2006). Electric double layer capacitance of multi-walled carbon nanotubes and B-doping effect. *Applied Physics A: Materials Science and Processing*, 82(4), 585-591.

[80] Wang, D., , W., et al. (2008). Synthesis and Electrochemical Property of Boron-Doped Mesoporous Carbon in Supercapacitor. *Chemistry of Materials*, 20(22), 7195-7200.

[81] Gao, Y., et al. (2012). High power supercapacitor electrodes based on flexible TiC-CDC nano-felts. *Journal of Power Sources*, 201(0), 368-375.

[82] Zhou, J., et al. (2009). Mesoporous carbon spheres with uniformly penetrating channels and their use as a supercapacitor electrode material. *Materials Characterization*, 61(1), 31-38.

[83] Korenblit, Y., et al. (2010). High-Rate Electrochemical Capacitors Based on Ordered Mesoporous Silicon Carbide-Derived Carbon. *ACS Nano*, 4(3), 1337-1344.

[84] Fu, C., et al. (2011). Supercapacitor based on electropolymerized polythiophene and multi-walled carbon nanotubes composites. *Materials Chemistry and Physics*, 132(2), 596-600.

[85] Wu, Z. S., et al. (2012). Graphene/metal oxide composite electrode materials for energy storage. *Nano Energy*, 1(1), 107-131.

[86] Bhandari, S., et al. (2009). PEDOT-MWNTs commposite films. *J. Phys. Chem B*, 113, 9416-9428.

[87] Zhang, X, et al. (2011). Ultralight conducting polymer/carbon nanotube composite aerogels. *Carbon*, 49(6), 1884-1893.

[88] Crispin, X., et al. (2006). The Origin of the High Conductivity of Poly(3,4-ethylenedioxythiophene) Poly(styrenesulfonate) (PEDOT/PSS) Plastic Electrodes. *Chemistry of Materials*, 18(18), 4354-4360.

[89] Antiohos, D., et al. (2011). Compositional effects of PEDOT-PSS/single walled carbon nanotube films on supercapacitor device performance. *Journal of Materials Chemistry*, 21(40), 15987-15994.

[90] Kim, K. S., & Park, J. S. (2011). Influence of multi-walled carbon nanotubes on the electrochemical performance of graphene nanocomposites for supercapacitor electrodes. *Electrochimica Acta*, 56(3), 1629-1635.

[91] Hu, Y., et al. (2012). Defective super-long carbon nanotubes and polypyrrole composite for high-performance supercapacitor electrodes. *Electrochimica Acta*, 60(0), 279-286.

[92] Myoungki, M., et al. (2006). Hydrous RuO2/Carbon Black Nanocomposites with 3D Porous Structure by Novel Incipient Wetness Method for Supercapacitors. *Journal of The Electrochemical Society*, 153(2), A334-A338.

[93] Zhai, Y., et al. (2011). Carbon materials for chemical capacitive energy storage. *Advanced Materials*, 23(42), 4828-4850.

[94] Shan, Y., & Gao, L. (2007). Formation and characterization of multi-walled carbon nanotubes/Co3O4 nanocomposites for supercapacitors. Materials Chemistry and Physics (2â€"3): , 103, 206-210.

[95] Jayalakshmi, M., et al. (2007). Hydrothermal synthesis of SnO2-V2O5mixed oxide and electrochemical screening of carbon nano-tubes (CNT), V2O5, V2O5-CNT, and SnO2-V2O5-CNT electrodes for supercapacitor applications. *Journal of Power Sources*, 166(2), 578-583.

[96] Li, Q., et al. (2011). Structural evolution of multi-walled carbon nanotube/MnO2 composites as supercapacitor electrodes. *Electrochimica Acta*, 59(0), 548-557.

[97] Li, J. J., et al. (2012). Graphene/carbon nanotube films prepared by solution casting for electrochemical energy storage. *IEEE Transactions on Nanotechnology*, 11(1), 3-7.

[98] Dong, X., et al. (2011). The formation of a carbon nanotube-graphene oxide core-shell structure and its possible applications. *Carbon*, 49(15), 5071-5078.

[99] Wartanowicz, T. (1964). The theoretical analysis of a molten salt thermocell as a thermoelectric generator. *Advanced Energy Conversion*, 4(3), 149-158.

[100] Quickenden, T. I., & Mua, Y. (1995). A Review of Power Generation in Aqueous Thermogalvanic Cells. *Journal of The Electrochemical Society*, 142(11), 3985-3994.

[101] Kuzminskii, Y. V., Zasukha, V. A., & Kuzminskaya, G. Y. (1994). Thermoelectric effects in electrochemical systems. Nonconventional thermogalvanic cells. *Journal of Power Sources*, 52(2), 231-242.

[102] Wartsila. (2012). The world's most powerful reciprocating engine. Available from:, http://www.wartsila.com/en/engines/low-speed-engines/RT-flex96C.

[103] Dincer, I. (2002). On thermal energy storage systems and applications in buildings. *Energy and Buildings*, 34(4), 377-388.

[104] Bell, L. E. (2008). Cooling, Heating, Generating Power, and Recovering Waste Heat with Thermoelectric Systems. *Science*, 321(5895), 1457-1461.

[105] Ujihara, M., Carman, G. P., & Lee, D. G. (2007). Thermal energy harvesting device using ferromagnetic materials. *Applied Physics Letters*, 91(9), 093508-093503.

[106] Vining, C. B. (2009). An inconvenient truth about thermoelectrics. *Nat Mater*, 8(2), 83-85.

[107] Hertz, H. G., & Ratkje, S. K. (1989). Theory of Thermocells. *Journal of The Electrochemical Society*, 136(6), 1698-1704.

[108] Romano, M., et al. (2012). Novel carbon materials for thermal energy harvesting. *Journal of Thermal Analysis and Calorimetry*, 1-7.

[109] Goncalves, R., & Ikeshoji, T. (1992). Comparative studies of a thermoelectric converter by a thermogalvanic cell with a mixture of concentrated potassium ferrocyanide and potassium ferricyanide aqueous solutions at great temperature differences. *J. Braz. Chem. Soc*, 3(3), 4.

[110] Quickenden, T. I., & Vernon, C. F. (1986). Thermogalvanic conversion of heat to electricity. *Solar Energy*, 36(1), 63-72.

[111] Artjom, V. S. (1994). Theoretical study of thermogalvanic cells in steady state. *Electrochimica Acta*, 39(4), 597-609.

[112] Mua, Y., & Quickenden, T. I. (1996). Power Conversion Efficiency, Electrode Separation, and Overpotential in the Ferricyanide/Ferrocyanide Thermogalvanic Cell. *Journal of The Electrochemical Society*, 143(8), 2558-2564.

[113] Kang, T. J., et al. (2011). Electrical Power From Nanotube and Graphene Electrochemical Thermal Energy Harvesters.

[114] Hu, R., et al. (2010). Harvesting Waste Thermal Energy Using a Carbon-Nanotube-Based Thermo-Electrochemical Cell. *Nano Letters*, 10(3), 838-846.

[115] Salazar, P. F., Kumar, S., & Cola, B. A. (2012). Nitrogen- and Boron-Doped Carbon Nanotube Electrodes in a Thermo-Electrochemical Cell. *Journal of The Electrochemical Society*, 159(5), B483-B488.

[116] Quickenden, T. I., & Mua, Y. (1995). The Power Conversion Efficiencies of a Thermogalvanic Cell Operated in Three Different Orientations. *Journal of The Electrochemical Society*, 142(11), 3652-3659.

[117] Antiohos, D., et al. (2010). Electrochemical investigation of carbon nanotube nanoweb architecture in biological media. *Electrochemistry Communications*, 12(11), 1471-1474.

[118] Landi, B. J., et al. (2009). Carbon nanotubes for lithium ion batteries. *Energy & Environmental Science*, 2(6), 638-654.

[119] Wang, C., et al. (2003). Proton Exchange Membrane Fuel Cells with Carbon Nanotube Based Electrodes. *Nano Letters*, 4(2), 345-348.

[120] Baughman, R. H., Zakhidov, A. A., & de Heer, W. A. (2002). Carbon Nanotubes--the Route Toward Applications. *Science*, 297(5582), 787-792.

[121] Nugent, J. M., et al. (2001). Fast Electron Transfer Kinetics on Multiwalled Carbon Nanotube Microbundle Electrodes. *Nano Letters*, 1(2), 87-91.

[122] Inoue, S., et al. (1998). Capillary Condensation of N2 on Multiwall Carbon Nanotubes. *The Journal of Physical Chemistry B*, 102(24), 4689-4692.

[123] Eswaramoorthy, M., Sen, R., & Rao, C. N. R. (1999). A study of micropores in single-walled carbon nanotubes by the adsorption of gases and vapors. *Chemical Physics Letters*, 304(3-4), 207-210.

[124] Banks, C.E, et al. (2004). Investigation of modified basal plane pyrolytic graphite electrodes: definitive evidence for the electrocatalytic properties of the ends of carbon nanotubes. *Chemical Communications* [16], 1804-1805.

[125] Chun, K. Y., Lee, H. S., & Lee, C. J. (2009). Nitrogen doping effects on the structure behavior and the field emission performance of double-walled carbon nanotubes. *Carbon*, 47(1), 169-177.

[126] Charlier, J. C., et al. (2002). Enhanced Electron Field Emission in B-doped Carbon Nanotubes. *Nano Letters*, 2(11), 1191-1195.

[127] Deng, C., et al. (2009). Electrochemical detection of l-cysteine using a boron-doped carbon nanotube-modified electrode. *Electrochimica Acta*, 54(12), 3298-3302.

[128] Panchakarla, L. S., Govindaraj, A., & Rao, C. N. R. (2007). Nitrogen- and Boron-Doped Double-Walled Carbon Nanotubes. *ACS Nano*, 1(5), 494-500.

[129] Gong, K., et al. (2009). Nitrogen-Doped Carbon Nanotube Arrays with High Electrocatalytic Activity for Oxygen Reduction. *Science*, 323(5915), 760-764.

[130] Terrones, M, et al. (2002). N-doping and coalescence of carbon nanotubes: synthesis and electronic properties. Applied Physics A: Materials Science & Processing , 74(3), 355-361.

[131] Lee, Y. T., et al. (2003). Growth of Vertically Aligned Nitrogen-Doped Carbon Nanotubes: Control of the Nitrogen Content over the Temperature Range 900–1100 °C. *The Journal of Physical Chemistry B*, 107(47), 12958-12963.

[132] Banks, C. E., & Compton, R. G. (2005). Exploring the electrocatalytic sites of carbon nanotubes for NADH detection: an edge plane pyrolytic graphite electrode study. *Analyst*, 130(9), 1232-1239.

[133] Deng, C., et al. (2008). Boron-doped carbon nanotubes modified electrode for electroanalysis of NADH. *Electrochemistry Communications*, 10(6), 907-909.

[134] Strmcnik, D., et al. (2009). The role of non-covalent interactions in electrocatalytic fuel-cell reactions on platinum. *Nat Chem*, 1(6), 466-472.

[135] Novoselov, K. S., et al. (2004). Electric Field Effect in Atomically Thin Carbon Films. *Science*, 306(5696), 666-669.

[136] Bolotin, K. I., et al. (2008). Ultrahigh electron mobility in suspended graphene. *Solid State Communications*, 146(9-10), 351-355.

[137] Stoller, M. D., et al. (2008). Graphene-Based Ultracapacitors. *Nano Letters*, 8(10), 3498-3502.

[138] Coleman, J. N. (2009). Liquid-Phase Exfoliation of Nanotubes and Graphene. *Advanced Functional Materials*, 19(23), 3680-3695.

[139] Park, S., et al. (2009). Colloidal Suspensions of Highly Reduced Graphene Oxide in a Wide Variety of Organic Solvents. *Nano Letters*, 9(4), 1593-1597.

[140] Qiu, L., et al. (2010). Dispersing Carbon Nanotubes with Graphene Oxide in Water and Synergistic Effects between Graphene Derivatives. *Chemistry- A European Journal*, 16(35), 10653-10658.

[141] Romano, M. S., et al. (2012). Novel Carbon Nanomaterials for Thermal Energy Converters. in International Society of Electrochemistry 10th Spring Meeting. Perth, Australia.

[142] Bergin, S.D, et al. (2009). Multicomponent Solubility Parameters for Single-Walled Carbon Nanotube–Solvent Mixtures. *ACS Nano*, 3(8), 2340-2350.

[143] Bergin, S.D, et al. (2008). Towards Solutions of Single-Walled Carbon Nanotubes in Common Solvents. *Advanced Materials*, 20(10), 1876-1881.

[144] Israelachvili, J. (1991). Intermolecular and Surface Forces. *Academic Press, London.*

[145] White, B., et al. (2007). Zeta-Potential Measurements of Surfactant-Wrapped Individual Single-Walled Carbon Nanotubes. *The Journal of Physical Chemistry C*, 111(37), 13684-13690.

[146] Hunter, R. (1994). Introduction to Modern Colloid Science. *Oxford: Oxford Science Publications.*

[147] Sun, Z., et al. (2008). Quantitative Evaluation of Surfactant-stabilized Single-walled Carbon Nanotubes: Dispersion Quality and Its Correlation with Zeta Potential. *The Journal of Physical Chemistry C*, 112(29), 10692-10699.

[148] Liu, J., et al. (1998). Fullerene Pipes. *Science*, 280(5367), 1253-1256.

[149] Geng, H. Z., et al. (2008). Absorption spectroscopy of surfactant-dispersed carbon nanotube film: Modulation of electronic structures. *Chemical Physics Letters*, 455(4-6), 275-278.

[150] Chen, J., et al. (2007). Flexible, Aligned Carbon Nanotube/Conducting Polymer Electrodes for a Lithium-Ion Battery. *Chemistry of Materials*, 19(15), 3595-3597.

[151] Wei, B. Q., et al. (2003). Assembly of Highly Organized Carbon Nanotube Architectures by Chemical Vapor Deposition. *Chemistry of Materials*, 15(8), 1598-1606.

[152] Chen, J., et al. (2008). Direct Growth of Flexible Carbon Nanotube Electrodes. *Advanced Materials*, 20(3), 566-570.

[153] Myounggu, P., et al. (2006). Effects of a carbon nanotube layer on electrical contact resistance between copper substrates. *Nanotechnology*, 17(9), 2294.

[154] Cola, B. A., et al. (2007). Photoacoustic characterization of carbon nanotube array thermal interfaces. *Journal of Applied Physics*, 101(5), 054313-054319.

[155] Wilson, I. A. G., Mc Gregor, P. G., & Hall, P. J. (2010). Energy storage in the UK electrical network: Estimation of the scale and review of technology options. *Energy Policy*, 38(8), 4099-4106.

[156] Hall, P. J. (2008). Energy storage: The route to liberation from the fossil fuel economy? *Energy Policy*, 36(12), 4363-4367.